A student's guide to the seashore

Third Edition

This unique, concise and beautifully illustrated guide allows students to identify over 650 of the common, widespread animals and seaweeds of the shore. User-friendly dichotomous keys are supported by details of diagnostic features and biology of each species. Now enhanced with 32 pages of colour, this much-acclaimed guide is invaluable to students of marine biology at any level. Questions such as: *How does the species reproduce? What is its life-cycle? How does it feed?* are answered in the notes accompanying each species to give a fascinating insight into the diversity and complexity of life on the shore. The text is supported by an extensive glossary of scientific terms, and a comprehensive bibliography is included to aid further study. The third edition builds on the excellent reviews of earlier editions and will continue to appeal to a wide readership, including students, teachers and naturalists.

The authors have many years' experience of research and teaching in marine biology. Susan Fish is a Research Associate and John Fish the former Director of the Institute of Biological Sciences and Dean of the Faculty of Science at Aberystwyth University. They have published extensively on intertidal ecology, and the *Student's Guide* draws on this personal research and their substantial experience of leading field courses in marine biology for undergraduate and extramural students.

A student's guide to the
seashore

Third Edition

J. D. Fish and **S. Fish**
Institute of Biological, Environmental and Rural Sciences, Aberystwyth University

CAMBRIDGE UNIVERSITY PRESS
Cambridge, New York, Melbourne, Madrid, Cape Town, Singapore,
São Paulo, Delhi, Dubai, Tokyo, Mexico City

Cambridge University Press
The Edinburgh Building, Cambridge CB2 8RU, UK

Published in the United States of America by Cambridge University Press, New York

www.cambridge.org
Information on this title: www.cambridge.org/9780521720595

First published by Unwin Hyman 1989

Second edition published by Cambridge University Press 1996

Third edition 2011

Printed in the United Kingdom at the University Press, Cambridge

A catalogue record for this publication is available from the British Library

Library of Congress Cataloguing-in-Publication Data

Fish, J. D.
   A student's guide to the seashore / John Fish, Susan Fish. – 3rd ed.
      p.   cm.
   ISBN 978-0-521-72059-5 (Pbk.)
   1. Seashore biology–Great Britain.   2. Seashore biology–Europe, Northern.   I. Fish, S. (Susan)
II. Title.
   QH137.F52 2010
   578.769′90941–dc22

                                                                        2010017271

ISBN 978-0-521-72059-5 Paperback

Additional resources for this publication at www.cambridge.org/9780521720595

(Pastel sketch by B.M. Jones)

'What branch of Natural History shall I begin to investigate, if it be but for a few weeks, this summer?'
To which I answer, Try 'the Wonders of the Shore.'

There are along every sea-beach more strange things to be seen, and those to be seen easily, than in any other field of observation which you will find in these islands. And on the shore only will you have the enjoyment of finding new species, of adding your mite to the treasures of science.

Charles Kingsley
Glaucus, or The Wonders of the Shore
Macmillan and Co., 1890
(By kind permission of Macmillan Publishers Ltd)

Contents

The colour plates are situated between pages 372 and 373.

Preface

The fusion of identification guide and biological text, the hallmark of *A Student's Guide to the Seashore,* has been widely acclaimed since the publication of the first edition in 1989 and over the past 20 years we have been greatly encouraged by the support and encouragement received from users of the *Guide*. It was therefore a pleasure to accept the invitation from Cambridge University Press to prepare a third edition.

The publication of a third edition has given us the opportunity to review the list of species included in the light of our field experience and that of colleagues. Additional species have been added, some of which are introduced species in the sense that they have now become established in Britain and north-west Europe, bringing the total coverage to over 650 species. At the same time, the biological information on individual species has been updated on the basis of the latest research and expanded wherever possible and this, together with the extended bibliography of primary sources of data, will encourage wider lines of enquiry and help those readers who wish to investigate further this fascinating habitat. One of the most significant changes since the publication of the earlier editions has been in taxonomy following the increased use of molecular techniques and there is no doubt that important advances in our understanding of taxonomic relationships will continue to be made through the application of molecular phylogeny. Taxonomy and systematics are at the heart of ecology and when considering changes in taxon names we have erred on the side of caution; where there is ongoing research and debate on the status of particular species and groups of species, we have drawn attention to this. Taxonomic change can be confusing to the reader and where changes in species names have been made we have included for ease of reference the widely used former names.

The design and layout of the book follows that of previous editions and is described in detail following the Introduction. The Illustrated Guide to the Plants and Animals of the Shore, which is designed to help those readers who are unable initially to assign an organism to its phylum or class, has been extensively reorganized and is now more user-friendly and easier to navigate. The inclusion of colour plates makes the identification of those organisms for which black and white drawings will always fail to convey key diagnostic features, that much easier and more reliable. We have also taken the opportunity to illustrate aspects of the biology of some species in colour.

As in previous editions, colleagues have been generous in giving guidance and advice. Dr Joanne Porter helped us unravel the bryozoan genus *Alcyonidium*; Dr Graham Oliver gave advice on aspects of bivalve taxonomy; Dr Robert Brownlow was a source of much helpful discussion; Mr Keith Wilson provided specimens of *Eriocheir sinensis* from the Thames estuary; and Mr Gareth Owen provided a steady supply of fresh specimens where additional drawings were needed. Ms Gwen Jenkins and Mr Anthony Pugh gave invaluable professional advice on all aspects of photo-imaging, and Ms Catherine Flack and Ms Joanna Endell-Cooper, Cambridge University Press, guided us sympathetically through the final stages of preparation and design. We are also grateful to staff of the Hugh Owen Library, Aberystwyth University, the National Library of Wales, Aberystwyth and the Natural History Museum, London. Unfortunately, it has not been possible to cite individually in the text all the references used and, once again, we wish to record our indebtedness to those whose research has contributed to the 'wonders of the shore' and made this edition possible.

Aberystwyth, December 2009   **J. D. Fish**
                                                                                              Susan Fish

Acknowledgements for illustrations

We are grateful to the following authors and publishers for kind permission to use the figures listed:

Edward Arnold and J. R. Lewis for Figure 17 and Plate 38C in *The ecology of rocky shores*. Hutchinson Educational Ltd and A. D. Boney for Figures 19, 32, 40(e) in *A biology of marine algae*. British Museum (Natural History) and N. Tebble for Figures 52A, B, C, 54A, B, 95C, 101B in *British bivalve seashells*; British Museum (Natural History) for Figures 70B, 239B in Newton, L., *A handbook of the British seaweeds*. Conseil International pour L'Exploration de la Mer and W. Greve for Figure 5 in Zooplankton Sheet 146, *Fiches d'identification du zooplancton*. The Ray Society for Plate 2, Figures 1, 2; Plate 4, Figure 1; Plate 5, Figure 1; Plate 9, Figures 5, 6 in Allman, G. J., *A monograph of the gymnoblastic or tubullarian hydroids*; and Plate 4, Figure 2, Vol. 1; Plate 22, Figure 3, Vol. 2, in Stephenson, T. A., *The British sea anemones*; and Plate 1, Figures 1, 4; Plate 3, Figures 4, 6; Plate 5, Figure 1, Vol. 1, Part 1; Plate 25, Figure 1; Plate 26, Figure 1, Vol. 1, Part 2; Plate 114, Figure 2, Vol. 4, Part 1; Plate 116, Figures 2, 3, 5, Vol. 4, Part 2 in McIntosh, W. C., *A monograph of the British marine annelids*; and Fam. 1, Plate 5, Figure 1; Fam. 1, Plate 10, Figure 2; Fam. 1, Plate 11, Figures, 1, 8 in Alder, J. & Hancock, A., *A monograph of the British nudibranchiate Mollusca*; and Figures 97A, 98L, 102A, 103G, 107, 108I, 109A, 110E in Tattersall, W. M. & Tattersall, O. S., *The British Mysidacea*; and Plate 11, Figure 3, Vol. 1; Plate 51, Figure 1; Plate 52, Figure 3; Plate 54, Figure 6, Vol. 3 in Alder, J. & Hancock, A., *The British Tunicata*. Cambridge University Press for Plates 3 and 4 in Russell, F. S., *The medusae of the British Isles*, Vol. 2; Plate 14, Figure 61; Plate 21, Figure 100a in Eales, N. B., *The littoral fauna of Great Britain*. Cambridge University Press and The Marine Biological Association of the United Kingdom for Plate 16 in Smith, J. E., *'Torrey Canyon' pollution and marine life*. Academic Press, The Estuarine and Brackish-water Sciences Association, The Linnean Society of London and R. L. Manuel for Figures 70, 78A, B in *British Anthozoa. Synopses of the British fauna*, No. 18. Academic Press, The Linnean Society of London and P. E. Gibbs for Figures 2A, 13A in *British sipunculans. Synopses of the British fauna*, No. 12. Academic Press, The Linnean Society of London and E. Naylor for Figure 10D, G. I in *British marine isopods. Synopses of the British fauna*, No. 3. Academic Press, The Linnean Society of London and R. H. Millar for Figures 10A, 55A in *British Ascidians. Tunicata: Ascidiacea. Synopses of the British fauna*, No. 1. E. J. Brill/ Dr W. Backhuys, The Estuarine and Brackish-water Sciences Association, The Linnean Society of London and P. J. Hayward for Figure 8A in *Ctenostome bryozoans. Synopses of the British fauna*, No. 33. S. A. Masson for Figure 669 in *Traité de Zoologie*, Vol. 5. The Linnean Society of London for Plate 44, Figures 1, 3 in Dyster, F. D., *Transactions of the Linnean Society of London*, Vol. 22. Penguin Books for Plate 5 in Stephenson, T. A., *Seashore life and pattern*. Editions Lechevalier for Figures 18i, 21b in Fauvel, P., *Faune de France 5. Polychètes errantes* and Figures 46b, 120a in Fauvel, P., *Faune de France 16. Polychètes sédentaires*. The Malacological Society of London for Figure 6B, D, G in Matthews, G., *Proceedings of the Malacological Society of London*, Vol. 29. The Freshwater Biological Association for Figure 15b, c in *A key to the British fresh- and brackish-water gastropods*, F. B. A. Scientific Publication No. 13. The Field Studies Council and J. H. and M. Crothers for Figure 9 in *Field Studies*, Vol. 5. The Zoological Society of London for Plate 1, in Bramble, F. W. R. & Cole, H. A., *Proceedings of the Zoological Society of London*, Vol. 109. V. Driver (Du Heaume) for Figures on pp. 365, 424, 438 in Wheeler, A., *The fishes of the British Isles and north-west Europe*. Cammermeyers Forlag, Oslo, The Institute of Marine Research, Bergen and The Zoologisk Museum, Bergen, for Plate 9 (part), Plate 10 (part), Plate 212 (part), Plate 225 (part), Plate 236 (part) in Sars, G. O., *An account of the Crustacea of Norway*, Vol. 1, *Amphipoda*; and Plate 31 (part), Plate 34 (part), Plate 43 (part) in Sars, G. O., *An account of the Crustacea of Norway*, Vol. 2 *Isopoda*. Collins for Figure 53 in Yonge, C. M., *The sea shore*. HMSO and The Natural History Museum, London for Figure 17E in Irvine, L. M. & Chamberlain, Y. M., *Seaweeds of the British Isles*, Vol. 1 *Rhodophyta* Part 2 B. G. Rogerson for Figures 61A, C in Burrows, E. M., *Seaweeds of the British Isles*, Vol. 2. Cambridge University Press for Figure 2A, B in Cadman, P. S. & Nelson-Smith. A., *Journal of the Marine Biological Association of the United Kingdom*, Vol. 73 and Figure 50A in Gibson, R., *British nemerteans. Synopses of the British fauna*, No. 24. Academic Press for Figures 4E, 5B in *British marine isopods. Synopses of the British fauna*, No. 3. Academic Press and The Linnean Society of London for Figures 4C, 5C in Gibbs, P. E., *British sipunculans. Synopses of the British fauna*, No. 12 and Figures 4E, 5B in Naylor, E., *British marine isopods. Synopses of the British fauna*, No. 3. The Ray Society for Plate VI, Figure 3 in McIntosh, W. C., *A monograph of the British marine annelids*, Vol. 1, Part 1 and Figures 26A, 1, 70G, 73F, 36A, 35 in Tattersall, W. M. & Tattersall, O. S., *The British Mysidacea*. Editions Lechevalier for Figures 30c, 411, 150c in Fauvel, P., *Faune de France 5. Polychètes errantes* and Figures 72d, 73a in Fauvel, P., *Faune de France 16. Polychètes sédentaires*. The Malacological Society of London for Figure 3D in Matthews, G., *Proceedings of the Malacological Society of London*, Vol. 29. Estuarine and Coastal Sciences Association and the Linnean Society of London for Figures 26B, 49 in Thompson, T. E., *Molluscs: benthic opisthobranchs. Synopses of the British fauna*, No. 8 (2nd edn).

Introduction

The seashore

One of the most striking features of the shore is the rich diversity of plant and animal life to be found there. A wide range of invertebrates, some highly mobile, others fixed or sedentary, and shore fishes, are a characteristic feature. Brightly coloured lichens often form distinct bands on the high shore; seaweeds may be present in abundance, and on mud-flats flowering plants often dominate. Physical factors change rapidly and it is here that the student has the opportunity to observe and study the fascinating adaptations to this environment shown by both plants and animals.

Tides

The dominating force on the shore is the rise and fall of the tide. Tides result from the gravitational forces between the Moon and Sun and the seas and oceans on the Earth's surface. The tides with which we are most familiar in north-west Europe are semi-diurnal, that is, there are usually two high tides and two low tides each day. This can be appreciated if the Earth is pictured revolving on its axis during the course of a day and passing through a water envelope which has been distorted by the gravitational forces of the Moon and Sun as in Fig. 1.1. Because of the constantly changing positions of the Earth, Sun and Moon, the length of time between successive high tides is not exactly 12 hours but is about 12 hours 25 minutes. Subsequently, the tides are approximately 50 minutes later each day and there will not necessarily be two high tides and two low tides every day. Superimposed on the regular daily pattern of the tides are changes brought about by the relative positions of the Earth, Sun and Moon based on a monthly and a yearly cycle. The Moon revolves around the Earth in approximately 28 days. At the times of full and new moon, the Earth, Moon and Sun are in line and the combined gravitational attraction of the Moon and Sun on the seas produces tides with a large range. Such tides are referred to as spring tides (Fig. 1.2). At times of half moon when the Sun and Moon are at right angles to one another and their forces are in opposition, the gravitational pull on the surface of the Earth is less, giving rise to tides with a smaller range. These tides are known as neap tides (Fig. 1.2). The height of the tide is measured in metres above a datum level known as chart datum (Fig. 1.3).

The revolution of the Earth around the Sun in an elliptical orbit during the course of a year also affects the range of the tides. When the Sun is closest to the Earth

Figure 1.1   The Earth sectioned through the equator is rotating on its own axis every 24 hours. At 0 h high water is experienced; 6 h later low water; 12 h later high water; and 18 h later another low water. For detail see text.

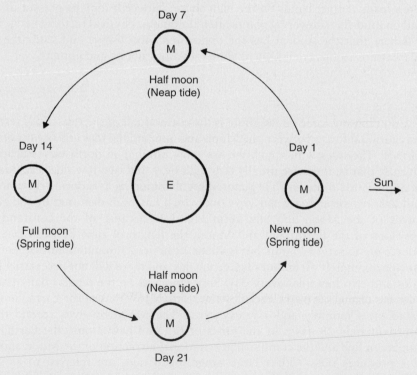

Figure 1.2   Orbit of the Moon around the Earth showing the times of spring and neap tides.

Figure 1.3   Shore profile showing tidal levels. EHWS, extreme high water level of spring tides; MHWS, mean high water level of spring tides; MHWN, mean high water level of neap tides; MTL, mean tide level; MLWN, mean low water level of neap tides; MLWS, mean low water level of spring tides; ELWS, extreme low water level of spring tides.

its gravitational pull is greatest, and in March and September the combined pull of Sun and Moon results in very large spring tides, the spring and autumn equinoctial tides. Atmospheric pressure variation and wind speed can have a marked impact on the height of the tide and in extreme cases meteorological conditions can lead to storm surges during which the predicted height of the tide is greatly exceeded. This can lead to natural disasters such as that in the North Sea in January 1953 which resulted in extensive flooding of the coastlines of England and the Netherlands and serious loss of life. The configuration of the coastline also has a substantial effect on tidal range. In the Bristol Channel, for example, where the tide is funnelled between narrowing headlands, a spring tide range of over 12 m is recorded, compared with less than 1 m in some areas.

The daily rise and fall of the tide results in different levels of the shore being covered (submersed) and uncovered (emersed) for varying periods of time. Specific tidal levels are often referred to and these prove to be useful reference points for shore ecologists. These are average levels and five such reference points are extensively used; these are: mean high water level of spring tides (MHWS), mean low water level of spring tides (MLWS), mean high water level of neap tides (MHWN) and mean low water level of neap tides (MLWN) (Fig. 1.3). Mean tide level (MTL) is the average of these four tidal heights. Reference is also made to extreme spring tide levels (Fig. 1.3), these being the highest and lowest tides of the year. Animals living at MHWS are covered by the sea for only a short period of time at high water of spring tides and are not covered during neap tides. Animals at MLWS are uncovered for only a brief period at low water of spring tides and are permanently submersed at neap tides. These varying periods of submersion and emersion lead to the development of a gradient of physical conditions such as temperature and desiccation from high to low shore and the response of animals and plants to this gradient, together with the impact of biological interactions between groups of organisms such as competition for space and food, lead to zonation (Plate 6).

Zonation is the occurrence of different species of seaweeds and animals at different levels on the shore and can be seen on rocky shores all around the world.

Rocky shores

In his book on the ecology of rocky shores, Lewis (1964) divided the shore into three major zones marked by the presence of conspicuous and widespread seaweeds and animals. The nomenclature proposed by him is shown in Fig. 1.4 and it is important to note that the zones are not defined by reference to the tidal levels described above but by reference to biological factors, that is, the distribution of specific organisms or groups of organisms. The highest zone on the shore is the littoral fringe, the upper limit of which is marked by the upper limit of the periwinkles (p. 214) and black lichens (p. 70). The middle zone is the eulittoral zone, the upper limit of which is marked by the upper limit of the acorn barnacles (p. 305). The lowest zone is the sublittoral zone which extends below low water but its upper limit, marked by the presence of the large laminarian seaweeds (p. 57), can be explored at low water of spring tides (Plate 5b). The littoral fringe and the eulittoral zone together make up the littoral zone.

The upper limits of the zones on rocky shores are extended on coasts that are exposed to heavy wind and wave action (Fig. 1.4) as a result of spray being carried far up the beach. In very exposed situations, the upper limit of the littoral fringe is raised by many metres forming a broad zone extending well above the level of the highest tide. On sheltered shores the littoral fringe is narrow. Exposure to wave action also affects the range of seaweeds and animals found on the shore as some species are more

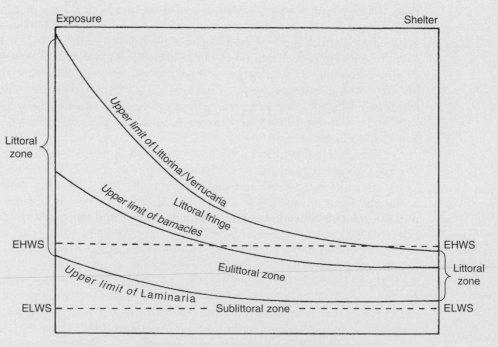

Figure 1.4   Rocky shore zonation. ELWS, extreme low water level of spring tides; EHWS, extreme high water level of spring tides (after Lewis, 1964).

tolerant of exposure than others and the density, diversity and form of the seaweeds on rocky shores are useful indicators of exposure to wave action (see Ballantine, 1961). The upper reaches of the shore are harsh physical environments and it is not surprising to find that the number of different species is lower there compared with the middle and lower reaches where conditions are less stressful. Some organisms may be restricted to one zone on the shore but the distribution of the majority of species extends across more than one zone.

The distribution of littoral organisms rarely coincides with the boundaries of the zones defined above. As a result, the terms lower shore, middle shore and upper shore are often used by ecologists to note relative positions on the shore. These terms have been used extensively in this text: the lower shore extends from the upper sublittoral to include the lower part of the eulittoral zone, the middle shore refers to the middle region of the eulittoral, and the upper shore includes the upper eulittoral and littoral fringe.

Sandy and muddy shores

Rocky shores offer a stable, solid surface on which organisms can attach, and anyone walking or browsing on such shores will immediately be aware of a rich diversity of seaweeds and animals. This is in contrast to sandy beaches where there is a lack of stable surfaces on which organisms can attach and here the fauna generally lives buried in the sediment for protection. The size of individual particles making up the sediment is of great importance as it determines the stability of the shore and factors such as the amount of water that can be held by the sediment when the tide is out. Coarse pebble and shingle deposits are unstable habitats and when waves break on such beaches, the pebbles roll against one another and are likely to crush the fauna. When the tide retreats, the sea water drains quickly and little is retained in the spaces between individual particles. These are harsh physical environments and very few animals are adapted to living in these conditions. Sand deposits are largely made up of quartz particles, often with substantial amounts of calcareous material, and the characteristic yellow-brown colour results from iron deposits on the grains. Sand deposits have a much smaller particle size than pebble and shingle deposits and retain greater quantities of water during emersion. As a result, they offer more stable habitats for burrowing animals and sheltered sandy beaches support large numbers of animals but the number of different species, the species diversity, is generally less than that found on rocky shores.

The finest particles, known as silt and clay, are deposited in sheltered areas giving rise to mud-flats. Here, the increased stability of the sediment allows seaweeds and salt-tolerant flowering plants to gain anchorage and their presence further stabilizes the mud surface. Muddy shores are found most frequently in estuaries where there is a heavy load of particulate matter in suspension, mainly brought down by rivers as runoff from the land. When fresh and salt water meet, the fine particles tend to flocculate and settle to form mud banks. These are rich in organic matter but there is little circulation of water through such fine deposits and this can lead to stagnant and anaerobic conditions. Just under the surface of mud deposits, oxygen may be completely lacking and in such situations anaerobic bacteria flourish, producing the characteristic smell of hydrogen sulphide so often detected when the mud is disturbed. Anaerobic conditions are also found in sandy deposits but these are generally encountered a few centimetres beneath the surface and are marked by a conspicuous black layer (Plate 24b) in which the sediment is discoloured by sulphide deposits produced through the metabolism of anaerobic bacteria. The presence of this black layer is a useful ecological indicator. The fauna of mud-flats survives anaerobic sediments by a variety of morphological and physiological adaptations and can therefore take advantage of the rich supplies of food available in the form of organic debris.

Organisms living in estuaries experience water of constantly changing salt concentration or salinity. Sea water of the open oceans away from the influence of freshwater drainage from the land generally has a salinity of 3.5 grams of salt per 100 grams, and this has traditionally been expressed in parts per thousand as 35‰, i.e. 35 grams of salt per litre of solution. In 1978 salinity was redefined using a Practical Salinity Scale and salinities are now expressed in Practical Salinity Units (PSU) simply as numerical values. The constantly changing salinities in estuaries mean that organisms can experience values ranging from almost 0 to above 30. These are harsh physical conditions and as a result, diversity of species is low compared with adjacent marine and freshwater habitats but abundance of individual species can be very high. Productivity of estuaries is high and the high density and biomass of invertebrates make intertidal mud-flats important feeding grounds for a wide variety of birds, often seen feeding as the tide recedes, and a range of fishes some of which migrate into the estuary on the flood tide to feed, retreating as the tide ebbs. The drainage channels so conspicuous on mud-flats (Plate 27c) are home to a number of resident invertebrates and fishes and in some estuaries they may be important nursery grounds for the young of commercial species.

Collection of specimens

One of the primary objectives of all who use the shore should be to cause little disturbance to the organisms living there. Wherever possible, specimens should be identified in the field, where necessary with the aid of a good-quality hand lens. If specimens are removed from the shore, few should be taken and it should be borne in mind that many species will survive if returned to the beach from which they were collected after careful handling and observation in a laboratory. Before going to the shore it is essential to have information on any peculiarities of the rate of ebb and flow of the tide and the times of high and low water for that area. Remember that tidal conditions, including the time and height of high and low waters, vary from one location to another. Tidal data are available on the internet and from *Admiralty Tide Tables* published by the Hydrographer of the Navy, but most coastal authorities now produce inexpensive tide tables covering the coastline within their district. Sampling should begin before the time of low water, starting on the lower shore and moving up well in advance of the incoming tide. Special care is required when working in estuaries, where tidal currents can be very fast and quicksands can be encountered. Always seek local knowledge before working on sand- and mud-flats, and inform appropriate contacts of your whereabouts and programme of activities.

The type of shore being studied will determine the method of sampling, and survey methods are described in a number of texts listed in the introductory section of the Bibliography (p. 475). Many animals are immediately visible on rocky shores, but careful searching of crevices and rocky overhangs will reveal many more specimens. The moist conditions under a dense cover of seaweed offer refuge for a diversity of species, while the surface of the weed is often colonized by hydroids, bryozoans and tubiculous polychaete worms. The holdfasts of the large laminarian seaweeds provide shelter for a surprising variety of small invertebrates, and boulders and large rocks can be lifted (*but always returned to their original position*), to reveal a wide range of fauna. Although the need for conservation on the shore cannot be emphasized too strongly, it can be very rewarding to take back to the laboratory small amounts of scrapings of, for example, acorn barnacles and the holdfasts of laminarians for examination in sea water under a microscope.

On sandy shores, where the fauna lives buried in the sediment (infauna), sampling has to be carried out by digging. The surface few centimetres of sediment are removed and wet-sieved through a brass sieve, generally of mesh diameter of 0.5 mm, and animals above this size, known as the macrofauna, are retained for examination. Smaller animals, members of the meiofauna, pass through the sieve. These are not included in this book; they are adapted to living in the tiny spaces between the sand grains and different sampling techniques are required for their capture.

Many macrofaunal species live in the surface 50 mm or so of sediment where the interstitial water circulates freely between the grains and is oxygenated. Others live deeper in the sediment. For example, the lugworm (p. 147) lives at depths of 200 mm or more and betrays its presence by worm casts on the surface (Plate 24c). A wide range of bivalve molluscs also lives in anaerobic conditions deep in the sediment and like the lugworm shows a range of adaptations to survival in these conditions. Animals living in mud often construct permanent burrows and the practised eye can detect the openings of these on the surface. Such specimens can be dug for selectively but it must be borne in mind that much damage can be done to sand- and mud-flats by indiscriminate and reckless digging.

Nomenclature and classification

All organisms described in this text have been given their scientific name. The system of scientific nomenclature in use today stems from the work of the Swedish naturalist Linnaeus and the publication of the 10th edition of his *Systema Naturae* in 1758. This marked the beginning of the binomial system of nomenclature by which plants and animals are given two names. The first (the generic name) is the name of the genus to which the organism belongs and begins with a capital letter. The second is the species or specific name, which begins with a small letter. The name of the person (or persons) who assigned the scientific name follows the name of the organism. This person is known as the author. If, since the naming of the animal it has been transferred to a different genus, the name of the author is in parentheses. Two names are often written after the scientific name of a plant. This means that subsequent to its original description by the first author (whose name is in parentheses), the plant has been transferred to a new genus by the second named author. It should be noted that in most texts the names of the authors of plants are given in abbreviated form. The internationally accepted system of nomenclature using the scientific name followed by the author ensures that there is no ambiguity regarding the species under consideration. In some cases, however, taxonomic research results in changes to scientific names and this can lead to confusion. In this text the former, sometimes more familiar, scientific name(s) has also been included. Common names in wide usage are also given. Rules governing nomenclature are given by the International Commission on Zoological Nomenclature (1999) and by McNeill *et al.*, International Code of Botanical Nomenclature (2006).

Organisms are grouped according to their similarities and affinities and are thus arranged in a system of classification. Closely related species are grouped into genera and genera into families. Families are grouped into orders; orders into classes and related classes are grouped into phyla. Morphological and anatomical features are used by taxonomists when considering classification and in some cases the creation of subphyla, subclasses and suborders has proved useful.

The nomenclature and classification of organisms is constantly being researched and revised and for this reason information given in different texts may vary. The nomenclature used here mainly follows Costello, Emblow & White (2001) and The Marine Biodiversity and Ecosystem Functioning EU Network of Excellence (Mar BEF). Details of other sources used are given in the relevant chapters. The system of classification is based largely on Brusca & Brusca (2003).

Design and layout of the book

This book has been designed both as a guide to the identification of the common and widespread organisms of the shore and as a biological text giving information on their biology and ecology. It is intended for use by students, lecturers and teachers, and also by naturalists and others who might have little or no formal scientific training.

In the following pages, the different groups of organisms are arranged in phylogenetic sequence. Readers who are unable initially to assign an organism to its phylum or group are referred to the illustrated guide on p. 11. When the phylum or group to which an organism belongs is known, reference can be made directly to the appropriate chapter where the reader will find an introduction giving the main characters of the phylum or group together with a full classification in which those classes, subclasses, orders, etc. included in this text are highlighted in bold type. A brief statement on the morphology and biology of the group is also included. For most groups a simple dichotomous key is provided. In some cases this is to the level of families and leads to a full statement of family characteristics, followed where appropriate by a key to species. In other cases the phylum or group has been keyed directly to species. Morphological features are the basis on which identification is made but in many cases the type of substratum on which the organism is found, position on the shore and distribution are important characteristics. This information is included in the text where appropriate.

The entry for each species is presented as follows:

Genus species Authority (*former name(s)*) Common name(s) **Fig.**

Generic and specific names are shown in bold italic script followed by the name of the person(s) who described the species, the authority, which may or may not be in brackets (see p. 7). This is followed by any former names by which the species has been known and referred to in earlier/other texts. Finally, common names in wide usage are included, followed by reference to the figure number in bold type. In some cases there is also reference to a plate number.

For each species there is a statement of diagnostic features, in italic script, a drawing and notes on its biology. Scientific terminology used in the description of species has been kept to a minimum and is explained in the introductory sections at the beginning of each chapter and the glossary at the end of the book. Important diagnostic features for a particular group or individual species are labelled on the appropriate drawings. A scale has not been included on the drawings but for guidance the size of each species is given in the diagnostic features. For ease of reference, drawings are interspersed in the text.

The inclusion of family and species keys followed by a full statement of diagnostic features for each species has led to some repetition but this enables those readers more familiar with the identification of shore organisms, and using the book as a reference text, to turn directly to an appropriate page reference where their provisional identification of an organism can be confirmed and where they will find detail on the biology of the species in question.

At the end of each chapter there is a short list of specialist references, useful in further identification, covering the phylum/group in question. Some of these have been referred to in the text where they are in bold type. Other references, cited in the text but not in bold type, are to be found in an extensive bibliography of research papers and books (p. 475) from which information on the biology of the species has been gathered. This will allow readers with specific interests to develop further lines of enquiry. For ease of reference the bibliography is broken down into phyla or in some cases classes.

The keys cover only those organisms included in this text and readers will undoubtedly come across species not included here. For this reason, every effort has been made to prevent false identification. Attention is drawn to cases where confusion might arise and where appropriate reference to specialist texts and further reading is given. Where identification to species is beyond the scope of this book, identification has been made to genus. When an organism has been identified to a family, it should be checked against the family characteristics before proceeding further. When identification to species has been made, this should be checked carefully against the diagnostic features, the drawing and the habitat characteristics given for that species.

Illustrated guide to the plants and animals of the shore

The following guide has been designed to help those readers who are unable initially to assign an organism to its correct phylum or class. The guide is divided into nine groups, each recognized by the general characters given, for example, encrusting, worm-like, etc. Each of these groups is broken down into subdivisions which are described and illustrated, and for each subdivision a page reference is given where further information can be found. In a few cases the guide will indicate that an organism belongs to one of two or more subdivisions, and for each a page reference is given.

   To design a guide covering a broad range of plants and animals for use by readers with widely differing background knowledge is difficult as its success will to some extent be affected by subjective assessment of the characters listed. Representative sketches, not to scale, of each group of organisms have been included as a further aid to identification.

Plants and plant-like animals	group 1 p. 12
Worm-like animals, varying in shape from long and rounded to flat and leaf-like	group 2 p. 14
Animals with jointed appendages; mobile	group 3 p. 16
Soft and slug-like or leathery and cucumber-shaped; creeping	group 4 p. 17
Animals with spiny skin, hard outer skin, hard shell, shell plates or hard tube; without jointed appendages (or jointed appendages not obvious)	group 5 p. 18
Attached at base, lacking tentacles; gelatinous. Not plant-like	group 6 p. 20
Encrusting, without hard shell, shell plates or hard tube	group 7 p. 20
Soft and gelatinous; swimming or floating	group 8 p. 22
Fishes	group 9 p. 22

group 1 | Plants and plant-like animals

Plant-like in form, generally attached at one end. Includes obvious plants and plant-like animals.

1. Green, brown or red; branched, unbranched, thread-like, tubular, membranous, leaf-like, strap-like, globular. Delicate, tough, leathery or calcareous. Few millimetres to several metres in length. Mainly on rocky shores, also in estuaries. Widespread; all levels of the shore.
.... **Seaweeds** p. 23

2. Upright or pendulous, often in tufts, normally brittle, dry. Brownish-black or greyish-green. Small, usually no more than 10 mm in height. Rocky shores, from terrestrial fringe to middle shore.
.... **Lichens** p. 69

3. Green, with flowers, seeds and roots. Typically in estuaries.
.... **Flowering plants (Anthophyta)** p. 75

4. Tubular, vase- or urn-like; may be branched. Single, conspicuous opening at free end. Spongy. Attached to rocks, stones, seaweeds. Lower shore.
.... **Sponges (Porifera)** p. 79

5. Delicate, branching growths; sometimes tufted. Whitish, light brown or buff colour. Tentacles retractile or non-retractile. Usually no more than 50 mm height. Attached to stones, shells, seaweed, etc. Middle shore and below.
.... **Sea-firs, hydroids (Cnidaria: Hydrozoa)** p. 99
.... **or Sea-mats (Bryozoa)** p. 382

6. Delicate, inverted bell/trumpet shape. Eight groups of short tentacles around margin. On stones and seaweed, lower shore.
.... **Stalked jellyfish (Cnidaria: Scyphozoa)** p. 125

7. Soft bodied, attached at base. Many tentacles at free end, best seen when animal undisturbed under water. Red, green, purple, white, orange. Middle and lower shore.
.... **Sea-anemones (Cnidaria: Anthozoa)** p. 87

8. Fleshy lobes, sometimes finger-like. When undisturbed, surface covered with many small, pinnate tentacles. White, yellow, orange, pink. Attached to rocks, stones and shells. Lower shore. *Do not confuse with sea-mats under 10 below.*
.... **Dead men's fingers (Cnidaria: Anthozoa)** p. 88

9. Soft body, many long tentacles. Withdraws into hard, ridged case when disturbed. Attached to rocks, stones, shells. Lower shore (see also Group 5.1, p. 18).
.... **Stony corals (Cnidaria: Anthozoa)** p. 97

10. Irregular, lobed gelatinous mass, may be branched. Smooth with large number of tiny openings through which retractile tentacles project. Colour varied, yellowish-brown, brown, grey. Lower shore. Often washed ashore after storms. *Do not confuse with dead men's fingers under 8 above.*
.... **Sea-mats (Bryozoa)** p. 382

11. Flattened fronds, brittle; buff colour. May be mistaken for dried seaweed. Strandline.
.... **Sea-mats (Bryozoa)** p. 382

12. Flower-like, 5 pairs feathery arms. Attached to rocks and walls of crevices, lower shore and sublittoral.
.... **Feather-stars (Echinodermata: Crinoidea)** p. 397

group 2

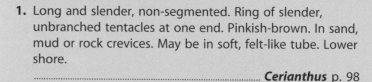

Worm-like animals, varying in shape from long and rounded to flat and leaf-like

A large assemblage of animals varying widely in size but all 'worm-like'. Found under stones and rocks, burrowing into sediments and in tubes made of various materials.

1. Long and slender, non-segmented. Ring of slender, unbranched tentacles at one end. Pinkish-brown. In sand, mud or rock crevices. May be in soft, felt-like tube. Lower shore.

.... ***Cerianthus*** p. 98

.... **and burrowing sea-anemones (Cnidaria: Anthozoa) p. 97**

2. Thin and leaf-like, non-segmented. Moves over substratum with gliding motion. Small, generally less than 30 mm length. Eye spots and tentacles often visible. Under stones, seaweed, etc. Widely distributed on shore.

.... **Flatworms (Platyhelminthes) p. 129**

3. Long, ribbon-like, non-segmented, often slimy; varying in length from few millimetres to several metres. Yellow, black, brown. In crevices, under stones and seaweeds; buried in mud and sand. Lower shore.

...... **Ribbon worms, proboscis worms (Nemertea) p. 133**

4. Flattened, leech-like, with posterior ventral sucker. On gills of bivalve molluscs such as the cockle.

.... ***Malacobdella*** **(Nemertea) p. 138**

5. Segmented body, lateral bristles, often borne on conspicuous paddle-shaped outgrowths. Dorsal surface sometimes covered with dense bristles or scales. Anterior end often with eyes and tentacles. Shape varies from broad and oval to long and slender. Under stones and rocks, and burrowing in mud and sand. Some live in tubes of mucus, mud, sand and shell fragments. Widespread on shore.

.... **Bristle worms (Annelida: Polychaeta) p. 141**

6. Cylindrical, superficially segmented body with anterior proboscis covered in spines and posterior tassle-like appendage. Pinkish-brown. Burrowing in mud and muddy-gravel on lower shore.
.... **Priapulans (Priapula)** p. 140

7. Unsegmented and cylindrical. Narrow anterior region and somewhat wider posterior region. Lobes or tentacles at anterior. Yellow, brown, grey. In burrows in sand or gravel, or in empty shells. Lower shore.
.... **Sipunculans (Sipuncula)** p. 377

8. Body long, unsegmented. Anterior end covered with small, whitish shell plates. Animal secretes calcareous tube. Burrows into submerged timbers such as piles, piers, wooden ships.
.... **Shipworms (Mollusca: Bivalvia)** p. 296

9. Unsegmented, narrow, cylindrical. Bears horseshoe- or crescent- or oval-shaped group of anterior tentacles. Greyish, yellowish. In chitinous tube, either encrusting, or burrowing into rocks, shells, sand and gravel. Lower shore.
.... **Phoronids (Phoronida)** p. 394

10. Unsegmented. Narrow anterior region with ventral groove and cylindrical, wider posterior region. Yellow, grey, blue, pink. In burrows in mud and sand, and in rock crevices. Lower shore.
.... **Echiurans (Echiura)** p. 380

11. Body unsegmented. Divided into 3 regions, conical anterior, short cylindrical collar-like region and long, narrow posterior region. In U-shaped burrows in sand and muddy-sand; under rocks and stones. Lower shore.
.... **Acorn worms (Hemichordata)** p. 427

12. Elongate, unsegmented. Branched tentacles at anterior. May be tacky to the touch. Pinkish-brown. In sand and muddy-sand, and crevices. Lower shore.
.... **Sea-cucumbers (Echinodermata: Holothurioidea)** p. 421

group 3 · Animals with jointed appendages; mobile

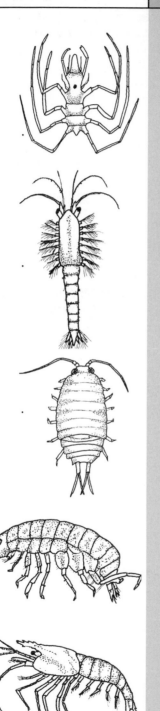

Includes a wide range of animals in the Arthropoda (p. 298).

1. Four pairs of legs; body elongate and divided into sections; spider-like. Among stones, seaweeds, sea-firs, sea-anemones, etc. Middle and lower shore.
.... **Sea-spiders (Pycnogonida)** p. 299

2. Small, delicate, often transparent, shrimp-like. Legs with 2 branches, lacking pincers. Free-swimming, often in swarms in areas of reduced salinity. *Do not confuse with prawns and shrimps under 5 below.*
.... **Opossum shrimps, mysids (Crustacea: Mysida)** p. 312

3. Body shape resembles woodlouse. Seven (rarely 5) pairs of similar legs. Wide variety of habitats; in crevices, under stones and seaweed, burrowing in sand, mud, old timbers. Widely distributed.
.... **Sea-lice, sea-slaters (Crustacea: Isopoda)** p. 317

4. Body slender, laterally compressed. Seven (rarely 5) pairs of legs, not all alike. Under stones and among seaweed; when disturbed often seen to move on one side. Also burrow in sand, mud, gravel. Occasionally on undersurface of jellyfishes. Widely distributed.
.... **Includes sand-hoppers (Crustacea: Amphipoda)** p. 329

5. Shrimp-like, often transparent. One or more pairs of legs have pincer tips. Usually free-swimming in rock pools, some burrow into sand. Middle and lower shore. *Do not confuse with mysids under 2 above.*
.... **Prawns, shrimps (Crustacea: Decapoda)** p. 349

6. Body elongate, strongly built, powerful pincers. Posterior region extended (lobsters) or curled underneath (squat lobsters). Some (hermit crabs) live in empty shells. Among rocks, stones, in crevices, etc. Lower shore.
.... **Lobsters, squat lobsters, hermit crabs (Crustacea: Decapoda)** p. 355

7. Body rounded, or triangular; strongly built, powerful pincers. Legs may be short or very long and spider-like. Posterior region tightly folded under main body. Among seaweed, rocks, stones, crevices and in sand and mud. Widely distributed.
.... **Porcelain crabs, true crabs (Crustacea: Decapoda)** p. 358

8. Body divided into 3 regions, head, middle region with 3 pairs of legs, and posterior region without legs. Found floating in surface film of rock pools, and in crevices. Upper shore. Up to about 10 mm length, but some 2–3 mm.
Includes bristle tails (Arthropoda: Hexapoda) p. 374

group 4

Soft and slug-like or leathery and cucumber-shaped; creeping

1. Body soft and slug-like, head with tentacles. Surface may be covered with finger-like or branched outgrowths. Often very brightly coloured. In rock pools, among seaweed; sometimes on saltmarshes. Lower shore.
.... **Sea-slugs (Mollusca: Opisthobranchia)** p. 235
.... **but see *Lamellaria perspicua* (Mollusca: Prosobranchia)** p. 228

2. Cucumber- or sausage-shaped; skin may be rough and leathery. Ring of branched tentacles at anterior end. Brown, black or greyish colour. On rocks, mud and sand. Lower shore.
.... **Sea-cucumbers (Echinodermata: Holothurioidea)** p. 421

group 5

Animals with spiny skin, hard outer skin, hard shell, shell plates or hard tube; without jointed appendages (or jointed appendages not obvious)

1. Soft body, many long tentacles; withdraws into hard, ridged case when disturbed. Attached to rocks, stones, shells. Lower shore (see also Group 1.9, p. 13).
.... **Stony corals (Cnidaria: Anthozoa)** p. 97

2. Tube coiled, sinuous or almost straight; attached to rocks and seaweed. When undisturbed, many tentacles project from anterior. Tube whitish colour. Lower shore.
.... **Tubicolous worms (Annelida: Polychaeta)** p. 141

3. Oval-shaped with 8 overlapping shell plates. Large muscular foot on undersurface. Dull red, olive, brownish. Under stones, in crevices. Middle and lower shore.
.... **Coat-of-mail shells, chitons (Mollusca: Polyplacophora)** p. 192

4. Body housed in single shell; spiral, cone-shaped or flattened. On rocks, stones, in crevices, among seaweed; also on sand and mud. Widespread on shore.
.... **Limpets, top-shells, winkles, whelks (Mollusca: Prosobranchia)** p. 198
.... **but see *Acteon tornatilis* and *Retusa obtusa*** p. 237
and pulmonates p. 248

5. Body housed in shell of two parts hinged together. Attached to rock, stones, also in gravel, sand and mud. Widespread on shore, often in dense beds.
.... **Mussels, cockles, clams, scallops (Mollusca: Bivalvia)** p. 250

6. Soft brown stalk with 'head' bearing number of hard, whitish-grey plates. When under water jointed appendages protrude from between plates. Often washed ashore on floating driftwood or spongy float.
.... **Stalked barnacles (Arthropoda: Cirripedia)** p. 304

7. Shell plates form cone attached to rocks and shells. Whitish colour. When under water apex of cone opens and jointed appendages protrude from between plates. At all shore levels; in dense aggregations on upper shore.
.... **Acorn barnacles (Arthropoda: Cirripedia)** p. 305

8. Star-shaped, 5 or more arms not sharply marked off from central area. Surface rough, spiny. Lower shore. On rocks, also sand and mud.
.... **Starfishes (Echinodermata: Asteroidea)** p. 399

9. Star-shaped, usually 5 slender arms, sharply marked off from central area. Surface rough, spiny. Lower shore. In crevices, under rocks; also on sand and mud.
.... **Brittle-stars (Echinodermata: Ophiuroidea)** p. 407

10. Globular, heart-shaped, oval or sometimes rather flattened, rigid and covered with spines. Lower shore. On rocks, also in sand and mud.
.... **Sea-urchins (Echinodermata: Echinoidea)** p. 413

group 6 — Attached at base, lacking tentacles; gelatinous. Not plant-like

When emersed and contracted the tentacles of some sea-anemones are not obvious and as a result they might be wrongly assigned to this group (see Group 1.7 p. 13).

1. Rounded or club-shaped with one to several conspicuous openings on surface. May be covered with sand or mud. Attached to a variety of substrata on lower shore. Up to about 50 mm height.
   .... **Sea-squirts (Chordata: Ascidiacea)** p. 431

2. Cylindrical or oval-shaped, attached at base or along one side. Usually gelatinous but may have adhering sand and shell particles. Two openings at or close to free end. Occur singly or in groups attached to variety of substrata on lower shore. May be up to 120 mm height.
   .... **Sea-squirts (Chordata: Ascidiacea)** p. 431

group 7 — Encrusting, without hard shell, shell plates or hard tube

Firmly attached to rock, seaweed, shells, etc. May be difficult to remove without damage.

1. Red, pink or purplish, irregular, hard patches on rock surfaces and on sides of rock pools. Sometimes as thin, paint-like film; sometimes thick and folded. Middle and lower shore. *Do not confuse with lichens under 2 below.*
   .... **Encrusting red seaweeds** p. 35

2. Irregular patches on rocks and stones; sometimes as very thin, paint-like film. Tough, brittle texture. Often with conspicuous, saucer-shaped structures on surface. Colour varied, black, yellow, orange, white. Extending from terrestrial fringe to middle shore. *Do not confuse with encrusting red seaweeds under 1 above.*
   .... **Lichens** p. 69

3. Irregular, spongy patches on rock, especially under overhangs and in crevices, and among seaweed holdfasts. Sometimes with raised, volcano-like openings on surface. Yellow, green, red. Lower shore.
.... **Sponges (Porifera)** p. 79

4. Irregular patches on rocks, stones, shell and especially seaweeds. Made up of large number of regular compartments, sometimes with 'hairy' appearance. Each compartment houses a tiny individual which has retractile tentacles. Usually brownish colour. Lower shore.
.... **Sea-mats (Bryozoa)** p. 382

5. Irregular, gelatinous patches on rocks, seaweed holdfasts, etc. Surface may be marked with flower-like or wavy line patterns. Brightly coloured, blue, yellow, brown, orange. Lower shore.
.... **Sea-squirts (Chordata: Ascidiacea)** p. 431

group 8

Soft and gelatinous; swimming or floating

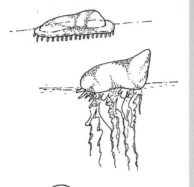

Normally found floating or swimming in the open sea but often stranded in rock pools and along strandline.

1. Floating in surface film with sail or gas-filled float projecting above surface. Tentacles, sometimes very long, on undersurface.
.... **Jack sail-by-the-wind, Portuguese man o'war (Cnidaria: Hydrozoa)** p. 119

2. Just below surface of sea. Saucer- or bell-shaped with arms and tentacles trailing from undersurface.
.... **Jellyfishes (Cnidaria: Scyphozoa)** p. 120

3. Just below surface of sea. Globular or thimble-shaped. May have pair of long, trailing tentacles.
.... **Sea-gooseberries (Ctenophora)** p. 126

group 9

Fishes

1. Shape varies from eel-like to that of typical fish with conspicuous fins, gill covers, head and tail. In rock pools and under stones. Also on sandy shores and in estuaries. Middle and lower shore.
.... **Shore fishes (Chordata: Actinopterygii)** p. 443

Seaweeds

Seaweeds are often one of the most conspicuous features of rocky shores, and in areas where there is little wave action they may be present in such high densities that they dominate the shore (Plate 1b). They photosynthesize and are thus restricted to the shore and shallow and surface waters where there is adequate light. Most are firmly anchored to the substratum by a holdfast, although in very calm waters some species have lost the attachment and survive as unattached forms.

Seaweeds are divided into three divisions, the greens (Chlorophyta), the reds (Rhodophyta) and the browns (Heterokontophyta).

Division Chlorophyta	Usually green in colour, may be bleached white. Chlorophylls *a* and *b* are the main pigments and starch is the main storage product. In this respect they resemble the higher plants.

Division Rhodophyta	Colour varies from pink to reddish-brown, to purple. Chlorophyll *a* is present but is masked by the phycobilin pigments phycoerythrin (red) and phycocyanin (blue-green), the former being the dominant pigment in the majority of red seaweeds. The main storage product is floridean starch which, chemically, is closely related to glycogen.

Division Heterokontophyta	Colour varies from olive-green to brownish-black. Chlorophylls *a* and *c* are present but the green colour is masked by the carotenoid pigment xanthophyll, which gives the brown or olive-green colour. The main storage product is a polysaccharide known as laminarin.

Seaweeds have traditionally been regarded as plants but research in molecular systematics and cell morphology suggests that the browns, unlike the greens and reds, are not closely related to land plants and should be assigned to a separate group, the Chromalveolata. This is one of five supergroups in the tree of life used to illustrate phylogenetic relationships, and also includes the diatoms and dinoflagellates. The green and red seaweeds, along with the land plants, make up the supergroup Archaeplastida.

The life histories of seaweeds are complex and varied and may involve vegetative propagation, sexual and asexual reproduction, and an alternation of generations involving haploid and diploid phases. Only a general outline can be given here. Alternation of generations usually involves a diploid sporophyte that produces haploid spores, which on germination give rise to haploid gametophytes producing male and female gametes. Fusion of gametes gives rise to a new

Figure 2.1   Life-cycle of *Ulva lactuca* illustrating the isomorphic condition (not to scale). (a) Diploid sporophyte, (b) haploid spores, (c) germination of spores, (d), (e) haploid gametophytes, (f) fusion of gametes to restore the diploid condition, (g) germination to produce new sporophyte (after Boney, 1966).

sporophyte. Where sporophyte and gametophyte are morphologically similar they are known as isomorphic; when different as heteromorphic. The common green seaweed *Ulva lactuca* (p. 30) is an example of an isomorphic life history where sporophyte and gametophyte are, to all practical purposes, indistinguishable on the shore (Fig. 2.1).

The laminarians (p. 57) are good examples of the heteromorphic life history. The conspicuous seaweed seen on the shore is the diploid sporophyte. This produces haploid spores which, on germination, give rise to microscopic male and female gametophytes. These produce male and female gametes, and after fertilization the sporophyte develops as a filamentous outgrowth from the female gametophyte (Fig. 2.2). In some of the most common brown seaweeds, such as *Fucus* (p. 63) and *Pelvetia* (p. 64), however, there is a single morphological form. This can be likened to the sporophyte of the laminarians but, unlike the latter, these species do not undergo alternation of generations. Instead, the gametes are borne in reproductive bodies known as receptacles seen as conspicuous, swollen areas on the seaweed and containing many flask-shaped invaginations opening by tiny pores onto the surface. These are the conceptacles; cells lining the wall of the conceptacle develop into male and female gametes and, when ripe, these are discharged in a drop of mucus onto the surface of the frond, usually as the seaweed dries out during low water. When the tide comes in the gametes are mixed and fertilization occurs. Germination of the zygote and subsequent growth produce the macroscopic seaweed (Fig. 2.3).

In the red seaweeds some species have two morphological forms, while others have three, only two of which are free-living, and to add to the complexity these may

Figure 2.2 Life-cycle of *Saccharina latissima* illustrating the heteromorphic condition (not to scale). (a) Diploid sporophyte, (b) male gametophyte, (c) female gametophyte, (d) new sporophyte developing on female gametophyte after fertilization by male gamete to restore diploid condition (partly after Boney, 1966).

be isomorphic or heteromorphic. As a generalization, the macroscopic sporophyte produces spores known as tetraspores, because they are produced in groups of four, and on germination these produce separate male and female gametophytes. The male gametophyte produces gametes and these fuse with the female reproductive organ which develops on the female gametophyte. This fusion gives rise to a carposporophyte which produces carpospores and is often described as being parasitic on the female gametophyte. Germination of the carpospores and subsequent growth restore the sporophyte generation. In view of this complexity, it is not surprising that there are examples of different stages in the life history of a single species which are morphologically distinct and have in the past been classified as different species.

Zonation of seaweeds is well documented and distinct bands or zones of different species, such as the fucoids and the laminarians, can be recognized on the shore. It has also been noted that as a rough generalization, green seaweeds are most common on the upper shore, the browns on the middle and lower shore and the reds on the lower shore. This distribution is often accounted for in terms of the relationship between the efficiency of photosynthesis of the different groups of seaweeds and the changing spectral composition of light with increasing depth of water during high tide. Red light, which penetrates only the surface layers of water, stimulates rapid photosynthesis in the green seaweeds, while the additional pigments of the brown and red seaweeds enable them to utilize the blue and green

Figure 2.3   Stages in the development of *Fucus* sp. (not to scale). (a) Mature plant with receptacle (b) containing many conceptacles (c). fg, female gametes; mg, male gametes (partly after Boney, 1966).

light, which penetrates deeper. It is clear, however, that this interesting hypothesis of chromatic adaptation cannot be applied as a general rule to the distribution of seaweeds on the shore. The lower limit to which seaweeds extend must, of course, be related to the penetration of light, but it may be a function of the quantity rather than the quality of light. It is also worth noting that red seaweeds are often found quite high on the shore in shaded crevices. Furthermore, there are interesting examples in which the lower limit of distribution of a seaweed is controlled by competitive interactions between the seaweed and associated fauna. For example, the lower limit of *Laminaria hyperborea* (p. 57) in some situations is controlled by the grazing activity of sea-urchins. Although the upper limit of many seaweeds is determined by physico-chemical factors such as temperature and salinity associated with tidal emersion, the lower limits are believed to be determined by an inter-action of factors, including competition with other species of seaweed and with animals.

Seaweeds are poorly represented on sedimentary shores where there is generally a lack of solid substrata for attachment, but thrive on rocky shores where their distribution and density are useful indicators of the degree of exposure of the shore to wave action. Sheltered rocky shores are dominated by seaweeds, notably *Fucus* (p. 63) and *Ascophyllum* (p. 61), while in exposed areas seaweed cover is much reduced (Plate 1a). In some species, such as *Fucus spiralis* (p. 63) and *Fucus vesiculosus* (p. 64), the morphology of the thallus varies in response to exposure to

Receptacle (reproductive body)

Air bladder

Lamina

Midrib

Stipe

Holdfast

Frond

Thallus

Figure 2.4   Generalized seaweed illustrating structures referred to in the text.

wave action and different ecotypes have been described from shores of different degrees of exposure. The density of seaweeds has a marked effect on the ecology of the shore. Many animals seek refuge in the moist conditions under seaweeds and take advantage of the niches provided there; the holdfasts of laminarians support a rich diversity of fauna and the seaweed thallus itself provides a substratum on which invertebrates can settle and interesting seaweed/fauna associations have developed. In sheltered situations some seaweeds carry large numbers of such epiphytes which reduce their photosynthetic capacity but others produce antibiotics which deter settlement of invertebrate larvae on at least part of the thallus; some periodically shed the outer layer of cells, so removing attached organisms. Seaweeds are grazed extensively by herbivorous invertebrates but in some cases this is minimized by the possession of tough, sometimes calcareous, cell walls, rapid growth and even the production of deterrent chemicals, including acids and polyphenols which are produced when the seaweed is damaged by the grazer. Harvesting of seaweeds for human consumption and agricultural and industrial uses is a long-established practice in many countries, and the large-scale culture of commercially important seaweeds is now widely practised, particularly in the Far East.

On some shores there is an abundance of different species of seaweeds and although some 90 species are described in the following pages, this represents no more than the most common. This is particularly so of the red seaweeds. Identification to species is often difficult. The gross morphological characters important in identification are shown in Fig. 2.4. The main part of the seaweed is known as the frond, and the region of attachment is called the holdfast which, in some species, is an important taxonomic feature. In some groups, notably the laminarians, there is a conspicuous stalk or stipe between holdfast and frond; holdfast and frond together are referred to as the thallus.

The nomenclature used in this text is based on **Hardy & Guiry (2003)**.

Key to the seaweeds

The separation of seaweeds into green, brown and red is an important first step in identification but it must be stressed that the distinction on the basis of colour is not always clear in the field. Brown seaweeds vary in colour from olive-green to brown and brownish-black, while the reds vary from pink, reddish-brown to purple and almost black. It must also be borne in mind that in some cases colour changes rapidly after death and care must be taken to examine fresh material.

Greens – usually a definite shade of green, but note that some specimens may show patches which are bleached white .... **Chlorophyta** p. 28

Reds – pink to reddish-brown in colour .... **Rhodophyta** p. 35

Browns – olive-green to brownish-black in colour .... **Heterokontophyta** p. 52

Division Chlorophyta – green seaweeds

There are about 100 species of green seaweeds in the British Isles and although this number is much lower than that for the browns and reds, green seaweeds are often a conspicuous feature on the shore. Some tolerate low salinities and extend into estuaries where in very sheltered areas they can cover extensive areas of ground. On rocky shores they are frequently abundant in areas of freshwater drainage and thrive in areas of nutrient enrichment. Identification to genus can generally be made on the basis of clear morphological features, but species identification can be difficult and often requires microscopic examination. Some species are microscopic and bore into shells and soft rocks. They are frequently recorded from the shells of a wide range of common shore molluscs but identification requires specialist techniques and they are not included in this text.

Key to species of green seaweeds

1 Frond prostrate, encrusting; a spongy, felt-like mat on rocks; 5–10 mm thick.
Dark green in colour .... ***Codium adhaerens*** p. 33
Frond not so **2**

2 Frond upright, dichotomously branched, up to 500 mm in length. Solid, spongy
texture. Dark green in colour .... ***Codium* spp.** p. 33
Frond not so **3**

3 Frond thin, flat membranous; with or without a stipe **4**
Frond not so **5**

4 Frond resembles a lettuce leaf; with wavy margin. With or without short stipe.
Generally bright green; up to 300 mm or more in length .... ***Ulva lactuca*** p. 30
Frond wedge-shaped; margins curled; arising from distinct stipe half as long as
frond. Up to 10 mm in length. Dark green in colour .... ***Prasiola stipitata*** p. 29

5 Frond tubular; elongate, often inflated; sometimes branching. Green to bright
green in colour, but often bleached white .... ***Ulva*** **spp.** p. 30
Frond filamentous, unbranched or branching **6**

6 (a) Frond unbranched; filaments of a single string of oblong cells which may
be visible to naked eye. Attached by a large basal cell ... ***Chaetomorpha*** **spp.** p. 31
or
(b) Frond much branched; tufted. Light green to dark green in colour.
Characteristically coarse to the touch .... ***Cladophora*** **spp.** p. 32
or
(c) Frond branching to give side branches each of which bears smaller branches
arranged as in a feather; glossy; limp to the touch .... ***Bryopsis plumosa*** p. 32

Prasiola stipitata Suhr ex Jessen **Fig. 2.5a**
*Frond thin, membranous; wedge-shaped to oval or linear. Attached by holdfast from
which stipe widens out into the frond; margins of frond curled. Dark green in colour.
Up to 10 mm in length, 3 mm wide.*

(a)

(b)

Figure 2.5 (a) *Prasiola stipitata.* (b) *Ulva intestinalis.*

P. stipitata is widely distributed in north-west Europe and around Britain on rocks and stones on the upper shore in shelter and exposure; it extends into estuaries and thrives in areas of nutrient enrichment. The species is generally more abundant in winter and there are records of its disappearance from some shores in southern Britain during summer.

The genus *Ulva* (*Enteromorpha*)

Research using molecular techniques has led to a review of the taxonomy of two of the most widely distributed and common groups of green seaweeds in north-west Europe formerly described under the separate genera of *Ulva* and *Enteromorpha*. Known under the common names of sea-lettuce and gut weed, respectively, they are readily separated on the basis of gross morphological characters such as a thin membranous frond of two cell layers thick, compared with an elongate, tubular, hollow frond one cell thick. Despite this, molecular phylogenetic analysis shows that they should not be assigned to separate genera but to the single genus **Ulva**, thirteen species of which have been recorded in Britain. The separation of the species can be difficult and sometimes depends on cell detail with reference to advanced texts but those described here can generally be identified on the basis of gross morphological characters. Nevertheless, care is required in identification as intraspecific variation in response to environmental conditions such as salinity is common. Reference to **Brodie** *et al.* (**2007**) is recommended.

Ulva intestinalis Linnaeus (*Enteromorpha intestinalis*) **Fig. 2.5b**

Frond elongate, tubular, inflated; may be compressed; may be branched when living in areas of reduced salinity; may be constricted at intervals and twisted; attached by disc-like holdfast or free-floating. Light to dark green in colour; up to 300 mm in length.

U. intestinalis is widely distributed in north-west Europe. It is found at all levels of the shore and, being tolerant of very low salinity, is particularly abundant where there is freshwater drainage and on the margins of pools on the upper shore where salinity is often reduced by rainfall. The impact of this is often a striking area of bright green running down the beach highlighting freshwater seepage. It is also common in estuaries and on saltmarshes, and in sheltered situations the fronds often become detached and survive as large, floating masses. It is an indicator of eutrophication and in areas where there is an abundant supply of nutrients, growth can be prolific, leading to fouling problems when the seaweed dies and decays.

U. intestinalis is sometimes difficult to separate from *Ulva compressa* Linnaeus (*Enteromorpha compressa*) in which the frond is similar but not inflated. A useful guide to separation is the fact that *U. compressa* has a much-branched frond while that of *U. intestinalis* is usually unbranched, except in plants living in reduced salinity. *Ulva compressa* does not tolerate low salinity but is widely distributed and common in north-west Europe on both sheltered and exposed shores attached to rocks and stones and in rock pools. It is also found as an epiphyte on other seaweeds. Like *U. intestinalis* it is an important fouling organism.

Ulva lactuca Linnaeus Sea-lettuce **Fig. 2.6a; Plates 2c, 5a**

Thin, membranous frond, lettuce-like. May have short, solid stipe; attached by disc-like holdfast Green to dark green in colour. Usually 100 to 300 mm in length but specimens up to 1 m have been recorded.

Figure 2.6   (a) *Ulva lactuca*. (b) *Chaetomorpha* sp. (c) *Ulva linza*.

U. lactuca is cosmopolitan in distribution and common on the shore, extending into the sublittoral. It is found on stones and rocks and other seaweeds and is common in brackish water habitats. It is commonly found on mud-flats attached to stones and shells, and in sheltered waters the thallus often attains a large size, sometimes becoming detached, the plants surviving as large floating masses. It thrives in areas of nutrient enrichment. The life-cycle is illustrated in Fig. 2.1.

Ulva rigida C. Agardh is morphologically very similar to *U. lactuca*. It is widely distributed on rocky shores in north-west Europe and like *U. lactuca* is often found free-floating in sheltered areas. The frond has a distinct stipe and is up to 300 mm in length but feels stiffer than *U. lactuca*.

Monostroma grevillei (Thuret) Wittrock also has a morphology similar to that of *U. lactuca*. The frond is up to 100 mm in length but more delicate than that of *Ulva* and usually funnel-shaped. It is found attached to rocks, most often in rock pools but also as an epiphyte on the red seaweed *Corallina* (p. 43).

Ulva linza Linnaeus (*Enteromorpha linza*) **Fig. 2.6c**
Elongate, unbranched, with hollow basal region which widens out into flattened frond up to 80 mm wide; shape variable, sometimes spiral-like with wavy margins. Pale green in colour; up to 300 mm in length.

U. linza is widely distributed in north-west Europe on the middle shore attached to rocks and in rock pools. It is also found as an epiphyte and is most abundant in fully marine, open coast situations, less so in areas where there is reduced salinity.

Chaetomorpha spp. **Fig. 2.6b**
Frond fine, filamentous, unbranched; made up of single string of oblong cells, which are sometimes visible to the naked eye. Usually attached by large basal cell but

sometimes unattached. Green in colour. Frond up to 300 mm in length but usually much smaller.

There are several species of *Chaetomorpha*, some of which are common and widespread in north-west Europe. Three species are widely distributed around Britain attached to rocks and seaweeds such as *Cladophora* (below) and *Codium* (p. 33), and the angiosperm *Zostera* (p. 75). Unattached forms are often found as large, interwoven masses of filaments or entwined with other seaweeds. Attached and unattached forms are common in both marine and brackish water.

Cladophora spp. **Fig. 2.7a**

Frond filamentous, usually much branched, often tufted; attached by a branched or disc-like holdfast, or unattached. Light to dark green in colour; up to 300 mm in length.

There are a number of species of the genus *Cladophora* in Britain and north-west Europe, but they are difficult to separate. They are widely distributed and common and occur at all shore levels, often in rock pools; some are tolerant of reduced salinity and extend into estuaries. In sheltered areas they occur as detached, floating masses. **Cladophora rupestris** (Linnaeus) Kützing is common all around Britain, occurring from the upper to lower shore in rock pools and under other seaweeds such as *Ascophyllum* (p. 61) and *Fucus* (p. 63). The frond is densely tufted, dark green to bluish-green in colour, characteristically coarse to the touch and up to 200 mm in length.

Acrosiphonia arcta (Dillwyn) Gain and **Spongomorpha aeruginosa** (Linnaeus) van den Hoek may be confused with *Cladophora*. Both are widely distributed around the British Isles.

A. arcta is found on the middle to lower shore on rocks and in rock pools. During winter the thallus is prostrate while branched filamentous fronds (up to 150 mm in length) are found through spring and summer, often in entangled masses.

S. aeruginosa is found on the lower shore. The erect, branched filaments (up to 50 mm in length) are often epiphytic, particularly on red seaweeds such as *Polyides* (p. 38). It is most abundant on the shore in spring and summer.

Figure 2.7 (a) *Cladophora rupestris*. (b) *Bryopsis plumosa*.

Bryopsis plumosa (Hudson) C. Agardh **Fig. 2.7b**

Frond erect, filamentous; branching to give side branches which each bear smaller branches arising in two opposite rows, the small branches becoming shorter distally, such that each side branch resembles a feather. Light to dark green in colour; glossy, limp to the touch. Up to 100 mm in length.

B. plumosa is widely distributed in north-west Europe and around Britain, where it can be found on the sides of deep pools and under rocky overhangs on the middle and lower shore, and into the sublittoral. It has also been recorded on stones and shells in sheltered areas. It is usually dioecious.

A second species, **Bryopsis hypnoides** J. V. Lamouroux, is widely distributed in the British Isles. The fronds are similar to those of *B. plumosa* but are more branched and the branches are often arranged spirally. It is found from the middle to lower shore in pools and attached to stones and shells, often in silt.

Codium spp.

Six species/subspecies of *Codium* have been recorded from around Britain and Ireland and although the form of the different species varies from spongy patches to erect, branching fronds up to 500 mm in length, the genus is readily identified by a number of diagnostic characters. Identification is based on microscopic examination of the frond and the shape of internal microscopic structures called utricles. These are about 1 mm in length and project onto the surface of the frond; they are clearly visible if the filaments are gently teased apart under water (Fig. 2.8). Different species/subspecies of *Codium* are known to coexist on some shores.

1	Thallus encrusting, spongy, grows in irregular patches .... ***Codium adhaerens*** p. 33	
	Thallus erect, dichotomous branching	2
2	Utricles terminate in point	3
	Utricles not pointed; club-shaped and bluntly rounded ... ***Codium tomentosum*** p. 34	
3	Utricles terminate in a short point ... ***Codium fragile*** subsp. *atlanticum* p. 33	
	Utricles terminate in an elongate point ... ***Codium fragile*** subsp. *fragile* p. 34	

Codium adhaerens C. Agardh **Fig. 2.8b**

Thallus grows as an encrusting spongy, felt-like mat 5–10 mm thick on rocky surfaces. Grows in irregular patches up to 1 m across; surface smooth or ridged, dark green in colour. Utricles terminate in flat apex with rounded edges.

C. adhaerens has been recorded from the south coasts of England and Ireland, extending as far south as Spain. It is found from the lower shore into the sublittoral and is tolerant of a wide range of exposure to wave action. It appears to thrive in areas where there is strong current flow.

Codium fragile (Suringar) Hariot **subsp. atlanticum** (A. D. Cotton) P. C. Silva **Fig. 2.8c**

Frond erect, formed of many interwoven branches arising from spongy disc-like holdfast formed of many fine threads; dichotomously branched, branches round in cross-section but may be flattened below point of branching. Solid, spongy, felt-like texture with many colourless hairs best seen when frond held under water. Utricles terminate in a short point. Dark green in colour. Up to 500 mm in length.

C. fragile subsp. *atlanticum* has been widely recorded around Britain and Ireland, extending as far north as Orkney. There are no reliable records for its distribution outside this area and this has led to the conclusion that, unlike *C. fragile* subsp. *fragile* (below), this subspecies is native to Britain and Ireland. The upright fronds do not

Figure 2.8 (a) *Codium fragile* ssp. *fragile*. Detail of utricles of (b) *C. adhaerens*, (c) *C. fragile* ssp. *atlanticum*, (d) *C. fragile* ssp. *fragile*, (e) *C. tomentosum*. ((c) and (e) after Burrows, 1991.)

survive well during winter and may be absent at this time, reappearing in spring and growing rapidly during summer.

Codium fragile (Suringar) Hariot **subsp. *fragile*** (*Codium fragile* subsp. *tomentosoides*) **Fig. 2.8a,d**
Frond as in C. fragile *subsp.* atlanticum (*above*). *Utricles terminate in an elongate point. Dark green in colour; up to 500 mm in length.*

C. *fragile* subsp. *fragile* is not native to Britain and is believed to have been introduced from Japan. It is now widely distributed in Europe and the Mediterranean and has been recorded from the Channel Islands, the Scilly Isles, Ireland, the south coast of England and the west coast of Scotland as far north as Argyll. It is found throughout the year on rocks and in rock pools on the middle and lower shore, into the sublittoral. It often carries epiphytic brown and red seaweeds.

Codium tomentosum Stackhouse **Fig. 2.8e**
Frond erect, formed of many interwoven branching filaments arising from broad, spongy holdfast; dichotomously branched, branches tend to taper towards apex but may be flattened below point of branching. Solid, spongy, felt-like texture

with many colourless hairs, best seen when frond held under water. Utricles somewhat club-shaped, apices bluntly rounded. Dark green; up to 450 mm in length.

C. tomentosum has been recorded from Britain and Ireland, extending northwards to Orkney. It appears to be distributed mainly in south-west Britain and in the Channel Isles has been recorded as abundant on the middle and lower shore. Although recorded on the shore throughout the year, it is most abundant in winter, reproducing in autumn and winter.

Division Rhodophyta – red seaweeds

The red seaweeds show wide variation in colour from bright red, greenish red to almost black, depending on the quantity of the red pigment phycoerythrin in the tissues. They are not as conspicuous on temperate shores as the greens and browns (being most abundant in tropical waters) and, being largely intolerant of desiccation, many are restricted to rock pools and the moist conditions under the seaweed canopy on sheltered rocky shores. Transects down the shore will show that some species are zoned but this is not as apparent to the casual observer as the conspicuous zonation patterns of the brown seaweeds. The morphology of the thallus of red seaweeds varies widely from delicate and membranous to tough and leathery, and in some of the common rock pool species the cell walls are heavily impregnated with calcium carbonate. Many species are epiphytic. Identification of red seaweeds is often difficult, requiring microscopic examination. In some species the morphology changes with the onset of reproduction and this may lead to difficulties in identification.

Key to species of red seaweeds

1 Thallus encrusting on surface of rock, stones, shell **2**
Thallus not encrusting **3**

2 Thallus hard, calcareous, usually thin but may be thrown into folds and bumps; encrusting rock, shells; in rock pools. Pink-red, purple, yellow in colour .... ***Lithophyllum incrustans*** p. 44
Thallus not calcareous, very thin; encrusting stones and rocks. Rose-red to dark red in colour .... ***Hildenbrandia rubra*** p. 46

3 Frond upright, much branched; made up of calcified segments; coarse to the touch; purplish, reddish, pinkish in colour **4**
Frond not made up of calcified segments **5**

4 Branching mostly pinnate; branches stiff, characteristically growing in rock pools; up to 120 mm in length. Purplish, reddish, pink, often bleached white .... ***Corallina officinalis*** p. 43

Branching dichotomous; small up to 25 mm in length; growing on other
seaweeds, especially *Cladostephus* (p. 60) .... ***Jania rubens*** p. 43

5   Frond low growing, creeping, moss-like or carpet-like in appearance; covering rock
surface **6**
Frond not so **7**

6   Frond of very fine, sparingly branched filaments; bound with sand to form a
spongy carpet over rocks; often extensive .... ***Rhodothamniella floridula*** p. 38
Frond much branched, constricted at intervals into different-sized segments;
moss-like in appearance; purple or black in colour   ... ***Catenella caespitosa*** p. 38

7   Frond with well-developed midrib **8**
Frond without midrib **11**

8   Frond without lateral veins. Ends of fronds pointed; branches arise from the
midrib .... ***Hypoglossum hypoglossoides*** p. 48
Frond with lateral veins which are conspicuous, or fine and difficult to see;
frond narrow or leaf-like **9**

9   Frond narrow; lateral veins very fine, seen with the aid of hand lens. Branches
arise from margin of frond (cf. *Hypoglossum* p. 48). Irregular dichotomous
branching .... ***Membranoptera alata*** p. 49
Frond leaf-like, veins conspicuous **10**

10 Margin of frond wavy, not serrated; with the appearance of a beech leaf; bright
red in colour .... ***Delesseria sanguinea*** p. 48
Margin of frond divided and serrated; with the appearance of an oak leaf;
crimson-brown in colour .... ***Phycodrys rubens*** p. 50

11 Frond broad, flat, not branching but may be divided; tough and opaque or very
thin and membranous **12**
Frond not so **16**

12 Frond flattened, mostly tough, may be leathery; undivided, or divided into broad
segments. Margin with or without outgrowths **13**
Frond flattened, very thin, membranous **15**

13 Frond widening from branching holdfast; undivided or irregularly divided;
margin with many small (5 mm) outgrowths forming fringe around
edge .... ***Calliblepharis ciliata*** p. 41
Frond not so **14**

14 Frond broad, widening from small disc-like holdfast and branching into broad
segments; no stipe. Margin of frond often with some small outgrowths.
Common epiphyte on stipes of laminarians .... ***Palmaria palmata*** p. 45
Frond broadening from small, disc-like holdfast, on short, cylindrical stipe;
undivided or split into a number of broad segments. Margin without
outgrowths. Found on stones and rocks .... ***Dilsea carnosa*** p. 45

15 Frond with irregular folded margin; central disc-like holdfast; dark purple,
reddish in colour .... ***Porphyra* spp.** p. 51
Frond divided only slightly or deeply into strap-like sections;
disc-like holdfast; surface of frond spotted; clear rose-pink to brownish-red
colour .... ***Nitophyllum punctatum*** p. 49

16 (a) Frond fine, tough and wiry in a tangled mass; often partly buried in sand .... ***Ahnfeltia plicata*** p. 40
or
(b) Frond narrow, tubular, inflated and twisted; dark brown-red in colour; in rock pools .... ***Dumontia contorta*** p. 45
or
(c) Frond very shiny, constricted at intervals into bead-like segments usually bright red in colour. On rocks, in rock pools .... ***Lomentaria articulata*** p. 46
or
(d) Frond shows alternate light and dark bands. When viewed under microscope seen to be made up of a row of large cells – monosiphonous. Usually with pincer-like tips. On rocks, stones, *Fucus* spp. (p. 63), laminarians (p. 57) .... ***Ceramium* spp.** p. 47
or
(e) When viewed under microscope frond seen to be made up of columns of cells – polysiphonous (Fig. 2.24b,c). On rock, stones, *Ascophyllum* (p. 61) .... ***Polysiphonia* spp.** p. 51
or
(f) Frond not fitting any of the above **17**

17 Frond flat, channelled or cylindrical, not constricted into segments; dichotomous branching **18**
Frond flat or cylindrical, not constricted into segments; branching alternate, opposite or irregular, may be much branched **21**

18 Frond cylindrical **19**
Frond flat or channelled **20**

19 Frond rising from much-branched holdfast; elongate reproductive bodies at ends of branches .... ***Furcellaria lumbricalis*** p. 38
Frond rising from disc-like holdfast; reproductive bodies as oval swellings along sides of branches .... ***Polyides rotundus*** p. 38

20 Frond flat, axils of branches rounded; dark red-purple in colour, edges often bleached green-yellow; tips of frond iridescent under water .... ***Chondrus crispus*** p. 42
Frond channelled, margins thickened, papilla-like reproductive bodies on surface of frond; dark reddish-brown, purple in colour .... ***Mastocarpus stellatus*** p. 42

21 Frond flattish, branching alternate or irregular, very distinctive; small branches in alternate groups, each small branch bearing tiny branches on one side only, so resembling teeth on a comb; rose-pink to dark red in colour. Turf-like or bushy growths .... ***Plocamium* spp.** p. 40
Branching not so **22**

22 Frond flattened, branching alternate in one plane only; tips of branches bluntly rounded; holdfast of intertwined creeping stolons; reddish-brown in colour often bleached to yellowish-green .... ***Osmundea pinnatifida*** p. 50
Frond not so **23**

23 Branching mostly alternate; branches become shorter towards apex; ends of branches bluntly rounded; dark purple to yellow in colour .... ***Osmundea hybrida*** p. 51
Frond not so **24**

24 Branching opposite or alternate, many smaller branches; branches narrow at base and tip; bright pink-red in colour .... *Lomentaria clavellosa* p. 47
Branching irregular, alternate. Much branched, many branchlets; soft and limp; brown, purplish in colour .... *Plumaria plumosa* p. 40

Rhodothamniella floridula (Dillwyn) Feldmann (*Audouinella floridula*, *Rhodochorton floridulum*) **(not illustrated)**

Frond of fine, sparingly branched filaments up to about 30 mm in length; some upright, some creeping. Brownish-red in colour. Filaments bound with sand to form a spongy, carpet-like covering up to about 30 mm thick over the rock surface; often extensive.

There are many species of *Rhodothamniella* and although microscopic examination of the filaments is required for species separation, *R. floridula* can be recognized by its habit of binding sand particles into a spongy, turf-like mass. It is a southern species extending to all parts of Britain, where it is widely distributed on the lower shore, often covering large areas of rock. It is perennial. The polychaete *Fabricia stellaris* (p. 172) often occurs among *Rhodothamniella*.

Furcellaria lumbricalis (Hudson) J. V. Lamouroux (*Furcellaria fastigiata*) **Fig. 2.9a**

Frond cylindrical, rising from much-branched holdfast (cf. Polyides, below); dichotomous branching; elongate reproductive bodies at ends of branches. Red-brown to brownish-black in colour (blackish-brown in transmitted light), glossy; sometimes bleached green. Up to 300 mm in length.

F. lumbricalis is widely distributed and common in north-west Europe on rocks and in pools on the lower shore, often in situations where the holdfast is buried in coarse, sandy deposits. Unattached, floating plants are found in sheltered localities. It is perennial and dioecious, living for up to seven years.

Polyides rotundus (Hudson) Greville **Fig. 2.9b**

Frond cylindrical rising from disc-like holdfast (cf. Furcellaria, above): dichotomous branching. Reproductive bodies seen as oval-shaped swellings along sides of branches. Dark red to black in colour (red in transmitted light). Up to 200 mm in length.

P. rotundus is widely distributed in north-west Europe on rocks and in pools on the lower shore, in similar situations to *Furcellaria lumbricalis* (above). It is perennial and dioecious.

Catenella caespitosa (Withering) L. M. Irvine (*Catenella repens*) **Fig. 2.10a**

Frond irregular and much branched, creeping; constricted at intervals into different-sized segments: holdfast of tangled fibres. Moss-like in appearance. Dull purple or black in colour. Up to 20 mm in height. Not to be confused with the lichen Lichina pygmaea (p. 73).

C. caespitosa is widely distributed and common in north-west Europe on rocks on the upper shore, and is sometimes abundant in sheltered situations such as saltmarshes, forming dense, matted growths. It is perennial and hermaphroditic.

Figure 2.9 (a) *Furcellaria lumbricalis.* (b) *Polyides rotundus.*

Figure 2.10 (a) *Catenella caespitosa.* (b) *Plocamium cartilagineum.* (c) *Ahnfeltia plicata.*

Plocamium spp. Cockscomb **Fig. 2.10b**

Frond compressed or nearly flat, much branched; branching alternate or irregular; larger branches have smaller branches arising alternately; smaller branches characteristically with tiny branchlets arising from one side only, so resembling the teeth on a comb. Rose-pink to dark red in colour, often bleached. Ranging in size from small and turf-like, 25 mm or less in length to bushy growths up to 120 mm in length.

Although the genus *Plocamium* is readily identified and widely distributed, research on molecular and morphological characters suggests that there are four species in northern Europe. Of these, ***Plocamium cartilagineum*** (Linnaeus) P. S. Dixon is the most widely distributed. It occurs on the lower shore on rocks and as an epiphyte, especially on the stipe of *Laminaria hyperborea* (p. 57), but is largely sublittoral. It is extensively branched and the pink to red fronds grow to a length of up to 120 mm. It is perennial and dioecious and rounded reproductive bodies are frequently seen on the fronds. It is often washed up after storms.

Plocamium subtile Kützing is similar in form to *P. cartilagineum* but the fronds are smaller, generally up to 100 mm in length, and of a darker red colour. The fronds are less frequently branched than those of *P. cartilagineum* and the branching is sparse in the proximal regions, increasing distally. It has been recorded from around Ireland and northern France on stones on the lower shore and as an epiphyte; it extends into the sublittoral.

By comparison, ***Plocamium nanum*** G. W. Saunders & Lehmkuhl and ***Plocamium maggsiae*** G. W. Saunders & Lehmkuhl can be described as dwarf species, reaching no more than 60 mm in length. Their distribution is poorly known and requires further study but both have been recorded intertidally and have distinctive habitats. *P. nanum* grows on rocks under fucoid seaweeds while *P. maggsiae* forms turf-like mats on rocks in moist areas where sea water drains from intertidal pools. Reference to Saunders & Lehmkuhl (2005) is recommended for species separation.

Ahnfeltia plicata (Hudson) Fries **Fig. 2.10c**

Frond very fine, tough, wiry; branching irregular or dichotomous. Ends of branches blunt. Disc-like holdfast. Purple when moist, almost black when dry. Growing in tangled masses; up to 150 mm in length.

A. plicata is widely distributed and common in north-west Europe in rock pools on the middle and lower shore, often partly buried in sand. It is perennial and dioecious.

Plumaria plumosa (Hudson) Kuntze (*Plumaria elegans*) **Fig. 2.11**

Frond arising from small disc-like holdfast. Main branch with irregular, alternate branching; very much branched with many branchlets themselves branched; grows in tufts. Soft and limp, brownish-purplish in colour. Up to 130 mm in length.

P. plumosa is widely distributed in north-west Europe and occurs all round Britain. It is found on the lower shore and is often abundant on shaded vertical rock faces where it hangs limply. It is rarely found as an epiphyte.

P. plumosa may be confused with ***Ptilota gunneri*** P. C. Silva, Maggs & L. M. Irvine which is also much-branched and grows from a small holdfast. Unlike *Plumaria*, however, the frond of *Ptilota* is more rigid and grows to 300 mm in length. There is a major difference in the habitats of the two species, *Ptilota* being epiphytic on the stipe and holdfast of *Laminaria hyperborea* (p. 57) and rarely found on rock. It is essentially a sublittoral species extending from the extreme low shore to depths of 10 m, occasionally deeper.

Calliblepharis ciliata (Hudson) Kützing **Fig. 2.12a**

Frond tough, cartilaginous; flattened, widening to about 70 mm across from a narrow, cylindrical stipe. Undivided or irregularly divided into strap-like sections. Margin of frond with many outgrowths, 5 mm or so in length, and giving the impression of a fringe around the margin; outgrowths sometimes also arise from surface of frond. Conspicuous branching holdfast. Dark red, purplish-red in colour, dull. Up to 300 mm in length.

C. *ciliata* is a conspicuous red seaweed reaching its northern limit in south and west Britain, becoming rare in Scotland. It is essentially sublittoral but also occurs in pools on the lower shore. It is dioecious and annual; the plants reach maturity in winter and are often washed ashore in the early part of the year.

Calliblepharis jubata (Goodenough & Woodward) Kützing (*Calliblepharis lanceolata*) is similar to C. *ciliata* but has a narrower, much more divided frond, up to 300 mm in length, and the margins are fringed with numerous long (up to 30 mm), narrow outgrowths (Fig. 2.12b). These often become entwined with one another and with other seaweeds. C. *jubata* is found in pools on the lower shore attached to rock and the stipes of laminarians.

Figure 2.11 *Plumaria plumosa.*

Figure 2.12 (a) *Calliblepharis ciliata.* (b) *Calliblepharis jubata.*

Chondrus crispus Stackhouse **Fig. 2.13a,b**

Frond flat, widening from narrow, unbranched stipe; dichotomous branching, axils of branches rounded. Disc-like holdfast. Dark red, purple in colour, edges of frond often bleached green-yellow. Tips of frond iridescent under water. Up to about 200 mm in length.

C. crispus is widely distributed in north-west Europe and often abundant on rocks and in pools on the middle and lower shore. It tolerates some reduction in salinity and is found in estuaries. Morphologically it is a highly variable species and the differences in the appearance of plants collected from shores of differing exposure, and even from different levels on the same shore, can lead to confusion in identification. *Chondrus* is easily confused with *Mastocarpus stellatus* (below). Together with the latter species it is harvested commercially as Irish moss or carragheen, used in the medical and food industries. It is perennial and dioecious, living for up to six years.

Figure 2.13   (a), (b) *Chondrus crispus*; note variation in form. (c) *Mastocarpus stellatus*.

Mastocarpus stellatus (Stackhouse) Guiry (*Gigartina stellata*) **Fig. 2.13c**

Frond channelled, particularly at base, with thickened margin; widening from narrow stipe; dichotomous branching. Disc-like holdfast. Mature plants with conspicuous, papilla-like reproductive bodies on surface of frond. Dark reddish-brown, purple in colour, may be bleached. Up to about 170 mm in length.

M. stellatus is widely distributed and common in north-west Europe on rocky shores, particularly those exposed to wave action, where it grows among barnacles and mussels. It is found on the lower shore and in rock pools and in less exposed conditions it is often abundant under fucoids. Like *Chondrus* (above), with which it is easily confused, it is variable in form. The channelled frond and the papilla-like reproductive bodies of *Mastocarpus* are important features in separating the species. It is a perennial, living for several years, and is probably dioecious.

Corallina officinalis Linnaeus **Fig. 2.14; Plate 2a**

Frond made up of calcified segments, longer than wide. Coarse to the touch. Branching mostly pinnate; branches stiff. Reproductive bodies seen as slightly swollen segments. Hard, encrusting holdfast. Colour varied, purple, red, pink and shades of yellow; extremities often bleached white. Up to 120 mm in length.

C. officinalis is widely distributed and often abundant in north-west Europe. It is one of the characteristic seaweeds of rock pools on the middle and lower shore and provides a habitat for a wide range of invertebrates including *Spirorbis corallinae* (p. 179), *Janua pagenstecheri* (p. 179) and small *Mytilus edulis* (p. 254). It is dioecious.

On south and west coasts of Britain and Ireland, *Corallina elongata* J. Ellis & Solander may be locally common in rock pools and gullies on the lower shore. The fronds are calcareous and up to 200 mm in length, but unlike those of *C. officinalis* they are rather limp and the segments are almost as wide as long. The reproductive bodies often have small, horn-like branches; branching of the reproductive bodies is much less frequent in *C. officinalis*.

A third species, *Corallina caespitosa* R. H. Walker, J. Brodie & L. M. Irvine, has recently been described from rock pools in south-west England. The fronds are stiff, up to 45 mm in length, smaller than those of *C. officinalis* and *C. elongata*, and much more densely branched.

Figure 2.14 *Corallina officinalis*, with detail of frond.

Jania rubens (Linnaeus) J. V. Lamouroux **Fig. 2.15**

Frond made up of calcified segments. Dichotomous branching; branches fine (cf. Corallina spp., above). Small encrusting, sometimes branching holdfast. Reproductive bodies as rounded swellings at fork of dichotomous divisions (females) or at ends of branches (males). Colour pink, red. Up to 25 mm in length.

J. rubens is widely distributed on shores of a wide range of exposure in north-west Europe growing in dense tufts on other seaweeds on the middle and lower shore, especially on *Cladostephus spongiosus* (p. 60) and *Cystoseira* (p. 66).

In addition to *Jania rubens*, **Jania rubens var. corniculata** (Linnaeus) Yendo also occurs in Britain, but is only found in the south. The habitat of these is similar and they are difficult to separate. *Corallina* and *Jania* can be confused with **Haliptilon squamatum** (Linnaeus) H. W. Johansen, L. M. Irvine & A. Webster, another species in which the fronds are made up of calcified segments. It is found on wave exposed shores, usually as an epiphyte on *Cystoseira* (p. 66), and is restricted to south and west coasts of Britain.

Figure 2.15   *Jania rubens* on *Cladostephus spongiosus* (detail of frond after Irvine & Chamberlain, 1994).

Lithophyllum incrustans   Philippi   **Plate 2a**

Thallus in the form of a hard, smooth, calcareous patch; chalky texture when dry; encrusting rock surfaces, shells and sometimes seaweed. Usually very thin, less than 1 mm in height, thicker at edges; but may be several millimetres in height and thrown into folds and bumps. Pink, red, purple, grey-yellow in colour.

There are several species of *Lithophyllum*. *L. incrustans* is very common and widely distributed in north-west Europe particularly on exposed shores and in rock pools on the middle and lower shore, often covering large areas of rock. It extends into the sublittoral. On death, the white, calcareous skeleton of the plant remains attached to the rock for some time, often giving a white fringe to rock pools. *Lithophyllum* is long-lived, reaching 12 or 13 years of age. Spores are released mainly during the winter months. On the west coast of Wales, the plants reproduce when about two years old.

Other genera of encrusting red seaweeds are commonly found in north-west Europe. These include **Lithothamnion** and **Phymatolithon** (well known because they form the rare maërl beds of shallow water) and **Mesophyllum**. These genera can be difficult to identify and reference to **Irvine & Chamberlain (1994)** is recommended.

Figure 2.16 *Dilsea carnosa.*

Dilsea carnosa (Schmidel) Kuntze   Red rags
Fig. 2.16
Frond flat, thick, opaque; tough and leathery; broadening from a short, cylindrical stipe; small, disc-like holdfast. The entire frond of small specimens is pear-shaped but in larger specimens the margin is split and the frond divided into broad segments; margin without outgrowths. Dark red in colour. Up to about 1 mm thick, 500 mm in length and 200 mm or so broad; smaller specimens more common.

D. carnosa is widely distributed and common on the lower shore attached to stones and rocks, and in the shallow sublittoral to depths of about 25 m. It is perennial; the young plants are generally seen on the shore in autumn and reach maturity the following spring.

Dumontia contorta (S. G. Gmelin) Ruprecht
(*Dumontia incrassata*) **Fig. 2.17a**
Frond tubular, narrow and rounded at base and at origin of branches, becoming wider and twisted; inflated in older specimens. Alternate or irregular branching. Disc-like holdfast. Dark brownish-red in colour, sometimes bleached. Up to 230 mm in length.

D. contorta is widely distributed and common on rocky shores in north-west Europe. It is found on the middle and lower shore on rocks and, particularly, in rock pools. The plants are small in winter and grow rapidly in spring and summer before dying back in the autumn. It is dioecious.

Palmaria palmata (Linnaeus) Kuntze (*Rhodymenia palmata*) Dulse **Fig. 2.17b**
Frond broad flat, very varied in form, generally widening from base and dividing into broad segments. Membranous, tough, opaque. Margins of frond often with small outgrowths. Disc-like holdfast. No obvious stipe. Purplish-red in colour. Up to 500 mm in length, about 80 mm across, but usually much smaller.

P. palmata is widely distributed and very common on rocky shores in north-west Europe. It is found on the lower shore attached to rocks or other seaweeds, particularly the laminarians, and is often abundant on the stipe of *Laminaria hyperborea* (p. 57).

In the south and west of Britain, a somewhat similar species, **Rhodymenia pseudopalmata** (J. V. Lamouroux) P. C. Silva (*Rhodymenia palmetta*) is often found growing on the stipe of *L. hyperborea* (p. 57). The pinkish-red frond arises by a short stipe from a disc-like holdfast and is up to about 50 mm in length. The frond is

Figure 2.17 (a) *Dumontia contorta.* (b) *Palmaria palmata.*

membranous, tough, opaque and much divided to become fan-shaped in outline. Unlike *P. palmata*, the margin of the frond does not have outgrowths.

Hildenbrandia rubra (Sommerfelt) Meneghini **Plate 2b**

Thallus very thin, encrusting on stones and rock, not calcareous; may cover large areas of rock surface. Reproductive bodies in pits on surface of thallus. Rose-red to dark red in colour; dull when dry.

H. rubra is widely distributed on rocky shores in north-west Europe and is common over most of the shore, particularly in shaded, moist situations where it forms conspicuous patches up to about 150 mm across.

      Hildenbrandia crouaniorum J. Agardh is found in similar locations and often with *H. rubra*. The two species are difficult to separate but colour can be a useful distinguishing feature; *H. crouaniorum* is brownish-red compared with the rose-red to dark red of *H. rubra*.

Lomentaria articulata (Hudson) Lyngbye **Fig. 2.18a**

Frond constricted at intervals into conspicuous bead-like segments, hollow; branches occur at constrictions. Disc-like holdfast. Dark brown to red in colour, very shiny; bleaching to pink and orange. Up to 100 mm in length, occasionally much longer.

L. articulata is widely distributed and common in north-west Europe. It is found on rocks and in pools on the middle and lower shore, sometimes occurring in profusion. It has been recorded growing on laminarians and is perennial and dioecious.

Figure 2.18   (a) *Lomentaria articulata*. (b) *Lomentaria clavellosa*.

Lomentaria clavellosa   (Turner) Gaillon   **Fig. 2.18b**

Frond cylindrical, not constricted into segments; main axis hollow, narrow at base. Branches opposite or alternate, with many smaller branches. Branches narrow at base and tip. Disc-like holdfast. Bright pinkish-red colour. Up to 400 mm in length.

L. clavellosa is widely distributed in north-west Europe on the lower shore on rocks and in pools. The extent and nature of branching varies with exposure to wave action. It is dioecious and probably annual.

Ceramium spp.   **Fig. 2.19**

Frond filamentous, made up of row of large cells (mono-siphonous) overlain by smaller cells at intervals to give a banding effect. Irregular, dichotomous branching; branches narrower near tips which are usually pincer-like. Reddish-brown, red or purple in colour, usually with alternate light and dark bands. Up to 300 mm in length.

The genus is widely distributed in north-west Europe and as many as 14 species

Figure 2.19   *Ceramium* sp., with detail of tip of frond (after Newton, 1931).

Figure 2.20 (a) *Delesseria sanguinea.* (b) *Hypoglossum hypoglossoides.*

of *Ceramium* have been described from Britain. '***Ceramium rubrum***' is often referred to in the literature but this is now known to be an aggregate of species which are very difficult to separate. Identification to the level of genus is recommended.

Ceramium spp. are found intertidally and in the sublittoral attached to rocks and stones, and epiphytically on other seaweeds such as *Fucus* spp. (p. 63), *Masto-carpus* (p. 42) and laminarians, particularly *Laminaria hyperborea* (p. 57).

Delesseria sanguinea (Hudson) J. V. Lamouroux Sea-beech **Fig. 2.20a**

Frond leaf-like; prominent midrib with lateral veins; margin wavy, not serrated, similar to beech leaf in appearance. Disc-like holdfast. Stipe tough, branched. Bright red in colour. Up to 250 mm in length.

D. sanguinea is widely distributed in north-west Europe extending from northern Spain to Norway and Iceland. It is a conspicuous seaweed found in deep pools on the lower shore, particularly in shade. The leafy fronds are best seen in spring and summer; reproductive bodies are borne on the stipe after the fronds have been cast in winter. It is perennial, probably living five or six years.

Hypoglossum hypoglossoides (Stackhouse) F. S. Collins & Hervey (*Hypoglossum woodwardii*) **Fig. 2.20b**

Frond narrow, leafy, usually up to 2 mm wide; with prominent midrib, without lateral veins. Often much branched, the branches arising from the midrib. Ends of fronds pointed. Rose-pink colour. Up to 200 mm in length.

H. hypoglossoides is widely distributed and common. It is found on rocks and stones, and on seaweeds, particularly the stipe of *Laminaria hyperborea* (p. 57). It has been recorded in pools on the lower shore and sublittorally to depths of 30 m. It is dioecious.

Hypoglossum is similar in appearance to ***Apoglossum ruscifolium*** (Turner) J. Agardh, which is found in the same type of habitat but is not as common as

Figure 2.21 (a) *Membranoptera alata*. (b) *Nitophyllum punctatum*.

Hypoglossum. The two are distinguished by the presence of microscopic lateral veins on the frond of *Apoglossum*, which reaches a length of 100 mm and is bright red in colour.

Membranoptera alata (Hudson) Stackhouse **Fig. 2.21a**
Frond flat, narrow, with midrib and many fine, lateral veins seen clearly under a hand lens. Branches arise from margin of frond. Irregular, dichotomous branching; branches narrower towards tip. Red or reddish-brown in colour. Up to 200 mm in length.

M. *alata* is widely distributed and common in north-west Europe extending from Spain to the Arctic. It is found in pools on the middle and lower shore, frequently growing on other seaweeds, particularly laminarians. The lamina is often damaged and torn during winter. It is perennial.

Nitophyllum punctatum (Stackhouse) Greville **Fig. 2.21b**
Frond flat, very thin (one cell thick over most of frond); delicate, without veins; no midrib. Short stipe or stipe absent; disc-like holdfast. Frond divided only slightly or deeply divided into strap-like sections. Surface of frond with many elongate reproductive bodies giving spotted appearance. Clear rose-pink to brownish-red in colour. Up to 500 mm in length.

N. *punctatum* is widely distributed in north-west Europe, and in Britain is commonest on the west coast where it occurs on rocks and stones on the lower shore and, particularly, in the sublittoral as an epiphyte on *Laminaria hyperborea* (p. 57). The shape of the frond varies considerably and while this is believed to be related to habitat care must be taken in identification as N. *punctatum* is difficult to separate from other similar species of red seaweeds growing epiphytically on the stipes of laminarians.

Figure 2.22 *Phycodrys rubens.*

Phycodrys rubens (Linnaeus) Batters **Fig. 2.22**
Frond leaf-life, with midrib and lateral veins; margin divided and serrated, giving appearance of oak leaf. Disc-like hold-fast. Stipe tough, elongated. Crimson-brown in colour. Up to 300 mm in length.

P. rubens is widely distributed in north-west Europe extending from Portugal to the Arctic. It is common in pools on the lower shore and sublittorally. It is a perennial, frequently found growing on the stipes of laminarians. As in *Delesseria* (p. 48), much of the lamina is lost in winter.

Osmundea pinnatifida (Hudson) Stackhouse (*Laurencia pinnatifida*) Pepper dulse **Fig. 2.23a**
Frond flattened, branches alternate, decreasing in length to apex; tough, cartilaginous. Branching in one plane only. Tips of branches bluntly rounded. Holdfast of intertwined creeping stolons. Reddish-brown in colour bleaching to yellow-green. Up to 150 mm in length.

O. pinnatifida is widely distributed all around Britain and extends south to the Mediterranean. It is found on the middle and lower shore and shows considerable variation in form and this, to some extent, is dependent on the position on the shore. Plants at the higher shore levels are often small, rarely more than 30 mm in length, and yellow-green in colour as a result of prolonged exposure to light. Large areas of rock are often covered by a dense growth of *O. pinnatifida*, which forms a turf supporting a wide range of invertebrates. It is perennial and dioecious.

(a) (b)

Figure 2.23 (a) *Osmundea pinnatifida.* (b) *Osmundea hybrida.*

Osmundea hybrida (A. P. de Candolle) K. W. Nam (*Laurencia hybrida*) **Fig. 2.23b**

Frond cylindrical, with alternate, occasionally opposite, branches; tough, cartilaginous. Branches become shorter towards apex; ends of branches bluntly rounded; disc-like hold-fast. Dark purple to yellowish in colour. Up to 200 mm in length.

O. *hybrida* is a southern species occurring all around Britain. It is found on the middle and lower shore on shells, rocks and in rock pools. It is dioecious and annual or perennial.

Polysiphonia spp. **Fig. 2.24a,b,c; Plate 4b**

Frond filamentous, polysiphonous (central cell is surrounded by a number of long, siphon-like cells). Generally dark reddish-brown colour. Some species up to 350 mm in length but most are much smaller. On seaweeds, rocks and stones.

The genus *Polysiphonia* is widely distributed and common and some 19 species have been described from British waters. One of the characters used to identify the species is the number and arrangement of cells, or siphons, seen in transverse section to surround a central siphon. *Polysiphonia* spp. are found on the middle and lower shore on rocks, stones and the shells of snails and limpets, and are epiphytic on a wide range of seaweeds, including fucoids, *Corallina*, the stipe of laminarians and *Chorda*, and are also on *Zostera*. **Polysiphonia lanosa** (Linnaeus) Tandy grows as conspicuous, dark brown-red filamentous tufts on *Ascophyllum nodosum* (p. 61), less frequently on *Fucus* spp. It has 12–24 siphons around a central siphon. Root-like fibres penetrate the tissues of *Ascophyllum*. P. *lanosa* is rarely found on stones. It is dioecious.

Figure 2.24   (a) *Polysiphonia lanosa* on *Ascophyllum nodosum*, with (b) diagrammatic representation of frond when viewed under microscope, and (c) when seen in transverse section, to show polysiphonous nature. (d) *Porphyra* sp.

Porphyra spp. **Fig. 2.24d**

Frond membranous, irregular folded margin; widening from narrow base; delicate, gelatinous texture. Small, central disc-like holdfast. Colour varies, dark olive, purple, red; blackens when dry and becomes brittle. Up to 200 mm in length, occasionally longer.

The genus *Porphyra* is widely distributed and includes a large number of species, some seven of which have been described from Britain. The genus is readily identified on the shore but separation of the species is difficult and identification to genus level is recommended. *Porphyra* spp. are found attached to rocks and boulders on the upper and middle shore, some species extending into the shallow sublittoral. They are found on a wide range of substrata including rocks, boulders and mollusc shells and are often abundant on man-made concrete structures and wooden jetties. Some species tolerate lowered salinity and are found in estuaries. *Porphyra* is harvested in Britain, particularly in Wales, and marketed as laver bread. This is likely to include a number of different species. In Japan, *Porphyra* is grown on nets in shallow water and dried on bamboo screens to provide the sheets of 'nori' used as wrappings in the sushi industry.

Division Heterokontophyta – brown seaweeds

Nearly 200 species of brown seaweeds have been recorded from the British Isles where they are often the most dominant seaweeds on rocky shores, extending from the high shore into the sublittoral. They are most abundant in cool temperate waters where nutrients are plentiful. Some, such as the laminarians (p. 57), grow to a considerable size and on the middle shore in sheltered waters canopy-forming species are common. These give protection and shelter to other seaweeds and invertebrates, and the fronds often carry a wide range of epiphytic species. They often show clear zonation of species down the shore and a commonly observed sequence is: *Pelvetia canaliculata* (p. 64) on the high shore below which is a belt of *Fucus spiralis* (p. 63) followed by a mixture of *Ascophyllum nodosum* (p. 61) and *Fucus vesiculosus* (p. 64), below which is *Fucus serratus* (p. 63), with laminarians marking the upper limit of the sublittoral zone. The cell walls of brown seaweeds comprise cellulose and gel-forming polysaccharides and this reduces the rate of water loss during emersion. They are generally intolerant of low salinity and as a result few species are found in estuaries.

Brown seaweeds are widely used in production of pharmaceuticals, and algin extracted from the cell walls is used in the food industry as a thickening agent in a range of processed foods, including ice cream.

Key to species of brown seaweeds

1 Thallus encrusting on rocks and shells; margins can be raised. Surface knobbly, dark brown, reddish-brown. (Do not confuse with lichens and encrusting red seaweeds.) .... *Ralfsia verrucosa* p. 56
Thallus not encrusting **2**

2 Frond with midrib **3**
Frond without midrib **8**

3 Frond not branching; long, strap-like with wavy margin; clusters of leaf-like
reproductive bodies on either side of stipe .... *Alaria esculenta* p. 60
Frond branching; reproductive bodies on frond, not on stipe .... *Fucus* **spp**. **4**

4 Margin of frond serrated .... *Fucus serratus* p. 63
Margin of frond not serrated **5**

5 Frond with paired spherical air bladders on either side of
midrib .... *Fucus vesiculosus* p. 64
Frond without spherical air bladders **6**

6 Frond fan-shaped, often inflated on either side of midrib; reproductive bodies
at ends of branches, narrow pointed. In brackish water and where freshwater
streams run onto beach .... *Fucus ceranoides* p. 62
Frond and habitat not as above **7**

7 Frond usually twisted; reproductive bodies at ends of branches, almost
round, surrounded by narrow rim of frond, the sterile margin. Upper
shore .... *Fucus spiralis* p. 63
Frond not twisted. Reproductive bodies without sterile margin; often forked
and pointed. Found on rocky shores, generally with a high degree of exposure
to wave action .... *Fucus vesiculosus* **forma** *linearis* p. 64

8 Frond large, may be well over 1 m in length; strap-like or wide and deeply
divided (digitate); not branched. Stipe and holdfast well developed. On the
lower shore and below; the kelps **9**
Frond not so **12**

9 Frond long, undivided, strap-like; surface wrinkled; margins wavy. Often in rock
pools on the middle and lower shore .... *Saccharina latissima* p. 58
Frond divided (digitate); surface smooth **10**

10 Stipe broad and flat with frilled margin; holdfast large, bulbous,
warty .... *Saccorhiza polyschides* p. 59
Stipe and holdfast not so **11**

11 Stipe narrow, smooth, flexible; oval in section. Frond wide, deeply divided
(digitate) .... *Laminaria digitata* p. 57
Stipe narrow, rough, stiff; rounded in section; often with red seaweed attached.
Frond wide, deeply divided .... *Laminaria hyperborea* p. 57

12 Frond globular or lobed and irregular; often hollow; attached to other seaweeds
and shells **13**
Frond not so **14**

13 Frond globular, lobed, irregular; solid when young, hollow later; thick walled and
shiny; growing on seaweeds, especially *Corallina* .... *Leathesia difformis* p. 55
Frond globular, thin walled and hollow; smooth and papery, not shiny;
growing on seaweeds and shells .... *Colpomenia peregrina* p. 55

14 Frond fine, filamentous and much branched, interwoven, slimy, *or* frond
cord-like, unbranched, tangled and slimy **15**
Frond not so **16**

15 Frond fine, filamentous and much branched; slimy, usually growing on other seaweeds; microscopic to 300 mm in length .... ***Ectocarpus* spp.** p. 55
Frond cord-like, unbranched; slimy, covered with colourless hairs. Up to 8 m in length .... ***Chorda filum*** p. 56

16 Frond button-like with very short stipe; with long, strap-shaped 'thong' up to 2 m in length, growing from centre of button. *Note:* the button-like fronds are without the thong-like outgrowth in the first year .... ***Himanthalia elongata*** p. 65
Frond not button-like **17**

17 Frond with large egg-shaped air bladders at intervals along it; rounded reproductive bodies on short stalks arising from frond. Up to 2 m or more in length; widespread on middle shore .... ***Ascophyllum nodosum*** p. 61
Frond not so **18**

18 Frond branching, bushy, stiff with whorls of tiny branches *or* frond branching, bushy, spiny and nodular with small air bladders. Up to 450 mm in length **19**
Frond branching; not bushy or spiny; may be flattened, cylindrical or channelled, with or without air bladders **20**

19 Frond stiff, branches with whorls of tiny branches. Dirty brown in colour; up to 250 mm in length .... ***Cladostephus spongiosus*** p. 60
Frond spiny, rough texture. Reproductive bodies with spines and nodules, at ends of branches; small air bladders below reproductive bodies. Iridescent when under water. Reminiscent of heather plant. Up to 450 mm in length .... ***Cystoseira tamariscifolia*** p. 66

20 Frond delicate, flat; tips of branches rounded, usually divided into two lobes. Reproductive bodies as fine specks over surface of frond .... ***Dictyota dichotoma*** p. 60
Frond flat, channelled or cylindrical, with or without air bladders **21**

21 Frond curled to form a marked channel; without air bladders; reproductive bodies at ends of branches. Upper shore. Black and brittle when dry .... ***Pelvetia canaliculata*** p. 64
Frond flattened or cylindrical; with or without air bladders. Not curled to form a marked channel **22**

22 Frond erect, hollow, may be constricted; without air bladders ....
.... ***Scytosiphon lomentaria*** p. 56
Frond with air bladders, not as above. In rock pools on the middle and lower shore **23**

23 Frond flattened, with pod-shaped air bladders divided by transverse septa. Reproductive bodies pod-shaped, on short stalks at ends of branches. Main stem zigzag .... ***Halidrys siliquosa*** p. 67
or
Frond cylindrical; elongate reproductive bodies at ends of branches; rounded air bladders may be present below the reproductive bodies .... ***Bifurcaria bifurcata*** p. 66
or
Frond with alternate branching; large number of lance-shaped, branchlets, spherical air bladders and reproductive bodies on short stalks ....
.... ***Sargassum muticum*** p. 68

Figure 2.25 *Ectocarpus* sp. with detail of reproductive bodies (after Newton, 1931).

Ectocarpus spp. **Fig. 2.25**
Frond of very fine much-branched filaments; tufted; grows as mass of interwoven filaments; often slimy in texture. Swollen reproductive bodies at ends of branches. Olive, brown, yellow in colour. Up to about 300 mm in length.

Two species of *Ectocarpus* occur around Britain, **Ectocarpus fasciculatus** Harvey and **Ectocarpus siliculosus** (Dillwyn) Lyngbye. Separation of the species is difficult and requires microscopic examination. Both are widely distributed and sometimes abundant in north-west Europe from the middle shore to the shallow sublittoral where they are found on rocks and on other seaweeds, including laminarians (p. 57), *Himanthalia* (p. 65) and *Chorda* (p. 56).

Leathesia difformis (Linnaeus) J. E. Areschoug **Fig. 2.26a**
Frond globular, or lobed and irregular; gelatinous; solid and rounded in young specimens, becoming hollow and lobed; thick walled and shiny (cf. Colpomenia, below). Yellow-brown in colour and up to 50 mm across.

L. difformis is widely distributed in north-west Europe growing on other seaweeds, especially *Corallina* (p. 43), on the middle and lower shore. The plants make their appearance in spring and grow rapidly to reach maturity during summer. *Leathesia* is an annual.

Colpomenia peregrina (Sauvageau) Hamel Oyster thief **Fig. 2.26b; Plate 4a**
Frond globular, thin walled (cf. Leathesia, above) and hollow; attached by fine filaments. When mature dotted with conspicuous brown specks. Smooth but not shiny; papery. Often torn and collapsed. Green, olive colour; up to 90 mm across.

C. peregrina is widely distributed in north-west Europe and in Britain is most common in the south-west. It grows attached to rocks, seaweeds and shells on the lower shore

(a) (b)

Figure 2.26 (a) *Leathesia difformis* on *Corallina*. (b) *Colpomenia peregrina*.

in sheltered situations. The hollow frond sometimes becomes filled with air during low water and the plant floats on the next tide and has been known to carry away an attached mollusc, hence the common name. *Colpomenia* was first recorded in Britain in the early 1900s possibly as an introduced species from the Pacific.

Ralfsia verrucosa (J. E. Areschoug) J. Agardh **(not illustrated)**

Thallus encrusting as thickish, irregular patches on rocks and shells; up to 100 mm across; surface knobbly. Quite firmly attached to substratum but margins can be raised. Olive to dark brown in colour. Not to be confused with lichens (p. 69) and encrusting red seaweeds (p. 44).

R. verrucosa is the only species of the genus found in Britain. It is widely distributed and abundant on exposed and sheltered shores where it extends from high to low shore but has not been recorded from the sublittoral. It is found in damp situations and in rock pools, and is commonly recorded on the shells of barnacles and molluscs, but not on seaweeds.

Scytosiphon lomentaria (Lyngbye) Link **Fig. 2.27a**

Erect, unbranched tubular fronds; may be inflated or collapsed; constricted at intervals. Small disc-like holdfast. Yellow to dark brown in colour; up to 400 mm in length.

S. lomentaria is widely distributed and common in north-west Europe where it is found on rocks, shells and stones from the high shore into the sublittoral, most often in rock pools. It is an annual species, most abundant in the spring and summer. The constrictions on the fronds are characteristic of mature plants.

A second species, *Scytosiphon dotyi* M. J. Wynn, has also been described from Britain but its distribution is restricted to the east coast of England where it is found on the high shore. The fronds are sometimes twisted but lack the constrictions characteristic of *S. lomentaria*. It is up to 140 mm in length.

Figure 2.27 (a) *Scytosiphon lomentaria*. (b) *Chorda filum*.

Chorda filum (Linnaeus) Stackhouse   Sea-lace, dead men's ropes, mermaid's tresses **Fig. 2.27b**

Frond cord-like, unbranched, round in section and hollow; 5–6 mm across; densely covered with short, colourless hairs during summer; rising from a tiny, disc-like holdfast. Slimy texture; brown to dark brown in colour. Up to 8 m in length.

C. filum is widely distributed and common in north-west Europe and in some areas is abundant. It is most frequently found in sheltered bays attached to stones and shells on the lower shore, often with the holdfast buried in sand and gravel. Sublittorally it is found to depths of about 20 m; occasionally it occurs higher on the shore in rock pools but is intolerant of emersion. The plants often grow together in a tangled mass of entwined fronds, sometimes supporting rich populations of epiphytes. *Chorda* tolerates some reduction of salinity. It is a summer annual, disappearing during the winter months. It is often washed ashore in large quantities.

The kelps

The large brown seaweeds found in the shallow sublittoral and on the lower shore of rocky shores are commonly known as kelps. In temperate waters these include the genera *Laminaria*, *Saccharina*, *Saccorhiza*, *Alaria* and *Undaria*, but are best known through the giant kelps of the Pacific belonging to the genus *Macrocystis* which can reach lengths of over 100 m. In temperate waters their presence marks the shoreward extension of the sublittoral zone (see Fig. 1.4). In many areas they form kelp forests and it has been estimated that there are more than 3 million tonnes of kelp in the shallow waters around Britain. They often dominate the ecology of the shallow sublittoral, reducing light penetration and providing refuges for a wide variety of fauna. They also provide an extensive surface for colonization by a diverse range of epiphytes. Research on the molecular phylogeny of the group has led to taxonomic reorganization and review.

Laminaria digitata (Hudson) J. V. Lamouroux Oarweed, tangle **Fig. 2.28 Plate 5b**
Stipe smooth, flexible, oval in section; expanding into a digitate frond. Without midrib. Attached by much-branched, slightly dome-shaped holdfast. Glossy, golden-brown in colour. Up to 2 m in length.

L. digitata is widely distributed in north-west Europe and is a conspicuous and characteristic species, forming a distinct zone at low water on most rocky shores. The blue-rayed limpet (*Ansates pellucida*, p. 204) is often found on the stipe and frond where it can do considerable damage, weakening the stipe. The holdfast offers a refuge for a wide variety of fauna. New growth takes place at the base of the frond where it joins the stipe and, although more or less continuous, is fastest in spring. Older parts of the frond are progressively worn away. It is a perennial.

Figure 2.28 *Laminaria digitata.*

Laminaria hyperborea (Gunnerus) Foslie
Fig. 2.29a
Stipe stiff, rough texture, rounded in section (cf. L. digitata, above); narrowing towards frond, which is expanded and deeply divided into strap-like sections. Without midrib. Holdfast large, much-branched and cone-like. Glossy, golden-brown in colour. Up to 3.5 m in length.

L. hyperborea is widely distributed in north-west Europe where it is common on the extreme lower shore, often forming a distinct zone below *L. digitata*. It extends into the sublittoral and where conditions are favourable forms dense forests. The rough stipe is often heavily clothed with a wide range of epiphytes such as *Palmaria* (p. 45), *Lomentaria* (p. 46) and *Delesseria* (p. 48), and is often weakened by the browsing activities of the blue-rayed limpet, *Ansates pellucida* (p. 204). *L. hyperborea* is a perennial. The frond almost ceases growth in summer, new growth

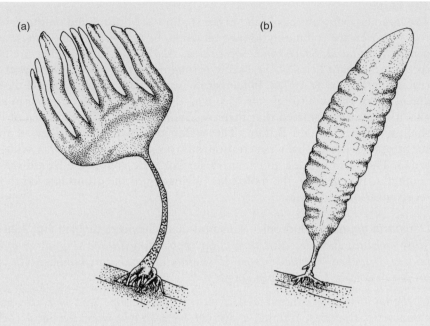

Figure 2.29　(a) *Laminaria hyperborea.* (b) *Saccharina latissima.*

commencing the following spring at the base of the old frond which is suddenly shed. The age of individual plants during the first few years of life can, with experience, be determined by counting growth lines laid down in the stipe and revealed by sectioning. Longevity up to about 20 years has been recorded in sheltered situations but is reduced on exposed shores where the older plants are susceptible to being torn away by wave action. *L. hyperborea* reproduces over winter.

Saccharina latissima　(Linnaeus) C. E. Lane, C. Mayes, Druehl & G. W. Saunders (*Laminaria saccharina*)　Sea-belt, sugar kelp　**Fig. 2.29b**
Stipe smooth, flexible, broadening to a long, undivided frond with wrinkled surface and wavy margins. Without midrib. Branching holdfast. Yellowish-brown in colour. Up to 4 m in length.

S. latissima is widely distributed in north-west Europe and is often abundant on sheltered rocky shores where large plants can be found and growth rates of over 20 mm per day have been recorded. It occurs on the lower shore and in the sublittoral while small, isolated plants are found in rock pools on the middle shore. The specific name refers to a sweet-tasting, whitish powder which forms on the dried frond. *S. latissima* is a perennial and although new growth occurs more or less continuously at the base of the frond, growth is most rapid in the first half of the year. Seasonal growth of the stipe results in alternate rings of light and dark tissue being laid down during periods of fast and slow growth, respectively. These rings can be seen in sectioned material and give an indication of the age of the plant. Length of life is up to three years. *S. latissima* reproduces over winter. The bryozoan *Celleporella hyalina* (p. 392) is often found in the depressions on the frond.

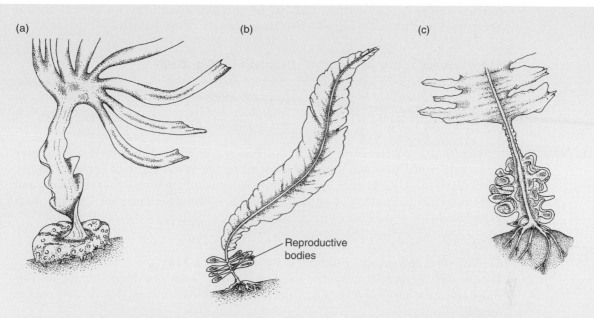

Figure 2.30 (a) *Saccorhiza polyschides*. (b) *Alaria esculenta*. (c) *Undaria pinnatifida*, detail of stipe.

Saccorhiza polyschides (Lightfoot) Batters Furbelows **Fig. 2.30a**

Stipe broad, flat, twisted at base and with conspicuously frilled margin; broadens into a wide frond often deeply divided into ribbon-like sections. Without midrib. Holdfast disc-like at first but grows into large, bulbous structure with warty appearance. Golden-brown in colour. Up to 4 m in length.

S. polyschides is widely distributed in north-west Europe and is often common at extreme low water and in the shallow sublittoral, particularly on sheltered shores. The morphology of the frond shows considerable variation depending on the nature of the habitat. Plants in areas of weak current flow have broad, undivided fronds while those in areas of strong current have long fronds deeply divided into many sections. Plants growing in areas without current flow but subjected to wave action have short fronds divided into relatively few sections. *Saccorhiza* is a very fast-growing annual and a growth rate of more than 2 m in two months has been recorded. Reproduction occurs during autumn and winter after which the fronds decay and are shed leaving the holdfasts attached to the rock. Eventually these become detached, many becoming washed up on the strandline.

The invasive kelp **Undaria pinnatifida** (Harvey) Suringar (Fig. 2.30c), a native of Japan, Korea and China has greatly extended its range worldwide over the past 20 years and was first recorded in Britain from the Hamble estuary in 1994. It is believed to have been transported from Brittany on the hulls of boats, and since that time has become established in a number of harbours and marinas on the south coast of England where in some places it has out-competed a few native species. *U. pinnatifida* grows up to 2 m and more in length. It has a branched holdfast, a stipe with distinctive corrugated margins and a broad frond with a midrib, giving superficial resemblance to *Alaria esculenta* (p. 60) but can be separated from the latter by the morphology of the stipe. Although it has spread rapidly since its introduction to Britain, *Undaria* is restricted to ports and marinas and has not become established on natural coastal habitats; research to date suggests that its spread into these habitats will be slow.

Alaria esculenta (Linnaeus) Grevill Dabberlocks **Fig. 2.30b**

Frond elongate, with wavy margins; distinct midrib, which is a continuation of the short stipe. Clusters of leaf-life reproductive bodies, each up to 100 mm in length, on either side of stipe of mature plant. Branching holdfast. Yellowish to olive-green in colour. Up to 1 m in length.

A. *esculenta* is widely distributed in north-west Europe and in Britain occurs most frequently in the west and north. It is found on the lower shore of exposed shores, particularly on vertical rock faces. The fronds are often torn by wave action and sometimes amount to little more than midrib. A. *esculenta* is a perennial, new growth being produced at the base of the frond.

Cladostephus spongiosus (Hudson) C. Agardh **Fig. 2.31a**

Frond stiff; branched, irregular or dichotomous; branches with whorls of tiny branches. Disc-like holdfast. Characteristic dirty brown colour. Up to 250 mm in length.

C. *spongiosus* is widely distributed in north-west Europe on rocks and in pools on the middle and lower shore. It sometimes supports dense tufts of the red seaweed *Jania rubens* (p. 43).

(a)   (b)   (c)

Figure 2.31   (a) *Cladostephus spongiosus.* (b) *Dictyota dichotoma.* (c) *Cutleria multifida,* tips of frond.

Dictyota dichotoma (Hudson) J. V. Lamouroux **Fig. 2.31b**

Frond delicate (3 cells thick), flat, no midrib; dichotomously branched, rising from disc-shaped holdfast. Tips of branches rounded, usually divided into 2 lobes. Reproductive bodies seen as fine specks on surface of frond. Yellow-brown, olive in colour, lighter towards tips; iridescent when under water. Up to about 150 mm in length.

D. dichotoma is widely distributed in north-west Europe and is sometimes abundant on south and west coasts of Britain in pools on the lower shore and in the sublittoral. It is much less common in the north and east. Gametes are released fortnightly on spring tides. *Dictyota* is an annual.

Cutleria multifida (J. E. Smith) Greville, widely distributed in the south and west of Britain, is a sublittoral seaweed often washed up on the shore. It bears some resemblance to *Dictyota* (above) but is up to 400 mm in length and the tips of the branches are divided several times to give a fringe-like effect (Fig. 2.31c). Like *Dictyota*, the reproductive bodies are seen as small dots over the surface of the frond.

Ascophyllum nodosum   (Linnaeus) Le Jolis   Knotted wrack
Fig. 2.32a; Plates 1b, 4b,c
Frond narrow, without midrib. Large, swollen, egg-shaped air bladders at intervals along middle of frond. Reproductive bodies rounded, on short stalks rising from margins of frond. Dichotomously branched. Olive-green in colour; reproductive bodies golden-yellow. Up to 2 m or more in length.

Figure 2.32   (a) *Ascophyllum nodosum* with the epiphyte, *Polysiphonia*. (b) *Ascophyllum nodosum* var. *mackayi*.

A. nodosum is widely distributed in north-west Europe. It is found on the middle shore below *Fucus spiralis* (p. 63) and is abundant on sheltered shores where it has a major impact on the ecology, offering refuge for a wide variety of flora and fauna in the moist conditions under the canopy. Wherever a suitable substratum is present it extends into estuaries. On more exposed shores scattered plants are found but the fronds are broken, torn and only a few centimetres in length; it is absent from the most exposed shores. *A. nodosum* is dioecious. Gametes are discharged in spring after which the reproductive bodies and their stalks are shed. After one or two years' growth, each branch produces a single air bladder per year, providing a means of estimating the age of undamaged plants. A length of life of about 12 years has been suggested but there are estimates as high as 25 years. Differences in morphology and age have been recorded in fronds growing at different shore levels; fronds on the higher shore levels have smaller air bladders and reach an age of about six years compared with fronds from lower on the shore where longevity of about 17 years has been recorded. *Ascophyllum* often bears tufts of the filamentous, epiphytic red alga *Polysiphonia lanosa* (p. 51) and occasionally epiphytic *Ectocarpus* or *Ectocarpus*-like seaweeds (see p. 55).

An unattached form of *Ascophyllum*, **Ascophyllum nodosum var. mackayi** (Turner) Børgesen & Jónsson (Fig. 2.32b), is widely distributed in north-west Europe. It is found on very sheltered shores, salt marshes and in sea lochs, and is common on the west coasts of Ireland and Scotland. The frond is flat or rounded with extensive dichotomous branching and bears few, if any, small air bladders. It has low fertility or is sterile and slow growing reaching a length of 400 mm. It drifts in large, spherical masses in the sheltered waters.

(a)   (b)

Figure 2.33   (a) *Fucus ceranoides.* (b) *Fucus serratus.*

Fucus ceranoides   Linnaeus   **Fig. 2.33a; Plate 3b**
Frond thin with smooth margin; prominent midrib; without air bladders but frond on either side of midrib may be inflated. Dichotomous branching, repeated frequently to give fan-shaped frond. Reproductive bodies narrow, pointed, at ends of branches. Pale olive-green in colour. Up to 600 mm in length, but usually much smaller.

F. ceranoides is widely distributed in north-west Europe but is common only in brackish water, and is often abundant where freshwater streams run onto the beach. It is attached to rocks and stones on the middle and lower shore and is one of the characteristic seaweeds of estuaries. It is usually dioecious but some plants are hermaphroditic. Gametes are released on daytime high tides at about the time of the new and full moon.

Fucus serratus Linnaeus Toothed wrack, serrated wrack **Fig. 2.33b**
Frond with serrated margin and conspicuous midrib; without air bladders. Dichotomously branched. Reproductive bodies at ends of fronds and surrounded by narrow rim of frond, the sterile margin; regions of frond bearing reproductive bodies flattened. Surface of frond with many tiny 'hairs'. Olive-brown in colour. Up to 1 m or more in length.

F. serratus is widely distributed and often abundant in north-west Europe but absent from the most exposed shores. It is found on the lower shore, often in a distinct zone below *Ascophyllum nodosum* (p. 61) and *Fucus vesiculosus* (p. 64). On sheltered shores the lamina of the frond is much broader than that from exposed shores and supports a wide range of epiphytes, including sponges, hydroids, bryozoans and the polychaete *Spirorbis spirorbis* (p. 178). It has been shown that fronds with heavy encrustations of bryozoans have a reduced rate of photosynthesis. *Lacuna pallidula* (p. 213) and the flat periwinkles *Littorina fabalis* and *L. obtusata* are also found on *F. serratus* together with their characteristic oval- or kidney-shaped, whitish egg masses (p. 216). *F. serratus* is dioecious and has a lifespan of up to about three years.

(a)   (b)

Figure 2.34 (a) *Fucus spiralis*. (b) *Fucus vesiculosus*.

Fucus spiralis Linnaeus Spiral wrack **Fig. 2.34a**
Frond with smooth margin; prominent midrib; without air bladders; often twisted. Dichotomously branched. Reproductive bodies at ends of branches; each almost round in outline, surrounded by narrow rim of frond, the sterile margin. Olive-brown in colour. Up to 400 mm in length but usually much smaller.

F. spiralis is widely distributed and common on sheltered rocky shores. It is found on the high shore just below *Pelvetia* (p. 64) and often forms a distinct zone between *Pelvetia* and the two common brown seaweeds of the middle shore, *Fucus vesiculosus* (p. 64) and *Ascophyllum nodosum* (p. 61). It is usually absent from the most exposed shores; in such situations the typical plant is sometimes replaced by a dwarf form,

Fucus spiralis forma *nanus* which in addition to being shorter is less branched and has narrower fronds. The two forms often occur on the same shore, but forma *nanus* is generally found higher on the shore. *Fucus spiralis* is hermaphroditic and lives for up to about four years.

Fucus vesiculosus Linnaeus Bladder wrack **Fig. 2.34b**

Frond with prominent midrib and almost spherical air bladders. Air bladders usually in pairs but may be absent in very small plants. Margin of frond smooth. Dichotomously branched. Reproductive bodies at tips of branches; often forked and pointed. Olive-brown in colour. Up to 1 m or more in length.

F. vesiculosus is widely distributed in north-west Europe and in sheltered situations is abundant. It is found on the middle shore, along with *Ascophyllum nodosum* (p. 61), in a zone above *Fucus serratus* (p. 63). It is tolerant of wide degrees of exposure to wave action, and the morphology of the plant varies in response to environmental conditions. On sheltered shores the frond has many air bladders, the number decreasing in exposed situations. Where exposure to wave action is high, plants without bladders may be common and these are known as *F. vesiculosus* forma *linearis*. They are no more than about 200 mm in length. The presence of air bladders, which in sheltered waters buoy up the fronds and help keep them in the illuminated surface waters, would be disadvantageous on wave-swept beaches, increasing the likelihood of the seaweed being torn away. On exposed shores the frond is much torn and the basal regions consist of little more than midrib. It is dioecious and lives for up to about three years.

The genus *Fucus* includes some of the best-known brown seaweeds of temperate waters. Those described here generally possess a number of distinctive, morphological features and occupy definite zones on the shore, but the morphology of the frond varies in response to environmental conditions, leading in some cases to difficulty in identification. This is clearly illustrated by *Fucus vesiculosus* forma *linearis* and *Fucus spiralis* forma *nanus*. Difficulty in identification also arises as a result of interbreeding between species to give hybrids that are believed to be common on some shores. Unattached fucoids are found on saltmarshes and very sheltered shores. Also occurring in these habitats is **Fucus cottonii** M. J. Wynne & Magne (*Fucus muscoides*) which takes the form of a dense mossy growth no more than 4 cm in height on mud and amongst saltmarsh vegetation, particularly on the west coasts of Scotland and Ireland.

Dictyopteris polypodioides (A. P. de Candolle) J. V. Lamouroux has a prominent midrib and has some similarity to a fucoid seaweed. The frond is, however, thinner and more delicate and is dotted with numerous clusters of tiny 'hairs'. It is up to 300 mm in length. Specimens just collected have an unpleasant, pungent smell. *Dictyopteris* is a southern species extending northwards to south-west Britain but is very local in distribution. It is essentially sublittoral under the canopy of laminarians, but is sometimes found on the lower shore.

Pelvetia canaliculata (Linnaeus) Decaisne & Thuret Channelled wrack
Fig. 2.35a; Plates 1b, 3a

Frond curled to form marked channel. Dichotomously branched; reproductive bodies at ends of branches. Without air bladders and midrib. Dark olive-green colour, becoming black and brittle as the frond dries out. Up to 150 mm in length.

(a)   (b)

Strap-like
reproductive
bodies

Figure 2.35   (a) *Pelvetia canaliculata.* (b) *Himanthalia elongata*, with early stages in the development
of the vegetative thallus.

P. canaliculata is common on rocky shores in north-west Europe and in some
situations forms a distinct band on the upper shore above *Fucus spiralis* (p. 63).
It is one of the highest zoned seaweeds and it has been estimated that it can spend
as much as 90% of its time out of water, surviving water loss of up to 65%. The
ability to survive such harsh conditions is believed to be related to the high levels
(20%) of fucoidan, a sulphated polysaccharide, in the cells. Fucoidan has the
property of taking up water. Fucoids and laminarians living on the lower shore
contain much lower levels (5%). Reproduction occurs during summer and autumn
when the reproductive bodies become orange coloured, often giving a distinct
band of colour on the upper shore. *Pelvetia* is hermaphroditic and lives for up to
five years.

Pelvetia is also found on saltmarshes growing among the angiosperms found
there. In these situations the thallus is sterile and has a distinct morphology from that
seen on rocky shores. On saltmarshes the thallus is stunted and twisted and not
attached.

Himanthalia elongata   (Linnaeus) S. F. Gray   Thong weed   **Fig. 2.35b; Plate 3c,d**
*Frond button-like with concave upper surface, up to about 30 mm across, with very
short stipe. Strap-shaped, dichotomously branched reproductive bodies grow from
centre of button; dotted with brown spots when ripe. Olive-green in colour. Up to 2 m
in length.*

H. elongata is widely distributed in north-west Europe and may be locally abundant.
It is found on the lower shore on wave-swept beaches and often forms a distinct zone
just above the laminarians. The young fronds are conical or club-shaped and broaden

Figure 2.36   (a) *Bifurcaria bifurcata.* (b) *Cystoseira tamariscifolia.*

as they grow to form a characteristic button shape with a concave upper surface. This is the vegetative thallus from which the strap-like reproductive bodies grow, generally in the following year. These reach maturity after about one year and die after the gametes are shed, usually in summer and autumn. *Himanthalia* is dioecious and lives for two to three years. The 'buttons', 1 or 2 cm across, are sometimes very numerous on the shore and often have a range of epiphytes on the undersurface (including the tube worm *Spirorbis inornatus*, p. 179), the upper surface apparently having some property which prevents this, presumably a deterrent against settling of spores and larvae. The reproductive straps support a rich flora and fauna.

Bifurcaria bifurcata   R. Ross   **Fig. 2.36a**

Frond cylindrical; unbranched near base then branching alternately or dichotomously. Rounded air bladders sometimes present. Elongate reproductive bodies at ends of branches. Yellow-olive in colour, much darker when dry. Up to about 500 mm in length.

B. bifurcata is a southern species found on the Atlantic coast of France and extending to the south and west coasts of England and the west coast of Ireland, where it is sometimes common. It is restricted to rock pools on the middle and lower shore, particularly on exposed beaches. *Bifurcaria* is perennial and hermaphroditic.

Cystoseira tamariscifolia   (Hudson) Papenfuss   Rainbow bladder weed   **Fig. 2.36b**

Frond cylindrical, much-branched, branches not swollen at base; bearing many small, spine-like structures giving rough texture and bushy appearance, reminiscent of heather plant. Air bladders small, usually occur just below reproductive bodies. Reproductive bodies with nodules and spines, borne on ends of branches. Olive-green in colour, almost black when dry; blue-green iridescence when under water. Up to 450 mm in length.

C. tamariscifolia is a southern species found on the Atlantic coast of France and south-west Britain and, although it has been recorded from the Hebrides, is rare in the north. It is found in rock pools on the lower shore, and is often common among shell and gravel deposits on sheltered shores. It extends into the sublittoral. *C. tamariscifolia* is hermaphroditic and perennial.

Other species of *Cystoseira* are found on south-west coasts, but *C. tamariscifolia* is readily distinguished by the iridescence seen when the plant is under water.

Halidrys siliquosa (Linnaeus) Lyngbye **Fig. 2.37**

Frond flattened, with alternate branches. Main stem of zigzag appearance. Ends of some branches with pod-shaped air bladders, divided by transverse septa. Reproductive bodies similar in appearance to air bladders but without the septa and borne on short stalks at ends of branches. Young plants olive-green in colour, older plants rich brown and leathery. Up to 2 m in length but usually much smaller.

Figure 2.37 *Halidrys siliquosa.*

H. siliquosa is widely distributed in north-west Europe on the middle and lower shore in rock pools and in the sublittoral. It often supports a range of epiphytes, including hydroids, bryozoans and ascidians. *Halidrys* is hermaphroditic and perennial.

Sargassum muticum (Yendo) Fensholt Japan weed, wireweed **Fig. 2.38**

Frond arises as a single stem from a conical disc-like holdfast; alternate branching; large number of lance-shaped branchlets, spherical air bladders and reproductive bodies borne on short stalks. Dark green or olive-green in colour. Up to 5 m in length.

S. muticum is an invasive species, native to Japan, which has spread rapidly in Europe and since the 1940s become established from south-west Portugal to the west coast of Norway. It was first recorded in Britain on the Isle of Wight in 1973 and has spread rapidly along the Channel coast. Its spread in Europe has been related to the importation of the Japanese oyster *Crassostrea gigas* (p. 259). *S. muticum* has now been recorded from Wales, Ireland and the west coast of Scotland. It is found on the lower shore and into the sublittoral, attached to stones, rocks and shells, and thrives in rock pools where at low water the large fronds are often conspicuous floating on the surface of the pool. It has also become established in soft sediments where it is attached to stones and shells buried in the sediment. When the fronds are large, their buoyancy can cause them to drift away taking the stone or shell to which they are attached with them and it is believed that dispersal in this way is important in the spread of the species. Unattached fronds are common, and fragments of

Figure 2.38 *Sargassum muticum.*

fronds continue to discharge spores as they drift, thus aiding distribution. Growth can be rapid and rates of 40 mm per day have been recorded in May and June in the English Channel. Where growth is fast it can out-compete and sometimes displace indigenous species and research suggests that it might pose a threat to *Zostera* sea-grass beds (see p. 75). There is evidence to suggest that the indigenous herbivores prefer the native seaweeds, giving *S. muticum* a competetive advantage. *S. muticum* carries a wide range of epiphytes, including ascidians, hydroids and polychaetes. It does not tolerate very low salinity but is believed to survive and reproduce in salinities down to about 20.

References

Brodie, J. A. & Irvine, L. M. (2003). *Seaweeds of the British Isles*. Vol. 1. *Rhodophyta*. Part 3B. *Bangiophycidae*. Andover: Intercept.

Brodie, J., Maggs, C. A. & John, D. M. eds. (2007). *The green seaweeds of Britain and Ireland*. London: British Phycological Society.

Bunker, F., Brodie, J., Maggs, C. & Bunker, A. (2010). *Seasearch guide to seaweeds of Britain and Ireland*. Ross-on-Wye: Marine Conservation Society.

Burrows, E. M. (1991). *Seaweeds of the British Isles*. Vol. 2. *Chlorophyta*. London: British Museum (Natural History).

Dixon, P. S. & Irvine, L. M. (1977). *Seaweeds of the British Isles*. Vol. 1. *Rhodophyta*. Part 1. *Introduction, Nemaliales, Gigartinales*. London: British Museum (Natural History).

Fletcher, R. L. (1987). *Seaweeds of the British Isles*. Vol. 3. *Fucophyceae (Phaeophyceae)*. Part 1. London: British Museum (Natural History).

Hardy, F. G. & Guiry, M. D. (2003). *A check-list and atlas of the seaweeds of Britain and Ireland*. London: British Phycological Society.

Hiscock, S. (1979). A field key to the British brown seaweeds (Phaeophyta). *Field Studies*, 5, 1–44.

Irvine, L. M. (1983). *Seaweeds of the British Isles*. Vol. 1. *Rhodophyta*. Part 2A. *Crypto-nemiales* (sensu stricto), *Palmariales, Rhodymeniales*. London: British Museum (Natural History).

Irvine, L. M. & Chamberlain, Y. M. (1994). *Seaweeds of the British Isles*. Vol. 1. *Rhodophyta*. Part 2B. *Corallinales, Hildenbrandiales*. London: HMSO.

Maggs, C. A. & Hommersand, M. H. (1993). *Seaweeds of the British Isles*. Vol. 1. *Rhodophyta*. Part 3A. *Ceramiales*. London: HMSO.

Newton, L. (1931). *A handbook of the British seaweeds*. London: British Museum (Natural History).

Lichens

Although the majority of lichens are terrestrial, some grow in areas subject to salt spray and are a conspicuous feature of the upper reaches of many rocky shores. Some withstand daily submersion by the tide and are found in the eulittoral zone, where they compete for space with other organisms, for example, barnacles and seaweeds. Lichens are complex plants composed of fungal and algal components existing in what is believed to be a mutually beneficial association known as symbiosis in which photosynthesis of the algal cells provides the fungus with organic nutrients, while the fungus provides the alga with support and protection. The exact nature of the relationship is still debated. Each species of lichen has its own species of fungus but the algal component is not specific to a particular fungus, and it is believed that the algae can exist independently, unlike the fungus which survives only in the lichen association.

The main part of the lichen, known as the thallus, is typically seen in one of three forms; crustose, foliose and fruticose. The flattened, crust-like thallus of crustose lichens is firmly attached to the substratum and very difficult to remove without damage to the lichen. Foliose lichens have a horizontal, leaf-like thallus usually loosely attached to the substratum and often divided into lobes. In the fruticose lichens the thallus stands upright or hangs downwards from the base.

Reproduction of lichens is complex and either sexual or asexual. In sexual reproduction, fungal spores, ascospores, are formed in groups, usually of eight, in special cells called asci. Asci are grouped on the surface of the lichen in apothecia or ascocarps some of which may be saucer-shaped and conspicuous (Plate 8a). The form of the ascocarps is important in identification. In other species, flask-shaped reproductive bodies known as perithecia are embedded in the thallus and open onto the surface via a pore through which the spores escape. Perithecia are generally black in colour. The fungal spores are dispersed by wind and after germination must combine with suitable algal cells to re-establish the partnership, a process known as lichenization. Asexual or vegetative reproduction occurs in a variety of ways. Fragmentation of the thallus, particularly when dry and brittle, with survival of the fragments leads to spread of the species, but in some cases distinct structures comprising both fungal and algal elements detach from the thallus and are dispersed.

Lichens grow best on stable surfaces and do not flourish on friable slates and shales or in areas where there is atmospheric pollution. They are affected by the chemical nature of the substratum, different species growing on acidic rocks compared with calcareous rocks. Lichens are generally regarded as slow-growing and long-lived, and while some show almost no growth, others attain growth rates of 10–30 mm per year. Longevity in excess of 50 years has been recorded for some species. Lichens found intertidally and at the littoral-fringe–terrestrial boundary often form distinct zones according to their ability to withstand submersion in seawater and salt spray. On rocky shores, the most striking of these is a zone of the black lichen *Verrucaria maura* (p. 71) above which is a zone of the orange lichens, *Caloplaca* spp. (p. 73) and *Xanthoria parietina* (p. 73) together with the grey-green coloured *Ramalina* spp. (p. 72) (see Plate 6b).

Key to species of lichens

Although some lichens are difficult to identify, those common on the shore can generally be recognized on the basis of external characters seen under a good-quality hand lens. For more advanced studies, examination under a microscope is essential and chemical testing may be necessary.

1 Thallus crustose – crust-like, firmly attached to the substratum; difficult to remove **2**
Thallus foliose – horizontal leaf-like thallus often divided into lobes, usually only loosely attached to the substratum **8**
Thallus fruticose – thallus upright or hangs downward from the base; much branched and shrubby or strap-like **9**

2 Thallus black, smooth, very difficult to remove from rock; crossed by a meshwork of fine cracks; covers large areas; resembles an oil stain on rock surface; at upper limit of littoral fringe often forming distinct band .... ***Verrucaria maura*** p. 71
Thallus not resembling a black oil stain **3**

3 Thallus dark green, smooth, almost gelatinous, very difficult to remove from rock; in irregular patches up to 300 mm across on stones and rock; not a distinct band; littoral fringe to middle shore .... ***Verrucaria mucosa*** p. 71
Thallus not dark green **4**

4 Thallus orange **5**
Thallus as small black dots embedded in barnacle or mollusc shells, particularly barnacles, and calcareous rocks ***or*** thallus grey, encrusting **6**

5 Thallus rusty-orange, granular, margin irregular; just above littoral fringe .... ***Caloplaca marina*** p. 73
Thallus shiny, bright orange colour, margin lobed; just above littoral fringe .... ***Caloplaca thallincola*** p. 73

6 Thallus as small black dots embedded in shells of barnacles, molluscs and calcareous rocks; as blackish patches on harder substrata
.... ***Collemopsidium foveolatum*** p. 74
Thallus grey, encrusting **7**

7 (a) Thallus thick, cracked warty surface; apothecia with black centres and pale irregular margins; above littoral fringe .... ***Tephromela atra*** p. 72
or
(b) Similar to above, but apothecia with smooth margins .... ***Lecanora gangaleoides*** p. 72
or
(c) Thallus thick, warty, edged with white; apothecia with pinkish-grey centres and thick margins; above littoral fringe .... ***Ochrolechia parella*** p. 72

8 Thallus bright orange, lobed, leafy; loosely attached to substratum; just above littoral fringe .... ***Xanthoria parietina*** p. 73
Thallus dark green, spongy when wet, brown and stiff when dry; thick with branching overlapping lobes; just above littoral fringe ..... ***Anaptychia runcinata*** p. 72

9 Thallus light grey-green, strap-shaped; upright or hanging downwards; upper limit of littoral fringe and above .... ***Ramalina siliquosa*** p. 72
Thallus dark brown, black; upright branching lobes; shrubby; may be reminiscent of small dried seaweed **10**

10 Thallus up to 10 mm in height; flattened, branching lobes; grows in fairly open tufts, often covering large areas; from lower littoral fringe to mid-shore. Reminiscent of small dried seaweed .... ***Lichina pygmaea*** p. 73
Thallus up to 5 mm in height; rounded, much-branched lobes, grows in dense tufts; not covering extensive areas; upper littoral fringe and above .... ***Lichina confinis*** p. 74

Verrucaria maura Wahlenberg **Plate 6a,b**
Thallus crustose, smooth; adheres closely to substratum and very difficult to remove. Surface of thallus crossed by meshwork of fine cracks. Black in colour. Perithecia visible as black spots.

V. maura is a very common lichen, widely distributed in north-west Europe. Where conditions are favourable it forms a broad, black band on the upper reaches of the littoral fringe, contrasting sharply with the grey-white barnacles below and the bright orange-yellow *Xanthoria* (p. 73) and *Caloplaca* (p. 73) above. It is often described as being similar in appearance to an oil stain on the rocks. The upper limit of *Verrucaria* forms the upper limit of the littoral fringe (see p. 4). In the sheltered sea lochs of the west coast of Scotland, it is one of the most characteristic organisms of the upper shore.

Verrucaria mucosa Wahlenberg **Plate 9c**
Thallus crustose, smooth, almost gelatinous; in irregular patches up to 300 mm across; adheres closely to surface of substratum, very difficult to remove. Green, olive in colour, becoming darker in bright sunlight. Perithecia visible as black spots.

V. mucosa is widely distributed and often abundant on the west coast of Britain on rocks and boulders and is also found on very sheltered shores, often on loose stones a few centimetres across. On some shores large patches are very conspicuous, particularly when moistened, because of the green colour. It extends from the littoral fringe to the middle shore, but does not form a distinct band so characteristic of *V. maura* (above).

Tephromela atra (Hudson) Hafellner (*Lecanora atra*) **Plate 8a**

Thallus crustose, thick with cracked, often warty surface; grey in colour. Apothecia with black centres and pale, irregular margins. When apothecium is cut open it is seen to be purple-brown in colour.

T. atra is a common terrestrial lichen of north-west Europe found on stones and walls. It withstands salt spray and extends into the lower reaches of the terrestrial fringe where it is common on rocky outcrops.

Lecanora gangaleoides Nylander **(not illustrated)**

Thallus crustose, thick with cracked, often warty surface; grey or greenish-grey in colour. Apothecia with black centres and pale, smooth margins. When apothecium is cut open it is seen to be green-brown in colour.

L. gangaleoides is common on rocks close to the sea. It is similar in appearance to *Tephromela atra* (above) and is often confused with that species. The tendency to a greenish-grey colour of the thallus in *L. gangaleoides* is a useful character in the separation of the species, as is the colour difference seen when the apothecia are cut open.

Ochrolechia parella (Linnaeus) A. Massalongo **Plate 8b**

Thallus crustose, thick and warty; grey in colour, edged with white. Apothecia with pinkish-grey centres and thick margins; with powdery surface. In patches up to 200 mm across.

O. parella is a common terrestrial lichen of north-west Europe, found on stones, walls and trees. It extends into the lower reaches of the terrestrial fringe where it is sometimes abundant on rocks exposed to salt spray.

Ramalina siliquosa (Hudson) A. L. Smith Sea-ivory **Plate 7a**

Thallus fruticose, upright or hanging downwards; may be sparsely branched; strap-shaped; up to 100 mm in length. Light grey-green colour. Apothecia usually at distal ends of branches, white or pale brown in colour.

R. siliquosa is widely distributed in north-west Europe. It is found in patches at the upper limit of the littoral fringe and above, and in some situations forms a distinct band above *Xanthoria* (p. 73) and *Caloplaca* (p. 73). It often grows with the moss *Grimmia maritima* (p. 73), and in parts of Britain is grazed by sheep.

A second species, ***Ramalina cuspidata*** (Ach.) Nylander, is similar to *R. siliquosa* and the two can be difficult to separate. As a guide, the base of *R. cuspidata* tends to be darker in colour than the rest of the thallus. It is found on rocks exposed to spray but is less common than *R. siliquosa* and is not found inland.

Anaptychia runcinata (Withering) J. R. Laundon (*Anaptychia fusca*) **Plate 9b**

Thallus foliose, thick, with branching, overlapping lobes. Dark green and spongy when wet; brown and stiff when dry. Apothecia with black centres and brown, wavy margins.

A. runcinata is widely distributed in north-west Europe, growing on rocks at the littoral-fringe–terrestrial boundary.

Caloplaca marina (Weddell) Zahlbruckner in Du Rietz **Plate 7b**
Thallus crustose, with many small granules, particularly in centre of thallus, together with apothecia. Thallus and apothecia rusty-orange colour. Margin of thallus irregular. May be confused with Xanthoria parietina *below, but thallus flatter and not leafy.*

C. marina is widely distributed and common in north-west Europe. Along with *Caloplaca thallinicola* (below) and *Xanthoria parietina* (below) it often forms a broad, yellow-orange band just above the littoral fringe.

Caloplaca thallincola (Weddell) Du Rietz **Plate 7c**
Thallus crustose, with apothecia in the centre of the thallus. Thallus and apothecia shiny, bright orange colour. Margin of thallus lobed.

C. thallincola is widely distributed in north-west Europe and on western coasts of Britain. It often occurs with *C. marina* (above) and is easily confused with that species, but is brighter orange in colour, the thallus is not as flat and the margin is lobed. It occurs in sheltered, shaded areas in contrast to the sunny positions preferred by *C. marina*.

Xanthoria parietina (Linnaeus) Th. Fries **Plate 9a**
Thallus foliose, lobed; broad and leafy. Loosely attached to substratum. In patches up to about 100 mm across. Bright orange colour changing to greenish-yellow in shaded, humid conditions. Apothecia deep orange in colour. May be confused with Caloplaca marina *and* C. thallinicola *(above) but thallus leafy.*

X. parietina is widely distributed in north-west Europe, growing in profusion on rocks, walls, trees and roofs, especially in areas enriched by nitrogenous compounds. It is also abundant on rocks on the high shore and along with *Caloplaca* spp. often forms a distinct orange-yellow band just above the littoral fringe. In some places it is covered by the highest spring tides.

The maritime moss ***Grimmia maritima*** Turner often occurs with *Xanthoria parietina* and *Ramalina siliquosa*, especially on the west coast of Britain, but rarely on calcareous rocks. It is dark, brownish-green in colour and grows as small, compact clumps up to 40 mm in height.

Lichina pygmaea (Lightfoot) C. Agardh **Plate 8c**
Thallus fruticose, formed of upright, flattened, branching lobes. Up to 10 mm in height. Dark brown, black; brittle when dry. Rounded apothecia at ends of branches. Grows in fairly open tufts, often covering large areas. Not to be confused with the red seaweed Catenella caespitosa *(p. 38).*

L. pygmaea is widely distributed and common in north-west Europe. It has been recorded from the lower littoral fringe to about the middle shore where it is regularly covered by the tide and tolerates exposure to wave action. It is usually found in association with barnacles and in favourable situations such as steep slopes in open sunny positions on exposed shores it can cover several square metres of rock surface. The lichen harbours large numbers of invertebrates, particularly the isopod *Campecopea hirsuta* (p. 326), which feeds on *Lichina*, and the bivalve *Lasaea adansoni* (p. 266).

Lichina confinis (O. F. Müller) C. Agardh **(not illustrated)**

Thallus fruticose, formed of upright, rounded, much-branched lobes. Up to 5 mm in height. Dark brown, black. Rounded apothecia at ends of branches. Grows in dense tufts; does not cover extensive areas of rock.

L. confinis is widely distributed and common in north-west Europe. It occurs higher on the shore than *L. pygmaea* (above) and is found on the upper littoral fringe in the *Verrucaria maura* zone and above, often with *Caloplaca marina* (p. 73). It is most abundant on sheltered shores in sunny positions.

Collemopsidium foveolatum (A. L. Smith) F. Mohr *(Pyrenocollema halodytes, Arthopyrenia halodytes)* **Plate 9d**

Thallus crustose. On hard rocks seen as small, blackish-brown patches; on calcareous rock and shells of barnacles and molluscs is embedded in the shell and appears as small black dots. Obvious on the plates of acorn barnacles.

C. foveolatum is widely distributed and common in north-west Europe. It occurs on the upper shore on rocks and on the shells of a number of common intertidal invertebrates, particularly acorn barnacles but also on mussels and limpets.

References

Alvin, K. L. (1977). *The observer's book of lichens*. London: Frederick Warne.

Dalby, D. H. & Dalby, C. (2005). *Shetland lichens*. Lerwick: Shetland Amenity Trust.

Dobson, F. S. (2005). *Lichens. An illustrated guide to the British and Irish species*, 5th edn. Slough: Richmond Publishing.

Dobson, F. S. & Dalby, C. (1997). *Lichens of rocky shores*. Slough: Richmond Publishing.

Fletcher, A. (1975). Key for the identification of British marine and maritime lichens. I. Siliceous rocky shore species. *Lichenologist*, 7, 1–52.

Fletcher, A. (1975). Key for the identification of British marine and maritime lichens. II. Calcareous and terricolous species. *Lichenologist*, 7, 73–115.

Smith, C. W., Aptroot, A., Coppins, B. J., Fletcher, A., Gilbert, O. L., James, P. W., & Wolseley, P. A. eds (2009). *The lichens of Great Britain and Ireland*. London: Natural History Museum Publications.

Anthophyta

The Anthophyta, commonly known as the angiosperms, are flowering plants and include the trees, shrubs, herbs and grasses so conspicuous in the terrestrial environment. They are characterized by seeds which develop and ripen inside an ovary. While the vast majority are terrestrial, some live in freshwater ponds, lakes and rivers, but few are adapted to the marine environment. Of these, some are tolerant of periodic submersion in sea water and are often common on the shore, while others are tolerant of salt spray and are characteristic of the littoral-fringe–terrestrial boundary. The angiosperms included here tolerate daily submersion by the tide, and together with seaweeds such as *Ulva* (*Enteromorpha*) (p. 30), are important colonizing and stabilizing species of sand and mud deposits. Almost invariably, the presence of these plants slows down water movement, resulting in the deposition of fine particles and an increase in the level of the beach leading to a plant succession typical of saltmarshes and dominated by angiosperms such as thrift (*Armeria maritima* Willd.), sea aster (*Aster tripolium* L.) and seablite (*Suaeda maritima* (L.) Dumort.); these are not included in this text. The three genera described below are readily identified and common on sheltered shores and estuaries.

Salicornia europaea Linnaeus Common glasswort, Marsh samphire, Pickle-weed
Fig. 4.1c; Plates 27b, 28a
Plant branched, fleshy and shiny; leaves opposite; stem cylindrical. Bright green in colour changing to reddish colours in autumn. Up to 150 mm in height. Tiny flowers in axils of leaves. Dies back in winter.

There are several species of *Salicornia* in north-west Europe and of these *Salicornia europaea* is widely distributed. It is one of the first angiosperms to colonize mud-flats but is not able to colonize soft sands. The plants withstand regular submersion by the tide and are often found along with the green seaweed *Ulva* (p. 30) and isolated plants of *Spartina* (p. 77). The seeds of *Salicornia* are an important component in the diet of wildfowl such as the teal, and the plants themselves are occasionally harvested and marketed for human consumption. *S. europaea* is an annual.

Zostera marina Linnaeus Eel-grass, sea-grass, grass-wrack **Fig. 4.1a**
Long, narrow, flattened leaves with rounded tips; grass-like; creeping rhizome. Leaves 5–10 mm wide, up to 1 m in height, but usually much less. Dark green in colour. Flowers borne on one side of much-branched stem. Seeds ribbed.

In the early 1930s much of the *Zostera* population, particularly the sublittoral populations, on the Atlantic coasts of North America and Europe was destroyed by a wasting disease seen as dark necrotic patches on the leaves. Various factors including fungal infection, bacterial infection, weakening of the plants due to lack of sunshine and increased water temperature due to unusually warm summers and mild winters

Figure 4.1 (a) *Zostera marina* (seed inset). (b) *Spartina anglica*. (c) *Salicornia europaea.*

have been suggested as possible causes of the disease, which has been seen as recurrent outbreaks in some localities. It has now been established that a marine slime mould, *Labyrinthula zosterae* Porter & Muehlstein is responsible. It may be that *Labyrinthula* is less active at low salinities and this may account for the survival of some intertidal populations. *Zostera* recovered to some extent and by the 1950s had returned to many areas. It is now widely distributed in north-west Europe but occurs only sporadically around Britain in sheltered bays and estuaries growing on mud, sand and gravel from the lower shore into the shallow sublittoral where it often covers extensive areas. The plants flower in summer and ripe shoots become detached and are pollinated in the water by rafts of floating pollen. The ripe seeds sink to the bottom and germinate. *Zostera* beds are important and sensitive habitats. They support many species, including hydroids, bryozoans, crustaceans, polychaete worms, gastropod molluscs and fishes, and are grazed extensively by wildfowl such as brent geese and widgeon. The plants have in the past been used as packing material for mattresses and cushions, and were used in the building of sea dykes in the Netherlands during the nineteenth century, but do not appear to have been used commercially since the 1930s.

A second species, *Zostera noltii* Hornemann, the dwarf eel-grass, is also widely distributed in north-west Europe. It grows in sheltered bays and estuaries in mud and sand, usually between mean high and mean low water of neap tides and is generally found higher on the shore than *Z. marina*, rarely extending into the sublittoral. The leaves are up to 200 mm in height, much narrower (1 mm in width) than those of *Z. marina* and have divided tips. The flowering stems are unbranched and the seeds smooth. *Z. noltii* is less susceptible to the slime mould *Labyrinthula* than is *Z. marina*.

Spartina anglica C. E. Hubbard Common cord-grass, rice-grass
Fig. 4.1b; Plates 27a,b,c, 28b
Tall, tough grass with stiff, pointed leaves. Large, creeping rhizome. Greyish-green in colour. Usually up to about 1 m in height. Yellowish flowers in spikes during summer. Dies back in winter.

There are many species of *Spartina* colonizing the saltmarshes of north-west Europe and the history and development of *S. anglica* is of considerable interest. It is now believed that a North American species, *Spartina alterniflora* Loiseleur, was accidently introduced by shipping into the area around Southampton Water in the English Channel in the early 1800s. It was first recorded there in 1829. As a result of a cross between this species and the endemic British species, *Spartina maritima* (Curtis) Fernald, *Spartina* × *townsendii* H. & J. Groves was formed and this was first recorded in 1870 at Hythe in Southampton Water. *S.* × *townsendii* is a sterile hybrid and spreads to new areas by fragmentation and plants carried by currents. As a result of a natural doubling in the number of chromosomes of *S.* × *townsendii*, a new vigorous plant, *S. anglica*, was formed. It was first recorded in 1892 on the south coast of England at Lymington and spread rapidly by seedlings colonizing large areas of mud-flats, and is now widely distributed in north-west Europe.

S. anglica, like *Salicornia* (p. 75), is a pioneer colonizer of bare mud. Where the two species grow together, *Salicornia* occurs lower on the shore but in areas where the sediment is very unstable, *Salicornia* cannot become established and the deeper-rooting *Spartina* is the main colonizer. The deep roots are extremely effective in binding mud and stabilizing the sediment while the shallower, more delicate roots are mainly absorptive. The plants slow down water movement and cause fine particles to be deposited with a consequent accretion and increase in the level of the sediment. *Spartina* withstands long periods of tidal submersion. Characteristically, the initial colonization of bare mud results in the development of 'islands' of *Spartina* plants which spread vegetatively and by seedlings, and dense swards soon become established (Plate 27a,c). The decaying detritus associated with such areas, much of which is derived from the dying back of the grass at the end of the growing season, forms an important component of estuarine food webs and once established, *Spartina* has a major impact on the ecology of the estuary. The characteristic infauna of mud-flats is lost and large populations of periwinkles, *Littorina littorea* (p. 215) and *L. saxatilis* (p. 218), the mud snail, *Hydrobia ulvae* (p. 222) and juveniles of the shore crab, *Carcinus maenas* (p. 366) are often found in association with *Spartina*. Shoals of small fishes, including the sand goby, *Pomatoschistus minutus* (p. 472), occur in the creeks and drainage channels among *Spartina* and on the west coast of Wales, the drainage channels in estuaries colonized by cord-grass are important nursery grounds for young bass (p. 457) and mullet (p. 458).

The spread of *Spartina* in north-west Europe was hastened by planting of the grass to prevent sediment erosion and as a means of reclaiming land for grazing. In 1991 it was claimed that *S. anglica* covered about 10 000 hectares of intertidal saltmarsh. This has resulted in the loss of the natural mud-flat habitat and its characteristic infauna of polychaete worms, bivalve molluscs and crustaceans which are so important in the diet of wildfowl. Such has been the rate of spread of *Spartina* in some areas that the loss of these important feeding grounds is giving cause for concern and control of the spread of the species by the application of herbicides has been considered. Interestingly, die-back of dense stands of *Spartina* has been recorded on the south coast of England and this is attributed to rotting of the rhizome, a phenomenon which may in some way be age-related as older populations have suffered most. *S. anglica* also suffers from infection by the ergot fungus, **Claviceps purpurea** (Fr.) Tulasne, which prevents the seed from setting.

References

Den Hartog, C. (1970). *The sea-grasses of the world*. Amsterdam: North-Holland Publishing Company.
Stace, C. (1997). *New flora of the British Isles*, 2nd edn. Cambridge: Cambridge University Press.

Porifera

Sponges are frequently found on the lower shore in gullies and under rocky overhangs. Here they avoid desiccation and display a range of colours and growth forms giving them a superficial resemblance to plants. Indeed, it was not until the eighteenth century that they were classified as animals and as late as 1825 before the issue was put beyond doubt by the research of Dr R. E. Grant of Edinburgh.

Sponges are sessile, some forming encrusting, irregular growths covering large areas of rock, while others are vase-like and attached to the substratum at the base. They exhibit the cellular grade of organization, in which the cells do not form tissues and organs so characteristic of higher animals, and as a result they are often referred to as primitive animals. Despite this simplicity, they exhibit a range of complexity in body structure and have colonized both freshwater and marine habitats, extending from the shore to the greatest depths.

The basic design of a simple sponge is that in which the body wall is perforated by many tiny pores, ostia, through which water enters the sponge, and a larger opening, the osculum, through which water leaves. The cavity of the sponge is lined by flagellated cells known as choanocytes which maintain a flow of water through the body. A simple system is seen in Fig. 5.1c. In many sponges, however, the system is more complex because of the folding, sometimes extensive, of the body wall and this leads to a series of cavities and canals called the aquiferous system. In some species the body wall is supported by a skeletal network of calcareous or siliceous spicules which are of various shapes and sizes. Others have a skeleton of protein fibres known as spongin, while some have a combination of siliceous spicules and spongin fibres. These skeletal arrangements are important in taxonomy.

As water flows through the body of the sponge, suspended food particles are filtered out by the flagellated cells but, interestingly, in some species from deep water where suspended food is in short supply, the system is modified and the sponge is carnivorous. Hook-shaped spicules project onto the outer surface of the body where they trap prey such as small crustaceans.

Regeneration of sponges is well documented, particularly through the now famous experiment in which a sponge is forced through a fine mesh and within a short time the dissociated cells aggregate to form many new sponges. Regeneration from fragments is an important form of asexual reproduction, others being the production of internal or external buds and the release of clusters of cells known as gemmules which develop into a new sponge.

Sponges also reproduce sexually. Most are hermaphroditic but cross-fertilization generally takes place. Sperm are discharged through the osculum and enter a neighbouring individual in the inhalant current. In some cases, fertilized eggs are discharged into the sea where they develop into ciliated larvae known as planulae, but in the majority the larva develops within the sponge and is then released. Larval life varies from a few hours to a few weeks and the adult lives for one to several years, some species showing regression and fragmentation during the winter months. Sponges often have a rich associated fauna of, for example, amphipod crustaceans and are preyed on by a variety of animals including sea-slugs and echinoderms.

Sponges are divided into three classes, which are distinguished by taxonomists according to the nature of the spicules.

Phylum PORIFERA

Class Calcarea	**Sponges with calcium carbonate spicules. Restricted to firm substrata. Most species found in shallow water.**

Class Hexactinellida	Sponges with siliceous spicules mainly of six rays. Many colonize soft substrata. Found in deep water. Commonly known as the glass sponges.

Class Demospongiae	**Sponges with siliceous spicules, spongin fibres or both. Found on a wide range of substrata from rock to soft mud; some bore into calcareous material. Found in shallow and deep water; includes the majority of the common shore species.**

Key to species of sponges

Accurate identification of sponges is difficult and sometimes impossible without dissection and microscopic examination of the skeleton and spicules. A small number of common shore species is included here and reliable identification of these can generally be made by reference to the external features described. However, it should be noted that some of these species show considerable morphological variation from one habitat to another in response to factors such as water depth and speed of current flow, and this can lead to difficulty in identification.

1 (a) Flattened, vase-shaped or tubular .... **purse sponges 2**
or
(b) Encrusting, may be in thin sheets or substantial cushion-like growths several centimetres thick. Sometimes lobed .... **encrusting sponges 4**
or
(c) Limestone rocks and shells, especially oyster shells, often riddled with small round holes about 2 mm in diameter. Yellowish-orange projections may be seen at entrance to holes. .... **boring sponge *Cliona celata*** p. 85

2 Delicate, tubular; upright, often branching tubes arise from encrusting network. Osculum at end of each branch. Up to 20 mm in height. On seaweeds. Grey-white in colour .... ***Leucosolenia* spp.** p. 83
Vase-shaped, not branching **3**

3 Flattened, vase-shaped; up to 20 mm in height. Conspicuous terminal osculum, occasionally more than 1. Often hanging from rocks and seaweeds .... ***Grantia compressa*** p. 81
More cylindrical; conspicuous terminal osculum surrounded by stiff spicules. Rough, 'hairy' outer surface. Usually to about 30 mm in height. On rocks and seaweeds .... ***Sycon ciliatum*** p. 82

4 Large, thick fleshy sponge, up to 50 mm thick. May be in the form of rounded lobes. Surface channelled or smooth. Oscula conspicuous **5**
Not as above **6**

5 Usually rounded in shape; up to 200 mm across. Surface smooth. Oscula few in number but conspicuous .... ***Suberites ficus*** p. 83
Thick, irregular lobes; surface marked with channels, slimy. Oscula conspicuous, unevenly distributed, often aligned along ridges .... ***Myxilla incrustans*** p. 84

6 Flat, thin, usually about 3 mm thick; surface smooth. Oscula well defined, with small collar. Blood-red, orange-red in colour .... ***Ophlitaspongia papilla*** p. 85
Not as above **7**

7 Surface relatively smooth. Oscula well developed, often raised above surface of sponge and likened to 'miniature volcanoes'. In some situations may be up to 20 mm thick. Usually olive-green colour. When broken has distinctive smell of seaweed
.... ***Halichondria panicea*** p. 84
Surface rough, often grooved. Relatively thin encrustations or papillate. Scattered, small oscula, not always obvious. Orange-red colour .... ***Hymeniacidon perlevis*** p. 84

Class Calcarea

Grantia compressa (Fabricius) (*Scypha compressa, Spongia compressa*) Purse sponge
Fig. 5.1b; Plate 11a
Vase-shaped, sometimes rather elongate. Conspicuous terminal osculum; occasionally more than one. Up to 20 mm in height, sometimes larger; 10 mm wide. Thin and flat. Colour varied, white, grey, yellow.

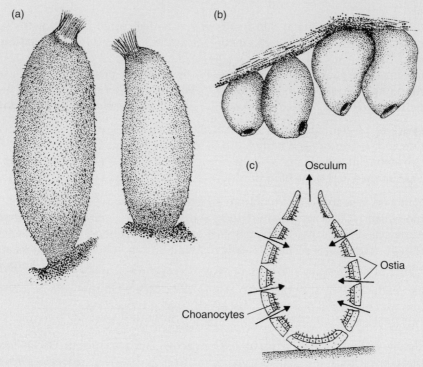

Figure 5.1   (a) *Sycon ciliatum*. (b) *Grantia compressa*. (c) Diagrammatic section of sponge to show direction of flow of water (arrows).

G. compressa is widely distributed in north-west Europe and around Britain in sheltered crevices and under boulders on the lower shore where it occurs singly or in groups attached to a variety of substrata, often hanging from the roofs and walls of crevices. It is an annual species. Overwintering specimens die after releasing larvae in spring and summer; sponges settling in spring reproduce in late summer and autumn and survive the winter to reproduce in the following spring. Specimens settling later in the year are believed to reproduce only once, in the spring.

Sycon ciliatum (Fabricius) (*Scypha ciliata, Sycon coronatum, Spongia ciliata*)
Fig. 5.1a

Body upright, vase-shaped or tubular; more cylindrical than G. compressa *(above). Height about 30 mm but much larger specimens have been recorded. Conspicuous terminal osculum fringed with stiff spicules (cf.* G. compressa*). Rough, 'hairy' outer surface, sometimes smooth. Yellow-grey colour.*

S. ciliatum is widely distributed in north-west Europe and around Britain where it is common on the lower shore. It is attached singly or in groups to stones and seaweeds, the largest specimens being found in sheltered situations. It is an annual. Overwintering specimens release larvae in the spring and some of these develop rapidly to reproduce in the autumn of the same year, before overwintering.

Figure 5.2   (a) *Leucosolenia* sp. growing on red seaweed. (b) *Suberites ficus*. (c) *Myxilla incrustans*.

Leucosolenia spp. Fig. 5.2a
Encrusting network from which upright, often branching tubes arise. Oscula conspicuous, terminal. Surface rough. Up to about 20 mm in height, but usually much smaller, grey-white in colour.

Leucosolenia is widely distributed in north-west Europe and around Britain. The species are difficult to identify. They are attached to seaweed and stones on the lower shore and in the sublittoral, sometimes with the purse sponge, *Grantia compressa* (p. 81). *Leucosolenia* is believed to be an annual, breeding twice, once in autumn and again the following summer.

Class Demospongiae

Suberites ficus (Johnston) Sulphur sponge, sea-orange Fig. 5.2b
Large, thick, fleshy sponge with smooth, velvety surface. Usually rounded in shape and up to 200 mm across. Oscula large, conspicuous, few in number. Orange, yellow or brownish in colour.

S. ficus is widely distributed in north-west Europe attached to a variety of hard substrata on the lower shore and in the sublittoral, sometimes encrusting gastropod shells and scallops. It has been suggested that *S. ficus* is a complex of three species.

Other species of *Suberites* are common in the sublittoral, including ***Suberites domuncula*** (Olivi) which is often found on shells occupied by hermit crabs, the sponge completely enclosing the shell which is sometimes dissolved away but the hermit crab is protected by the sponge. Separation of the species of *Suberites* is difficult and requires microscopic examination of the spicules.

Halichondria panicea (Pallas) Breadcrumb sponge **Plate 10a**

Encrusting, from relatively thin sheets to thick encrustations. Relatively smooth surface. Oscula well developed, sometimes raised well above surface of sponge. Large encrustations may be 20 mm thick. Very variable in form. Characteristically olive-green in colour but shades of yellow common. When broken has a distinctive smell of seaweed.

H. panicea is widely distributed in north-west Europe and around Britain and is sometimes abundant. It is one of the most common sponges on the lower shore where it is found in gullies, under rocky overhangs and among the holdfasts of seaweeds. It covers extensive areas of rock and often overgrows other encrusting invertebrates. The form of the sponge varies, particularly in relation to current flow, and in areas of strong currents it is much thicker and the oscula are more raised than in sheltered waters, the faster current presumably enhancing the flow of water through the sponge. The typical olive-green colour of the species is due to the presence of green algal symbionts in the tissue; specimens from deeper water lack the symbionts and are yellow in colour. In thick specimens, only a thin outer layer is green, the deeper tissues being yellow. Large patches of *Halichondria* often have a diverse associated fauna including crustaceans, annelids, echinoderms and fishes. The variegated scallop *Chlamys varia* (p. 261) is often overgrown by *Halichondria* forming an association which reduces predation on the scallop by starfishes, while the sponge takes food from the inhalant current created by the scallop. There is evidence to suggest that some amphipods, e.g. *Caprella* (p. 346), are attracted to chemicals given off by the sponge. *Halichondria* reproduces in spring and summer when larvae are released into the plankton. Lifespan is believed to be about three years and in some localities breeding has been recorded annually.

   Halichondria bowerbanki Burton is similar in appearance to *H. panicea* and like *H. panicea* it is very variable in form. The two species are easily confused but useful guidelines for separation are that *H. bowerbanki* is mostly sublittoral and rarely found on the shore, is more tolerant of muddy and silty conditions such as in harbours, is creamish in colour, not green, and when broken does not have the distinctive smell of seaweed that is so characteristic of *H. panicea*.

Myxilla incrustans (Johnston) **Fig. 5.2c**

Thick, encrusting, up to 200 mm across and 50 mm thick. Oscula well developed, unevenly distributed and often aligned along ridges. Surface marked with channels, slimy. Yellow to orange in colour.

M. incrustans is widely distributed in north-west Europe, encrusting rocks and boulders on the lower shore and in the sublittoral. Specimens with embryos have been recorded during August and September.

Hymeniacidon perlevis (Montagu) (*Hymeniacidon perleve, Hymeniacidon sanguinea*) **Plate 11c**

Encrusting, relatively thin or papillate intertidally, sublittoral specimens thicker and cushion-like; sometimes buried in mud. Surface rough and often grooved. Small, scattered oscula, randomly arranged, not always obvious. Orange-red to deep red in colour; colour, form and texture vary widely.

H. perlevis is widely distributed in north-west Europe and around Britain, and like *Halichondria* (p. 84) is one of the commonest intertidal sponges. The largest encrustations are found on the lower shore and cushion-like growths have been recorded from the shallow sublittoral. The species extends to the middle shore where it is found as small fragments. It is found on a variety of substrata including rocks, seaweed holdfasts and also on muddy gravel, where only the oscula can be seen on the surface of the sediment. Growth in the summer months is often followed by a reduction in size and by fragmentation of the sponge in winter. Off the south coast of England embryos have been recorded in the sponge from July to October, with larval release presumably taking place in late summer. Longevity is believed to be three or more years.

Ophlitaspongia papilla Bowerbank (*Ophlitaspongia seriata*) **Plate 10b**
Encrusting; flat and thin, usually about 3 mm thick; growing in patches 50–100 mm across. Surface smooth. Oscula well defined, mainly regular in distribution; with small collar. Blood-red, orange-red in colour.

O. papilla is found on west and south-west coasts of Britain, as far north as the Hebrides and may be locally common. It has been recorded as far south as Portugal. It is found on the lower shore and sublittorally to depths of 1–2 m. Embryos are found in the sponge during summer and larval life is about two days.

Cliona celata Grant Boring sponge **Plate 11b**
Limestone rocks and stones on the lower shore, and the empty shells of bivalve molluscs such as the oyster, are often riddled with small, round holes about 2 mm in diameter. These have probably been made by *C. celata*, a boring sponge sometimes seen as yellowish-orange projections at the entrance to the holes. The sponge bores a complex of cavities in the substratum and the extent of this can be seen if the rock or shell is broken open. Boring is started by the larva and is brought about by secretions, possibly including carbonic anhydrase, from special cells which eat into the rock. A small chip of the substratum is surrounded by the cell and eventually undercut and removed in the exhalant current, the sponge gradually penetrating deeper. An unidentified species of *Cliona* has been recorded in the shells of living *Littorina littorea* (p. 215) and this weakens the shell and makes the periwinkle more vulnerable to predation by shore crabs.

Cliona is also seen in a non-boring form as large growths up to 1 m across, with many rounded, raised oscula. These occur sublittorally to 40 m and may show considerable morphological variation. Recent research suggests that *C. celata* is probably a complex of sibling species.

References

Ackers, R. G., Moss, D. & Picton, B. E. (1992). *Sponges of the British Isles (Sponge V)*. Ross-on-Wye: Marine Conservation Society.
Hooper, J. N. A. & van Soest, R. W. M. (2002). *Systema Porifera. A guide to the classification of sponges*, Vols. 1 and 2. New York: Kluwer Academic/Plenum Publishers.

Cnidaria

The phylum Cnidaria, formerly known as the Coelenterata, includes the hydroids or sea-firs, the sea-anemones, jellyfishes and corals. These apparently diverse forms have in common a basic radial symmetry and a relatively simple body structure with a body wall made up of an outer epidermis (ectoderm) and an inner gastrodermis (endoderm), separated by a jelly-like mesogloea. Cnidarians therefore have two layers of cells and are described as diploblastic, compared with the triploblastic grade of organization shown by the majority of animals, in which there is a third cell layer, the mesoderm. The body cavity has a single opening to the exterior called the mouth and this is usually surrounded by tentacles, which are without cilia, but have special cells known as cnidocytes (nematocytes). These contain capsule-like structures, the cnidae, that are unique to the phylum, and from which the phylum gets its name. The cnidae contain a thread which is discharged on stimulation; in some (the nematocysts) the thread penetrates the prey and injects a toxin; in others (the spirocysts) the thread is adhesive. A few cnidarians are suspension feeders but most are carnivorous and when the tentacles make contact with the prey, it is immobilized and passed to the mouth. The majority of cnidarians are dioecious and both asexual reproduction by budding and sexual reproduction are common.

A striking feature of cnidarians is the high degree of polymorphism often seen in a single species. Two basic types exist; these are a sessile polyp (e.g. a hydroid or sea-anemone) and a free-swimming medusa (e.g. a jellyfish). The presence or absence of polyp or medusa in the life-cycle forms the basis of the classification of the phylum but the relationships between the different groups of cnidarians remains controversial. Both polyp and medusa forms have been suggested as ancestral cnidarians but molecular evidence suggests that the anthozoans are the most primitive and that they arose from the sponges.

Phylum CNIDARIA

Class Anthozoa	Without medusae; includes sea-anemones, soft corals, stony corals sea-fans and sea-pens. Marine.

Subclass Octocorallia (Alcyonaria)	Colonial; polyps generally small, with 8 pinnate tentacles. Includes the soft corals, sea-fans and sea-pens. Marine.

Subclass Hexacorallia (Zoantharia)	Colonial or solitary; polyps usually with many smooth tentacles. In temperate waters best known through the sea-anemones. Also includes the stony corals. Marine.

Class Hydrozoa	Usually both polyp and medusoid stages in the life-cycle, the polyp generally being the most obvious. Mainly marine; includes the hydroids or sea-firs, floating colonial forms such as the Portuguese man o'war and a few freshwater species such as *Hydra*.

Class Scyphozoa	The medusa is the most conspicuous stage in the life-cycle. Includes the free-swimming jellyfishes and some bottom-dwelling species. Marine.

Class Cubozoa	Tall, box-shaped medusae; largely tropical and subtropical and noted for their dangerous sting which can be lethal to humans. Commonly known as sea-wasps. Marine.

Class Anthozoa

The anthozoans are the largest group of cnidarians and include the corals, sea-fans, sea-pens and sea-anemones. They exist only as polyps: there is apparently no medusa in the life-cycle. The anthozoan polyp is generally much bigger and more complex than the hydrozoan polyp and both sexual and asexual reproduction are common. Asexual reproduction occurs by budding and also by longitudinal or transverse fission, or by basal laceration. During basal laceration, small pieces of the anemone break away from the basal region of the polyp and develop into new anemones. In sexual reproduction, both internal and external fertilization are known to occur. When fertilization is internal, the eggs are often retained inside the parent and released as planula larvae or as tiny anemones. Such is the complexity of reproduction in sea-anemones that in some species the method differs from one population to another.

Cnidae (see p. 86) are found in all anthozoans, being used in defence and food capture, and in some cases they aid in attachment to the substratum. Some anemones possess structures known as acontia. These are long and thread-like, and attached to internal mesenteries. When the anemone contracts they are discharged through the mouth and through pores in the wall of the column, and being armed with nematocysts they presumably play a role in defence. They can be withdrawn into the anemone. The presence of discharged acontia can be an important aid to identification. Anemones have few predators, the most notable being sea-slugs, some of which prey on single species of anemones. The pycnogonids *Nymphon* spp. (p. 300) and *Pycnogonum litorale* (p. 300) are often found on sea-anemones but are not believed to be serious predators.

Subclass Octocorallia (Alcyonaria)

Colonial anthozoans in which the polyps are small and have eight pinnate tentacles. This group includes the sublittoral sea-fans and sea-pens, and the soft corals, one species of which is sufficiently common on the lower shore in north-west Europe to be included here.

Alcyonium digitatum Linnaeus Dead men's fingers **Fig. 6.1**
Colonial, much-lobed thick, irregular mass; usually with stout, finger-like lobes; up to 250 mm in height and usually more than 20 mm across. Colour varied, grey, white, yellow, orange or pink. When under water and undisturbed it is covered by many small polyps, each with 8 pinnate tentacles giving a 'feathery' appearance. Polyps retractable.

A. digitatum is widely distributed in north-west Europe and around Britain and, although recorded at low water of spring tides, the most substantial colonies are found sublittorally down to about 50 m, especially in areas where there is much water movement. The majority of colonies are either male or female, a few being hermaphroditic. Colonies as small as 10–20 mm in height can be sexually mature, and in the Irish Sea spawning occurs in December and January. Fertilization is external and after five to seven days a free-swimming planula larva develops and normally metamorphoses in one or two days, but has the potential for a prolonged pelagic life. Sexual maturity is reached in the second or third year. *Alcyonium* feeds on both phytoplankton and zooplankton but the polyps remain retracted and non-feeding from about July to November, when the surface of the colony becomes covered with encrusting

Figure 6.1   *Alcyonium digitatum*, with detail of polyps.

seaweed and hydroids. When the colony recommences feeding in December the surface film, together with the outer epithelium, is shed. *Alcyonium* is long-lived and believed to survive for up to 25 years and more.

A second species, **Alcyonium glomeratum** (Hassall), is found in deeper water but is occasionally seen on the very low shore in sheltered situations. It can be distinguished from *A. digitatum* by the more slender lobes and red-orange coloration.

Subclass Hexacorallia (Zoantharia)

The Hexacorallia includes both colonial and solitary polyps, some of which reach a considerable size. All have smooth, undivided tentacles surrounding the mouth, which is situated at the free end of the body or column. The most familiar examples in temperate waters are the sea-anemones of rocky shores, but anemones are also adapted to living in mud, sand and gravel and some are capable of burrowing into the sediment. Although commonly referred to as sessile and attached to the substratum by a basal disc, most sea-anemones are able to change position by a sliding movement and, indeed, when being attacked by predators and during aggressive encounters, some species, including the beadlet anemone, *Actinia* (p. 91), move away slowly. Sea-anemones are potentially long-lived and specimens are known to have lived for more than 65 years in aquaria. The subclass also includes the stony corals, characterized by an external calcareous skeleton into which the polyps can almost completely withdraw when disturbed; a few species are found in British waters.

Key to species of Hexacorallia

1 Polyp elongate, with about 70 long, slender tentacles surrounding many short tentacles. Inhabiting soft, mucous, felt-like, vertical tube buried in sand, mud or gravel .... **Cerianthus lloydi** p. 98
Polyp not so **2**

2 Polyp without external, calcareous skeleton; tentacles without terminal knob **3**
Polyp with external calcareous skeleton and knobbed tentacles or without a calcareous skeleton but with knobbed tentacles **16**

3 Polyp attached to prosobranch shell occupied by hermit crab **4**
Not as above **5**

4 Polyp cloak-like, enclosing prosobranch shell occupied by hermit crab. Tentacles and mouth lie underneath shell. Usually with distinctive red or mauve spots
.... **Adamsia carciniopados** p. 96
Polyp firmly attached to shell by base; not enclosing shell
.... **Calliactis parasitica** p. 96

5 Polyp firmly attached to substratum (rocks, stones, piers, pilings, sometimes in soft sediments with base attached to small stone or piece of shell) by a basal disc. More than 48 tentacles **6**

Polyp long and narrow with rounded base, not firmly attached to substratum. Fewer than 36 tentacles. Found in holes and cracks in rocks or burrowing in sand and gravel **17**

6 Column smooth without protuberances or suckers; without attached sand or gravel **7**
Column with protuberances or suckers, with or without attached sand, shell, gravel **11**

7 Up to several thousand short, fine tentacles giving a 'feathery' appearance. Often white in colour .... ***Metridium senile*** p. 94
Up to about 200 longish, slender tentacles, without a 'feathery' appearance (but see *M. senile*, p. 94) **8**

8 Tentacles long, slender, snake-like, almost always extended, even on contact; tips of tentacles usually purplish .... ***Anemonia viridis*** p. 92
Tentacles contract readily when anemone is disturbed **9**

9 Without conspicuous blue spots (acrorhagi) round inside of top of column. Column very narrow and elongate when fully extended; usually orange in colour; tentacles long and slender .... ***Diadumene cincta*** p. 93
Conspicuous blue spots (acrorhagi) round inside of top of column **10**

10 Column red, green or brown. Basal diameter up to 50 mm ...... ***Actinia equina*** p. 91
Column reddish covered with many green spots and resembling a strawberry in appearance. Basal diameter up to 100 mm .... ***Actinia fragacea*** p. 92

11 Column with 6 longitudinal rows of large white protuberances without attached shell or gravel. When contracted resembles denuded test of a sea-urchin .... ***Aulactinia verrucosa*** p. 93
Column not so **12**

12 Column with many protuberances with or without attached shell, gravel. Tentacles up to 200, short, thick. Large sea-anemones, basal diameter up to 250 mm; very firmly attached to substratum **13**
Column with white or greyish suckers, with or without attached sand, shell, gravel. Basal diameter about 50 mm or so. Up to 200 longish tentacles or several hundred very short tentacles **14**

13 Column with attached shell gravel. Colour of column varied, reddish, green; with grey protuberances. Tentacles banded red, white, green, blue. Lower shore and below .... ***Urticina felina*** p. 92
Column without attached shell and gravel. Colour varied but paler than *U. felina*. Usually sublittoral .... ***Urticina eques*** p. 93

14 Column with greyish suckers near free end, usually with attached shell gravel; free end wider than base and convoluted in outline. Several hundred very short tentacles. Found in crevices filled with coarse sand and gravel into which the anemone withdraws; also in sand and mud .... ***Cereus pedunculatus***[1] p. 95
Column with suckers, with or without attached sand, shell, gravel; free end wider or narrower than base. Up to about 200 longish tentacles **15**

15 Column with large white suckers, especially near free end, rarely with attached shell, gravel; free end wider than base. Found on rocky shores, often in holes in the rock. Rarely on sand or mud .... ***Sagartia elegans***[1] p. 94

Column with small suckers and attached shell, gravel; free end narrower than base. Typically found buried in sandy, muddy shores with the base attached to a small stone or piece of shell .... ***Sagartia troglodytes***[1] p. 94

16 Polyp with external, white calcareous skeleton; up to 80 tentacles, each ending in a white or brown knob .... ***Caryophyllia smithii*** p. 97
Polyp without external calcareous skeleton; column smooth, brilliantly coloured, often bright green; up to 100 tentacles, each ending in bright red knob .... ***Corynactis viridis*** p. 97

17 Most of column covered with thick brownish cuticle; usually 28 long slender tentacles; translucent, pink. Small anemone, up to 30 mm in length, 4 mm in diameter. Lives in holes and cracks in rocks .... ***Fagesia carnea*** p. 97
Column translucent, pink, pale brown with brown marks; 12 translucent tentacles patterned brown and cream. May be up to 300 mm in length, 25 mm diameter. Burrowing in sand and gravel .... ***Peachia cylindrica*** p. 97

[1] These three species may be difficult to separate.

Actinia equina (Linnaeus) Beadlet anemone **Plate 12a**

Column smooth; basal diameter up to 50 mm, broader than column. Up to 192 tentacles arranged in 6 circles. Column red, green, brown or orange, sometimes striped or with yellow, green or blue spots randomly distributed, commonest column colours shades of dark red or brown. Usually with narrow blue line round edge of base. Conspicuous blue spots, acrorhagi, each containing nematocysts, occur round inside of top margin of column. These increase in number, up to 48, with age of specimen.

A. equina is widely distributed and very common on rocky shores in north-west Europe. It extends from the level of mean high water of neap tides to depths of about 20 m, and is highly adapted to the intertidal zone, tolerating high temperature and desiccation. Interestingly, intraspecific aggression has been recorded between neighbouring individuals. During these encounters, which involve tentacle contact and discharge of nematocysts, one of the animals is eventually forced to move away from the other. This behaviour pattern is perhaps related to the spacing out of individuals on the shore. *Actinia* also shows high levels of aggression to a range of common shore species of anemones. The sexes of *Actinia* are separate but cannot be identified without microscopic examination of the gonad. Adults of both sexes can be found brooding young throughout the year and it is believed that the broods are produced by internal budding, although parthenogenesis has been suggested. The extent and significance of sexual reproduction are not fully understood but it has been suggested that *Actinia* reproduces sexually in the warmer months of the year. Embryos are brooded by the female for a few weeks and then discharged into the overlying sea water. No details of fertilization are known. Length of life of *Actinia* has been estimated to be about three years but many specimens are believed to live considerably longer.

The morphological variability of *Actinia equina* has attracted much research, and it has been suggested that the variation recorded reflects cryptic speciation. It has been concluded that the green form in which the column and basal disc are light green in colour is a separate species, **Actinia prasina** Gosse. This species is found predominantly on the lower shore, particularly under rocks and stones, and has been recorded from the Isle of Man and Morecambe Bay. The other colour morphs described here may encompass a number of species and it has been suggested that within the red-brown morphs there may be separate upper, middle and lower shore species.

Actinia fragacea Tugwell Strawberry anemone **Plate 12b**

Column smooth and generally similar to A. equina *but red to reddish-brown in colour (not green) and covered with many greenish spots, resembling a strawberry in appearance. More plump than* A. equina *and up to 100 mm basal diameter. Conspicuous blue, occasionally red, white or pink spots (acrorhagi), round inside of top of column.*

For many years, *A. fragacea* was considered a variety of *A. equina*. It has a restricted distribution and has been recorded from north-west France and south-west Britain. Its intertidal distribution is more restricted than that of *A. equina* (above), being limited to rocks and stones on the lower shore and when present is never as abundant as *A. equina*. Apparently, it does not brood young but little is known of reproduction in this species.

Anemonia viridis (Forskål) (*Anemonia sulcata*) Snakelocks anemone **Plate 14a**

Column smooth; basal diameter broader than column and up to 70 mm. About 200 long, snake-like tentacles; almost always extended. Two colour varieties predominate, yellow-brown, and green; cream or white specimens sometimes occur. Tips of tentacles purple.

A. viridis is a southern species recorded as far north as the west of Scotland. It is common on the south-west coast of England but absent from the east coast and the easterly half of the English Channel. It is found in rock pools on the middle shore and below, extending into the sublittoral to depths of about 20 m. *Anemonia* often occurs in well-lit situations and this is probably related to the presence of photosynthesizing, symbiotic algae in the tissues. Asexual reproduction by longitudinal fission has been recorded. When groups of *A. viridis* are placed close together in aquaria, they make contact with their tentacles, in what has been described as aggressive encounters, and eventually separate from one another, resulting in spacing out. Interestingly, when the brown and green colour varieties are close together, it is apparently always the green that move away. On the shore there is some separation of the colour varieties.

Urticina felina (Linnaeus) (*Tealia felina*) Dahlia anemone **Plate 13a,b**

Column with many protuberances with (in intertidal specimens) pieces of shell and gravel stuck to them making the contracted animal sometimes difficult to detect. Basal diameter up to 120 mm; may be much larger in sublittoral specimens; very firmly attached to substratum. About 160 short, thickish tentacles. When fully extended, central area around mouth is free of tentacles. Column very varied in colour, frequently reddish with green markings; protuberances grey; tentacles grey, with bands of red, white, green or blue. The largest British intertidal anemone.

U. felina is widely distributed in north-west Europe and around Britain, where it is found from the lower shore to depths of about 100 m. It thrives in areas exposed to wave action and is often abundant in gullies and crevices, being readily identified in the contracted state by the presence of shell and gravel stuck to the column. Gametes are shed throughout the year and give rise to pelagic planula larvae.

A second species, **Urticina eques** (Gosse), is widely distributed in north-west Europe and around Britain. It is essentially a sublittoral species growing up to 250 mm basal diameter, but is sometimes found on the extreme lower shore on rocks and shells. It is varied in colour, usually red or orange and generally paler than *U. felina*, but may be off-white, yellow, or lilac, often with a pale reticulate pattern on the lower part of column. Unlike the intertidal specimens of *U. felina*, it does not have bits of shell and gravel stuck to the column.

Aulactinia verrucosa (Pennant) (*Bunodactis verrucosa*) Wartlet anemone, gem anemone **Fig. 6.2**

Column pink; up to 48 longitudinal rows of protuberances, those in 6 of the rows large and white in colour. Protuberances not covered with shell or gravel. Base up to 25 mm across; broader than column; 48 banded tentacles. Closes tightly when disturbed giving a hemispherical shape resembling denuded test of Echinus *(p. 415).*

Figure 6.2   *Aulactinia verrucosa*, extended and contracted specimens.

A. verrucosa has a southerly distribution and apparently reaches its northern limit in the British Isles where it is most common on south and west shores, extending to the Hebrides. It occurs in rock pools mainly on exposed shores, often well camouflaged among the red seaweed *Corallina* (p. 43) and extends to depths of about 20 m. It is also often found in sandy deposits in rock pools. Young are brooded in the parent.

Figure 6.3 *Diadumene cincta* (after Stephenson, 1935).

Diadumene cincta Stephenson **Fig. 6.3**

Column smooth, narrow and elongate when fully extended. Basal diameter up to 15 mm; column up to 60 mm in height. Tentacles long, slender, up to 200 in number. Column and tentacles usually orange in colour, occasionally fawn. Collar visible below tentacles in contracted specimens, 'disappears' in expanded specimens. Bears some resemblance to small specimens of Metridium senile *(below) but can be distinguished by the absence of a collar below the tentacles in expanded specimens and by the absence of a white band on the tentacles.*

D. cincta has been recorded from the southern North Sea, the west coast of France and around Britain, where it is widely distributed. It occurs on stones, rocks and bivalve molluscs such as *Mytilus* (p. 254), from the lower shore to depths of about 20 m. When disturbed it contracts in what has been described as 'jerky' movements. Reproduction is by basal laceration.

Metridium senile (Linnaeus) Plumose anemone **Plate 14b**

Column smooth, narrower than base, which may be deeply lobed and up to 150 mm basal diameter, height up to 300 mm. Tentacles slender, up to several thousand in number, giving 'feathery' appearance. When fully relaxed prominent collar visible below tentacles. Colour varied, most often white or orange; may be brown, red or yellow; tentacles usually have a whitish band near the base.

M. senile is widely distributed in north-west Europe and around Britain where it is commonly found in crevices, on piers, pilings, stones and under rocky overhangs and other situations where there is good water flow. It is found on the lower shore, extending sublittorally to depths of 100 m and is occasionally found in brackish water. The fine tentacles trap plankton on which the anemone feeds. Sexual reproduction with pelagic larvae and asexual reproduction by basal laceration have been recorded, the latter often leading to a much-lobed basal disc.

Although adult specimens of *M. senile* have a number of distinctive characters and can be readily identified, the species is very varied in form. Two distinct morphological varieties are generally recognized from British waters; the variety ***dianthus*** as described above and a smaller variety, ***pallidus***, in which the tentacles are long and much fewer in number (about 200) than in the variety *dianthus*, have a conspicuous white band near the base and do not have a 'feathery' appearance. The variety *pallidus* is up to 25 mm in basal diameter and up to 25 mm in height. Specimens intermediate between the two varieties have been described.

Sagartia elegans (Dalyell) **Fig. 6.4**

Base often indented, wider than column, but generally not as wide as free end; up to 30 mm basal diameter. Column with rounded white suckers; larger and more numerous towards free end; rarely with shell or gravel attached to them. Up to about 200 tentacles. Very varied in colour and patterning, from red, brown, orange to white; 5 colour varieties have been described. Thread-like acontia are freely discharged by the anemone when disturbed.

Figure 6.4 *Sagartia elegans.*

S. elegans is widespread in north-west Europe and around Britain from the middle shore to depths of about 50 m. It occurs in pools, crevices and holes in the rock. Large numbers are often found together in crevices and overhangs. It has also been recorded from soft sediments. Reproduction is by basal laceration; sexual reproduction also occurs.

Sagartia troglodytes (Price in Johnston) (*Sagartia troglodytes* var. *decorata*) **Fig. 6.5**

Similar to S. elegans (above) but less brightly coloured. Base wider than column and generally wider than anterior region with tentacles expanded; up to 50 mm basal diameter and 120 mm in height. Column with small suckers; often with attached shell, gravel. Up to about 200 tentacles. Colour and patterning highly varied; generally dull shades of yellow, green, grey; basal part of column with pale vertical stripes; tentacles plain or banded, generally with a B-shaped mark at the base. Thread-like acontia are not freely discharged when the anemone is disturbed.

Figure 6.5 *Sagartia troglodytes* (a) extended specimen on rock, (b) contracted specimen from sandy shore, attached to piece of shell, (c) specimen in sediment.

S. troglodytes is widely distributed in north-west Europe and common around Britain with the exception of the English Channel. It is found intertidally, and sublittorally to 50 m, generally in soft sediments such as sand and mud with the base attached to stones, pebbles and pieces of shell well below the surface, but does also occur on rocky shores. The sexes are separate and sexual reproduction is believed to occur, with planktonic larvae.

Until recently, *S. troglodytes* was described as existing in two forms, variety *decorata* and variety *ornata*. It is now known that these two varieties are, in fact, two separate species, *S. troglodytes* (formerly variety *decorata*) described above, and ***Sagartia ornata*** (Holdsworth), formerly variety *ornata*. *S. ornata* is widespread and common intertidally but seldom occurs in soft sediments, being more typical of exposed rocky shores where it is found in cracks and crevices, and with mussels and barnacles. It has a basal diameter of up to 15 mm, is usually without attached shell and gravel, and is less varied in colour than *S. troglodytes*, being greenish or reddish-brown. The tentacles are fewer in number (up to 100) and are banded. Populations consist only of females and these brood young. It has been suggested that the broods are produced parthenogenetically.

Figure 6.6 *Cereus pedunculatus.*

Cereus pedunculatus (Pennant) Daisy anemone **Fig. 6.6**

Base up to 40 mm diameter; wider than lower part of column. Column widening out towards free end when fully relaxed, particularly in young specimens. Up to 120 mm in height. Greyish suckers near free end, generally with attached shell, gravel. Tentacles very short and numerous, always at least 200, and more than 700 in fully grown specimens. Region from which tentacles arise convoluted in outline and wider than base. Column very varied in colour, usually buff-grey, orange, darkening towards free end, often with a purplish tinge; tentacles light brown, may be banded or mottled. Relatively wide area round mouth without tentacles. Thread-like acontia freely discharged when anemone disturbed.

C. pedunculatus is a southern species extending north to Britain where it commonly occurs on south and west coasts but is rare on the east coast. It lives in rock pools, typically in inaccessible crevices, sometimes as high as the middle shore, and extends into the sublittoral to depths of about 50 m. Large numbers are often found attached to shell debris and buried in mud and sand with the tentacles exposed on the surface. *Cereus* is sometimes found at the mouths of estuaries. Young are brooded in the parent.

Figure 6.7 *Calliactis parasitica* on shell of *Buccinum*.

Calliactis parasitica (Couch) **Fig. 6.7**

Base wider than column; up to 80 mm diameter and 100 mm in height. Column cream to light brown with faint longitudinal lines; often mottled with dark brown or red. Up to 700 translucent, yellowish to orange tentacles with longitudinal reddish-brown lines. Thread-like acontia discharged from lower part of column when anemone disturbed. Usually on prosobranch shell occupied by hermit crab.

C. parasitica is a southern species extending as far north as the west coast of Wales and the south of Ireland. It occurs sublittorally down to about 60 m, only rarely being found on the lower shore. The anemone usually lives on shells of the whelk *Buccinum undatum*, occupied by the hermit crab *Pagurus bernhardus* (p. 360), although it has been recorded on other substrata. Both the anemone and the crab can survive alone. When associated with the hermit crab, some protection against predators must be gained by the crab, while the anemone gains some advantage in food collection. The anemone establishes this commensal relationship by a series of complex manoeuvres through which it eventually attaches to the shell while the crab remains passive. It is believed that the anemone uses a chemical sense to recognize suitable shells and, in aquaria, they have been seen to transfer to the shells of living *Buccinum* but in such cases are shaken off by the whelk. A single shell may provide a home for more than one anemone. Young are brooded in the parent.

Figure 6.8 *Adamsia carciniopados.*

Adamsia carciniopados (Otto) (*Adamsia palliata*) Cloak anemone **Fig. 6.8**

Base and column enclose prosobranch shell occupied by hermit crab in such a way that mouth and tentacles come to lie under crab. Base up to 700 mm diameter. Column white to fawn; usually with red or mauve spots. Tentacles white. Thread-like acontia, pinkish-white in colour, discharged when disturbed.

A. carciniopados is widely distributed around Britain and in north-west Europe, particularly in the sublittoral to depths of about 200 m, but is less common on the east coast of England. It is found at low water on sandy shores and is readily identified by its coloration and the distinctive way it folds itself round the prosobranch shell. Almost without exception around Britain, it lives in association with the hermit crab *Pagurus prideaux* (p. 362), without which the anemone apparently does not reach sexual maturity. As the crab grows a chitinous membrane is secreted from the base of the anemone and this protects the crab and overcomes the need for it to move to a larger shell as it grows.

Peachia cylindrica (Reid) (*Peachia hastata*) **Fig. 6.9**

A burrowing form. Varied in shape, often long and slender; base not adhesive, rounded. May be up to 300 mm in length, and 25 mm diameter. Small lobed structure (the conchula) projects from one side of mouth. Column translucent pink or pale brown flecked with brownish marks. Twelve translucent tentacles, conspicuously patterned with brown and cream. In buried specimens the tentacles can extend to become very long.

Figure 6.9 *Peachia cylindrica* (after Stephenson, 1928).

P. cylindrica is widely distributed, but not abundant, in northwest Europe and around Britain, extending from the extreme lower shore to depths of about 100 m. It is well known for its characteristic habit of burrowing in sand or gravel where it remains free and not attached to the substratum. Its life-cycle is unusual in that the planulae are parasitic on a number of species of planktonic medusae. They have been recorded in the plankton from about April to August.

Fagesia carnea (Gosse) (*Edwardsiella carnea*) **Fig. 6.10a**

Column narrow, elongate; base not adhesive, rounded. Up to 30 mm in length, 4 mm diameter. Up to 36 (usually 28) long, slender tentacles. Column translucent, pale pink, orange, yellow; tentacles translucent, pink. Most of column covered with thick, brownish cuticle. Contracts rapidly.

F. carnea has been recorded from southern Norway to the Atlantic coast of France and occurs all around Britain. Its distribution is very localized but it may be present in high densities. It extends from the lower shore to the shallow sublittoral living in cracks and holes in rocks; it does not burrow. Little is known of its life-cycle.

Corynactis viridis Allman Jewel anemone **Fig. 6.10b**

Column smooth; base up to 10 mm diameter, height up to about 15 mm. Up to about 100 tentacles, each ending in a small knob. Brilliantly coloured in a variety of colours, often bright green with bright red terminal knob on each tentacle.

C. viridis is a southern species, common on south and west coasts of Britain, and reaching its northern limit in the north of Scotland. It occurs on the lower shore and into the sublittoral to depths of 50 m on rocks, in crevices and under overhangs, often in large numbers. Reproduction is by longitudinal fission.

Caryophyllia smithii Stokes & Broderip Devonshire cup-coral **Fig. 6.10c,d**

External whitish, calcareous skeleton with many curved ridges. Up to 15 mm in height and 25 mm basal diameter. Polyp varied in colour, red, pink, orange, white, green, brown, often with contrasting colour surrounding mouth. Up to about 80 tentacles, usually transparent, each ending in white or brown knob. Solitary coral.

C. smithii is a southern species found on south and west coasts of Britain, extending northwards to Shetland. It occurs on the lower shore on a variety of substrata and

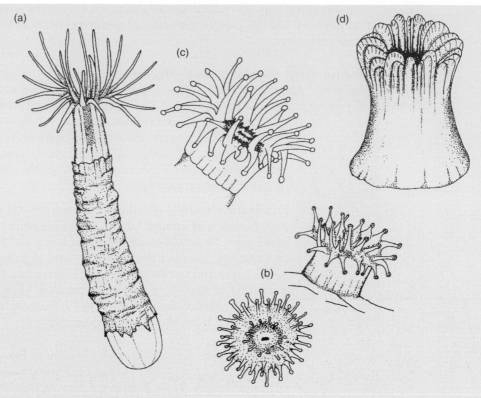

Figure 6.10   (a) *Fagesia carnea*. (b) *Corynactis viridis*. (c) *Caryophyllia smithii*, with (d) calcareous skeleton. ((a), (c), (d) after Manuel, 1988.)

Short inner tentacles

Figure 6.11   *Cerianthus lloydi* (modified after Manuel, 1988).

appears to favour vertical rock faces in exposed situations. It is relatively rare intertidally, but is often abundant in the sublittoral to depths of 200 m attached to rocks, stones and shells and in crevices. Growth rate is slow. A barnacle, **Megatrema anglicum** (Sowerby), is often attached to the calcareous skeleton, particularly on corals in the south and west. Several barnacles may be found on a single coral.

Cerianthus lloydi Gosse **Fig. 6.11**

Column elongate, base rounded; up to 150 mm in length. About 70 long, tapering outer tentacles surrounding many much smaller, smooth inner tentacles; tentacles cannot be retracted into column. Column brownish in colour, outer tentacles brown-green or white often banded with darker brown; inner tentacles not banded. Secretes a soft, felt-like, mucous tube up to 400 mm in length, in which it lives.

C. lloydi is widely distributed in north-west Europe and all around Britain and in some areas may be abundant. It is found on the lower shore and into the sublittoral to 100 m, buried in sand, mud and gravel. It inhabits a vertical tube, the anterior end of which projects a little way above the surface. *Cerianthus* is able to move freely in the tube and can contract rapidly. There is a pelagic larva, the arachnactis, in the life-cycle and this has been recorded in the plankton from January to August.

Class Hydrozoa

Hydrozoans exist as either polyp or medusa and the majority of species has both forms in the life-cycle. Most are colonial animals, a notable exception being the freshwater hydras which are solitary polyps without a medusa. Colonies are formed of groups of polyps arising from a common stolon linking them together and anchoring them to the substratum. The polyps are connected to one another by thread-like continuations of the body wall containing extensions of the body cavity. These may be in the form of horizontal stolons or upright stems which are supported by a stiff, protein–chitinous sheath, the perisarc. Hydrozoan colonies generally show division of labour, with some polyps (the hydranths or gastrozooids) specialized for feeding, some (the gonozooids or blastostyles) for reproduction and others (the nematophores and dactylozooids) for defence. Nematophores are equipped with many nematocysts and they are believed to deter settlement of larvae on the hydroid and to give some protection against predation. A few hydrozoans form floating colonies and are found in the open sea (p. 119).

Life-cycles are complex and varied. In species such as *Obelia* (p. 110), there is alternation of generations between the asexual polyp stage and the sexual medusa stage. The reproductive polyps of the colony bud off free-swimming medusae and at certain times of the year these are abundant in the plankton of inshore waters. The medusae reproduce sexually; sperm and eggs are liberated into the plankton where fertilization takes place to produce free-swimming larvae known as planulae. After a pelagic existence of a few hours to a few days, the planulae settle on the bottom to establish new colonies (Fig. 6.12). In many hydrozoans the medusa is not free-swimming but stays attached to the hydroid where it reproduces sexually and a free-swimming planula larva escapes into the plankton. This eventually settles on the bottom to begin a new colony. In a few species such as *Tubularia* (p. 102), the planula stage is passed through in the gonozooid and an actinula larva is released. This moves about on its tentacles and eventually attaches to the substratum to establish a new colony (Fig. 6.13).

The majority of hydroids occur sublittorally but many are seen as delicate, plant-like growths at low water where they sometimes occur in profusion attached to the walls of gullies and overhangs, and attached to stones, shells and seaweed. Sublittoral species are often washed ashore during storms. Hydroids are preyed on by a variety of invertebrates, including sea-slugs, sea-spiders and ghost shrimps.

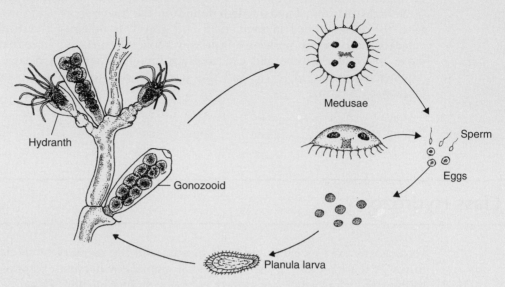

Figure 6.12 Life-cycle of *Obelia* (not to scale).

Figure 6.13 Life-cycle of *Tubularia* (not to scale) (adapted from Hincks, 1868). The medusae remain attached and the planula larvae develop into actinulae which are released and develop into new colonies.

Key to the groups of hydrozoans

The hydrozoans included here can conveniently be divided into three groups. The first two of these, the athecate and thecate hydroids, are the typical hydroids of the shore and shallow sublittoral and are usually of a small size, attached to rocks, shells and seaweeds. The remaining group comprises colonial hydrozoans characteristically found floating in open waters; they show a high degree of polymorphism.

1 **Athecate hydroids** – the perisarc stops short of the hydranth, gonozooids and nematophores and forms a protective covering only for the interconnecting strands. The hydranth cannot be withdrawn (Fig. 6.14a). Colonial or solitary .... p. 101

2 **Thecate hydroids** – the perisarc extends around the hydranth, forming a protective cup, the hydrotheca, into which the hydranth can be partially or wholly withdrawn, and around the gonozooid and nematophore to form a gonotheca and a nematotheca, respectively (Figs. 6.19, 6.29). Colonial .... p. 107

3 **Floating colonial hydrozoans** – species such as the Portuguese man o'war and Jack sail-by-the-wind. They belong to two separate orders within the Class Hydrozoa but are grouped together here for ease of identification. They are characteristic of the open sea but both are occasionally stranded on the shore, particularly in the south and west of Britain, sometimes in large numbers .... p. 119

Athecate hydroids

Key to families of athecate hydroids

1 Large, solitary, extensible hydranth covered with large number of small, blunt tentacles. Attached by a basal region with root-like processes .... **Candelabridae** p. 104
Hydroid not so **2**

2 Hydranth with knobbed tentacles, scattered or in several rings .... **Corynidae** p. 103
Hydranth with long narrow tentacles **3**

3 Tentacles scattered over hydranth .... **Hydractiniidae (*Clava*)** p. 105
Tentacles in one or more rings on hydranth **4**

4 Hydranths arise directly from an encrusting growth, especially on shells occupied by hermit crabs. Several polyp types and small spines present ...... **Hydractiniidae** p. 105
Hydranths borne on stems **5**

5 Tentacles arranged in 2 distinct rings; mouth on conical projection
.... **Tubulariidae** p. 102
Tentacles arranged in 1 ring; mouth on conical projection .... **Bougainvilliidae** p. 106

Family Tubulariidae

Hydranths on upright stems; with two rings of long, narrow tentacles, one ring near the mouth and the other near the base of the hydranth. Mouth on conical projection. Gonozooids usually in clusters on branched stalks between the two rings of tentacles.

Stems rarely branching, with very few, if any, annulations; up to 150 mm in height. Tentacles around mouth more numerous than those near base of hydranth .... *Tubularia indivisa* p. 102

Stems branched, often extensively; annulated at regular intervals; up to 35 mm in height. Hydranth with up to about 20 tentacles in each ring .... *Ectopleura larynx* p. 102

Tubularia indivisa Linnaeus **Fig. 6.14a**

Stems rarely branching, very few, if any, annulations; up to 150 mm in height; brown. Stems narrow at base and often twisted; rising from a stolon of tangled fibres. Hydranths pink to dark red with a ring of short, white tentacles around mouth more numerous than the ring of long, white, mostly drooping tentacles close to base. Gonozooids in clusters on branched stalks between the 2 rings of tentacles.

T. indivisa is widely distributed in north-west Europe and around Britain, attached to rocks, stones and shells at low water of spring tides, and extends into the sublittoral where it may be abundant. It is often found at the mouths of estuaries. Although breeding occurs throughout the year in some localities, peaks of recruitment have been recorded in spring and summer in Britain. There is no free-swimming medusa; a creeping actinula larva is produced. Longevity of the hydroid is believed to be about a year.

Ectopleura larynx (Ellis & Solander) (*Tubularia larynx*) **Fig. 6.14b**

Stems branched, often extensively; annulated at regular intervals; yellow-brown colour. Annulations sometimes difficult to see in living material. Generally up to 35 mm in height; sometimes more. Hydranths small, pale red; up to about 20 tentacles in each ring. Gonozooids as in T. indivisa *(above).*

E. larynx is similar in both habit and distribution to T. indivisa. The base of the colonies often accumulates mud in which small *Mytilus* (p. 254), *Jassa* (p. 345) and

Figure 6.14 (a) *Tubularia indivisa*; note athecate condition (after Hincks, 1868). (b) *Ectopleura larynx* (after Hincks, 1868).

other organisms can be found. In Britain, breeding has been recorded from spring to autumn and gonozooids have been recorded on individuals as early as seven days after settlement of the larva. Maturity can be reached in three weeks. There is no free-swimming medusa; an actinula larva is released from the hydroid.

Family Corynidae

Hydranths on upright stems; with knobbed tentacles scattered or in several rings. Gonozooids scattered over hydranth or below the tentacles.

Stems and branches annulated throughout length. Up to 25 mm in height .... *Coryne pusilla* p. 103

Much branched. Stems and branches without annulations, except at base of stem and above origin of branches. Up to 100 mm in height .... *Coryne eximia* p. 104

Coryne pusilla (Gaertner) **Fig. 6.15a**

Irregularly branched, brown stems rising from thread-like stolon; about 25 mm in height. Branches annulated throughout length. Hydranths slender, reddish, with 30 or more knobbed tentacles, either scattered or in several ill-defined rings. Spherical gonozooids scattered over hydranth.

C. pusilla is widespread in north-west Europe and around Britain, often attached to seaweeds in rock pools on the lower shore. Breeding has been recorded on the east coast of England from May to September. There is no free-swimming medusa; a planula larva is liberated.

Figure 6.15   (a) *Coryne pusilla* (after Hincks, 1868; inset after Allman, 1871). (b) *Coryne eximia* (after Hincks, 1868; inset after Allman, 1871).

Coryne eximia Gaertner (*Sarsia eximia*) **Fig. 6.15b**
Much branched, mainly from one side of stem; 50–100 mm in height. Yellow-brown colour. Annulations present at base of stem and above origin of branches. Reddish hydranths; 20–30 knobbed tentacles scattered over hydranth. Egg-shaped gonozooids borne on stalks rising from bases of tentacles.

C. *eximia* is widely distributed in north-west Europe and around Britain where it is found on a number of seaweeds, including laminarians (p. 57) and *Corallina* (p. 43), and on the sides of rock pools on the lower shore. Breeding has been recorded in spring and summer, and off the north-west coast of England medusae have been recorded in the plankton from April to July.

Family Candelabridae
Hydranth large, solitary, extensible, on root-like processes; covered with many small blunt tentacles.

Candelabrum cocksii (Cocks) (*Myriothela phrygia, Arum cocksii*) **Fig. 6.16a**
Hydranth solitary, cylindrical, with pointed apex; extensible, 40–100 mm in height. Apart from apical region, hydranth covered with large number of small blunt tentacles with reddish-brown tips. Attached to stones and rocks by a brown, basal region with root-like processes. Rounded, pink gonozooids borne on projections from base of hydranth.

C. *cocksii* is widely distributed in north-west Europe and around Britain where it is commonest in the south and west. It is found under stones and rocks on the lower shore and into the shallow sublittoral feeding on small crustaceans. On the south coast of England breeding has been recorded throughout the year. During development the fertilized eggs are held by specialized tentacles known as claspers. There is

Figure 6.16 (a) *Candelabrum cocksii* (after Hincks, 1868). (b) *Clava multicornis* (after Allman, 1871).

no free-swimming medusa; actinula larvae are released from the gonozooids and crawl about for several days before settling. *C. cocksii* has been confused in the literature with ***Candelabrum phrygium*** (Fabricius), a sublittoral Arctic species which lacks clasper tentacles.

Family Hydractiniidae

Hydranths arising directly from stolon or on upright stems or directly from encrusting growth, especially on gastropod shells occupied by hermit crabs. Tentacles scattered over hydranths or in one or two rings. Encrusting growths with several polyp types and spines present among polyps.

Clava multicornis (Forsskål) (*Clava squamata*) **Fig. 6.16b**

Pink, white or reddish hydranths rising directly from thread-like stolon; up to 25 mm in height. 30–40 tentacles scattered over hydranths. Spherical gonozooids below tentacles.

C. multicornis is widely distributed in north-west Europe and around Britain. It is commonly found on *Fucus vesiculosus* (p. 64), *F. serratus* (p. 63) and *Ascophyllum nodosum* (p. 61) on the middle and lower shore and there is some evidence to suggest that it is most abundant on seaweeds from sheltered shores. It is also found on shells and stones. The form of the colony ranges from a diffuse network of stolons with scattered hydranths, to a dense mat of stolons with many hydranths, and this variation is believed to be governed by environmental factors such as the nature of the substratum, water movement, food supply and tidal exposure. Reproduction takes place from spring to autumn. There is no free-swimming medusa; crawling planula larvae are released from the gonozooids during daylight hours. They are believed to seek out fronds of seaweeds such as *Ascophyllum*, where attachment and metamorphosis occur within a few days of their release.

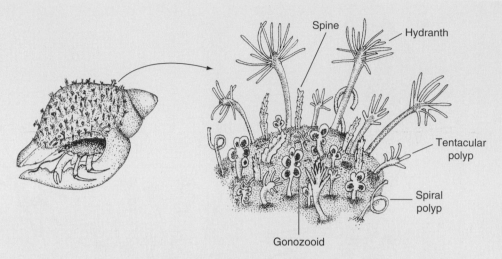

Figure 6.17 *Hydractinia echinata* on shell of hermit crab, with detail of colony.

Hydractinia echinata (Fleming) **Fig. 6.17**

Encrusting growth on prosobranch shells inhabited by hermit crabs; also occasionally recorded on other substrata such as rock, mussel shells, seaweed and barnacles. Large numbers of pink, white and brown polyps, some up to 10–15 mm in height. Four distinct types of polyp present – hydranths with 2 adjacent rings of slender tentacles; oval gonozooids; much-coiled spiral polyps and thin, extensible, tentacular polyps. Small serrated spines also present.

H. echinata is widely distributed in north-west Europe and around Britain where it is found on the lower shore extending well into the sublittoral. The spiral and tentacular polyps are believed to be associated with defence, the spiral polyps predominating on the edge of the colony towards the mouth of the shell. Interestingly, they disappear in a few days if the hermit crab is removed and reappear within a matter of days if the crab is returned. It appears that both the presence of the crab and the current systems it produces are essential for development of the spiral polyps. In Britain, breeding has been recorded throughout the year with peak activity in summer. *Hydractinia* colonies have separate sexes; sessile medusae are produced and these liberate gametes into the sea where fertilization and development of the planulae take place. Crawling planulae apparently detect moving shells occupied by hermit crabs. They crawl over the surface of the shell and become attached by nematocysts. Metamorphosis is induced by the presence of bacteria. Young colonies consist of hydranths and these extend at the periphery to form a dense network in the centre of which gonozooids develop. New colonies can also develop by asexual division. The presence of *Hydractinia* on shells occupied by hermit crabs is believed to offer some protection to the crab against attacks from other hermit crabs seeking to take over the shell.

Family Bougainvilliidae

Hydranths usually on upright stems; with a single ring of long, narrow tentacles; mouth on a conical projection. In addition to being found on rocks and stones, some members of the family are found on shells occupied by hermit crabs. They are readily separated from the Hydractiniidae (above).

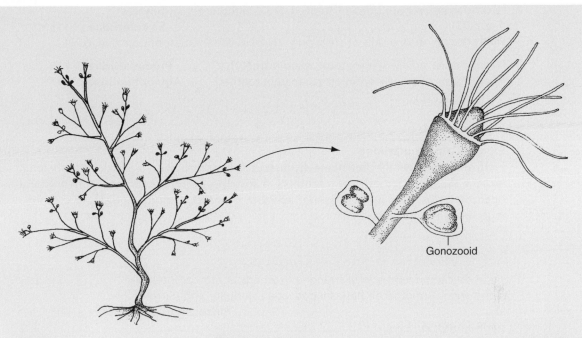

Figure 6.18 *Bougainvillia muscus* (after Allman, 1871).

Bougainvillia muscus (Allman) (*Bougainvillia ramosa*) **Fig. 6.18**
Main stem relatively thick, much branched; branches alternate. Height of 50 mm or more; yellow-brown colour. Hydranths pink to red, with single whorl of about 20 whitish tentacles. Stalked gonozooids just below hydranths.

B. muscus is widely distributed in north-west Europe and in Britain is probably commonest on southern shores. It is found on the lower shore attached to rock, shells and seaweed, and reproduces during spring and summer. At Plymouth, medusae have been recorded in the plankton from April to October. Other species of *Bougainvillia* are found in north-west Europe but *B. muscus* is probably the most widespread and common around Britain.

Thecate hydroids

The classification of the thecate hydroids follows **Cornelius (1995)**.

Key to families of thecate hydroids

1 Hydrothecae bell-shaped, on stalks (Fig. 6.19). Hydrothecae open, not closed by operculum; aperture rounded. Colonies upright or creeping
.... **Campanulariidae** p. 108
Hydrothecae not on stalks, arise directly from stem. Colonies upright **2**

2   Hydrothecae on both sides of stem. Hydroid without nematophores
(Fig. 6.25) .... **Sertulariidae** p. 113
Hydrothecae on one side of stem only. Hydroid with nematophores (Fig. 6.29) **3**

3   Hydrothecae radially symmetrical, margin smooth .... **Plumulariidae** p. 115
Hydrothecae bilaterally symmetrical, margin toothed .... **Aglaopheniidae** p. 117

Family Campanulariidae

Hydrothecae generally bell-shaped; on stalks; hydranth with a single ring of long, narrow tentacles; mouth on a trumpet-shaped projection. Aperture of hydrotheca open, not closed by an operculum; aperture rounded. Colony upright or creeping, without nematophores.

1   Hydrothecae on long stalks; arranged in whorls at regular intervals
along stem. Gonothecae flask-shaped, apex elongate and narrow
.... ***Rhizocaulus verticillatus*** p. 108
Hydroid not so **2**

2   Stem distinctly zigzag in appearance and 'jointed'; thickened below each joint,
especially on one side; occasionally branched .... ***Obelia geniculata*** p. 112
Stem not so **3**
Note: The following four species may be difficult to identify with certainty.
This key should be used in conjunction with the details given separately
for each species.

3   Rim of hydrotheca smooth or nearly so **4**
Rim of hydrotheca toothed (may be worn smooth by abrasion) **5**

4   Hydroid small, delicate (up to about 30 mm in height). Hydrothecae cup-shaped,
rims smooth. Gonothecae large, apex blunt .... ***Laomedea flexuosa*** p. 109
Hydroid small to large (up to 300 mm in height). Hydrothecae
bell-shaped, rim smooth to toothed. Gonothecae slender with raised apex
.... ***Obelia dichotoma*** p. 111

5   Hydroid with slender sinuous stem, dark brown to black, much-branched
branches, rising alternately. Up to 350 mm in height. Gonothecae with raised
apex .... ***Obelia longissima*** p. 111
Hydroid with stem slightly curved between branches, pale yellow-brown colour,
up to 30 mm in height. Gonothecae with up to 4 attached medusae protruding
from aperture .... ***Gonothyraea loveni*** p. 109

Rhizocaulus verticillatus (Linnaeus) (*Campanularia verticillata*) **Fig. 6.19**
Irregularly branched, thick stems up to about 60 mm in height. Hydrothecae bell-shaped, about 12 cusps on rim; on relatively long stalks sometimes annulated and arranged in whorls at regular intervals along stem. Gonothecae smooth, flask-shaped, apex elongate and narrow; borne on very short stalks.

Figure 6.19 *Rhizocaulus verticillatus*; note thecate condition (after Hincks, 1868).

R. verticillatus has been recorded in a wide range of localities from north-west France to Norway. It is essentially a sublittoral species extending to depths of 200 m and more, attached to stones and shells, particularly in sandy areas. Reproduction occurs in spring and summer. There is no free-swimming medusa; a planula is released.

Gonothyraea loveni (Allman) **Fig. 6.20**
Stems irregularly branched; annulated above origins of branches; up to about 30 mm in height; slightly curved between branches; pale yellow-brown colour. Hydrothecae alternate, elongate and rather slender; rim notched, with about 14 cusps to give castellated effect but may be worn smooth by abrasion; on short, annulated stalks. Gonothecae on short, annulated stalks. Up to 4 attached medusae, containing male or female gonads, protrude from aperture of gonothecae.

G. loveni is widely distributed in north-west Europe and around Britain in intertidal pools on seaweeds and stones, extending to depths of about 200 m. It tolerates salinities down to about 12. Breeding apparently occurs during most of the year. There is no free-swimming medusa; planula larvae are released and these have a pelagic life of about 36 hours.

Laomedea flexuosa Alder (*Campanularia flexuosa*) **Fig. 6.21**
Stems about 25–30 mm in height, rising from branching stolon; may be irregularly branched; annulated above origins of branches. Hydrothecae cup-shaped, alternate, on annulated stalks; rims smooth. Gonothecae large, elongate, smooth; apex blunt; on short, annulated stalks.

L. flexuosa is widely distributed in north-west Europe and is one of the commonest intertidal hydroids in Britain, extending from the middle shore into the sublittoral. It is found on seaweeds, such as *Ascophyllum nodosum* (p. 61) and *Fucus vesiculosus* (p. 64), and rocks and stones. Breeding has been recorded in spring and summer from

Figure 6.20 *Gonothyraea loveni* (after Hincks, 1868).

Figure 6.21 *Laomedea flexuosa* (after Hincks, 1868).

a number of localities. There is no free-swimming medusa; planula larvae are released which have a very brief pelagic life of about 30 minutes.

Obelia spp.

The genus *Obelia* contains a number of species three of which are described below but the validity of some of the others is questioned by taxonomists. The species described

Gonotheca

Figure 6.22 *Obelia dichotoma* (after Hincks, 1868).

here are distinctive and each is widely distributed in north-west Europe. The life-cycle of *Obelia* is illustrated in Fig. 6.12. Free-swimming medusae are liberated into the plankton where they are sometimes abundant in coastal waters. They have been recorded in the plankton throughout the year but identification to species is very difficult.

Obelia dichotoma (Linnaeus) **Fig. 6.22**

Slender, light brown stem, almost straight, irregularly branched. Annulated above origins of branches; up to about 40 mm in height but sometimes up to 300 mm. Hydrothecae bell-shaped with smooth to toothed rim; rising alternately on annulated stalks. Gonothecae in axils on annulated stalks; smooth, slender with raised apex.

O. dichotoma is widely distributed in north-west Europe and around Britain. It occurs on the lower shore and into the sublittoral attached to rocks, stones, shells and other hydroids. It breeds during the summer.

Obelia longissima (Pallas) **Fig. 6.23**

Slender, sinuous stem, with much-branched branches rising alternately. Stem dark brown, black; annulated above origins of branches; up to 350 mm in height. Branches sinuous with annulations above every side branch. Hydrothecae bell-shaped with 11–17 cusps on slightly toothed rim; on annulated stalks. Gonothecae in axils on annulated stalks; smooth, slender with raised apex.

O. longissima is widely distributed in north-west Europe and around Britain. It occurs sublittorally, usually in deep water attached to stones, shells and seaweed, often being washed ashore in large quantities. On the south coast of England breeding occurs in spring.

Figure 6.23 *Obelia longissima* (after Hincks, 1868).

Figure 6.24 *Obelia geniculata* (after Hincks, 1868).

Obelia geniculata (Linnaeus) **Fig. 6.24**

Stem only occasionally branched; distinctly zigzag in appearance and 'jointed'; thickened below each joint, especially on one side; up to 40 mm in height. Hydrothecae bell-shaped with smooth rim; alternate, 1 per joint on short, annulated stalks. Gonothecae slender, with narrow opening; rising from axils on short, annulated stalks.

O. geniculata has a worldwide distribution. It occurs usually on brown seaweed, less frequently on stones, and extends into the sublittoral. On the south coast of England, breeding has been recorded throughout spring and summer.

Family Sertulariidae

Hydrothecae not on stalks; arise directly from both sides of the stem; aperture oval. Hydranth can retract completely into hydrotheca; with a single ring of long, narrow tentacles; mouth on a conical projection. Hydrothecae usually closed by a flat operculum. Colony upright; without nematophores. A large family of hydroids, many species of which are sublittoral. Only three common and widespread species are described here.

1 Hydrothecae alternate. Gonothecae elongate, bell-shaped with a tooth on each side of aperture .... **Sertularia cupressina** p. 114

Hydrothecae opposite or nearly so. Gonothecae large, pear-shaped **2**

2 Hydroid with long, very slender stems alternately branched. Hydrothecae with outer edge prolonged into a spine, sometimes with 2 short lateral spines .... **Amphisbetia operculata** p. 113

Hydroid with straight stems, usually unbranched; if branched, then irregularly branched. Hydrothecae curve outwards .... **Dynamena pumila** p. 114

Amphisbetia operculata (Linnaeus) (*Sertularia operculata*) Sea-hair **Fig. 6.25**

Long, very slender stems alternately branched; up to 100 mm in height; creamish colour. Hydrothecae very small; arranged in opposite pairs. Aperture of cup faces towards stem; outer edge with long spine, sometimes with short, lateral spine on either side. Gonothecae large, pear-shaped or elongate.

A. *operculata* is found all around Britain but not the east coast of Scotland; it apparently reaches its northern limit at about Shetland. It has not been reliably

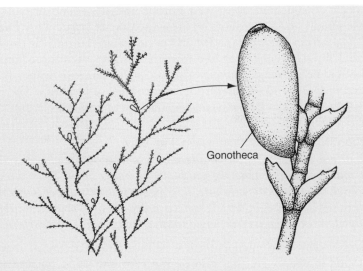

Figure 6.25   *Amphisbetia operculata* (after Hincks, 1868).

Gonotheca

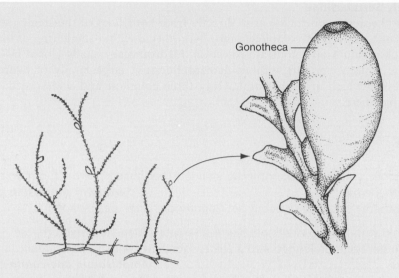

Gonotheca

Figure 6.26 *Dynamena pumila* (after Hincks, 1868).

recorded from Germany, Denmark or Scandinavia. It is found on the lower shore and in rock pools, attached to seaweeds, especially the stipe of *Laminaria digitata* (p. 57) and extends into the sublittoral to depths of about 70 m. Reproduction apparently takes place during summer. There is no free-swimming medusa.

Dynamena pumila (Linnaeus) (*Sertularia pumila*) **Fig. 6.26**
Straight, pale brown stems rising from stolon. Stems may be unbranched or sparsely branched and little more than 10 mm in height, or irregularly branched and up to 50 mm in height. Hydrothecae small, tubular, curving outwards; in opposite pairs on main stem and branches. Gonothecae large, pear-shaped.

D. *pumila* occurs throughout north-west Europe and all around Britain. It is found mainly on the middle and lower shore on brown seaweeds, particularly *Fucus serratus* (p. 63) and *Saccharina latissima* (p. 58) and occasionally on rock. It thrives in areas of strong tidal flow. Specimens on the lower shore are often more branched than those on the middle shore. *Dynamena* also occurs in brackish water. Breeding has been recorded in spring and summer from a wide range of localities. There is no free-swimming medusa; planula larvae have a pelagic life of about 36 hours.

Sertularia cupressina Linnaeus Whiteweed **Fig. 6.27**
Stout, more or less straight stem with alternate branches rising almost spirally; up to 350 mm or more in height, tapering gradually towards apex. Stem dark, branches brown in colour. Hydrothecae alternate, narrowing towards upper part which curves outwards only slightly; margin with 2 teeth usually of equal size. Gonothecae elongate, bell-shaped, most often with a single tooth on either side of aperture.

S. *cupressina* is widely distributed in north-west Europe and occurs all around Britain. It is often found in deep water but also on the very low shore on firm sand and in shallow pools attached to stone and shells. On the south coast of England breeding has been recorded from November to January. There is no free-swimming medusa;

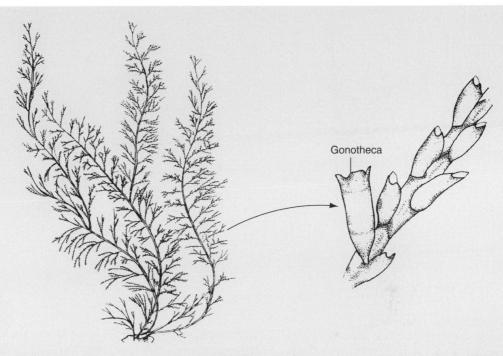

Figure 6.27 *Sertularia cupressina* (detail after Hincks, 1868).

planula larvae are released and have a free-swimming life of two to three days, possibly longer. When pressed and dyed whiteweed has some commerical value and is used for a variety of decorative purposes. A small industry existed round the Thames estuary where the weed was collected from the sea bed by raking from small boats. Commercial exploitation has also been carried out in Germany and the Netherlands.

Family Plumulariidae

Hydrothecae not on stalks, arise directly from stem on one side only, rim usually smooth. Hydrothecae radially symmetrical. Hydranth with single ring of long, narrow tentacles; mouth on a conical projection. Colony upright; branched or unbranched; with nematophores. Hydrotheca and nematotheca not joined. A large family, many species of which are sublittoral. Three common and widespread species are described here.

1 Hydroid large, up to 120 mm or more; white to yellowish in colour. Gonothecae roughly pear-shaped with spiny projections at top; on short stalks along both sides of main stem .... ***Kirchenpaueria pinnata*** p. 116
Hydroid usually small, up to about 30 mm in height. Gonothecae not as above **2**

2 Hydrothecae on every other joint of branches. Gonothecae smooth, flask-shaped, in axils between main stem and branches, often in a continuous row along one side of main stem .... ***Plumularia setacea*** p. 116
Hydrothecae separated from each other by 2, rarely 3, joints and so appear widely spaced on branches. Gonothecae large, on short stalks; with transverse grooves (female), smooth (male) .... ***Ventromma halecioides*** p. 116

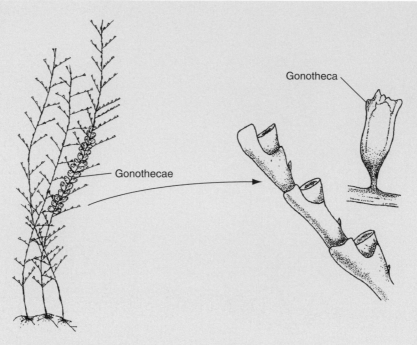

Figure 6.28 *Kirchenpaueria pinnata* (after Hincks, 1868).

Kirchenpaueria pinnata (Linnaeus) (*Plumularia pinnata*) **Fig. 6.28**

White to yellowish, delicate stems; fairly straight, rising in groups from stolon; up to 120 mm or more in height. Side branches alternate along main stem; hydrothecae on upper side of branches only, with smooth rim. Gonothecae on short stalks along both sides of main stem; roughly pear-shaped, with spiny projections at top.

K. pinnata is widely distributed in north-west Europe and around Britain in pools on the lower shore attached to stones, shells and seaweed. It extends into the sublittoral and is one of the invertebrates frequently recorded on the carapace of the spider crab *Hyas araneus* (p. 373). Breeding has been recorded throughout summer; there is no free-swimming medusa.

Plumularia setacea (Linnaeus) **Fig. 6.29**

Delicate stem with alternate branches; up to about 30 mm in height; pale yellow-brown colour. Branches with alternating long and short joints; hydranths on upper side of long joints. Hydrothecae small with smooth rim. Gonothecae smooth, flask-shaped, in axils, often forming a continuous row along one side of main stem.

P. setacea is a southern species infrequently recorded north of Britain. It is usually sublittoral and grows attached to rock, hydroids, sponges, ascidians, molluscan shells and the tubes of polychaete worms. On the south coast of England breeding has been recorded from January to June. There is no free-swimming medusa; planula larvae are released. Longevity is believed to be up to one year.

Ventromma halecioides (Alder) (*Plumularia halecioides*) **Fig. 6.30**

Delicate, irregularly branched stems up to 30 mm in height; branches arise alternately. Hydrothecae restricted to one side of branch; widely spaced and separated from each

Figure 6.29 *Plumularia setacea* (detail after Hincks, 1868); note gonotheca and nematothecae.

Figure 6.30 *Ventromma halecioides,* showing female gonotheca (after Hincks, 1868).

other by 2, rarely 3, joints. Hydrothecae with smooth rim. Gonothecae large, more or less egg-shaped, on short stalks; female gonothecae with transverse grooves; male gonothecae smooth.

V. halecioides is widely distributed around Britain on stones, seaweed and hydroids on the lower shore. There is no free-swimming medusa.

Family Aglaopheniidae
Hydrothecae bilaterally symmetrical; not on stalks, arise directly from stem on one side only. Rim of hydrothecae usually toothed. Hydranth with single ring of long, narrow tentacles; mouth on a conical projection. Colony upright, main stem branched

Figure 6.31 *Aglaophenia pluma.* (a) Branched colony, (b) single-stem colony, (c) detail. (Detail after Hincks, 1868).

or unbranched, feather-like. Nematophores present. Hydrotheca and nematotheca joined. Gonothecae usually protected by distinctive basket-like corbulae (Fig. 6.31). The family contains a number of genera, the genus *Aglaophenia* being the best known from around Britain. A single species of the genus is described here.

Aglaophenia pluma (Linnaeus) **Fig. 6.31**

Delicate, dark brown stem rising from a creeping stolon. Colony of two forms; a single stem on which side branchlets are alternate giving a feather-like appearance, or dichotomously branched with many feather-like side branches. Single stem colonies up to 40 mm in height, branched colonies 150 mm. Hydrothecae arranged on one side of branch; rim of hydrothecae with bluntly toothed rim of 9 cusps. Conspicuous nematotheca on front of hydrotheca. Stalked gonothecae protected by white, ribbed basket-like structures known as corbulae which arise from main stem in place of a branch.

A. pluma appears to have a patchy distribution around Britain, being most common in the south and west extending to the west coast of Scotland and eastwards to the Isle of Wight. It is found on the lower shore attached to stones and shells but usually grows on brown seaweeds, including laminarians and *Sargassum* and, in particular, *Halidrys siliquosa* (p. 67). Specimens found around Britain are mostly single-stemmed, while dichotomously branched colonies are more often found on the Atlantic coast of Europe and the western Mediterranean and it has been suggested that the much-branched colonies are produced in summer. This requires further investigation as it has also been suggested that the different growth forms are separate species. In south-west Britain breeding has been recorded in spring and summer. Male gonothecae can be identified by gaps between the ribs of the corbula, females by the fused ribs. There is no free-swimming medusa; planula larvae are released from the corbulae.

Although *A. pluma* is believed to be the commonest member of the genus in Britain, it is difficult to separate from other species and reference to **Cornelius (1995)** is recommended.

Floating colonial hydrozoans

In addition to the sea-firs, the class Hydrozoa also includes floating colonial animals, each colony being derived from a single larva. They are well known through descriptions of the much-feared Portuguese man o'war (below). These species have a high degree of polymorphism and some have a gas-filled float which supports the colony in the surface waters. They are oceanic and essentially tropical and semi-tropical. Two species described here are occasionally driven into British waters by prolonged south-westerly winds and in some years may be stranded on the shore in large numbers.

Figure 6.32 *Physalia physalis.*

Physalia physalis (Linnaeus) Portuguese man o'war
Fig. 6.32
Large, gas-filled float may be 300 mm in length, 100 mm wide; sky-blue or light green tinged with pink. Large crest along length of float. Many feeding, reproductive and defensive polyps beneath float, the defensive polyps often many metres in length.

P. physalis is the only species of the genus and belongs to a group known as the siphonophores. It is tropical and semi-tropical in distribution but occasionally found off south-west Britain, sometimes in large shoals. Specimens stranded on the shore are often damaged but they are easily recognized and contact with bare skin should be avoided as the tentacles possess a dangerous sting. The conspicuous float contains a gas, similar in composition to air but with up to 8% carbon monoxide. The float sits on the surface of the water and the crest acts as a sail. As in *Velella* (below), the position of the sail on the float determines whether the animal drifts to the right or left of wind direction. The long trailing tentacles are equipped with stinging cells which stun pelagic fishes and other prey coming into contact with them. *Physalia* is dioecious and fertilization external, leading to the development of a planula-like larva.

Velella velella (Linnaeus) Jack sail-by-the-wind **Plate 32a**
Oval disc, with deep blue to blue-violet margins; 80–100 mm in length. Erect sail set diagonally on disc. Central mouth on underside surrounded by gonozooids. Ring of fairly short tentacles round edge of disc.

V. velella is the only species of the genus and is classified with the athecate hydroids (p. 101). It is tropical and semi-tropical but in some years it occurs in shoals off southwest Britain and large numbers may be stranded on the shore. The disc contains a float and the stiff sail stands erect. The sail is set along either diagonal of the float and this line of insertion determines whether the individual drifts to the right or left of the wind direction. *Velella* is eaten by gastropod molluscs of the genus *Janthina*, warm water species often washed ashore with *Velella*. *Janthina*, in contrast to the snails of the shore, drifts in the surface waters on a bubble float secreted by the snail. *Velella* feeds on young fishes, crustaceans and other pelagic organisms caught by stinging cells on the tentacles. Free-swimming medusae are produced in large numbers and these are believed to sink into deep water. They develop ripe gonads and die after discharging the gametes. A planktonic larva develops into a new colony.

Class Scyphozoa

The scyphozoans include the well-known jellyfish. They are typically pelagic and exist for the greater part of their life as medusae. Most are dioecious and in the species described here, sperm are liberated through the mouth, which is on the undersurface, and drawn into a female, where internal fertilization takes place. A free-swimming planula larva is liberated and in most species this settles as a fixed polyp known as a scyphistoma (Fig. 6.33a). In the common jellyfish, *Aurelia* (p. 122), the scyphistoma reaches a height of about 14 mm. It eventually produces tiny medusae by a process of transverse fission. These medusae, also known as ephyrae (Fig. 6.33b), are no more than a few millimetres across. They become detached from the scyphistoma, feed and grow in the plankton and after a period of time which may be as long as two years, develop into sexually mature jellyfish.

Jellyfish are familiar inhabitants of the inshore waters of Britain and Europe, often occurring in such large numbers as to constitute shoals or blooms, the frequency of which has increased in recent years. They feed on a diverse range of plankton and in some situations may be significant predators of fish larvae. Scyphozoans have a wide range of natural predators, including other jellyfish and are a major item in the diet of sea turtles (p. 125). The benthic scyphistomae are preyed on by opisthobranch molluscs (p. 235). Jellyfish have long been exploited as a source of food for humans and commercial fisheries are now established in the tropical and subtropical waters of Asia, particularly for the large rhizostomatids (p. 124).

Six species are found in the inshore waters around Britain one of which, *Pelagia noctiluca* (Forsskål), is infrequently recorded; it is best described as oceanic. The remaining five species are described here. They are propelled through the water by gentle pulsations of the bell and almost inevitably many are driven ashore by the wind and become stranded and quickly perish.

Figure 6.33 Life-cycle of *Aurelia* (not to scale). (a) Scyphistoma attached to rock, (b) ephyra larva in plankton, (c) mature jellyfish, (d) planula larva.

Key to species of scyphozoans

1 Thick, dome-shaped. Margin of bell divided into large number of lobes; without tentacles. Up to 900 mm across .... ***Rhizostoma pulmo*** p. 124
Saucer-shaped. Margin of bell with tentacles **2**

2 Margin of bell with 16 lobes and 8 groups of tentacles, each group with 40–150 tentacles **3**
Margin of bell not as above **4**

3 Margin of bell with 8 groups of tentacles, each with 70–150 tentacles
.... ***Cyanea capillata*** p. 122
Margin of bell with 8 groups of tentacles, each with 40–60 tentacles
.... ***Cyanea lamarckii*** p. 122

4 Margin of bell with 8 small notches and numerous short tentacles.
Four violet-coloured semicircular or horseshoe-shaped patches on upper surface .... ***Aurelia aurita*** p. 122
Margin of bell with 32 semi-circular lobes, each spotted with brown; 24 tentacles in 8 groups of 3. Upper surface of bell with 16 V-shaped brown markings radiating from centre .... ***Chrysaora hysoscella*** p. 122

Figure 6.34 *Chrysaora hysoscella* (after Russell, 1970).

Chrysaora hysoscella (Linnaeus) Compass jellyfish
Fig. 6.34

Saucer-shaped; up to 300 mm diameter. Margin with 32 semi-circular lobes, each spotted with brown; 24 tentacles in 8 groups of 3, a sense organ situated between each group. Upper surface with 16 V-shaped, brown markings radiating from central brown spot. Varying degrees of brown pigment. Four arms surrounding mouth, longer than diameter of jellyfish.

C. hysoscella has been recorded in inshore waters all around Britain but is most commonly found in the south and west where it is generally recorded from May to September. It is a protandrous hermaphrodite. Fertilization is internal and planulae, with a free-swimming life of a few days, are liberated in summer and autumn. Ephyrae are produced from the scyphistomae the following spring and grow rapidly to produce large, mature individuals by the summer. Longevity is about one year. *Chrysaora* is omnivorous, feeding on the wide range of zooplankton coming into contact with the highly extensible, marginal tentacles from where it is transferred to the arms surrounding the mouth.

Cyanea capillata (Linnaeus) **Fig. 6.35**
Saucer-shaped, up to 500 mm diameter, occasionally much larger; yellow, brown or reddish in colour. Margin with 16 lobes and 8 groups of tentacles set a little distance in from margin; each group with 70–150 tentacles. Four large arms surrounding mouth, shorter than tentacles and much folded.

C. capillata is a northern species extending to the Arctic Circle. Commonly recorded in the North Sea and the Irish Sea, it is less frequently found off south-west Britain and is rare in the English Channel. It is not found in salinities below about 20. The life-cycle is similar to that of *Aurelia* (below); it feeds on plankton and small fishes that make contact with the tentacles. It has a powerful sting.

Cyanea lamarckii Péron & Lesueur **(not illustrated)**
Similar to Cyanea capillata *(above) but with 40–60 tentacles in each marginal group. Up to 150 mm in diameter but specimens as large as 300 mm have been recorded. Pale yellow to blue in colour.*

C. lamarckii is recorded all around Britain and has a more southerly distribution than *C. capillata*, extending from the Bay of Biscay to Iceland and the Faroes. The life-cycle is believed to be similar to that of *C. capillata*. The sting is much less powerful.

Aurelia aurita (Linnaeus) Common jellyfish, moon jellyfish **Fig. 6.36**
Saucer-shaped; almost colourless apart from 4 violet semicircular or horseshoe-shaped gonads. Up to 250 mm across, sometimes more. Margin with numerous short tentacles, smooth except for 8 small notches marking position of sense organs. Four thick arms surrounding mouth.

Figure 6.35 *Cyanea capillata* (after Russell, 1970).

Figure 6.36 *Aurelia aurita*. (a) Seen from above, (b) lateral view.

A. aurita is a cosmopolitan species widely distributed in the northern hemisphere. It is essentially an inshore species and is by far the most common jellyfish around the coasts of Britain. In some years it is so abundant during summer and autumn that it is a nuisance to fishermen and bathers, and swarms may be so dense that they block the water intake systems to power stations. *Aurelia* withstands salinities down to about 6 and extends into estuaries. The sexes are separate and size at onset of sexual maturity varies but is generally about 50 to 60 mm in diameter. Fertilization is internal with sperm being drawn in through the mouth. When the embryos leave the ovaries they become lodged in brood chambers on the arms surrounding the mouth where they remain until they reach the planula stage. Planulae are liberated in the autumn and after a brief free-swimming existence settle on hard substrata such as rock surfaces, seaweeds and barnacles. They develop into benthic scyphistomae (Fig. 6.33a) which produce ephyrae of about 1 mm diameter during winter and spring. These feed in the plankton by capturing prey with the aid of stinging cells on the tentacles and grow quickly to develop into fully grown jellyfishes in about three months. After spawning, the adults disappear from inshore waters during September and October. Most are believed to die but some survive the winter in deeper water. Adult *Aurelia* are plankton feeders, trapping a wide range of organisms in sheets of mucus transported across both upper and lower surfaces of the body by ciliary currents. The mucus collects round the margins of the bell and is removed by the arms and passed to the mouth. Both the marginal tentacles and the arms bear nematocysts which stun larger prey, including fish larvae, and swarms of *Aurelia* can have a significant impact on the density of commercial fish larvae, such as the herring, in the plankton.

Figure 6.37 *Rhizostoma pulmo.*

Rhizostoma pulmo (Macri) (*Rhizostoma octopus*)
Fig. 6.37

Thick, dome-shaped, up to 900 mm diameter; colour varied, from whitish to pale shades of green, blue, pink and brown. Margin without tentacles, finely divided into large number of lobes slightly darker in colour. Four pairs of very large arms on undersurface, each with many lobes bearing thousands of very small mouth openings. Four large pits on undersurface in which the amphipod Hyperia galba *(p. 347) is often found. Individuals up to 30 kg wet weight have been recorded.*

R. pulmo has been recorded as far north as southern Norway but is mainly found off south and west coasts of Britain, being rare off the east coast of Scotland and north-east England. Although it occurs throughout the year, it is most common in summer and autumn and is abundant in some years. The sexes are separate and the possibility of sexual dimorphism in colour has been suggested, generally the marginal lobes and gonads of males being bluish, those of the females reddish-brown. Fertilization is internal and the life-cycle is similar to that of *Aurelia* (above). It is believed that some

Figure 6.38 *Haliclystus auricula* (adapted from a number of sources).

specimens overwinter, probably in deep water, and that these are the large specimens recorded in the first half of the following year. It feeds on microscopic planktonic organisms, drawing in currents of water through the mouth openings.

Aerial surveys in the Irish Sea have revealed the presence of high densities of *Rhizostoma* during the summer months. These blooms attract leatherback turtles, **Dermochelys coriacea** (Vandelli), which undertake extensive feeding migrations into the area to feed on the jellyfish.

In addition to the jellyfishes described above, the scyphozoans include the Order Stauromedusae, the stalked jellyfishes. They are not pelagic but live attached to stones and seaweeds by a stalk with the mouth and bell uppermost. They are usually no more than about 50 mm in height and 30 mm across. They are not generally common on the shore, but one species, **Haliclystus auricula** (Rathke) (Fig. 6.38) is not uncommon in south-west Britain. It is up to 50 mm in height, green to reddish in colour and the margin of the bell has eight lobes, each with a group of short tentacles. It lives on the lower shore attached to laminarians and other seaweeds, and on *Zostera* (p. 75). Longevity is about one year.

References

Allman, G. J. (1871). *A monograph of the gymnoblastic or tubularian hydroids*. London: Ray Society.

Cornelius, P. F. S. (1995). *North-west European thecate hydroids and their medusae. Keys and notes for identification of the species*. Parts 1 and 2. Synopses of the British fauna (New Series), no. 50. Shrewsbury: Field Studies Council.

Hincks, T. (1868). *A history of the British hydroid zoophytes*. Vol. I. Text. Vol. II. Plates. London: John Van Voorst.

Kirkpatrick, P. A. & Pugh, P. R. (1984). *Siphonophores and velellids. Keys and notes for the identification of the species*. Synopses of the British fauna (New Series), no. 29. Leiden: E. J. Brill/Dr W. Backhuys.

Manuel, R. L. (1988). *British Anthozoa (Coelenterata: Octocorallia and Hexacorallia). Keys and notes for the identification of the species*. Synopses of the British fauna (New Series), no. 18, revised edn. Leiden: E. J. Brill/Dr W. Backhuys.

Russell, F. S. (1953). *The medusae of the British Isles. Anthomedusae, Leptomedusae, Limnomedusae, Trachymedusae and Narcomedusae*. Cambridge: Cambridge University Press.

Russell, F. S. (1970). *The medusae of the British Isles*. Vol. II. *Pelagic Scyphozoa with a supplement to the first volume on Hydromedusae*. Cambridge: Cambridge University Press.

Schuchert, P. (2008). The European athecate hydroids and their medusae (Hydrozoa, Cnidaria): Filifera Part 4. *Revue Suisse de Zoologie*, **115**, 677–757. (Part of a series on the athecate hydroids.)

Stephenson, T. A. (1928–35). *The British sea anemones*, Vols. I and II. London: Ray Society.

Wood, C. (2005). *Seasearch guide to sea anemones and corals of Britain and Ireland*. Ross-on-Wye: Marine Conservation Society.

Ctenophora

The ctenophores are often referred to as sea-gooseberries or comb-jellies. A few are able to creep on solid surfaces, some of these being ectoparasitic, but the majority are pelagic and at certain times of the year may be so abundant in inshore waters that they are trapped in rock pools by the ebbing tide. Body shape varies from ribbon-like, laterally flattened to spherical, and characteristically the surface is traversed by up to eight bands of cilia known as the comb-rows. Each comb-row consists of a series of plates bearing fused cilia or ctenes. The cilia beat in waves and the animal usually moves through the water with the mouth forward. During darkness bioluminescence is a characteristic feature of many ctenophores. In addition to comb-rows, some species have a pair of long, branched retractile tentacles bearing adhesive cells, the colloblasts, which are used in food capture. Ctenophores are carnivorous. Those lacking tentacles swallow the prey whole, or in some cases pieces of prey are cut away using specialized cilia. They do not possess nematocysts and although these have been recorded in one species, they are now believed to originate from the cnidarian medusae on which the ctenophore feeds. Most ctenophores are hermaphroditic and although eggs are brooded in some, the majority release gametes into the sea where fertilization takes place with the development of a free-swimming, spherical larva known as a cydippid.

The phylum comprises seven orders, species from two of which are described below.

Phylum CTENOPHORA

Order Cydippida

Pleurobrachia pileus (O. F. Müller) Sea-gooseberry **Fig. 7.1a**
Spherical body up to 30 mm in length; 8 comb-rows. A pair of branched tentacles up to 500 mm in length; completely retractile into pouches. Virtually transparent; comb-rows iridescent; some colour associated with gut.

P. pileus is widely distributed and almost cosmopolitan. It is typically found in the plankton of inshore waters, sometimes occurring in swarms. Around Britain it is most abundant in summer and autumn when it may be stranded in rock pools. *Pleurobrachia* breeds from spring to autumn. It is hermaphroditic; eggs and sperm are shed through

(a) (b)

Figure 7.1   (a) *Pleurobrachia pileus*. (b) *Beroe cucumis* (after Greve, 1975).

the mouth and fertilization is external. There is a free-swimming cydippid larva and some of the animals spawned in spring reproduce in the autumn of the same year. Most animals die after breeding. *Pleurobrachia* is a voracious carnivore and when present in high densities is a major competitor with pelagic and larval fishes for zooplankton. It feeds mainly on copepod crustaceans. The long, branched tentacles form an effective net and, once trapped, prey is held on the adhesive cells. *Pleurobrachia* is preyed on by fishes, medusae such as *Aurelia* (p. 122) and other ctenophores, particularly *Beroe* (below). Although *Pleurobrachia* has long been believed to be a bioluminescent species, research has shown that this is not so and the evidence on which this was based was anecdotal.

Order Beroida

Beroe spp. Fig. 7.1b
Body somewhat oval-shaped, 50 mm or more in length; 8 comb-rows; no tentacles. Mouth large. Body transparent or pale pink.

Two species of *Beroe* occur commonly in north-west Europe, **Beroe cucumis** Fabricius and **Beroe gracilis** Künne. Apparently young of the species cannot be separated and although adults of *B. cucumis* are generally pinkish in colour and those of *B. gracilis* whitish, they are very difficult to separate and reference to **Greve (1975)** is recommended. *B. cucumis* is widely distributed while *B. gracilis* is more restricted in its distribution and is found in the southern North Sea and off south and west Britain. Both species are voracious carnivores. The mouth has large, specialized cilia which are used to cut up prey too large to be swallowed whole. *Beroe* feeds extensively on other ctenophores.

Reference

Greve, W. (1975). *Ctenophora*. Fiches d'Identification du Zooplancton, no. 146. Copenhagen: Conseil International pour l'Exploration de la Mer.

Platyhelminthes

The platyhelminthes or flatworms are perhaps best known through their parasitic representatives, the flukes and tapeworms, but they also include a group of free-living worms, the turbellarians, with marine, freshwater and terrestrial representatives.

Phylum PLATYHELMINTHES

Class Turbellaria	**Free-living flatworms found in the terrestrial, marine and freshwater environments.**
Class Monogenea	Mainly ectoparasitic on fishes.
Class Trematoda	Endoparasites, mostly of vertebrates.
Class Cestoda	Endoparasites of vertebrates.

The classification of the phylum is under review. Research on the molecular phylogeny of the Platyhelminthes suggests that the flukes and tapeworms have characters in common that separate them from the Turbellaria. A new classification has been proposed.

Class Turbellaria

The turbellarians are the most primitive of the flatworms. The dorsoventrally flattened body has a bilateral symmetry and the digestive system has only one opening, the mouth, on the undersurface. Eye spots and tentacles are often visible at the anterior end. The animals move over the substratum with a gliding motion which results from the beating of cilia on the undersurface of the body but muscular contraction is also involved, particularly in the larger species. Most turbellarians are carnivorous, feeding on a wide range of invertebrates, but they also scavenge on dead and decaying material and some feed on microscopic algae. Turbellarians have considerable powers of regeneration and asexual reproduction by transverse fission has been recorded. Almost all are hermaphroditic but copulation and cross-fertilization generally take place. The eggs are usually laid in gelatinous masses and hatch after one to several weeks. The majority of species do not have a pelagic larva.

Flatworms extend from the intertidal into deep water. They are found in diverse habitats including the undersurface of stones and in mud, sand and shell deposits, and among seaweed, including laminarian holdfasts. They have to be searched for carefully and should be handled with care to avoid fragmentation.

The class is divided into a number of orders, the species included here belonging to two of these:

Order Tricladida The planarians. Mainly freshwater and terrestrial, with a few marine species. The gut has three branches, e.g. *Procerodes*.

Order Polycladida All polyclads in north-west Europe are marine. They have a central gut with many branches, e.g. *Leptoplana*, *Oligocladus*, *Prosthomeraeus*.

Key to species of turbellarians

Turbellarians are difficult to identify and in some cases accurate identification to species is based on microscopic detail of the anatomy. This key is based on distinctive morphological characters which can be seen with the aid of a hand lens in living specimens observed under water.

1 Animal without tentacles; 4 groups of eye spots at anterior .... ***Leptoplana tremellaris*** p. 131
Animal with 2 tentacles; head with 2 eyes, or tentacles with many eye spots **2**

2 Tentacles short, stumpy (Fig. 8.1a); head with 2 eyes ...... ***Procerodes littoralis*** p. 130
Tentacles distinct, well developed (Fig. 8.2), each with many eye spots **3**

3 Dorsal surface of body with a number of dark longitudinal lines, some of which merge in posterior region .... ***Prostheceraeus vittatus*** p. 131
Dorsal surface of body whitish, flecked with brownish-red .... ***Oligocladus sanguinolentus*** p. 132

Procerodes littoralis (Ström) (*Procerodes ulvae*) **Fig. 8.1a**
Body up to 9 mm in length, 1–2 mm wide. Dorsal surface olive-brown with pale streak down centre of body; 3 dark stripes on head; ventral surface pale. Anterior margin of head convex. Two short, stumpy tentacles. Two eyes on head, situated in the pale areas.

P. littoralis is widely distributed in north-west Europe and is common all around Britain. It is well known for its ability to tolerate large changes in salinity, even surviving in fresh water for one or two days. It lives under stones and boulders on the upper and middle shore and is often abundant in areas where freshwater streams run across the shore. *Procerodes* becomes sexually mature when about 5 mm in

Figure 8.1   (a) *Procerodes littoralis*. (b) *Leptoplana tremellaris*.

length. Fertilization is internal and yellowish-brown, oval-shaped egg capsules, about 1 mm in diameter, are attached to the underside of stones, each capsule containing on average five embryos. There is no larva in the life-cycle and the embryos hatch after about four to five weeks at a body length of 2 to 4 mm. On the south coast of England young are commonly found on the shore during spring and summer.

Leptoplana tremellaris (O. F. Müller) **Fig. 8.1b**
Elongate, delicate oval body up to 25 mm in length, wider at anterior than posterior. Pale brown or buff-coloured, sometimes with white patches. Margins of body frilly. Without tentacles. Four groups of black eye spots anteriorly; anterior 2 groups elongate with 20–25 eyes, posterior 2 groups rounded and with 6–12 larger eye spots.

L. tremellaris is widely distributed in north-west Europe and found all around Britain. It lives under stones, on mussel beds, in laminarian holdfasts, among seaweed and is associated with a variety of sessile invertebrates such as sponges and ascidians. There is no larva in the life-cycle.

Prostheceraeus vittatus (Montagu) **Fig. 8.2a**
Body up to 50 mm in length and 25 mm wide. Dorsal surface yellowish-white with number of dark, longitudinal lines, some of which merge in posterior region. Two distinct tentacles, each with a group of darkly pigmented eye spots. Further eye spots grouped in mid-dorsal region posterior to tentacles.

(a)   (b)

Figure 8.2   (a) *Prostheceraeus vittatus* (after Lang, 1884). (b) *Oligocladus sanguinolentus* (after Lang, 1884).

P. vittatus is widely distributed in north-west Europe and in Britain occurs mainly in the south and west. It is found on the lower shore and below, under stones and seaweed and among colonies of the sea-squirt *Ciona* (p. 437).

Oligocladus sanguinolentus (Quatrefages) **Fig. 8.2b**
Body up to 22 mm in length and 13 mm wide. Dorsal surface whitish, flecked with brownish-red. Two distinct tentacles each with many eye spots. Further eye spots grouped in mid-dorsal region posterior to tentacles.

O. sanguinolentus is widely distributed in north-west Europe. It occurs all around Britain under stones and among seaweed on the lower shore and below, and is often found in laminarian holdfasts.

References

Ball, I. R. & Reynoldson, T. B. (1981). *British planarians. Platyhelminthes: Tricladida. Keys and notes for the identification of the species.* Synopses of the British fauna (New Series), no. 19. Cambridge: Cambridge University Press. (Includes *Procerodes.*)

Prudhoe, S. (1982). *British polyclad turbellarians. Keys and notes for the identification of the species.* Synopses of the British fauna (New Series), no. 26. Cambridge: Cambridge University Press. (Includes *Leptoplana, Prostheceraeus, Oligocladus.*)

Nemertea

The nemerteans are elongate, often ribbon-like worms with bilateral symmetry. They are non-segmented and have a ciliated epidermis but unlike the flatworms have a separate mouth and anus. Dorsal to the gut, and separate from it, is a muscular, eversible proboscis used in the capture of food and in defence. It is from this structure that the animals get one of their common names, the proboscis worms. They are also known as ribbon worms. Nemerteans are carnivorous and feed on a wide variety of crustaceans, polychaete worms and molluscs, which are captured on the proboscis. In some cases the prey is ingested whole, while in others the body of the prey is torn by the proboscis, which in some species has one or more piercing stylets, and the nemertean feeds suctorially. Many species pursue their prey which is killed by powerful neurotoxins while others scavenge on dead and decaying animals. Despite the presence of toxins in the tissues, nematodes are themselves eaten by a wide range of fishes. Most nemerteans have separate sexes and fertilization is usually external. In some species the gametes are deposited in an egg sac where fertilization and development take place. Several different types of larvae have been described, including a ciliated pelagic larva known as a pilidium. Length of life is believed to be about one year but the larger species possibly live for several years.

Nemerteans are abundant in terrestrial, freshwater and marine habitats. On the shore they are found in crevices, under stones and seaweed, and in mud and sandy sediments. One genus lives commensally in the mantle cavity of bivalve molluscs and another is parasitic on the egg masses and gill lamellae of crabs. Most are no more than 300 mm in length but *Lineus longissimus* (p. 135) reaches a length of many metres. All show considerable extension and contraction and require careful handling if fragmentation is to be avoided. They are difficult to identify to species.

Phylum NEMERTEA

Class Anopla	Proboscis without stylets.

Class Enopla	Proboscis with one or more stylets, but note *Malacobdella* (p. 138).

Key to species of nemerteans

Accurate identification of nemerteans can be difficult and may depend on detail of the internal anatomy. The species described here can generally be identified according to morphological characters such as the shape of the head, the number and arrangement of eyes (when present) and the form of slits on the head, known as the cephalic grooves or slits. These are lined by ciliated epithelium and usually open to the cerebral organs.

1 Body with posterior, ventral sucker; leech-like. Living in the mantle cavity of bivalve molluscs .... *Malacobdella grossa* p. 138
Body without posterior ventral sucker. Not living in the mantle cavity of bivalve molluscs **2**

2 Body broad, very flattened, up to 150 mm in length; posterior part of body wider than anterior; truncate at posterior margin with single median cirrus .... *Cerebratulus fuscus* p. 135
Body not so **3**

3 Head with 4 eyes arranged in square or rectangular pattern (in some cases eyes may be difficult to see because of presence of pigment) **4**
Head with more than 4 eyes, may be arranged in groups or rows; sometimes difficult to see **6**

4 Conspicuous, large, rectangular black pigment patch on head; often obscuring anterior pair of eyes .... *Tetrastemma melanocephalum* p. 137
Head without rectangular black pigment patch **5**

5 Body flattened, narrow (0.5 mm wide); colour pale pink to white. Anterior pair of eyes the larger .... *Nemertopsis flavida* p. 137
Body rounded, up to 1–2 mm wide; colour varied; banded, striped and spotted forms common .... *Oerstedia dorsalis* p. 138

6 Eyes in 2 rows, 1 row on each side of head **7**
Eyes not in 2 rows **8**

7 Row of 10–20 dark-coloured eyes on each side of head; animal of very great length, up to 10 m. Brown-black in colour with purplish tints ....
.... *Lineus longissimus* p. 135
Row of 2–8 eyes in an irregular row on each side of head; up to 80 mm body length. Reddish-brown in colour .... *Lineus ruber* p. 135

8 Eyes in 4 conspicuous groups. Animal pink to white in colour; 2 pinkish-red blotches on head .... *Amphiporus lactifloreus* p. 136
Eyes tend to be arranged in 4 groups; may be difficult to see.
Dorsal surface of body with broken longitudinal brown streaks
.... *Paranemertes neesii* p. 137

Class Anopla

Figure 9.1
Cerebratulus fuscus (after McIntosh, 1873–4).

Cerebratulus fuscus (McIntosh) **Fig. 9.1**

Body very flattened, broad, up to 150 mm in length and 5 mm wide. Pale grey, brown, pinkish, yellowish in colour with reddish markings; transversely wrinkled. Head bluntly pointed with 4–13 eyes on each side; posterior part of body wider than anterior, truncate at posterior margin with single median cirrus. One pair of horizontal cephalic slits. Characteristically fragments when handled.

C. fuscus is widely distributed in north-west Europe and usually found in sandy deposits on the lower shore and sublittorally to depths of over 1000 m. It is a large, conspicuous nemertean, able to swim actively on being disturbed. It is known to prey on bivalve molluscs including *Mya arenaria* (p. 291) and *Ensis directus* (p. 275). The sexes are separate.

Lineus longissimus (Gunnerus) Bootlace worm **Fig. 9.2a**

Rounded body up to 10 m in length, 5 mm wide; larger specimens not uncommon. Brownish-black in colour, with purplish tints; anterior margin of head pale. Row of 10–20 dark-coloured eyes on each side of head, may be difficult to see because of brownish-black colour of body; behind each row a horizontal, cephalic slit.

L. longissimus is widely distributed in north-west Europe and because of its great length is probably the best-known species. It is the longest nemertean recorded and is widely distributed on the lower shore, lying as a coiled mass in crevices and under stones, particularly in areas where there is muddy-sand and shingle. When disturbed it produces large quantities of mucus. It feeds on a wide range of invertebrates, such as polychaete worms and ascidians, some of which are consumed whole. The sexes are separate.

Lineus ruber (Müller) **Fig. 9.2b**

Body slightly flattened, up to 80 mm in length, 3 mm wide. Reddish-brown, paler on ventral surface. Head slightly wider than rest of body; anterior margin rounded; 2 to 8 eyes in an irregular row on each side of head; a horizontal, cephalic slit behind each row.

L. ruber is common in north-west Europe and widely distributed around Britain under stones, among barnacles, mussels and seaweeds such as *Cladophora* (p. 32), *Ceramium* (p. 47) and in laminarian holdfasts. It tolerates salinities down to about 8 and is commonly found in estuarine sands and muds and on saltmarshes. It feeds on living and dead animals, such as polychaete worms and crustaceans captured by means of the proboscis. The sexes are separate and on the south coast of England and the Channel coast of France reproduction occurs from December/January to March. A male and a female become enclosed in a mucous cocoon secreted by the female and attached to the undersurface of rocks and seaweed. The eggs are fertilized and held in the cocoon where development takes place to give a creeping, non-pelagic larva, known as a Desor larva. Not all the eggs develop, some being eaten by the first larvae to hatch. After the eggs are fertilized, the males and females leave the cocoon and die a few weeks later. *L. ruber* breeds annually.

(a)   (b)

Figure 9.2   (a) *Lineus longissimus*. (b) *Lineus ruber* (after McIntosh, 1873–4).

Lineus viridis (Müller) is morphologically similar to *L. ruber* and was considered to be a colour morph of *L. ruber*. It is up to 80 mm in length, 3 mm in width, with 2–8 eyes in an irregular row on each side of the head. As the specific name suggests, it is olive-green to greenish-black, occasionally pale green in colour. It is found intertidally under rocks and stones in muddy sediments.

Class Enopla

Amphiporus lactifloreus (Johnston) **Fig. 9.3a**
Body rounded, up to 35 mm in length, 1.5 mm wide. Pale pink to white; 2 pinkish-red blotches on head. Head rounded; eyes in 4 groups; 2 pairs transverse cephalic slits.

A. lactifloreus is widely distributed in north-west Europe and around Britain from the upper shore into the sublittoral to depths of about 250 m. It is often found on laminarians and other seaweeds, also in sand and gravel. The amphipod crustacean *Gammarus locusta* (p. 334) is an important food item. It is trapped on the worm's proboscis and the worm then inserts its head into the body of the prey and feeds on the soft tissues. The sexes are separate and breeding occurs during spring and early summer. Groups of mature worms are found together and gametes are deposited in egg sacs on rocks and seaweed. There is no pelagic larva.

Figure 9.3   (a) *Amphiporus lactifloreus.* (b) *Paranemertes neesii.* (Both after McIntosh, 1873–4).

Paranemertes neesii (Ørsted) (*Emplectonema neesii*) **Fig. 9.3b**
Up to 500 mm in length, 4 mm wide; usually smaller. Dorsal surface pale to dark brown, with broken longitudinal brown streaks; ventral surface white to pale pink. Head rounded; many eyes, tend to be arranged in 4 groups but often difficult to see. Cephalic slits inconspicuous.

P. *neesii* is widely distributed in north-west Europe and around Britain. It is found under stones, among seaweed and in sand and gravel, from the middle shore into the sublittoral to depths of about 30 m. The sexes are separate and reproduction occurs from January to October. There is no pelagic larva.

Nemertopsis flavida (McIntosh) **Fig. 9.4a**
Body flattened, very narrow; up to 40 mm in length, 0.5 mm wide, narrowing to a point at posterior end. Pale pink to white. Head rounded; 4 small black eyes arranged in a rectangle, anterior pair the larger; 2 pairs transverse cephalic slits, one slit behind each eye.

N. *flavida* is widely distributed around Britain in pools on the lower shore, under stones and in laminarian holdfasts. It also lives in mud, sand and gravel and extends into the sublittoral. The sexes are separate and reproduction occurs during summer and autumn. There is no pelagic larva.

Tetrastemma melanocephalum (Johnston) **Fig. 9.4b**
Body up to 60 mm in length, 2.5 mm wide; yellowish-green in colour with conspicuous, large rectangular black patch on head. Head flattened; 4 eyes arranged in a rectangle, the anterior pair often obscured by black pigment patch. Without cephalic slits.

T. *melanocephalum* is widely distributed in north-west Europe and around Britain, where it is found on rocky shores in crevices, laminarian holdfasts and with seaweeds such as *Corallina* (p. 42). It also occurs on sand and mud-flats and saltmarshes, where

Figure 9.4 (a) *Nemertopsis flavida*. (b) *Tetrastemma melanocephalum*. (c) *Oerstedia dorsalis*, showing variation in body pattern (after McIntosh, 1873–4).

it has been recorded feeding suctorially on copepods and amphipods, especially *Corophium volutator* (p. 343) and *C. arenarium* (p. 344). It has been found in the sublittoral to depths of up to 40 m.

Oerstedia dorsalis (Abildgaard) **Fig. 9.4c**
Rounded, short body usually up to 15 mm in length, 1–2 mm wide. Colour varied, brown, red, green, yellow often with pale line along mid-dorsal surface; banded, striped and spotted forms common. Ventral surface pale. Head rounded; 4 eyes arranged in a square. Without cephalic slits.

O. dorsalis is widely distributed in north-west Europe and Britain, extending from the middle shore to depths of 80 m. It is commonly found among seaweeds, in rock pools and in laminarian holdfasts where it feeds suctorially on amphipods. The sexes are separate and breeding occurs from September to November. There is no pelagic larva; young develop within a gelatinous capsule attached to the substratum. In terms of colour and patterning *O. dorsalis* is one of the most varied nemerteans and research carried out on populations on the west coast of Sweden has shown that one of the colour varieties is a separate species, **Oerstedia striata** Sundberg.

Malacobdella grossa (Müller) **Fig. 9.5**
Body flattened, leech-like, with a posterior ventral sucker. Up to 40 mm in length and 15 mm wide. White or grey-coloured when immature; colour of gonad shows in mature specimens, males with pinkish coloration, females greenish.

Figure 9.5
Malacobdella grossa
(after Gibson, 1994).

M. grossa lives as an ectocommensal in the mantle cavity of bivalve molluscs. It occurs in a wide variety of species of hosts, including *Cerastoderma edule* (p. 269) and *Zirfaea crispata* (p. 295), and is widely distributed in north-west Europe. It is found between the gill lamellae of the host; usually there is only one nemertean in each host but as many as five have been recorded. The proboscis of *Malacobdella* is not armed with piercing stylets. It is a suspension feeder; the pharynx is alternatively expanded and contracted and water forced through a network of ciliated papillae, trapping food particles which are transported to the oesophagus.

References

Gibson, R. (1994). *Nemerteans. Keys and notes for identification of the species.* Synopses of the British fauna (New Series), no. 24, 2nd edn. Shrewsbury: Field Studies Council.

McIntosh, W. C. (1873–4). *A monograph of the British annelids.* Vol. I. Part 1. *The nemerteans.* London: Ray Society.

Priapula

Priapulans have been variously classified with different groups of invertebrates but most recent research assigns them to the phylum Priapula. They are worm-like, varying in length from a millimetre or so to about 200 mm. There are believed to be no more than about 16 living species. The body is divided into two distinct parts, an anterior proboscis, or introvert, and a posterior trunk. The proboscis is the shorter of the two regions and is retractile and covered with a large number of spines. The mouth is terminal. In some species, the posterior region of the body bears one or two much-branched processes. The trunk often has papillae on the surface and is annulated but the body is not segmented. The animals live in sandy and muddy sediments on the shore and in the sublittoral. They are carnivorous, feeding on polychaete worms and other invertebrates captured by means of the proboscis. The sexes are separate and the gametes are discharged into the sea where fertilization occurs. The larvae live in soft sediments.

Only one species is likely to be encountered on the shore.

Figure 10.1 *Priapulus caudatus* (after M'Intosh, 1875).

Priapulus caudatus Lamarck **Fig. 10.1**
Body elongate, cylindrical up to 150 mm in length. Large proboscis with rows of spines. Posterior of trunk with single, much-branched process. Pinkish-brown in colour.

P. caudatus is widely distributed and may be locally common in some areas of north-west Europe burrowing in mud, muddy-gravel and muddy-sand from the middle shore into the sublittoral. It is carnivorous, feeding on polychaete worms such as *Aphrodita* (p. 151) and brittlestars. The sexes are separate, fertilization external and breeding has been recorded in winter in specimens in the North Sea. The larvae inhabit the surface layers of sediment and are believed to feed on detritus. Development to the adult takes about two years.

References

M'Intosh, W. C. (1875). *The marine invertebrates and fishes of St. Andrews*. Edinburgh: A. & C. Black.

Stephen, A. C. (1960). *British echiurids (Echiuroidea), sipunculids (Sipunculoida) and priapulids (Priapuloidea) with keys and notes for the identification of the species.* Synopses of the British fauna (Series 1), no. 12. London: Linnean Society.

Annelida

The annelids include earthworms, bristle worms and leeches, and are characterized by the division of the body into a number of similar, externally visible segments. This is known as annulation, or metameric segmentation, and distinguishes them from other worm-like animals. In addition to being an important external feature, metamerism also affects the body cavity of the annelid, dividing it by septa into a number of units. In some annelids there is specialization and fusion of segments, particularly at the anterior end, and some segments have conspicuous lateral outgrowths. This is a large and diverse phylum with marine, freshwater and terrestrial representatives. It is divided into two classes.

Phylum ANNELIDA

Class Polychaeta	The bristle worms, characterized by paired, lateral outgrowths known as parapodia which bear chaetae (bristles). Marine.
Class Clitellata	Without paired lateral outgrowths. Marine, freshwater and terrestrial. Includes the earthworms (subclass Oligochaeta) and the leeches (subclass Hirudinoidea).
	Marine oligochaetes, known as sludge worms, are often found in sandy and muddy deposits. They are generally very small, rarely exceeding 20 mm in length and a few millimetres in width. In polluted estuaries they may be present in enormous densities, virtually to the exclusion of other macrofauna. Identification of species is difficult and beyond the scope of this book. The reader is referred to **Brinkhurst (1982)**.

Class Polychaeta

The polychaetes are widely distributed in the marine environment being commonly found on the shore under stones and rocks and buried in mud and sand. In addition to species such as the ragworms and lugworms, the class also includes a range of very small worms previously assigned to a separate phylum, the Archiannelida. Many of these tiny forms are interstitial and live in sedimentary deposits. They are members of the meiofauna (p. 7) and are not included in this text; **Westheide (2008)** should be consulted for identification.

Typically, the segmented body of the polychaete bears paired, lateral outgrowths, the parapodia (Fig. 11.1b). Each parapodium is divided into a dorsal notopodium and a ventral neuropodium, very often differing in structure. The parapodia usually bear chaetae which vary in shape and are very important in taxonomy.

The anterior segments of the body are often modified and fused, and carry a number of structures associated with feeding and sensory perception (Fig. 11.1a,c). The anus is terminal. Within the class there is wide diversity of form; this is related to whether the worm is active and mobile, often referred to as errant, or sedentary. Active, mobile forms usually have well-developed sensory structures such as eyes and antennae, well-developed parapodia and an eversible proboscis (referred to in some texts as a pharynx) often armed with jaws. In the sedentary forms, eyes and antennae are often reduced or absent, the parapodia are poorly developed, or very much modified, and

Figure 11.1 (a) Dorsal view of anterior end of a generalized polychaete with an active and mobile life-style (errant). (b) Diagrammatic transverse section through polychaete to show arrangement of parapodia. (c) Dorsal view of anterior end of a generalized polychaete with a sedentary life-style. (d) Dorsal view of a scale worm.

complex food-gathering structures may be present at the anterior end of the body. Some polychaetes burrow into sediments and usually live in a mucus-lined burrow, while others live in tubes made of sand grains, mud or calcium carbonate.

Polychaetes show a wide range of feeding habits including carnivorous, scavenging, deposit and suspension feeding. The sexes are usually separate and fertilization generally external. Eggs and sperm may be liberated through special ducts or the body wall of the worm may rupture, thereby releasing the gametes. An unusual form of reproduction known as epitoky occurs in some polychaetes where highly modified individuals, epitokes, are formed. These are packed with gametes and generally have enlarged eyes and large parapodia enabling them to swim in the surface waters where they swarm. Here synchronized release of gametes ensures cross-fertilization. In some cases the whole worm is modified for reproduction (epigamy) while in others chains of reproductive individuals are formed and these eventually break away and swim to the surface to spawn (schizogamy). Many polychaetes have a free-swimming pelagic larva, the trocophore; in others the eggs are either brooded by the parent or in the parental tube, or are laid in mucous egg sacs. Asexual reproduction by budding has been recorded in a few species.

The terminology used in the description of polychaetes is complex and there is often variation between authors. Figure 11.1 shows characters used in identification. It should be noted that the morphology of the palps varies and includes short palps, which are believed to have a sensory function, and longer palps used in feeding. The latter may be thread-like and numerous and are often referred to as tentacles. Care should be taken when handling specimens as many of the appendages such as palps are often lost. The scales (elytra) (Fig. 11.1d) of worms belonging to the Aphroditidae, Polynoidae and Sigalionidae may also be lost when the animal is handled, but the position of the scales, which is important in identification, can always be determined by the attachment scars left behind. The morphology of the proboscis is an important character in identification of several families; gentle pressure just behind the head is usually sufficient to evert the proboscis.

There is considerable debate as to the classification of polychaetes. Some authors make an initial distinction between errant and sedentary species and while this can be useful in that it highlights morphological differences that reflect the habit of the worm, it is now recognized that it has no basis in taxonomy. The systematic order followed here is based on Fauchald & Rouse (1997), Rouse & Fauchald (1997) and **Rouse & Pleijel (2001)** in which the polychaetes are grouped under two major divisions, **Scolecida** in which the head lacks appendages and **Palpata** in which palps and often other appendages are present on the head.

Characters used in the key are shown in Fig. 11.1.

Key to families of polychaetes

1 Dorsal surface of the body covered wholly or partially by overlapping scales (elytra) or with scales covered by a dense mat of chaetae giving the appearance of felt. *Note*: the scales may be lost when the worm is handled, leaving attachment scars as evidence of their position **2**
Dorsal surface of body without scales or mat of felt-like chaetae; worm inhabiting a tube or not **3**

2 (a) Short, broad body, not worm-like. Scales completely cover body and are often covered by a dense mat of felt-like chaetae. Head with single median antenna; eversible proboscis .... **Aphroditidae** p. 151
or
(b) Body long and narrow or short and broad, but worm-like; 12, 15 or 18 pairs of scales present at least on the anterior part of body, usually attached on every other segment. Head usually with 3 antennae; eversible proboscis ...... **Polynoidae** p. 152
or
(c) Body long, thin; many segmented, with many pairs of scales. Scales attached on every other segment anteriorly, and on all posterior segments. Head with 2 or 3 antennae; eversible proboscis .... **Sigalionidae** p. 155

3 Worm not inhabiting a tube **4**
Worm inhabiting a tube which may be calcareous, parchment-like, of mucus, mud particles, sand grains, shell fragments **17**

4 Body regularly divided into similar segments. Conspicuous parapodia all along body. Head generally square, rounded or heart-shaped; with 2 or 3 antennae, often with eyes, palps and tentacular cirri. With a large, eversible proboscis **5**
Body regularly divided into similar segments. Parapodia not always conspicuous. Head pointed or bluntly rounded and without appendages, or with 4 very small appendages set closely together at tip, or head or segments just posterior to head with 2 long, coiled palps or many long narrow palps (tentacles). With or without eversible proboscis **10**

5 Head small, squarish, with 2 short, pointed antennae, 1 at each anterior corner, giving some resemblance to the head of a cat; 2 short palps, no tentacular cirri; eyes not obvious. Large eversible proboscis with papillae and jaws .... **Nephtyidae** p. 164
Head usually with 2 or 4 eyes, 2 or 3 antennae and either or both palps and tentacular cirri **6**

6 Parapodia with large expanded paddle-shaped dorsal cirri. Head usually with 2 large eyes, 2 or 3 antennae; 2 short palps; 2–4 pairs of tentacular cirri. Large eversible proboscis sometimes with papillae but without jaws .... **Phyllodocidae** p. 165
Not so **7**

7 Parapodia divided into similar-sized notopodia and neuropodia. Head with 4 eyes, usually 2 antennae, 2 stout, knobbed palps, usually 4 pairs tentacular cirri. Large eversible proboscis with 2 large jaws and many paragnaths (Fig. 11.11a) .... **Nereididae** p. 157
Parapodia not clearly divided into similar-sized notopodia and neuropodia; dorsal cirri of parapodia slender and cylindrical, may be long and annulated **8**

8 Dorsal cirri of parapodia slender and cylindrical; not conspicuously long. Head usually with 2 eyes, 3 antennae, 2 palps; 1 pair of tentacular cirri or none. Eversible proboscis with complex jaws. First 2 segments of body without parapodia and chaetae .... **Eunicidae** p. 167
Dorsal cirri of parapodia mostly very long, often annulated. Head usually with 4 eyes **9**

9 Head usually with 2, 2-jointed palps; 2 or 3 antennae, usually 6–8 pairs of tentacular cirri. Large tubular, eversible proboscis; small jaws may be present .... **Hesionidae** p. 157

Head with 2, non-jointed palps (may be closely applied to ventral surface and not visible dorsally); 3 antennae, 2 pairs of tentacular cirri. Eversible proboscis often with 1 to many teeth .... **Syllidae** p. 162

10 Head pointed or bluntly rounded. Without appendages or with 4 very small appendages set closely together at tip **11**
Head or segment just posterior to head with 2 long, often coiled palps or 2 to many long narrow palps (tentacles) **15**

11 Head elongate, 4 very small appendages at tip. Body with many segments, each with 2 or 3 annulations. Very large eversible proboscis with 4 jaws .... **Glyceridae** p. 164
Head without appendages **12**

12 Long, many-segmented body, anterior region flattened; posterior region cylindrical. Large gills on dorsal surface of all segments except a few anterior ones. Tip of posterior region with 2 long cirri (Fig. 11.5) .... **Orbiniidae** p. 150
Body not as above **13**

13 Body solid, cylindrical with many segments, thicker anteriorly; almost all segments with 5 annulations. Large branching dorsal gills present on at least some segments .... **Arenicolidae** p. 146
Body and gills not as above **14**

14 Body shape reminiscent of an earthworm; parapodia and chaetae inconspicuous. Gills present or absent, but not large and branching .... **Capitellidae** p. 149
Body short and cylindrical with small number of segments; usually a deep ventral groove along part or all of length. Gills present on dorsal surface of some segments .... **Opheliidae** p. 150

15 Long, contractile thread-like gills present on at least some segments along sides of body. Two groups of 2 to many long, slender palps (tentacles) attached dorsally behind head .... **Cirratulidae** p. 180
Without long, contractile thread-like gills along sides of body **16**

16 Head with 2 long, often coiled palps, usually with 4 eyes, sometimes 2 frontal horns. Large flattened gills on dorsal surface of at least some segments.
Note: the palps are often lost when the worm is handled making identification difficult .... **Spionidae** p. 185
Head with many long, narrow intertwined palps (tentacles); usually 3 pairs of dorsal, branching gills just behind head .... **Terebellidae** p. 182

17 Worm inhabiting a calcareous tube; tube encrusting or upright; spirally coiled, straight or sinuous on seaweed, rocks, stones, shells, carapace of decapods
.... **Serpulidae** p. 173
Tube not calcareous **18**

18 Large, tough, parchment-like U-shaped tube. Body of worm of very distinctive appearance; divided into 2 or 3 regions with much-modified segments and parapodia; some parapodia fused to form fan-like structures ... **Chaetopteridae** p. 185
Tube and worm not as above **19**

19 Tube made of sand grains and shell fragments arranged in overlapping fashion. Head with a frilled membrane (Fig. 11.19a) .... **Oweniidae** *(Owenia fusiformis)* p. 167
Tube and worm not as above **20**

20 (a) Tube made of fine sand grains and shell fragments; rigid, narrow, cone-shaped, open at both ends. Worm as in Fig. 11.27 .... **Pectinariidae** p. 181
or
(b) Tube made of coarse sand grains; attached to rock and stones, often forming large colonies firm enough to walk on. Worm as in Fig. 11.20a. .... **Sabellariidae** p. 169
or
(c) Tube made of sand and shell fragments; anterior end flattened and extended into distinctive finger-like projections extending above surface of sediment. Worm as in Fig. 11.29a .... **Terebellidae (*Lanice conchilega*)** p. 183
Tube and body of worm not as above **21**

21 Mucous or membranous tube often covered with mud or sand particles. Head of worm with crown of pinnate projections (radioles) (Fig. 11.21a,b) which are often brightly coloured and banded, and sometimes bear eyes .... **Sabellidae** p. 170
Mucous tube; may be flimsy, or covered with mud particles, sand grains, shell fragments. Head of worm not as above **22**

22 Mucous tube often covered with mud particles, sand grains, shell fragments. Head of worm with 2 long, coiled palps or many long, narrow palps (tentacles) **23**
Mucous tube, may be rather flimsy; head of worm not as above **24**

23 Head of worm with 2 long, coiled palps (often lost when worm handled) .... **Spionidae** p. 185
Head of worm with many long, narrow intertwined palps (tentacles) .... **Terebellidae** p. 182

24 Parapodia divided into similar-sized notopodia and neuropodia. Head with 4 eyes, usually 2 antennae, 2 stout, knobbed palps, usually 4 pairs tentacular cirri. Large eversible proboscis with 2 large jaws and many paragnaths (Fig. 11.11a) .... **Nereididae** p. 157
Parapodia not divided into similar-sized notopodia and neuropodia, with mostly very long, dorsal cirri. Head with 4 eyes, 2 non-jointed palps (may be closely applied to ventral surface and not visible dorsally); 3 antennae, 2 pairs of tentacular cirri. Eversible proboscis often with 1 to many teeth .... **Syllidae** p. 162

Scolecida

Family Arenicolidae

Sedentary, in burrows in sand and mud. Body cylindrical, anterior (thoracic) region thicker than posterior (abdominal) region. Almost all segments with 5 annulations. Branching dorsal gills always present on at least some segments. Posterior region with or without chaetae. Head without eyes and appendages. Small, soft eversible proboscis.

Branching dorsal gills present on segments 7 to 19. Posterior region ('tail') narrow and without gills and chaetae .... *Arenicola marina* p. 147
.... see also *Arenicola defodiens* p. 147

Branching dorsal gills on all segments posterior to the 15th. No distinct 'tail' region .... *Arenicolides ecaudata* p. 149

Arenicola marina (Lamarck) Blow lug, lugworm **Fig. 11.2a,b,d; Plate 24c**

Body cylindrical, up to 200 mm in length. Anterior (thoracic) region of 19 segments, the posterior 13 of which have branching gills; narrower posterior (abdominal) region ('tail') of many segments lacking chaetae and gills. Reddish, greenish-yellow or black in colour.

A. marina is widely distributed in north-west Europe and occurs all around Britain on the middle and lower shore in sand and muddy-sand. It lives in J-shaped burrows, generally about 200 mm below the surface, and is tolerant of salinities down to about 12. In sheltered, estuarine sediments, densities as high as 100 to 150 per square metre have been recorded. *Arenicola* feeds on organic material such as microorganisms and detritus present in the sediment. When feeding, the worm moves forward to ingest sediment from the 'front' of the burrow and it has been suggested that the growth of microorganisms associated with the sediment in this area is enhanced by the flow of water through the burrow created by the worm. Sediment in this area may be further enriched by plankton contained in the water current becoming trapped in the sand just in front of the worm. Ingestion of sediment causes a shallow depression on the surface which, together with the characteristic worm casts, indicates the presence of *Arenicola* (Plate 24c). The sexes are separate and although breeding has been recorded in spring, late summer and autumn, it occurs over a well-defined, two- to three-week period in October and November at many localities around Britain and there is evidence of synchrony of spawning through the release of pheromones. Sperm are discharged from the burrows of the males onto the surface of the sediment, forming sperm puddles which are dispersed by the incoming tide. The female deposits eggs in the burrow where they are fertilized by sperm drawn in with the respiratory current. The larvae undergo their early development within the burrow and then move to the surface of the sediment and are transported by tidal currents to firmer areas, such as shingle and pebble, where they settle. Over the next few months they undergo development in mucous tubes attached to the substratum. At the end of this period the young worms, still in their mucous tubes, are found in the surface waters and are carried by the tide and eventually burrow into the sandy and muddy sediments where *Arenicola* is typically found. The worms become sexually mature after two or three years and spawn once a year, possibly for up to five or six years. *Arenicola* is preyed on by birds and flatfishes and is particularly prone to predation when it moves backwards in its burrow and comes close to the surface to deposit casts. Often only the 'tail' is cropped and in many cases the worm survives.

It has long been recognized that the lugworm exists in two varieties, the blow lug and the black lug. These are now recognized as two separate species. *A. marina* (above) and **Arenicola defodiens** (the black lug) described from South Wales by Cadman & Nelson-Smith (1993). The two species are similar in general appearance but *A. defodiens* is characterized by being generally darker (usually black) in colour and reaching a greater length (270 mm). In 90% or more of individuals of both species examined, the number of annuli between the first four consecutive pairs of chaetae differs as in Fig. 11.2b,c. *A. defodiens* has recently been recorded from the Wadden Sea and the Skagerrak. It occurs lower on the shore, living deeper in the sediment and seems to favour more exposed shores. *A. defodiens* is not found in estuaries. Where the species coexist, it has been suggested that *A. defodiens* spawns one or two months later than *A. marina*.

Figure 11.2   (a) *Arenicola marina*. (b) Anterior of *A. marina*, and (c) anterior of *Arenicola defodiens*
showing the number of annuli between consecutive pairs of chaetae. (d) Diagrammatic
representation of *Arenicola marina* in burrow; arrows indicate direction of flow of water.
(e) *Arenicolides ecaudata*. ((b) and (c) After Cadman & Nelson-Smith, 1993.)

Arenicolides ecaudata (Johnston) **Fig. 11.2e**

Similar to Arenicola marina *but lacks a distinct 'tail' region; external gills present on all segments posterior to the 15th. Up to 250 mm in length.*

A. ecaudata is widely distributed in north-west Europe but less commonly found than *A. marina* (above). It occurs on the lower shore burrowing in sand and mud, under stones and in crevices where sediment has been deposited.

Family Capitellidae

Sedentary; worms in burrows in sand and mud; anterior (thoracic) region of body short and thick; posterior (abdominal) region long and narrow. Parapodia small. Gills present or absent. Chaetae inconspicuous. Head conical, eyes present or absent; without appendages. Eversible proboscis. Resembles earthworm in body shape.

Capitella spp. **Fig. 11.3**

Body up to 100 mm or more in length, usually very much smaller. Up to about 100 segments. Narrowing gradually towards posterior. Gills absent. Blood-red in colour.

The different species of *Capitella* are very similar in appearance. The most commonly described species is **Capitella capitata** (Fabricius) which lives in muddy-sand, muddy-grit and fine sand on the middle and lower shore. It feeds on microorganisms and detritus and sometimes occurs in enormous numbers in areas of organic enrichment such as discharges from pulp mills and sewage farms. *C. capitata* is now known to be a species complex comprising a large number of sibling species. The life-cycle and mode of reproduction vary from species to species; some are dioecious, others hermaphroditic, and it appears that hermaphroditism is seen when population density is low. Females build tubes, made of fine material and open at both ends, in which the fertilized eggs are attached to the sides and held until the larval stage. Interestingly, in *C. capitata* more than one mode of development (poecilogony) has been recorded; eggs are either large and yolky leading to lecithotrophic development with benthic larvae, or small and non-yolky leading to planktotrophic development with pelagic larvae. Length of life varies widely in the species complex and estimates of nine months and one year have been made for *C. capitata* in Europe.

Figure 11.3 *Capitella* sp.

Family Opheliidae

Active, burrowing worms in sand and mud. Short, cylindrical body with small number (30 to 60) of segments; usually with deep ventral groove along part or all of length. Gills present on dorsal surface of some segments. Head pointed, eyes not visible; without appendages.

Ophelia bicornis Savigny **Fig. 11.4**

Body up to 45 mm in length. Up to about 32 segments. Thick anteriorly, narrowing towards posterior; ventral groove running from 10th segment with chaetae to posterior end of body; 15 pairs of gills on narrower part of body. Pinkish-purple in colour.

Ventral groove

Figure 11.4 *Ophelia bicornis* (after Fauvel, 1927).

O. bicornis is a southern species, extending as far north as the English Channel, but its distribution is often localized. It lives in fine, well-drained sand on the middle and lower shore where it feeds on detritus. The sexes are separate; during the breeding season the males are whitish in colour and the females dark green. The worms remain within the sediment and discharge the gametes onto the surface where fertilization takes place. The larvae are planktonic for about a week.

 O. bicornis is perhaps the best-known member of the genus but other species are found in sandy deposits in more northerly localities around Britain. Morphological and genetic studies on populations of *O. bicornis* from the Mediterranean suggest that it is a complex of two separate species.

Family Orbiniidae

Worms sedentary, burrowers in sand and mud. Head usually pointed; anterior (thoracic) region of body flattened; posterior (abdominal) region is longer, narrower and cylindrical. Large gills present on dorsal surface of all segments except a few anterior ones. Head without appendages. Tip of posterior region with 2 long cirri.

Scoloplos armiger (O. F. Müller) **Fig. 11.5**

Body up to 120 mm in length, tip of posterior region with 2 long cirri. Up to 200 or more segments. Gills present on dorsal surface from about 12th segment. Head

(a)

Posterior cirri

(b)

Figure 11.5 (a) *Scoloplos armiger.* (b) Egg cocoon on surface of sand.

cone-shaped, sharply pointed, with 2 eyes which may be difficult to detect. Red or reddish-orange in colour.

S. armiger is widely distributed in north-west Europe and around Britain on the lower shore and in the sublittoral, burrowing in muddy-sand. It is tolerant of reduced salinity and found in the lower reaches of estuaries where it feeds on organic deposits. The sexes are separate and when ripe, males are whitish in colour and the females brownish. On the south coast of England, *Scoloplos* first breeds when two years old and lives for about four years. Breeding occurs in early spring and is synchronized with spring tides. It is believed that the worms are in close association when spawning takes place and the eggs are fertilized as they leave the female. Greenish- to reddish-brown eggs are laid in pear-shaped, jelly-like cocoons, up to 15 mm in length, on the surface of the sediment. They are attached by a hollow stalk which penetrates up to 100 mm below the surface. The number of eggs per cocoon varies from about 80 to as many as 5000. They develop within the cocoon and after two to three weeks the larvae hatch at a length of about 1 mm, most of them migrating down the hollow stalk to emerge within the sediment and in this way avoid predators. There is no pelagic larval stage.

Research on intertidal and subtidal populations of *S. armiger* has revealed two different modes of development. The intertidal populations have egg cocoons and, as described above, lack a pelagic larva. In subtidal populations the eggs are not laid into cocoons but develop into larvae with a pelagic life of about 12 days. Both populations reproduce in spring but offshore populations also have an autumn spawning period. Intertidal and offshore populations are reproductively isolated and it has been concluded that they are sibling species.

Palpata

Family Aphroditidae

Active, mobile worms. Short, broad body; dorsal surface covered by large overlapping scales (elytra – expanded dorsal cirri of the parapodia) which are often covered by a dense mat of chaetae giving the appearance of felt. Head usually with 2 pairs of eyes, in some species on stalks; single median antenna; 2 palps; 2 pairs tentacular cirri. Eversible proboscis with jaws.

Aphrodita aculeata Linnaeus Sea-mouse **Fig. 11.6**

Wide, oval body up to 200 mm in length, usually smaller. Up to 40 segments. Dorsal surface rounded, ventral surface flat. Dorsal surface with 15 pairs flattened scales hidden by mat of felt-like chaetae. Head concealed, bears 2 pairs of sessile eyes. Dorsal surface with greyish-brown chaetae, lateral and ventral surfaces with iridescent green and gold chaetae and stout brown chaetae. Animal of distinctive and unmistakable appearance.

A. *aculeata* is widely distributed in north-west Europe on mud or muddy-sand at depths of 10–170 m, but occasionally large numbers are washed ashore. It normally feeds only when buried in the sediment and there are conflicting reports of the food taken; it is, however, believed to be carnivorous

Figure 11.6 *Aphrodita aculeata.*

and feeds on large polychaetes such as *Nephtys hombergii* (p. 165), *Hediste diversicolor* (p. 158), *Alitta virens* (p. 160) and *Arenicola marina* (p. 147). The sexes are separate and mature females can be recognized by the cream-coloured eggs visible through the walls of the parapodia. Little is known of the breeding cycle; mature animals have been recorded in spring and autumn on the south coast of England.

Family Polynoidae Scale worms

Mobile, active worms. Body long and narrow or short and broad; dorsal surface wholly or partly covered by scales (elytra – expanded dorsal cirri of the parapodia); depending on species there are generally 12, 15 or 18 pairs. The scales are present at least on the anterior part of the body; usually attached on every other segment. Some or all of the scales may be lost when the animal is handled, leaving attachment scars. Head with 4 eyes; usually 3 antennae; 2 long flexible palps; 2 pairs tentacular cirri. Eversible proboscis with concealed jaws.

1 Dorsal surface of body with 12 pairs of scales **2**
   Dorsal surface of body with 15 pairs of scales **3**

2 Scales overlap to some extent, but leave part of anterior mid-dorsal region
   uncovered; scales without a fringe of papillae .... *Lepidonotus clava* p. 152
   Scales overlap and completely cover dorsal surface; scales with tubercles and
   posterior fringe of papillae .... *Lepidonotus squamatus* p. 153

3 Scales cover only the anterior half of body; 50 or so posterior segments not
   covered. Worm long and narrow .... *Polynoe scolopendrina* p. 155
   Scales cover the whole length of the body or leave a relatively small part
   (posterior 8 or so segments) uncovered **4**

4 Posterior 8 or so segments not covered by scales ..... *Harmothoe extenuata* p. 154
   All of dorsal surface covered by scales **5**

5 Scales covered with tiny tubercles; quarter of scale margin with fine papillae.
   Worm longish, up to 65 mm in length .... *Harmothoe imbricata* p. 153
   Scales covered with many small and a few large tubercles; more than half of scale margin
   with papillae. Worm smallish, up to 25 mm in length .... *Harmothoe impar* p. 155

Lepidonotus clava (Montagu) **Fig. 11.7a**
Body up to 35 mm in length. Up to 26 segments. Dorsal surface rounded, ventral surface flattened; 12 pairs of rounded scales, leaving part of anterior mid-dorsal region uncovered. Scales patterned with shades of brown with lighter central spot.

L. clava is widely distributed in north-west Europe and in Britain is commonly found on south and west coasts where it occurs under stones on the lower shore. It is carnivorous, feeding on polychaetes, crustaceans and other invertebrates.

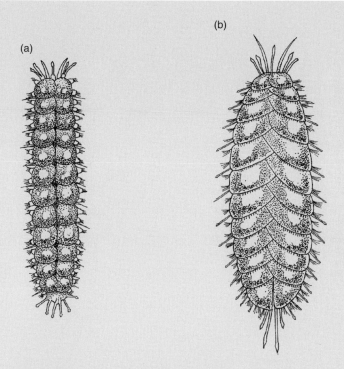

Figure 11.7   (a) *Lepidonotus clava* (after McIntosh, 1900). (b) *Lepidonotus squamatus* (after McIntosh, 1900).

Lepidonotus squamatus (Linnaeus) **Fig. 11.7b**

Body up to 50 mm in length. Up to 26 segments. Dorsal surface rounded, ventral surface flattened; 12 pairs of overlapping scales, oval to kidney-shaped, completely covering dorsal surface; scales covered with tubercles and with posterior fringe of papillae. Colour of scales varied, brown, yellow, grey, with clear central spot.

L. squamatus is widely distributed in north-west Europe and around Britain among stones and seaweed on the middle and lower shore and into the sublittoral. It is mainly carnivorous, feeding on small invertebrates, although it is also believed to feed on seaweed. The sexes are separate and breeding probably occurs during June and July. When disturbed the worm rolls up into a ball and relies on the scales, which are not easily dislodged, for some protection.

Harmothoe imbricata (Linnaeus) **Fig. 11.8a**

Body flattened; up to 65 mm in length. Up to 39 segments; 15 pairs of overlapping scales covering dorsal surface, the 1st pair almost circular; following pairs kidney-shaped, covered with tiny tubercles; quarter of scale margin fringed with fine papillae. Colour of scales varied, grey, blue or brown.

H. imbricata is widely distributed in north-west Europe and Britain, extending from the middle and lower shore to depths of over 3000 m. It has been described as the commonest polynoid of rocky shores and is found under stones and in laminarian holdfasts. It is an active predator, seizing prey such as amphipods and polychaetes by sudden eversion of the proboscis. The sexes are separate and on the north-east coast of

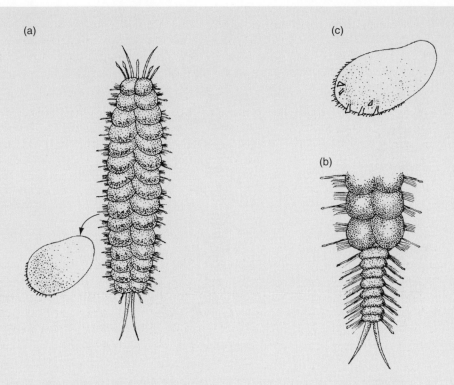

Figure 11.8 (a) *Harmothoe imbricata* (detail of scale after Fauvel, 1923). (b) *Harmothoe extenuata*, posterior region. (c) *Harmothoe impar*, detail of scale (after Fauvel, 1923).

England breeding occurs during March and April, the females spawning twice during these months. The adult worms are solitary but during the breeding season mature females release a pheromone that attracts males. The male lies across the dorsal surface of the female and fertilizes the eggs which are held under the scales of the female and here early development takes place. After two weeks, trocophore larvae are released into the plankton. Settlement of the young worms occurs during May and they breed during the following year. Length of life is probably up to four years.

Harmothoe extenuata (Grube) (*Lagisca extenuata*) **Fig. 11.8b**

Body flattened; up to 70 mm in length. Up to 50 segments; 15 pairs of scales, not covering posterior 8 or so segments; 1st pair of scales round, others kidney-shaped; scales covered with very small tubercles and with row of larger, round tubercles near posterior margin; less than quarter of scale margin with tiny papillae. Colour of scales varied, red, grey or brown, often with clear central spot.

H. extenuata is widely distributed in north-west Europe and Britain, under stones and among laminarian holdfasts extending into the sublittoral to depths of about 550 m. It is carnivorous. The sexes are separate and breeding has been recorded from November to January. The eggs are held by the female between the parapodia and also on the ventral surface of the body until they are released as free-swimming larvae.

Harmothoe impar (Johnston) **Fig. 11.8c**

Body flattened; up to 25 mm in length. Up to 40 segments; 15 pairs of overlapping scales covering dorsal surface; 1st pair of scales round, following pairs kidney-shaped, covered with many small and a few large tubercles: more than half of scale margin with papillae. Scales brownish in colour, often with central, yellow spot.

H. impar is widely distributed in north-west Europe and Britain and is common under stones and rocks and among laminarian holdfasts from the middle shore into the sublittoral. It is carnivorous. The sexes are separate and breeding takes place during spring. The worms come together in pairs; after fertilization the eggs are held under the scales of the female until they are liberated as free-swimming larvae.

Several other species of *Harmothoe* are recorded from around Britain. These are mostly rare and/or offshore.

Polynoe scolopendrina Savigny **Fig. 11.9a**

Body narrow, elongate; up to 120 mm in length with up to 100 segments. Dorsal surface rounded, ventral surface flattened; 15 pairs of scales covering only the anterior half of the body, 50 or so posterior segments not covered; scales rounded, with small tubercles anteriorly. Segments in posterior half of body each with 3 tubercles on the dorsal surface, giving 3 longitudinal rows along the length of the body. Anterior region reddish in colour; scales brownish with darker central area.

P. scolopendrina is widely distributed in north-west Europe under stones and in crevices on the lower shore. It is most often commensal with other polychaetes such as *Eupolymnia nebulosa* (p. 184).

Family Sigalionidae

Mobile, active worms. Body long and rather flattened, many segmented. Dorsal surface of body with many pairs of scales attached on every other segment anteriorly, and on all posterior segments. Head with 2 or 3 antennae, 2 palps; usually with 4 eyes; eversible proboscis with 4 jaws.

Outer edge of dorsal scales fringed with papillae (Fig. 11.9b) ...... **Sthenelais boa** p. 155

Outer edge of dorsal scales fringed with papillae with lateral branches (Fig. 11.9c) .... **Sigalion mathildae** p. 156

Sthenelais boa (Johnston) **Fig. 11.9b**

Body up to 200 mm in length. Up to 200 segments. Head with 1 median antenna and 2 lateral antennae; 2 palps; 2 pairs of eyes. Dorsal surface with about 180 pairs of overlapping scales completely covering surface; 1st pair rounded, others kidney-shaped; outer edge of scales fringed with papillae (cf. Sigalion, below). Colour varied, brown, yellow or orange, sometimes with darker transverse bandings.

Figure 11.9 (a) *Polynoe scolopendrina.* (b) *Sthenelais boa,* with detail of scale (after Fauvel, 1923). (c) *Sigalion mathildae,* anterior and detail of scale.

S. boa is widely distributed in north-west Europe burrowing in intertidal sands and gravel, and is commonly found under stones embedded in gravel and coarse sand. It is probably carnivorous feeding on small invertebrates.

Sigalion mathildae Audouin & Milne-Edwards **Fig. 11.9c**
Body up to 150 mm in length; up to 200 segments. Head with very small median antenna and 2 small lateral antennae which may be difficult to see; 2 palps; 4 small eyes, sometimes difficult to detect. Scales cover dorsal surface, 1st pair rounded, others kidney-shaped to rectangular; outer edge of scales fringed with papillae which have lateral branches giving feather-like effect (cf. Sthenelais, *above). Colour greyish-white, ventral surface iridescent.*

S. mathildae is widely distributed in north-west Europe in fine sand from the lower shore into the sublittoral. It has also been recorded from the Mediterranean. It lives buried in the sediment to depths of about 200 mm and when disturbed characteristically curls up into a spiral. *Sigalion* is believed to be carnivorous. Nothing appears to be known of its reproduction.

Family Hesionidae

Mobile, active worms. Cylindrical body; parapodia with long, usually annulated, dorsal cirri; not divided into similar-sized notopodia and neuropodia. Head with 4 eyes; 2 or 3 antennae; 2, 2-jointed palps; usually 6–8 pairs of tentacular cirri. Large tubular eversible proboscis, jaws present or absent.

Psamathe fusca Johnston (*Kefersteinia cirrata*) **Fig. 11.10**

Body up to about 75 mm in length. Up to 70 segments. Parapodia with long, annulated dorsal cirri; 2 antennae; 2 palps; 8 pairs of annulated tentacular cirri. Proboscis eversible, fringed with fine papillae. Colour varied, yellow, pink or red.

P. fusca is widely distributed in north-west Europe and around Britain on the lower shore and in the shallow sublittoral under stones, in coarse gravel, among laminarians and in mussel beds. It is carnivorous, feeding on small invertebrates. The sexes are separate and on the north-east coast of England breeding occurs during June and July. There is no epitokous stage; the worms swim for short periods at a time and release gametes.

Figure 11.10 *Psamathe fusca.*

Family Nereididae Ragworms

Mobile, active worms. Some species inhabit mucous tubes. Parapodia conspicuous, divided into similar-sized notopodia and neuropodia. Head with 4 eyes; usually 2 antennae; 2 stout, knobbed palps; usually 4 pairs tentacular cirri. Large, eversible proboscis with 2 large jaws and a number of small teeth known as paragnaths.

The distribution of the paragnaths on the dorsal surface of the proboscis is shown for each species, and although the number of paragnaths varies, the overall pattern of distribution can be an important aid to identification and where appropriate this has been highlighted in the figures. The proboscis of nereid worms can usually be everted by *gentle* pressure behind the head while the worm is alive. Colour and habitat are also useful aids to identification, but in some cases accurate identification can only be made following a detailed examination of the parapodia and chaetae. While the detail of individual parapodia may vary along the length of the body a useful character is the relative lengths of the dorsal cirrus and notopodial lobes in parapodia from the first 20 or so segments (excluding segments 1 and 2 which are rather atypical). Parapodia from this region of the body are illustrated for each species described here. For further detail reference to **Chambers & Garwood (1992)** is recommended.

1 Some of the paragnaths on the dorsal surface of the proboscis grouped to form 2 blackish crescents (Fig. 11.13a) .... ***Perinereis cultrifera*** p. 161
Paragnaths on the dorsal surface of the proboscis not grouped to form 2 crescents ... **2**

2 Paragnaths on the dorsal surface of the proboscis very small, arranged in short rows (Fig. 11.13c). Tentacular cirri long, 1 pair extending backwards to about segment 15 .... ***Platynereis dumerilii*** p. 161
Paragnaths and tentacular cirri not so **3**

3 (a) Paragnaths arranged as in Fig. 11.11c. Greenish-yellow, reddish in colour. In mud, muddy-sand; often in estuaries .... ***Hediste diversicolor*** p. 158
or
(b) Paragnaths arranged as in Fig. 11.12c. Yellowish-brown in colour with white band in middle of dorsal surface. Lives as a commensal in whelk shells occupied by hermit crabs .... ***Neanthes fucata*** p. 161
or
(c) Paragnaths arranged as in Fig. 11.11a. Reddish-brown, greenish-yellow in colour with an iridescent sheen. In crevices, laminarian holdfasts and under stones on the lower shore, usually in a mucous tube .... ***Nereis pelagica*** p. 158
or
(d) Paragnaths arranged as in Fig. 11.12a. Body large and thick; notopodia large and leaf-like. Greenish-blue in colour with iridescent sheen. In mud, muddy-sand on the lower shore .... ***Alitta virens*** p. 160

Nereis pelagica Linnaeus **Fig. 11.11a,b**
Body cylindrical; up to 120 mm or more in length. Up to 100 segments. Paragnaths on proboscis arranged as in figure. Antennae shorter than palps. Dorsal cirrus of parapodium longer than notopodial lobes. Reddish-brown, greenish, yellow in colour with an iridescent sheen.

N. *pelagica* is widely distributed in north-west Europe and Britain on the lower shore in crevices, among seaweed, in laminarian holdfasts and under stones. It extends into the sublittoral to depths of 120 m. The worms usually live in mucous tubes that are encrusted with sand particles and attached to stones and seaweed. They are omnivorous, feeding mainly on microorganisms. The sexes are separate and on the north-east coast of England the breeding season extends from December to April when the body changes into the epitokous condition and swarming occurs. Fertilization is external but there are conflicting views on the mode of development of the larva. Both pelagic and non-pelagic larvae have been recorded and laboratory studies suggest a free-swimming life of some 18 days for the pelagic larvae. Length of life is about three years. The worms breed once then die.

Hediste diversicolor (O. F. Müller) (*Nereis diversicolor*) Ragworm **Fig. 11.11c,d,e**
Body somewhat flattened; up to 200 mm in length. Up to 120 segments. Paragnaths arranged as in figure. Antennae shorter than palps. Dorsal cirrus of parapodium shorter than notopodial lobes. Colour varied, greenish-yellow, reddish.

H. *diversicolor* is widely distributed in north-west Europe and is one of the commonest intertidal polychaetes. It lives in a more or less permanent, often U-shaped, burrow

Figure 11.11 (a) *Nereis pelagica*, anterior end, (b) parapodium. (c) *Hediste diversicolor*, anterior end, (d) parapodium, (e) whole specimen. ((b), (d) Modified after Chambers & Garwood, 1992).

in mud and muddy-sand, most often in estuaries, and is tolerant of salinities down to 1. It is omnivorous, feeding on a variety of plant and animal material on the surface of the sediment, including grazing on algal mats, and also suspension feeds on phytoplankton contained in the water drawn in to irrigate the burrow. The phytoplankton is trapped on a mucous funnel secreted by the worm at the entrance to the burrow and periodically ingested. The main predators of *H. diversicolor* are fishes and birds. The sexes are separate and many populations have a high proportion of females. Ripe males can be distinguished by their bright green colour, compared to the much darker green of the ripe females. There is no free-swimming epitokous stage in *H. diversicolor*. Breeding has been recorded in spring at the times of new and full moon but there is some variation in the time of breeding between populations, even within a single estuary. Age at maturity can be anything from one to three years. During breeding the females remain in the burrow where the eggs are released, while mature males crawl about on the sediment surface and release sperm near the burrow entrance. Using the proboscis the female carries the sperm into the burrow where fertilization occurs and development of the larvae takes place. After 10–14 days the young worms leave the burrow; they are bottom-dwelling and not planktonic. Length of life is probably about three years. The worms breed once then die.

Figure 11.12 (a) *Alitta virens*, anterior end, (b) parapodium. (c) *Neanthes fucata*, anterior end, (d) parapodium. ((b), (d) Modified after Chambers & Garwood, 1992.)

Alitta virens (M. Sars) (*Neanthes virens, Nereis virens*) King ragworm **Fig. 11.12a,b**
Body large and thick; generally 200–300 mm but may reach 1 m or more in length. Up to 200 segments. Paragnaths arranged as in figure. Antennae a little shorter than palps. Notopodia large and leaf-like; dorsal cirrus small, much shorter than notopodial lobes. Greenish-blue in colour with iridescent sheen, often with purple tints.

A. virens is a northern species extending as far south as the English Channel; its distribution is sporadic around Britain, but it may be locally abundant. It is found on the lower shore in mud and muddy-sand up to 400 mm below the surface of the sediment, as well as under stones on rocky shores and in estuaries. It is, however, less tolerant of low salinity than *H. diversicolor* (above) and is rarely found in salinities less than 15. It is omnivorous and has a number of different feeding strategies, including deposit feeding, predation and scavenging. Its diet includes polychaetes such as *Nephtys hombergii* (p. 165). The sexes are separate and on the south-east coast of England breeding takes place during April and May when males and females become bright green in colour. The males become epitokous and swarm at high tide but the females do not have an epitokous stage and are believed to discharge eggs at high tide while remaining in, or close to, their burrows. The larvae are more or less benthic for the first five or six days after which they have a short pelagic phase of probably only a few hours. Length of life is believed to be two or three years. The worms breed once then die. *A. virens* has considerable economic importance as bait used by sea anglers and it is now reared in aquaria for the commercial market.

Neanthes fucata (Savigny) (*Nereis fucata*) **Fig. 11.12c,d**

Body up to 200 mm in length. Up to 120 segments. Paragnaths arranged as in figure. Antennae slightly longer than palps. Dorsal cirrus longer than notopodial lobes. Yellowish-brown in colour, with white band in middle of dorsal surface; parapodia white.

N. fucata is widely distributed in north-west Europe and Britain, occurring on the lower shore and in the sublittoral. It lives in a mucous tube as a commensal inside shells, often those of the whelk *Buccinum undatum* (p. 233), occupied by hermit crabs, especially *Pagurus bernhardus* (p. 360). It feeds on animal matter, usually scavenging on the remains of the hermit crab's prey and in some cases taking food from the mouthparts of the crab. The sexes are separate, and on the south coast of England breeding occurs during spring. Both sexes show epitokous changes, the males becoming whitish, the females lilac or blue, while the parapodia of both sexes become pinkish-red. The worms leave the host shells and are believed to swarm in the water while discharging gametes. After a pelagic life of four to six weeks, the larvae settle on the benthos where the young worms construct tubes from shell gravel and mucus in which they live for about four months, feeding on detritus and small invertebrates. At this time the worms become sensitive to vibrations and eventually locate a shell occupied by a hermit crab. The worm then leaves its tube and takes up position in the host shell where it constructs a mucous tube in which it lives. The worms breed once then die.

Perinereis cultrifera (Grube) **Fig. 11.13a,b**

Body up to 250 mm in length. Up to 125 segments. Paragnaths, some of which are grouped to form 2 crescents, arranged as in figure. Antennae shorter than palps. Dorsal cirrus of parapodium shorter than or equal to notopodial lobes. Green with bronze tints; parapodia reddish.

P. cultrifera is widely distributed in north-west Europe and around Britain on the lower shore in crevices, under stones and in laminarian holdfasts. It feeds mainly on seaweed and diatoms. The sexes are separate and both show epitokous changes and swarm in the surface waters, swimming in circles round each other. Breeding has been recorded in May and June in the English Channel; fertilization is external and there is a pelagic larva. Examination of populations on the north-west coast of Brittany suggests a length of life of three years.

Research on the number and morphology of paragnaths on the proboscis of different populations suggests that *P. cultrifera* is a complex of species.

Platynereis dumerilii (Audouin & Milne-Edwards) **Fig. 11.13c,d**

Body up to 60 mm or more in length. Up to 90 segments. Tentacular cirri long, 1 pair extend backwards to about segment 10–15. Paragnaths very small, in short rows, arranged as in figure. Antennae more or less the same length as palps. Dorsal cirrus of parapodium longer than notopodial lobes. Colour varied, iridescent, green, red or yellow.

P. dumerilii is widely distributed in north-west Europe and around Britain is found mostly on south and west coasts. It lives in a mucous or membranous tube on the lower shore in rock crevices among *Fucus* spp. and laminarian holdfasts. It is also found in shallow water on drifting seaweed, especially *Saccharina latissima* (p. 58). It feeds mainly on seaweed and associated microorganisms. On the west coast of Scotland breeding occurs from April to June. The sexes are separate and both become

Figure 11.13 (a) *Perinereis cultrifera*, anterior end, (b) parapodium. (c) *Platynereis dumerilii*, anterior end, (d) parapodium. ((b), (d) Modified after Chambers & Garwood, 1992.)

epitokous; in some populations males have been observed discharging sperm while swimming in circles round slowly moving females. After spawning the worms die. Fertilization is external and the larvae are pelagic for a few days.

A second species, ***Platynereis massiliensis*** (Moquin-Tandon), also occurs in north-west Europe and is apparently morphologically indistinguishable from *P. dumerilii*, but reproduces quite differently. The animals are protandrous hermaphrodites and eggs are deposited in the parental tube where they are fertilized and hatch as non-pelagic larvae. There is no epitokous stage.

Family Syllidae

Mobile, active worms, some species inhabiting mucous tubes. Small, slender body; parapodia not divided into similar-sized notopodia and neuropodia, often with well-developed dorsal cirri which may be distinctly annulated. Head usually with 4 eyes; 3 antennae; 2 palps, may be closely applied to ventral surface of head and not visible dorsally; 2 pairs of tentacular cirri. Antennae and tentacular cirri sometimes distinctly annulated. Eversible proboscis often armed with one to many teeth. Syllids are well known for the morphological changes that take place during reproduction, some showing epigamy and others schizogamy with the formation of stolons which become free and have a short pelagic life. Some species in the family show brood protection. A large group of polychaetes found in a wide variety of habitats, including laminarian holdfasts, among seaweeds, bryozoans and hydroids, and occasionally in muddy deposits. Only two of the most common genera are included here.

Figure 11.14 (a) *Autolytus* sp., detail of anterior and posterior, and (b) chain of individuals (stolons). (c) *Eusyllis blomstrandi*, detail of anterior and posterior. ((b) After Thorson, 1946).

Autolytus spp. **Fig. 11.14a,b**

Small, thin worms; about 3–15 mm in length. Number of segments varies, about 30–60 in the commoner species. Parapodia without ventral cirri. All appendages lack annulations. Palps not always visible dorsally. Proboscis armed with ring of teeth, visible through semi-transparent body wall when withdrawn. Colour varied, depending on species; some almost colourless.

Species of *Autolytus* are widespread in north-west Europe and Britain but are very difficult to separate. They are found on the lower shore and into the sublittoral, among hydroids, seaweeds, stones and shells where they build thin, transparent, mucous tubes. The worms are carnivorous and feed mainly on hydroids. Their reproductive biology is complex. Individuals are budded from the parent giving rise to chains of individuals (Fig. 11.14b), each individual being known as a stolon. These are specialized for sexual reproduction and are either male (the polybostrichus) or female (the sacconereis). They become pelagic and swarm together in the surface waters. In some species the male discharges sperm while swimming in a circle round the female. The fertilized eggs are carried by the female in a ventral egg sac until they hatch as free-swimming larvae.

Eusyllis blomstrandi Malmgren **Fig. 11.14c**

Body up to 30 mm in length. Up to 124 segments. Anterior parapodia with long, irregularly annulated dorsal cirri; posterior parapodia with short, more or less smooth dorsal cirri; ventral cirri of parapodia oval and short. Antennae and tentacular cirri irregularly annulated. Palps oval, visible dorsally. Median antenna longer than lateral ones. Proboscis eversible, with single large tooth and ring of many small teeth. Yellowish-orange in colour, tips of cirri brown.

E. blomstrandi is widely distributed in north-west Europe and Britain, often living in mucous tubes among stones, seaweeds, laminarian holdfasts, bryozoans and especially hydroids. It is carnivorous, feeding mainly on hydroids. The sexes are separate and on the east coast of England breeding occurs during spring and summer. The epitokous stages swarm in the surface waters, where the gametes are discharged. Breeding females of *Eusyllis* often shine with a green bioluminescence, presumably to attract males.

Family Glyceridae Bloodworms

Active, mobile worms. Body rounded, narrowing towards anterior and posterior. Head an elongate, pointed cone with annulations; 4 very small appendages set close together at tip. Extremely large, eversible proboscis with many papillae and 4 jaws. Body with many segments, each with 2 or 3 annulations.

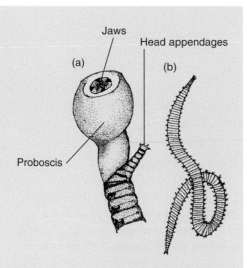

Figure 11.15 *Glycera convoluta*, (a) anterior end with proboscis everted (after Fauvel, 1923), (b) whole worm.

Glycera convoluta Keferstein (*Glycera tridactyla*) **Fig. 11.15**

Body round, narrowing markedly at anterior and posterior; up to 100 mm in length. Up to 180 segments but seems more as each segment has 2 annulations. Finger-shaped, non-retractile gills present on dorsal surface of parapodia from 25th segment to posterior of body. Head elongate, pointed cone with about 8 annulations. Reddish-pink in colour. Body of worm usually thrown into spiral coils.

G. convoluta is a southern species extending to the southern North Sea. It lives on the lower shore in mud and muddy-sand where it inhabits branching burrows. The worms are carnivorous, feeding on small invertebrates, and glands near the jaws are known to secrete a toxic substance. The sexes are separate and on the south coast of England breeding occurs from June to August; there is a pelagic larva.

Several other species of *Glycera* occur in European waters and reference to Böggemann & Fiege (2001) is recommended for further study.

Family Nephtyidae Catworms

Mobile, active worms. Body more or less flattened. Head small, squarish, with 2 short, pointed antennae, 1 at each anterior corner; 2 short palps. Large, eversible proboscis with papillae and 2 jaws. Parapodia with sickle-shaped gills between the 2 lobes; notopodia and neuropodia of similar size.

Figure 11.16 *Nephtys* sp., (a) anterior end, (b) posterior end, (c) proboscis everted, lateral view, (d) parapodium showing position of sickle-shaped gill.

Nephtys spp. Catworms **Fig. 11.16**

Length and segment number vary with species, often about 200 mm in length; up to 200 segments. Head sometimes with 2 small, visible eyes. White, pinkish, greyish, yellowish in colour, often with a pearly iridescence; gills red; chaetae yellowish.

The different species of *Nephtys* are difficult to identify, requiring detailed examination of the parapodia and chaetae; reference to Rainer (1991) is recommended. The species are widely distributed in north-west Europe and are found intertidally and sublittorally burrowing in sand, muddy-sand, mud and gravel. They are carnivorous on molluscs, crustaceans and other polychaetes although there are reports that they are also omnivorous. The sexes are separate and individuals breed several times over a number of years. On the north-east coast of England, two common species, ***Nephtys caeca*** (Fabricius) and ***Nephtys hombergii*** Savigny, mature at two or three years of age and breed during April and May. The worms remain in the sediment during spawning and eggs and sperm are released onto the surface of the sediment during low tide. Fertilization occurs when the gametes are mixed by the incoming tide. The larvae are pelagic. Both species live for several years and the presence of growth rings on the jaws enables accurate assessments to be made of the age of individual specimens; 12 years has been recorded for *N. caeca* and six years for *N. hombergii*.

Family Phyllodocidae Paddle worms

Mobile, active worms. Body long and slender with many segments. Parapodia with large, expanded paddle-shaped dorsal cirri. Head usually with 2 large eyes; 2 or 3 antennae; 2 short palps; 2–4 pairs tentacular cirri. First 2 or 3 segments without paddle-shaped dorsal cirri. Large eversible proboscis sometimes covered with papillae but without jaws.

A large group of polychaetes found in crevices and under stones on rocky shores, in laminarian holdfasts and in a wide range of sediments from shell gravel to mud and sand. The species can be difficult to separate. Only two genera are described and illustrated here; both are widely distributed and common.

Phyllodoce spp. (*Anaitides* spp.) **Fig. 11.17a**

Body up to 450 mm in length and with up to 700 segments, depending on species, the 2 described below being much smaller, up to 100 mm in length with up to 270 segments. Dorsal and ventral cirri of parapodia flattened; dorsal cirri very conspicuous, heart-shaped, kidney-shaped, oval. Head heart-shaped; 2 eyes; 2 antennae; 2 small palps; 4 pairs longish, slender tentacular cirri. Large, eversible proboscis with papillae. Colour varied depending on species: body often yellowish-green with brown markings on each segment, but sometimes whitish with brown spots; dorsal cirri often greenish-brown, but sometimes clear with brown spots.

There are several species of the genus *Phyllodoce*, two of which, *Phyllodoce maculata* (Linnaeus) and *Phyllodoce mucosa* Ørsted, are widely distributed in north-west Europe and around Britain, but the species are difficult to separate and reference to **Pleijel & Dales (1991)** is recommended. Both occur on the lower shore and shallow sublittoral in sand, muddy-sand and under stones, and are carnivorous, feeding on small invertebrates and a wide variety of carrion including molluscs, crabs and worms. The sexes are separate and tangled masses consisting of several males with a single female have been observed on the surface of the sediment. Eggs and sperm are shed into mucous cocoons secreted by the female and attached to sediment and stones. The eggs are green in colour and develop into pelagic larvae.

Expanded dorsal cirri

Figure 11.17 (a) *Phyllodoce maculata*. (b) *Eulalia clavigera*.

Eulalia clavigera (Audouin & Milne-Edwards) (*Eulalia viridis*) **Fig. 11.17b**
Body narrow; up to 95 mm in length. Head with 2 large eyes; 3 antennae; 2 small palps; 4 pairs tentacular cirri, the longest reaching back to about segment 7. Eversible proboscis covered with small, conical papillae. Bright green in colour; may have dark spots below parapodia.

E. clavigera is widely distributed around Britain and its range probably extends to the Atlantic coast of Spain and possibly into the Mediterranean. It is found from the middle shore into the shallow sublittoral in rocky crevices, among barnacles and mussels, and in laminarian holdfasts, being very conspicuous because of its bright green colour. It is a scavenger, feeding mainly on damaged barnacles and mussels. The sexes are separate and breeding has been recorded in July and August. Green gelatinous egg masses, usually found attached to seaweed, are often described as being those of *E. clavigera* but this requires confirmation. There is a planktonic larva. The worms probably live for several years, breeding in the second and subsequent years.

Prior to the 1990s, *Eulalia* from Britain and north-west Europe were recorded as *Eulalia viridis* (Linnaeus). However, detailed biochemical and morphological studies suggest that *E. viridis* in northern Europe is a complex of two species, *E. clavigera* as described above and *E. viridis* which has been recorded from Scandinavia and Germany. While geographic distribution may be a useful indicator of the species, reference to Bonse *et al.* (1996) is recommended.

Family Eunicidae

Active, mobile worms. Body cylindrical, flattened posteriorly. Head usually with 2 eyes, 3 antennae, 2 palps, 1 pair of tentacular cirri or none. Parapodia not divided into similar-sized notopodia and neuropodia; dorsal cirri slender and cylindrical. First 2 segments without parapodia and chaetae. Proboscis with complex jaws. Gills usually present.

Marphysa sanguinea (Montagu) Rock worm **Fig. 11.18**
Body cylindrical, flattened posteriorly; up to 400 mm in length. Up to 500 segments. Three antennae; 2 palps; no tentacular cirri. Gills present from segments 10–40 to within 12 segments of posterior. Most gills formed of several filaments. Pink, grey or orange-brownish in colour with a marked iridescence; gills red.

Gills

Figure 11.18 *Marphysa sanguinea*, (a) anterior end, (b) middle region, (c) posterior end.

M. sanguinea is widely distributed and has been described as a cosmopolitan species. It is found in mucus-lined burrows and galleries on the lower shore and in the shallow sublittoral in a variety of habitats such as clay, mud, gravel, rock crevices and under stones. It is also found in brackish water habitats. It is predominantly carnivorous. On the south coast of England breeding occurs in September. The sexes are separate and external fertilization is believed to take place in the burrow; there are unconfirmed reports that the eggs are deposited in masses of firm jelly. Early development takes place in the maternal burrow and there is a short pelagic phase. In some areas the worms are harvested commercially for use as bait in sea angling. Recent research suggests that *M. sanguinea* is a species complex.

Family Oweniidae

Worms sedentary, tube-dwelling; tube covered with sand grains and shell fragments. Cylindrical body with up to about 30 segments, those of anterior region longer than posterior ones. Head lobe-like, or with 2 palps, or as in the species described here, with a frilled membrane.

Owenia fusiformis Delle Chiaje **Fig. 11.19**

Body cylindrical, up to 100 mm in length. Up to 30 segments. Anterior 3 segments short, the next 5–7 much longer; posterior segments becoming progressively shorter. Head bears frilled membrane, with 2 eye spots at base. Greenish-yellow in colour with clear, transverse bands; frilled membrane reddish. Lives in tough, flexible tube

Figure 11.19 *Owenia fusiformis,* (a) removed from tube, (b) anterior end of tube.

covered with flat, usually whitish fragments of shell debris arranged in overlapping fashion. Larger specimens may have tubes up to 300 mm in length.

O. fusiformis is widely distributed in north-west Europe in sand and muddy sand on the lower shore and into the sublittoral. Population densities as high as 1000 per square metre have been recorded in the English Channel but overall distribution is patchy. The worm can move the tube and during feeding the anterior end is exposed above the surface of the sediment. *Owenia* feeds on suspended matter using the cilia on the frilled membrane, but is also able to bend over until the membrane touches the surface of the sediment and pick up sand grains and detritus. The sexes are separate and the larva, known as a mitraria, has a planktonic life of about four weeks. On the south coast of England, breeding occurs during June and July. Length of life is four years, the worms breeding each year.

  O. fusiformis has been described as a cosmopolitan species but detailed morphometric studies suggest that it is a complex of at least eight species.

Family Sabellariidae Honeycomb worms

Tube-dwelling, sometimes forming large colonies; tube made of sand grains attached to rocks and stones. Body short, divided into 3 regions; anterior with rings of stout chaetae (paleae) forming an operculum which closes the tube; middle with conspicuous parapodia and gills; posterior narrow, smooth and bent under body. Head with 2 palps and large number of oral filaments.

Stout chaetae of the outer ring with smooth, terminal teeth (Fig. 11.20b). Tubes usually forming large colonies .... ***Sabellaria alveolata*** p. 169

Stout chaetae of the outer ring with terminal teeth, the median one long and barbed (Fig. 11.20d). Tubes usually separate .... ***Sabellaria spinulosa*** p. 170

Sabellaria alveolata (Linnaeus) **Fig. 11.20a,b,c; Plate 23a,b**

Body up to 40 mm in length. Up to 37 segments. Paleae in 3 rings, those of outer ring with smooth, unbarbed terminal teeth. Reddish-brown in colour. Tube made of sand grains and fragments of shell, inner lining membranous; may occur singly but more usually forming large colonies or reefs.

S. alveolata reaches its northern limit in the North Sea and in Britain is commonest on south and west coasts. It is found on the lower shore and below on rocks close to sand deposits which provide the raw materials for tube building, such as coarse sand and pieces of shell. These are trapped by the oral filaments when the worms are submersed. Substantial reefs can be formed. Along with the materials used in tube building, food particles are also removed from suspension. The sexes are separate and during the breeding season males can be distinguished by their creamy colour, while the females are purplish in colour. On the south coast of England breeding occurs in July and the larvae are pelagic, some for about six weeks, while others have a long pelagic life of at least nine months. At settlement the larvae detect the tubes of adult *Sabellaria* and old colonies, and often settle on them in large numbers. Length of life is usually three to four years, but some colonies have been estimated to be nine years old.

3 Head with 2 lobes, each bearing 8–45 radioles, banded purple,
   brown, red. Tube covered with mud particles, extends well above
   sediment surface .... ***Sabella pavonina*** p. 172
   Head with 2 lobes, each bearing 5–15 radioles, some with up to 8 eye spots.
   Tube covered with mud or sand particles; free end rolls over when
   worm withdraws .... ***Pseudopotamilla reniformis*** p. 172

Sabella pavonina Savigny (*Sabella penicillus*) Peacock worm **Fig. 11.21a,b; Plate 24d**
Body up to 270 mm in length, 200 or more segments. Thoracic region of 8–12 segments. Head with 2 lobes, each bearing up to 40 or more long radioles. Greyish-purple or yellowish-orange in colour; radioles banded or patterned with purple, brown or red. Tube tough, membranous, covered with fine particles of mud at anterior end; posterior end more delicate; may be 500 mm or more in length.

S. pavonina is widely distributed in north-west Europe and around Britain. The tubes extend up to 100 mm above the surface of muddy and sandy sediments on the lower shore and below (Plate 24d). They are often found in large numbers, sometimes with seaweeds, sponges and ascidians attached to them. The worms feed on suspended particles collected by cilia on the radioles. Particles are sorted on the basis of size; large particles are rejected, small ones are used for food, and medium-sized ones mixed with mucus and added to the anterior end of the tube to increase its length. The sexes are separate and breeding has been recorded in spring and summer. The larvae are pelagic.

Pseudopotamilla reniformis (O. F. Müller) (*Potamilla reniformis*) **Fig. 11.21c,d**
Body delicate; up to 100 mm in length. Up to 300 segments. Thoracic region of 7–14 segments. Head with 2 lobes, each bearing 5–15 radioles, some with up to 8 dark eye spots; dorsally 2 pointed lobes at base of radioles. Body orange or red, often brownish anteriorly; radioles white or pink, often banded with purple or brown. Tube tough, often covered with mud or fine sand; free end flattened, rolls over when worm withdraws.

P. reniformis is widely distributed in north-west Europe and around Britain where it occurs on the lower shore and below in rock crevices and on stones and shells. It is a suspension feeder.

Fabricia stellaris (O. F. Müller) (*Fabricia sabella*) **Fig. 11.22a**
Body slender, up to 3 mm in length. Up to 12 segments. Thoracic region of 8 segments. Head with 2 lobes, each bearing 3 radioles; 2 (sometimes 4) eye spots; 2 eye spots on last segment of body. Yellowish or pinkish in colour, transparent. Tube of mucus, covered with mud.

F. stellaris probably has a worldwide distribution. It is found among empty barnacles, beds of mussels, the red seaweed *Rhodothamniella* (p. 38), laminarian holdfasts and other situations where silt accumulates. Where conditions are suitable, densities can exceed 1 million per square metre. It is a suspension feeder. The sexes are separate and sperm are taken into the tubes of females in the feeding current and stored in a seminal receptacle. Breeding occurs in spring. Eggs are laid in groups within the tube of the

(a)

(b)

Figure 11.22 (a) *Fabricia stellaris*. (b) *Myxicola infundibulum*. ((b) After Stephenson, 1944.)

female, each group containing one to seven eggs enclosed in a mucous cocoon. Development takes place within the parental tube; there is no pelagic larva.

Myxicola infundibulum (Montagu) **Fig. 11.22b**
Body up to 200 mm in length. Up to 150 segments, each with 2 annulations; 7–8 segments in thoracic region. Head with 2 lobes, each bearing 20–40 radioles united almost to their tips by a membrane; triangular projection between lobes of head on ventral surface. Body yellow or orange in colour; radioles with deep purple tips, inner surface purple or brown. Tube thick, mucilaginous.

M. infundibulum is widely distributed in north-west Europe and around Britain. The tubes are almost completely buried in sand and mud from the very low shore into the sublittoral. It is a suspension feeder. The sexes are separate and gametes are discharged into the sea, where fertilization occurs. The larvae have a pelagic life of one to two weeks.

Family Serpulidae
Tube-dwelling; tube calcareous; encrusting or upright, straight or sinuous, attached to rocks, stones and shells, or encrusting and spirally coiled on seaweed, rocks and stones. Often on carapace of decapod crustaceans. Body divided into 2 regions; a short, anterior

(thoracic) region with few segments enveloped in a membranous collar, and a posterior (abdominal) region of many segments. Head bears crown of 2 groups of pinnate projections called radioles which sometimes bear eyes, and which project from the tube when the worm is feeding. Usually 1 (rarely 0 or 2) of the radioles modified as an operculum (Fig. 11.23a) which closes the tube when the worm withdraws.

1 Tube small, spirally coiled, encrusting on seaweed, rocks, stones and shells **2**
   Tube straight, sinuous, twisted, encrusting or upright; may be several entwined
   together, on rocks, stones, shells **3**

2 Tube coiled sinistrally, i.e. tube opens clockwise; smooth or ridged
   .... ***Spirorbis* spp.** p. 178
   Tube coiled dextrally, i.e. tube opens anticlockwise; often with 3 longitudinal
   ridges .... ***Janua pagenstecheri*** p. 179

3 Tube very small, thin; large numbers often entwined together. Head with 2 small
   cup-shaped opercula .... ***Filograna implexa*** p. 176
   Tube encrusting or upright. Head without operculum or with 1 operculum **4**

4 Head without operculum. Radioles banded red, orange and with many red eye spots.
   Tube smooth, upright, cylindrical, attached at narrow end ..... ***Protula tubularia*** p. 178
   Head with 1 operculum. Radioles banded, blue, red, white. Tube encrusting or upright **5**

5 Stalk of operculum with winged expansions near tip (Fig. 11.23c,e). Tube
   encrusting, irregularly curved **6**
   Stalk of operculum without winged expansions near tip. Tube encrusting
   or upright **7**

6 Tube with single, longitudinal ridge on upper surface. Operculum cone-shaped.
   Usually sublittoral .... ***Pomatoceros triqueter*** p. 175
   Tube with 3 longitudinal ridges on upper surface. Operculum cup-shaped. Lower
   shore and below .... ***Pomatoceros lamarcki*** p. 176

7 Operculum funnel-shaped, edged with rounded teeth (Fig. 11.23a). Tube sinuous
   with 3 longitudinal ridges, or upright in older specimens and transversely ridged
   at intervals .... ***Serpula vermicularis*** p. 174
   Operculum funnel-shaped with centrally 10–20 toothed spines (Fig. 11.24a).
   Tube encrusting or upright, often several entwined ..... ***Hydroides norvegicus*** p. 176

Serpula vermicularis Linnaeus **Fig. 11.23a,b**
Body up to 70 mm in length. Up to 250 segments; 7 segments in thoracic region. Head with 2 lobes, each with 30–40 radioles united at base; single operculum, funnel-shaped and edged with rounded teeth. Varied in colour, body yellow to red; radioles red, pale pink or banded red and pink; operculum with red and white rays. Tube calcareous, pinkish-white; young animals with sinuous tubes usually with 3 longitudinal ridges; older animals with upright tubes, smooth and transversely ridged at intervals.

Figure 11.23 *Serpula vermicularis*, (a) removed from tube, (b) tube. (c) *Pomatoceros lamarcki*, removed from tube, (d) tubes. (e) *Pomatoceros triqueter*, operculum, (f) tube. ((a) After Fauvel, 1927.)

S. vermicularis has been described as cosmopolitan in distribution. It is found at extreme low water and sublittorally to depths of 210 m, encrusting rocks, stones and especially bivalve shells. It usually occurs as solitary individuals on the lower shore, particularly in the south and west of Britain, but in a few areas aggregation of individuals forms reef-like structures in the shallow sublittoral. *S. vermicularis* has a long breeding season and settlement of larvae on a reef habitat off the west coast of Scotland has been recorded from June to October. *Serpula* is a suspension feeder.

Pomatoceros triqueter (Linnaeus) **Fig. 11.23e,f**
Body up to 25 mm in length. Up to 100 segments; 7 segments in thoracic region. Head with 2 lobes, each bearing 18–20 radioles united at base; single operculum, cone-shaped and with winged expansions near tip of opercular stalk. Body varied in colour, yellow, red, brown or green; radioles banded with various colours, often blue and white, or red and white. Tube encrusting, calcareous, whitish; irregularly curved, with longitudinal ridge on upper surface often projecting as point over anterior end.

P. triqueter is widely distributed in north-west Europe and around Britain, encrusting stones, rocks and shells, and the carapace of some species of decapods. It is commonly found on the edible mussel (p. 254) and, along with **Pomatoceros lamarcki** (below), can have a serious economic impact as a fouling organism in areas where mussels are cultured. It is predominantly sublittoral, occurring to depths of about 70 m. It is a suspension feeder. The worms are protandrous hermaphrodites and breeding probably takes place throughout the year with a peak in spring and summer. The larvae have a pelagic life of two to three weeks in summer but about two months in winter. The newly settled worms first secrete a delicate semi-transparent tube and calcareous material is later added at the anterior end. Growth is rapid and sexual maturity can be reached in about four months. Length of life is about one and a half years and most worms die after breeding.

Pomatoceros lamarcki (Quatrefages) **Fig. 11.23c,d**
As P. triqueter *but with cup-shaped operculum, and tube with 3 longitudinal ridges on the upper surface.*

P. lamarcki is widely distributed in north-west Europe and in some places is abundant. It is found from the lower shore to depths of about 30 m, encrusting stones, rock, shell and the carapace of some species of decapods. Along with *P. triqueter* (above) it is a serious fouling organism on cultured mussels (p. 254). *P. lamarcki* is a protandrous hermaphrodite and has an extended breeding period.

There is confusion in the literature concerning the identification of the two species of *Pomatoceros* and it is now believed that some records referring to *P. triqueter* may in fact have been *P. lamarcki*, the species more likely to be encountered intertidally.

Hydroides norvegicus Gunnerus **Fig. 11.24a,b**
Body up to 30 mm in length. Up to 100 segments; 7 segments in thoracic region. Head with 2 lobes, each bearing 13–20 radioles united at base; single operculum, funnel-shaped, with centrally about 10–20 toothed spines. Body red in colour; radioles banded red and white. Tube calcareous, white; lying flat or upright; irregularly twisted; often many entwined together.

H. norvegicus has a worldwide distribution and although it does occur on the lower shore, it is mainly sublittoral on stones, shells and bryozoans, as well as causing problems as a fouling organism on ships' hulls. It is a suspension feeder. *Hydroides* is a protandrous hermaphrodite; on the south coast of England breeding has been recorded in August. There is a planktonic larva with, according to laboratory experiments, a pelagic life of eight to 21 days. The newly settled worm secretes a transparent tube and after a few days calcareous material is added at the anterior end.

Filograna implexa M. Berkeley **Fig. 11.24c,d**
Body up to 5 mm in length. Up to 35 segments; 6–9 segments in thoracic region. Anterior and posterior regions of body separated by non-segmented region without chaetae. Head with 2 lobes, each with 4 radioles; 2 opercula, very small, cup-shaped, transparent. Body greyish and transparent; radioles colourless. Tube very small, thin, calcareous and white; aperture sometimes slightly broader than tube; large numbers often entwined together and have superficial similarity to coral.

Figure 11.24 *Hydroides norvegicus*, (a) removed from tube, (b) tubes. (c) *Filograna implexa*, anterior projecting from tube, (d) tubes. (e) *Protula tubularia* removed from tube, (f) tube. ((a), (b), (c), (e) After McIntosh, 1923.)

F. implexa is widely distributed in north-west Europe and around Britain on the lower shore and sublittoral on pebbles, stones and shells. The sexes are separate but asexual reproduction by budding commonly occurs and the buds are most frequently seen during spring and summer.

Protula tubularia (Montagu) **Fig. 11.24e,f**

Body up to 50 mm in length. Up to 125 segments; 7 segments in thoracic region. Head with 2 lobes, each bearing 25–45 radioles arranged in spiral whorls; without operculum. Body red or orange in colour with greenish anterior region; radioles banded red and orange with many red eye spots. Tube smooth, white and cylindrical; attached at narrow end.

P. tubularia is widely distributed in north-west Europe and around Britain. It occurs mainly sublittorally attached to rocks, stones and shells; the tubes are often washed ashore. Gametes are released into the sea; there is a lecithotrophic larva.

Spirorbis spp. **Fig. 11.25a,b**

Small worms, no more than a few millimetres in length; number of segments varies with species, up to about 30. Usually 3 segments in thoracic region. Radioles few in number. Operculum flattish. Body often orange-red in colour. Embryos develop inside parental tube in egg string attached by thread to wall of tube. Tube opaque, calcareous, smooth or ridged; often white; usually coiled sinistrally, i.e. worm lies with ventral surface upwards and tube opens clockwise.

Figure 11.25 *Spirorbis spirorbis*, (a) removed from tube, (b) tubes on *Fucus serratus*. (c) *Janua pagenstecheri*, anterior projecting from tube.

There are several species of *Spirorbis* in north-west Europe, each of which is typically associated with a particular substratum, for example, *Spirorbis spirorbis* (Linnaeus) with *Fucus* spp., *Spirorbis corallinae* de Silva & Knight-Jones with *Corallina officinalis* (p. 43), *Spirorbis tridentatus* Levinsen with stones and rock crevices, *Spirorbis inornatus* L'Hardy & Quiévreux with red seaweeds, laminarian holdfasts and the undersurface of 'buttons' of *Himanthalia* (p. 65) and *Spirorbis rupestris* Gee & Knight-Jones with encrusting red seaweeds. For more detail, reference to Knight-Jones & Knight-Jones (1977) is recommended.

One of the most widespread and common spirorbids in north-west Europe and Britain is *Spirorbis spirorbis*, formerly known as *Spirorbis borealis*. This species occurs on *Fucus*, particularly *F. serratus* (p. 63), on sheltered shores. The worms live permanently in characteristic smooth, spiral tubes with a diameter of about 3 mm. The shape is believed to be an adaptation to withstand flexure of the seaweed frond. *Spirorbis* feeds on suspended particles trapped on the radioles. It is hermaphroditic with fertilization taking place inside the tube and, although self-fertilization may occur, cross-fertilization is more usual. Sperm are released into the water and are believed to be taken into the tubes of neighbouring individuals in the feeding current and then stored in a spermatheca. Breeding occurs during summer and on the coast of south Wales has been recorded from April to November. At some localities the worms show a lunar periodicity, with the larvae being released fortnightly at neap tides; up to 12 broods a year may be produced. The eggs develop in an egg string, each string containing about 20 to 35 eggs attached to the wall of the tube. The larvae have a pelagic life of only a few hours and show a marked preference for settling on seaweeds such as *F. serratus* on which adult spirorbids are already present. The first tube secreted is thin and made of mucus, and it has been suggested that release of larvae at neap tides ensures that the young worms avoid the hazards of desiccation in the period before they have had time to secrete a tough, calcareous tube. The worms are gregarious and large numbers can be found on a single piece of *Fucus*. *Spirorbis* breeds for the first time in the spring following settlement when the diameter of the spiral tube is about 1.5 mm; length of life is about one and a half years.

Janua pagenstecheri (Quatrefages) **Fig. 11.25c**

Body up to 2 mm in length. Up to 15 segments; 3 segments in thoracic region; 8 radioles. Operculum cup-shaped with uncalcified, transparent walls closed by thin, flat plate. Embryos brooded in operculum. Two colour forms exist, a brightly coloured orange-brown form with orange radioles, and a pale form which is greyish in colour. Tube coiled dextrally, i.e. worm lies with ventral surface upwards and mouth of tube opens anticlockwise. Tube opaque, calcareous; off-white or brown in colour, often, but not always, with 3 longitudinal ridges.

J. pagenstecheri is widely distributed in north-west Europe and Britain and it has been suggested that it is probably the commonest spirorbid in the world. It is found from about the middle shore to depths of 120 m on a wide range of substrata, including *Corallina officinalis* (p. 43) and other seaweeds, rocks, stones, shells and the carapace of crabs. The brightly coloured form occurs high on the shore, especially in shallow rock pools, while the pale form is found in more shaded situations and extends into the sublittoral. The worms are hermaphroditic and it is believed that both cross- and self-fertilization occur, the latter probably being more usual. On the coast of south Wales, breeding takes place from May to October and there appears to be no rhythmicity in larval release. Up to 12 broods a year are produced. About 10 to

20 embryos are brooded in the operculum and pelagic larvae are released when the wall of the operculum splits.

Family Cirratulidae

Sedentary worms with cylindrical body. Long, thread-like, contractile gills (Fig. 11.26a) present on at least some segments along sides of body. Head sometimes with eyes. Two to many long feeding palps (tentacles) in 2 groups attached dorsally just behind head.

Large number of palps (tentacles) in 2 groups arising behind head. Head without conspicuous eyes .... *Cirriformia tentaculata* p. 180

Two groups of up to 8 palps (tentacles) arising behind head. Head with 4–8 eyes in row on each side .... *Cirratulus cirratus* p. 180

Cirriformia tentaculata (Montagu) (*Audouinia tentaculata*) **Fig. 11.26a**
Body up to 200 mm in length. Up to 300 or more segments. Dorsal surface rounded, ventral surface flattened. Paired, long, thread-like gills present on 1st segment with chaetae and along body to near posterior. Head pointed, without eyes in adult; pigment spots present. Large number of feeding palps (tentacles) in 2 groups, arising posteriorly to the first few pairs of gills. Orange, brown or reddish in colour; gills and tentacles red.

C. tentaculata has been recorded in north-west Europe from as far north as Denmark and is widely distributed around Britain, but commonest in the south. It is found on the middle and lower shore in a variety of habitats including mud, muddy-sand and gravel, and under stones where it feeds on organic matter picked up from the surface of the sediment by the palps which are often seen extending over the surface of the sediment while the body of the worm remains buried. The sexes are separate and can be distinguished during the breeding season by the yellow colour of the males and the dull green colour of the females. On the south coast of England, breeding occurs during spring and summer. The gametes are discharged into the sea where external fertilization occurs. Pelagic larvae have been recorded and these settle after about 10 days. Length of life is about 15 months.

Cirratulus cirratus (O.F. Müller) **Fig. 11.26b**
Body up to 100 mm or more in length. Up to 150 segments; cylindrical. Head rounded. Paired long, thread-like gills present on 1st segment with chaetae and along body to near posterior end. Head with 4–8 eyes in row on each side. Two groups of up to 8 feeding palps (tentacles) on 1st segment with chaetae. Orange, red or brown in colour; gills and feeding palps red, yellow.

C. cirratus is widely distributed in north-west Europe and around Britain. It occurs on the lower shore in mud and muddy-sand, under stones and in rock crevices, where it feeds on organic material on the surface of the sediment. The worms generally live in aggregations, sometimes as many as 200 individuals together. The sexes are separate

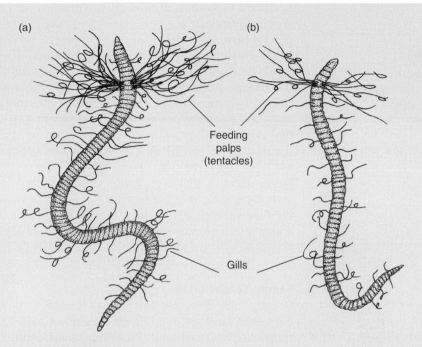

Figure 11.26 (a) *Cirriformia tentaculata.* (b) *Cirratulus cirratus.*

and during the breeding season can be recognized by the bright yellow colour of the females and the white colour of the males. On the north-east coast of England, breeding occurs throughout the year, each female usually spawning once a year for up to four or five years. Eggs and sperm are deposited in mucous egg masses found attached to the sediment and in rock crevices. They are yellowish in colour but become covered with mud particles and are therefore difficult to detect. After about a week or so benthic larvae emerge.

Aphelochaeta filiformis (Keferstein) (*Cirratulus filiformis*) is also found on the lower shore, usually among encrusting red seaweeds and in crevices. It can be distinguished from *C. cirratus* by the pointed head and lack of eyes, and in having only one or two feeding palps in each of the two groups.

Family Pectinariidae

Tube-dwelling; tube cone-shaped, smooth and rigid; made of sand grains and shell fragments, open at both ends. Body short, divided into 3 regions. Anterior with many stout, golden chaetae (paleae) and 2 pairs of gills; middle with conspicuous chaetae; posterior very short, flattened. Head with large number of feeding palps (tentacles); 2 pairs tentacular cirri.

Lagis koreni Malmgren (*Pectinaria koreni*) **Fig. 11.27**

Body short and wide, up to 50 mm in length. Middle region of 15 segments with chaetae. Body pinkish in colour; gills red. Tube rigid, slightly curved; made of fine sand grains and shell fragments; open at both ends, posterior end narrower than anterior.

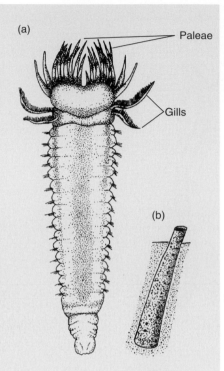

Figure 11.27 (a) *Lagis koreni*, removed from tube, (b) tube in sediment.

L. koreni is widely distributed in north-west Europe and around Britain in sandy deposits from the lower shore into the sublittoral. Large numbers of empty tubes are often washed ashore after storms. The tube lies diagonally or nearly upright in the sediment with the wide end below the surface and the narrow end projecting just above it. The paleae are used for burrowing and loosening the sediment and the palps for sorting the sediment particles which the animal ingests. The sexes are separate and breeding occurs during spring and summer when eggs and sperm are shed into the sea. Fertilization is external and the larvae have a planktonic life of about two weeks, but juvenile worms are able to secrete a sail-like structure onto the tube and become pelagic. Length of life is thought to be about one year; the animals breed once then die.

Family Terebellidae

Usually tube-dwelling but some members of the family burrow in sand and mud. Tube sometimes covered with shell fragments, mud and sand. Body long, divided into 2 regions; a thick anterior (thoracic) region, usually with 3 pairs of branching gills arising dorsally from separate segments; a narrow, posterior (abdominal) region with small neuropodia, without notopodia. Head bears large number of long, narrow sinuous feeding palps (tentacles).

1 Twenty-four anterior segments with conspicuous chaetae. Thin mucous tube .... ***Neoamphitrite figulus*** p. 182
   Seventeen anterior segments with conspicuous chaetae. Tube of sand, shell fragments **2**

2 Large, flat, lateral lobes on side of head. Worm up to 300 mm in length. Tube with anterior end flattened and extended into finger-like projections covered with sand, shell fragments .... ***Lanice conchilega*** p. 183
   Without large, flat lateral lobes on side of head. Worm delicate, up to 150 mm in length. Body with many white spots. Tube delicate, without anterior, finger-like projections .... ***Eupolymnia nebulosa*** p. 184

Neoamphitrite figulus (Dalyell) (*Amphitrite johnstoni*) **Fig. 11.28**
Body up to 250 mm in length. Up to 100 segments. Thoracic region of 24 segments with conspicuous chaetae. Body and palps (tentacles) brown, yellow, pink; gills red.

N. figulus is widely distributed in north-west Europe and around Britain in mud, muddy-sand and gravel, and in laminarian holdfasts. It tolerates low salinity and is found in estuaries. It lives in a thin, mucous tube in a vertical, U-shaped burrow and

Feeding palps
(tentacles)

Gills

Figure 11.28 *Neoamphitrite figulus.*

feeds by picking up organic detritus and diatoms from the surface of the sediment, using the long palps (tentacles).

Lanice conchilega Pallas Sand mason **Fig. 11.29a,b; Plate 23c**

Body delicate, up to 300 mm in length. Up to 300 segments. Thoracic region of 17 segments with conspicuous chaetae; large, flat lateral lobes on segment 3. Eyes sometimes visible. Body light pink or brown in colour, gills red. Tube made of mud, sand and shell fragments; anterior end flattened and extended into finger-like projections covered with sand and shell fragments; often extends well above surface of sediment.

L. conchilega is widely distributed in north-west Europe and around Britain, where it is found in sand and sandy grit on the lower shore and below. It commonly occurs in very large numbers and densities as high as 20 000 per square metre have been recorded. The anterior end of the tube projects above the surface of the sediment, making it very conspicuous. *Lanice* feeds on organic detritus picked up from the surface of the sediment by the long palps (tentacles) on the head. The anterior projections of the tube are set at right angles to the current flow and cause suspended organic material to be deposited behind them where the water is relatively still. It also feeds on suspended particles trapped in mucus on the palps. The sexes are separate and breeding occurs during spring and summer. Gametes are discharged into the sea where fertilizaton occurs. The larvae have a long pelagic life of about two months and on settlement, juveniles attach the tube to a variety of hard substrata found in the sediment but there appears to be a preference for attachment to the exposed parts of adult tubes. After about a month, the juveniles detach and resettle in another location. This behaviour would seem to reduce the risk of young worms being washed away. *Lanice* lives for two to three years.

Eupolymnia nebulosa (Montagu) (*Polymnia nebulosa*) **Fig. 11.29c,d**

Body delicate, up to 150 mm in length. Up to 100 segments. Thoracic region of 17 segments with conspicuous chaetae. Head with many eyes. Body orange, brown, with white spots, palps (tentacles) paler; gills red. Tube delicate and slimy, covered with sand and fragments of shell; without anterior projections.

E. nebulosa is widely distributed in north-west Europe and around Britain, where it is found on the lower shore and below. The tube is attached under stones and shells, and the worm feeds at night on detritus and microorganisms. The sexes are separate; on the Channel coast of France breeding occurs during May, June and July. Gametes are discharged into the sea and the larva has a short pelagic life. On the Mediterranean coast,

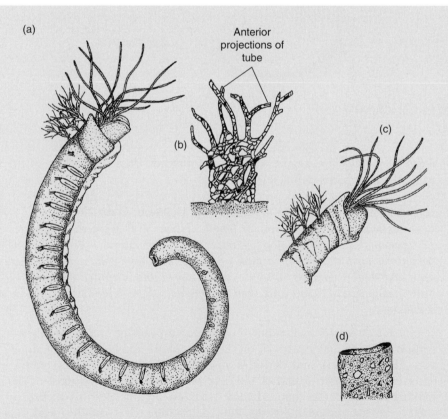

Figure 11.29 *Lanice conchilega*, (a) removed from tube, (b) anterior of tube protruding from sediment. (c) *Eupolymnia nebulosa*, anterior, (d) anterior end of tube. Feeding palps (tentacles) reduced in number and length in (a) and (c).

however, the pattern of reproduction is quite different and the eggs are deposited in mucous cocoons at the entrance to the tube of the female where development takes place.

Another species, *Eupolymnia nesidensis* (Delle Chiaje), also occurs on the lower shore and below, often in laminarian holdfasts. It is smaller (up to 60 mm in length) than *E. nebulosa* and the reddish-brown body is without white spots. The tube is thin and delicate and covered with sand.

Family Chaetopteridae

Tube-dwelling; tube parchment-like. Body divided into 2 or 3 regions with much-modified segments and parapodia. Head usually with 2 small eyes; 2 palps; sometimes 1 or 2 pairs tentacular cirri; large terminal mouth.

Chaetopterus variopedatus (Cuvier) **Fig. 11.30**

Body delicate, up to 250 mm in length. Up to 85 segments. Divided into 3 regions; anterior with 8–12 segments; middle with 5 segments, 1 segment with long, wing-like notopodia, 3 segments with large notopodia fused to form fan-like structures; posterior with many segments; 4th segment of body bearing conspicuous, dark-coloured chaetae. Head without tentacular cirri; palps short. Yellowish-green or whitish-yellow colour; middle region dark green. Bioluminescent. Very distinctive appearance. Tube parchment-like, U-shaped, up to 300 mm in length.

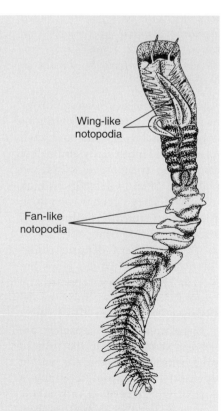

Wing-like notopodia

Fan-like notopodia

Figure 11.30 *Chaetopterus variopedatus.*

C. variopedatus is widely distributed in north-west Europe and around Britain but is not found on the southern North Sea coast of England. It lives in a parchment-like, U-shaped tube buried in sand. The ends of the tube are narrower than the main part and both ends are exposed above the surface of the sediment. It is normally sublittoral but is also found very low on the shore in sheltered areas. The tubes are sometimes cast up after storms. The worm remains permanently in the tube and feeds on suspended particles in the water current drawn through the tube by the action of the large fan-like notopodia of the middle region of the body. Food particles are trapped on a mucous bag suspended from the long, wing-like notopodia, and from time to time the bag is rolled up and passes along a dorsal ciliated groove to the mouth. With increase in size, the worm becomes too large for the tube but is able to tear it open with the modified chaetae on the fourth segment and add to it, thus increasing the size. The sexes are separate, fertilization is external and the larva apparently has a long pelagic life. *C. variopedatus* has been described as cosmopolitan but phylogenetic analysis using molecular techniques indicates that it is a species complex.

Family Spionidae

Burrowers in sand and mud or inhabiting sand encrusted mucous tubes, or living in rock and shell. Flattened gills usually present on dorsal surface of at least some segments. Head usually with 4 eyes, but maybe more; characteristically 2 long coiled palps which are often lost when the worm is handled. Tip of posterior region saucer-shaped, funnel-shaped or with several cirri.

1 Fifth segment of body enlarged, bearing stout, spine-like chaetae
.... ***Polydora ciliata*** p. 187

Fifth segment not so **2**

2 Series of paired gills present in middle region of body; tip of posterior region with
4 short conical projections; worm inhabits tiny, sandy tube. About 15 mm in length.
Often occurring in very high densities   .... ***Pygospio elegans*** p. 186
Series of paired gills present from 1st or 2nd segment of body to near posterior **3**

3 Head pointed anteriorly and posteriorly; tip of posterior region crenulate
.... ***Scolelepis squamata*** p. 187

Head pointed posteriorly, anteriorly with 2 small frontal horns; tip of posterior
region with 6–8 short cirri   .... ***Malacoceros fuliginosus*** p. 188

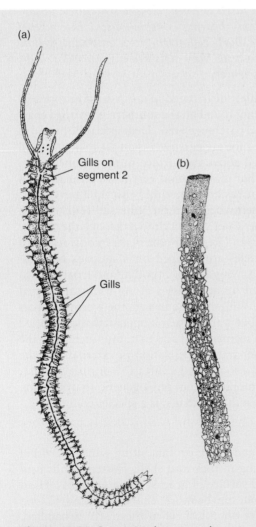

Figure 11.31 *Pygospio elegans*, male
(a) removed from tube, (b) tube.

(a)

Gills on
segment 2

(b)

Gills

Pygospio elegans Claparède **Fig. 11.31**

Body small, slight, up to 15 mm in length; tip of posterior region with 4 short, conical, projections, each with many papillae. Up to 60 segments. Gills present on dorsal surface, starting between segments 11 and 20; 7–9 pairs in the female, 20–28 pairs in the male; male also with 2 large gills on segment 2. Head bluntly divided anteriorly, pointed posteriorly and projecting backwards to segment 2. Eyes 2–8, irregularly placed. Yellowish-green anteriorly, brownish colour of gut visible posteriorly.

P. elegans is widely distributed in north-west Europe and around Britain. It lives in thread-like chitinous tubes covered with fine particles of sand and shell fragments. The tubes are found intertidally and below in muddy sediments among rock crevices and on sandy shores and mud-flats, often in densities as high as 60 000 per square metre. They occur in very high densities in areas that are organically polluted and also in brackish areas, the worms being tolerant of very low salinities. *Pygospio* is a deposit feeder picking up surface detritus with the paired palps but can also suspension feed. The sexes are separate and breeding has been recorded throughout most of the year. *P. elegans* shows more than one mode of development. In some populations egg capsules are deposited in rows, or egg strings, inside the tube of the female, the capsules being connected to the wall of the tube by a short stalk. Each capsule contains a number of eggs, some of which develop into larvae with a rich supply of yolk (lecithotrophic development) and are brooded in the capsules. Other eggs do not develop but serve as food for the developing embryos. In other populations, the eggs hatch into larvae which feed in the plankton (planktotrophic) for several weeks and then settle. There is no evidence to suggest that populations with different modes of development are reproductively isolated. In some populations asexual reproduction by body fragmentation has been reported.

Polydora ciliata (Johnston) **Fig. 11.32a,d**

Body up to 30 mm in length, tip of posterior region saucer-shaped. Up to 180 segments; 5th segment is enlarged and bears stout, spine-like chaetae. Gills present on dorsal surface from segment 7 to within 10 segments of posterior. Eyes often absent in adult. Yellowish-brown in colour.

P. ciliata is widely distributed in north-west Europe and around Britain. It usually burrows into substrata containing calcium carbonate such as limestone, chalk and clay as well as the shells of oysters, mussels and periwinkles. It is also found in muddy sediments, wood and laminarian holdfasts. It has been suggested that burrowing is achieved by mechanical action of the chaetae, especially those of the fifth segment, but this is open to some doubt; chemical action may also be involved. The U-shaped burrow is lined by mucus and particles of sand, and the two ends extend a few millimetres above the sediment surface. As a rule *Polydora* apparently feeds on detritus removed from the sediment by the two long palps. It also feeds on suspended particles in the water and on occasions has been observed to eat dead barnacles and other dead invertebrates. The sexes are separate and breeding has been recorded in spring at a number of localities. Sperm are drawn into the burrow of the female in the respiratory current and eggs are laid in a string of capsules. A single female produces many capsules, each containing up to about 60 eggs, the individual capsules being attached by two threads to the wall of the burrow. After a week, larvae emerge and are believed to have a pelagic life of from two to six weeks before settling. Length of life is no more than one year. *Polydora* is a serious pest of oysters and mussels, but invades only the shell and does not eat the soft tissue. When infestation is heavy, the shell is weakened and this makes the mollusc more susceptible to predation by crabs. Occasionally, the inner surface of the shell is damaged by the worms and the mollusc secretes a nacreous substance to seal the perforation. The resulting blisters are an indication of attack by *Polydora* and mussels heavily infested with the polychaete have reduced flesh content, possibly as a result of some impact on the metabolism of the mussel. It has been shown experimentally that periwinkles in which the shell has been attacked and weakened by *Polydora* are more vulnerable to predation by shore crabs.

Several other species of *Polydora* are recorded from north-west Europe. The enlarged fifth body segment is characteristic of the genus but the species are difficult to separate. They are found in calcareous rocks, encrusting red seaweed and molluscan shells; *P. ciliata* is likely to be the most widespread and common on the shore. This species, however, shows wide morphological variation and there is evidence to suggest that it is a species complex.

Scolelepis squamata (O. F. Müller) (*Nerine cirratulus*) **Fig. 11.32b**

Body up to 80 mm in length, tip of posterior region crenulate. Up to 200 segments. Dorsal gills present on 2nd segment with chaetae and on most segments to near posterior of body. Head pointed anteriorly and posteriorly. Bluish-green in colour, gills red.

Figure 11.32 *Polydora ciliata*, (a) removed from burrow, (d) in U-shaped burrows. (b) *Scolelepis squamata*. (c) *Malacoceros fuliginosus*.

S. squamata is widely distributed in north-west Europe and around Britain. It is found on the middle and lower shore mainly in sand, or sometimes muddy-sand, in vertical burrows lined with mucus. It feeds on particles of suspended organic material using the long palps. On the south coast of England, breeding occurs from March to at least July. The sexually mature worms are not pelagic. Fertilization is external and the larva is free-swimming for some five weeks before settlement.

A second species, **Scolelepis foliosa** (Audouin & Milne-Edwards) (*Nerine foliosa*), is found on the lower shore in fairly clean sand. It is distinguished from *S. squamata* by a small median tentacle at the base of the two palps and by the absence of gills on the posterior third of the body.

Malacoceros fuliginosus (Claparède) (*Scolelepis fuliginosa*) **Fig. 11.32c**
Body up to 60 mm in length; tip of posterior region with 6–8 short cirri. Up to 150 segments. Dorsal gills present from 1st segment with chaetae to near posterior end of body. Head pointed posteriorly and with 2 small frontal horns; 4 eyes. Reddish in colour.

M. fuliginosus is widely distributed in north-west Europe and around Britain. It is found on the middle shore in mucus-lined tubes under stones and in sand, muddy-sand and mud. It feeds on organic matter deposited on the surface of the sediment. The sexes are separate and on the north-east coast of England, breeding takes place in spring and summer. Fertilization is external and there is a free-swimming larva with a planktonic life of four to six weeks. It has been suggested that *M. fuliginosus* is a species complex.

References

Brinkhurst, R. O. (1982). *British and other marine and estuarine oligochaetes. Keys and notes for the identification of the species.* Synopses of the British fauna (New Series), no. 21. Cambridge: Cambridge University Press.

Chambers, S. (1985). *Polychaetes from Scottish waters.* Part 2. *Families Aphroditidae, Sigalionidae and Polyodontidae.* Edinburgh: Royal Scottish Museum Studies.

Chambers, S. J. & Garwood, P. R. (1992). *Polychaetes from Scottish waters.* Part 3. *Family Nereidae.* Edinburgh: National Museums of Scotland.

Chambers, S. J. & Muir, A. I. (1997). *Polychaetes. British Chrysopetaloidea, Pisionoidea and Aphroditoidea. Keys and notes for identification of the species.* Synopses of the British fauna (New Series), no. 54. Shrewsbury: Field Studies Council.

Fauvel, P. (1923). *Faune de France.* Vol. 5. *Polychètes errantes.* Paris: Lechevalier.

Fauvel, P. (1927). *Faune de France.* Vol. 16. *Polychètes sédentaires. Addenda aux errantes, archiannélides, myzostomaires.* Paris: Lechevalier.

George, J. D. & Hartmann-Schröder, G. (1985). *Polychaetes. British Amphinomida. Spintherida and Eunicida. Keys and notes for the identification of the species.* Synopses of the British fauna (New Series), no. 32. Leiden: E. J. Brill/Dr W. Backhuys.

Hartmann-Schröder, G. (1996). *Annelida, Borstenwürmer, Polychaeta. Die Tierwelt Deutschlands,* Vol. 58, 2nd edn. Jena: Gustav Fischer.

McIntosh, W. C. (1900–23). *A monograph of the British marine annelids.* Vols. 1 (part II), 2–4. London: Ray Society.

Pleijel, F. & Dales, R. P. (1991). *Polychaetes. British phyllodocoideans, typhloscolecoideans and tomopteroideans. Keys and notes for the identification of the species.* Synopses of the British fauna (New Series), no. 45. Oegstgeest, The Netherlands: UBS/Dr W. Backhuys.

Rouse, G. W. & Pleijel, F. (2001). *Polychaetes.* New York: Oxford University Press.

Tebble, N. & Chambers, S. (1982). *Polychaetes from Scottish waters.* Part 1. *Family Polynoidae.* Edinburgh: Royal Scottish Museum Studies.

Westheide, W. (2008). *Polychaetes. Interstitial families. Keys and notes for the identification of the species.* Synopses of the British fauna (New Series), no. 44, 2nd edn. Shrewsbury: Field Studies Council.

Mollusca

Molluscs are found in marine, freshwater and terrestrial environments and in terms of numbers of species are second only to the arthropods. Body form varies widely within the phylum but the basic molluscan plan shows a head, often well developed and bearing sensory structures, a muscular foot and a visceral hump. The delicate tissue covering the visceral hump is known as the mantle (or pallium). This projects beyond the edge of the visceral mass to enclose a cavity known as the mantle (pallial) cavity where gills and other structures are found. The mantle secretes a calcareous shell consisting of a number of layers. There is usually an outer proteinaceous layer known as the periostracum, beneath which are layers of calcium carbonate. In some molluscs an inner mother-of-pearl or nacreous layer is often exposed on the surface of the shell by the wearing away of the outer layers. Although described as one of the most characteristic features of the phylum, it is important to note that in some groups the shell is internal, while in others it is absent. The characteristic feeding organ is a ribbon-like, chitinous structure, the radula (Plate 16d), which comprises rows of teeth. In many species these have a rasp-like action and remove adhering microorganisms from the rock surface. In some gastropods, the radula is modified to enable it to bore into prey. The bivalves lack a radula and are deposit or suspension feeders using the gill to filter out fine particles of food. Most molluscs have separate sexes and while some shed the gametes and have external fertilization, in others fertilization is internal following copulation. Brood protection is well known but many species have a pelagic larva.

There is no universally accepted scheme of classification for the molluscs. Nomenclature follows **CLEMAM**: check list of European marine Mollusca.

Phylum MOLLUSCA

Class Aplacophora	Worm-like molluscs without a shell but may have chitinous cuticle and calcareous spicules or scales in epidermis.

Subclass Chaetodermomorpha	Worm-like molluscs without a shell. No foot. Body wall with chitinous cuticle and calcareous spicules. Live in marine sediments; deposit feeders.

Subclass Neomeniomorpha	Worm-like molluscs without a shell. Foot reduced. Body wall with calcareous spicules or scales. Carnivorous on cnidarians. Marine.
Class Monoplacophora	A primitive group with a single, symmetrical shell and replicated organs. Found in the deep sea.
Class Polyplacophora	**The chitons or coat-of-mail shells. The shell is formed of eight, transverse articulating plates (valves). Broad foot. Marine.**
Class Gastropoda	**The largest class of molluscs. The shell is a single unit, often coiled but may be cone-shaped or flattened. Shell sometimes absent. Found in marine, freshwater and terrestrial environments.**
Subclass Prosobranchia	**The winkles, whelks, limpets and top-shells. Well-developed shell. Marine; a few freshwater and terrestrial species.**
Subclass Opisthobranchia	**The sea-slugs and sea-hares. Shell usually reduced and internal, or absent, but primitive members retain external shell. Nearly all marine; a few freshwater species.**
Subclass Pulmonata	**Freshwater and land snails and slugs. Also includes a few marine representatives. Mantle cavity modified for air breathing. Shell present or absent.**
Class Bivalvia (Pelecypoda; Lamellibran-chiata)	**Includes the mussels, oysters and cockles. Shell with 2 valves joined by a ligament. Marine; a few freshwater representatives.**

Class Scaphopoda	The tusk shells. Shell a cylindrical tube, open at both ends. Sublittoral.

Class Cephalopoda	Squids, cuttlefishes and octopuses. Shell usually internal and reduced. Marine. Very occasionally stranded on the shore.

Class Polyplacophora

Chitons, or coat-of-mail shells, are dorsoventrally flattened molluscs which have retained some primitive features. The shell is made of eight articulating plates, also known as valves (Fig. 12.2a), enabling the animal to curl up when disturbed. The shell plates are surrounded by a fleshy mantle, the girdle (Fig. 12.1), and on the ventral surface the most conspicuous feature is a broad foot enabling chitons to creep slowly and adhere firmly to rock surfaces. Between foot and mantle is a groove, the mantle or pallial groove, in which there is a series of gills (Fig. 12.1). These may be restricted to the posterior part of the groove (merobranch condition) or distributed throughout the entire groove (holobranch condition), and their position may be important in identification. The head is generally not clearly distinguished and both eyes and tentacles are lacking but some species are believed to have specialized light-sensitive cells in the shell plates. Chitons graze on a variety of microorganisms and seaweeds by means of the radula some teeth of which are hardened with iron and silicate compounds, enabling the animals to feed on the tougher seaweeds, including encrusting calcareous forms. The sexes are usually separate and the gametes discharged into the sea. Fertilization is external, the eggs developing into trochophore larvae but in a few species they are brooded in the mantle cavity. Chitons are common on most rocky shores clinging to the undersurface of rocks and stones, especially those loosely buried in coarse sand and gravel. Crabs and fishes are believed to be their main predators.

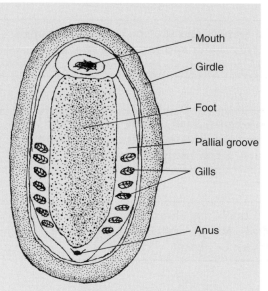

Figure 12.1 Ventral view of a chiton.

Labels: Mouth, Girdle, Foot, Pallial groove, Gills, Anus

Key to families of polyplacophorans

Of the four families of British chitons, representatives of three are widely distributed and common on the shore. They are described below using characters easily seen with the aid of a hand lens or low-power stereo-microscope. Fine detail of the shell plates, seen only when the plates are dissected out, is not used in the key.

1 Girdle with tufts (usually 18) of conspicuous bristles ... **Acanthochitonidae** p. 195
   Girdle without tufts of bristles **2**

2 Girdle narrow, covered in scales; fringed with spines ..... **Leptochitonidae** p. 193
   Girdle relatively broad, with granules varying from minute to coarse, widely
   separated to dense; fringed with spines .... **Ischnochitonidae** p. 194

Family Leptochitonidae
Small, oval-shaped chitons. Shell plates finely granular. Girdle narrow, covered in scales; fringed with spines. Gills in posterior region of mantle groove.

Leptochiton asellus (Gmelin) (*Lepidopleurus asellus*) **Fig. 12.2a**
Oval body. Shell plates with keel, finely granular; whitish or yellow colour often marked with darker patches. Girdle narrow, covered with longitudinally ribbed scales

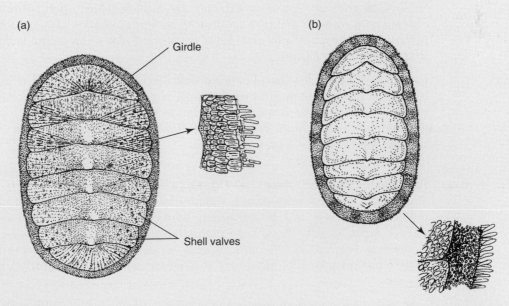

Figure 12.2 (a) *Leptochiton asellus*, with detail of girdle (detail after Matthews, 1953). (b) *Lepidochitona cinerea*, with detail of girdle (detail after Matthews, 1953).

and fringed with slender spines; 8–13 pairs of gills in posterior region of mantle groove. Body length up to 19 mm.

L. asellus is widespread and common in north-west Europe. It is found all around the coast of Britain where it is common on the lower shore and into the sublittoral to depths of 250 m.

Family Ischnochitonidae
Oval to elongate-shaped chitons. Shell plates smooth and shiny or with fine granules. Girdle relatively broad, granular; fringed with spines.

1 Up to 15 pairs of gills (usually 10–15), in posterior region of mantle groove; shell plates smooth. Girdle densely covered with oval granules ... ***Tonicella rubra*** p. 195
With 16–26 pairs of gills **2**

2 Girdle leathery, with a few well-spaced, minute granules; fringed with spatulate spines. Shell plates with minute granules but appear smooth
.... ***Tonicella marmorea*** p. 194
Girdle covered with rounded granules; fringed with large, golden spines. Shell plates granular .... ***Lepidochitona cinerea*** p. 194

Lepidochitona cinerea (Linnaeus) (*Lepidochitona cinereus*) **Fig. 12.2b**
Shell plates granular; colour varied, grey, green, blue, white; often patterned. Girdle covered with rounded granules and fringed with large golden, spines; banded colour pattern; 16–19 pairs of gills distributed throughout almost the entire mantle groove. Body length up to 28 mm.

L. cinerea is widely distributed in north-west Europe and common all around Britain. It is found throughout most of the intertidal region, generally on the undersurface of stones. The sexes are separate and fertilization is external, the larvae spending only a few hours in the plankton. Winter growth checks can clearly be seen on some of the shell valves. Longevity is up to five years.

Tonicella marmorea (O. Fabricius) **Fig. 12.3a**
Shell plates covered with minute granules but appear smooth; red to dark red in colour with cream and brown areas. Girdle leathery with a few well-spaced, minute granules and fringed with spatulate spines; purple, red or green in colour, sometimes banded; 17–26 pairs of gills, usually in posterior region of mantle groove but occasionally distributed throughout entire groove. Body length up to 45 mm.

T. marmorea is a northern species and in Britain is found as far south as North Wales on the west coast and Northumberland on the east coast. It occurs on rocks and stones from the lower shore to depths of about 200 m. The sexes are separate and fertilization external.

Figure 12.3 (a) *Tonicella marmorea*, with detail of girdle (detail after Matthews, 1953). (b) *Tonicella rubra*, with detail of girdle (detail after Matthews, 1953).

Tonicella rubra (Linnaeus) **Fig. 12.3b**
Shell plates smooth and shiny; pinkish-red in colour with cream areas. Girdle densely covered with oval granules and fringed with golden spines; with banded colour pattern; 10–15 pairs of gills in posterior region of mantle groove. Body length up to 21 mm.

T. rubra is widespread in north-west Europe and around Britain where it is found on stones on the lower shore extending to 300 m. The sexes are separate and fertilization external.

Family Acanthochitonidae
Shell plates granular, with keel; partly covered by girdle. Girdle broad, with (usually) 18 tufts of bristles.

Shell plates with large, pear-shaped or oval granulations, irregularly spaced .... ***Acanthochitona crinita*** p. 195

Shell plates with tiny, rounded granulations; granulations dense, evenly spaced .... ***Acanthochitona fascicularis*** p. 196

Acanthochitona crinita (Pennant) (*Acanthochitona crinitus*) **Fig. 12.4a**
Large, pear-shaped or oval, irregularly spaced granulations on shell plates; brown-yellow colour with varied patterning; white V-shaped area in centre of plates. Girdle

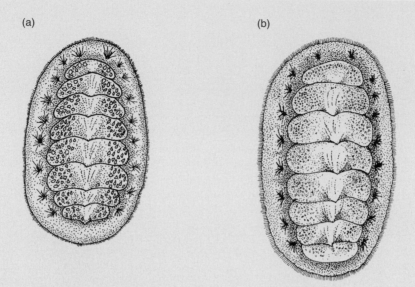

(a)   (b)

Figure 12.4   (a) *Acanthochitona crinita.* (b) *Acanthochitona fascicularis.*

broad, with 18 tufts of colourless bristles; fringed with spines; 10–15 pairs of gills in posterior region of mantle groove. Body length up to 34 mm.

A. *crinita* is widely distributed in north-west Europe, extending as far north as southern Norway. It is found all around Britain on rocky shores, extending from the lower shore to depths of 50 m. The sexes are separate and fertilization is external.

Acanthochitona fascicularis (Linnaeus) (*Acanthochitona communis*) **Fig. 12.4b**
Shell plates with very small, rounded granulations, closely packed and evenly spaced; varied coloration, whitish, olive, brown, often with whitish 'V' along mid-dorsal region. Girdle broad, with 18 tufts of colourless bristles; fringed with long spines; 10–15 pairs of gills in posterior region of mantle groove. Body length up to 60 mm. The largest British species of chiton.

A. *fascicularis* is a southern species extending to the Channel coasts, the west coast of Ireland and as far north as Anglesey. It is found from the lower shore to depths of about 50 m.

Class Gastropoda

A typical gastropod has a single, usually coiled shell, a large ventral foot and a well-developed head bearing eyes and conspicuous tentacles. In the opisthobranchs and pulmonates shell reduction and loss has occurred. A few marine gastropods are adapted to a pelagic habit but the most familiar species are benthic and include the often abundant limpets, periwinkles and top-shells belonging to the subclass

Prosobranchia, and the sea-slugs and sea-hares belonging to the subclass Opistho-branchia. During development gastropods undergo torsion, a process by which the visceral mass of the larva twists through 180° so that the mantle cavity, gills and anus come to lie just behind the head. This condition persists in the adult prosobranch and there is much controversy and debate as to its adaptive significance. In opisthobranchs and pulmonates, a process of detorsion has taken place. Torsion of the visceral mass is quite distinct from coiling of the shell which is believed to result from increase in body size during growth. While many gastropods have separate sexes, others are hermaph-roditic and the class shows a variety of reproductive patterns. Fertilization may be external, with the development of a pelagic larva, or internal. When internal the eggs may be brooded in an internal brood pouch, laid in egg masses or capsules attached to the substratum, or released into the sea and development is either direct or pelagic. In some gastropods, for example the limpets and top-shells, the typical larva is a trochophore, similar to that of the polychaetes, with a short free-swimming

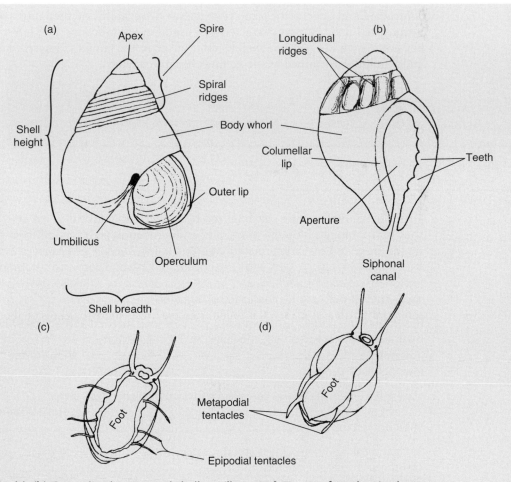

Figure 12.5   (a), (b) Generalized gastropod shells to illustrate features referred to in the text. (c), (d) Ventral view of animal showing (c) epipodial and (d) metapodial tentacles.

life. In others, such as the winkles, mud snails and sea-slugs, the typical larva is a veliger (Fig. 12.18). This has a tiny shell and swims by means of two ciliated lobes, the velum. Many veligers have a long pelagic life.

Subclass Prosobranchia

Prosobranch molluscs are among some of the most common members of the rocky shore fauna. They are primitive members of the Class Gastropoda from which the opisthobranchs (p. 235) and the pulmonates (p. 248) are believed to have evolved. They are characterized by a conical (e.g. the limpets) or coiled shell (e.g. the periwinkles) but in one family (p. 228) the shell is internal. The external shell is to varying degrees sculptured with a surface ornamentation of ridges, lines and ribs, which are often useful in identification. The aperture of the shell is often extended into a groove, known as the siphonal canal, which in life houses a fold of mantle tissue, the siphon. Most prosobranchs have a tough, proteinaceous disc, the operculum, which is sometimes impregnated with calcium. The operculum closes the shell when the animal is contracted and offers some protection against desiccation and predation when the tide is out. Shell and soft part characteristics useful in identification and referred to in the key are shown in Fig. 12.5. Shell length as used in the families Fissurellidae, Lottiidae and Patellidae is defined as the greatest basal diameter.

Key to families of prosobranchs

Prosobranch molluscs are characterized by a shell made up of a single unit. Variation in shell morphology within the group allows the recognition of the different families. It should be noted that some pulmonates (including *Auriculinella bidentata*, p. 248, *Myosotella myosotis*, p. 249, and *Otina ovata*, p. 249) and some opisthobranchs (including *Acteon tornatilis*, p. 237, and *Retusa obtusa*, p. 238) also possess an external shell. They should not be confused with prosobranchs.

1 Resembles sea-slug (Opisthobranchia, p. 235); shell internal and enclosed by mantle. Mantle covered with tubercles; siphon-like extension at anterior .... **Velutinidae** p. 228
External shell present **2**

2 Shell a flattened spiral with conspicuous row of rounded openings ....
.... **Haliotidae** p. 200
Shell not as above **3**

3 Shell conical (i.e. more or less straight-sided, without spiral whorls) or cap-shaped; without operculum **4**
Shell not conical, not cap-shaped; with or without operculum **8**

4 Shell conical with single apical aperture or a marginal slit ..... **Fissurellidae** p. 201
Shell conical or cap-shaped, without apical aperture or marginal slit **5**

5 Shell with internal shelf extending across part of aperture .... **Calyptraeidae** p. 225
Shell without internal shelf extending across part of aperture **6**

6 Shell conical to cap-shaped; smooth, may be transparent; oval outline, with 2–8
usually bright, blue-green lines. On stipe, frond and holdfasts of laminarians, also
on *Fucus, Himanthalia* and occasionally other seaweeds .... **Patellidae** p. 204
Shell conical, not as above **7**

7 Shell smooth; with tortoiseshell pattern or pink or brown rays. Firmly attached
to rock surface, stones, shells .... **Lottiidae** p. 203
Shell strong, usually with radiating ridges. Firmly attached to rock
surface .... **Patellidae** p. 204

8 Shell glossy, strongly ridged with ridges running into aperture, or smooth;
body whorl very large with narrow elongate aperture; may have very small spire.
Without operculum. Typical cowrie shape .... **Triviidae** p. 227
Shell not so. With operculum **9**

9 Shell with siphonal canal which may be open or closed; shell may have large
pointed expansion of outer lip **10**
Shell without siphonal canal **13**

10 Shell with large pointed expansion of outer lip to give appearance of pelican's
foot .... **Aporrhaidae** p. 225
Shell not so **11**

11 Shell usually large (about 100 mm shell height), stout; body whorl large; aperture
large; outer lip not usually thickened; siphonal canal not closed over. Shell smooth
or ridged, with spiral ridges, or spiral and longitudinal ridges. Predominantly
sublittoral on soft sediment; dead shells washed ashore .... **Buccinidae** p. 233
Shell up to 50 mm shell height, stout, with spiral or longitudinal ridges or both.
Outer lip may be thickened; sometimes with teeth in older specimens. Siphonal
canal may be closed over. Often common on rocky shores **12**

12 Siphonal canal open, short; in non-apertural or 'back view' the end of the
canal is seen as a notch. Groove runs from lower end of siphonal canal to
columella round 'back' of shell. .... **Nassariidae** p. 234
Siphonal canal open or closed, not seen as a notch in non-apertural or
'back view' .... **Muricidae** p. 230

13 Shell may be nacreous, especially at apex when worn, and within aperture; often
colourful with red, purple, green markings. Umbilicus obvious or closed. Rounded
operculum. Lateral lobes of foot with epipodial tentacles .... **Trochidae** p. 207
.... see also **Calliostomatidae** p. 209
Shell, operculum and foot not so **14**

14 Shell with conspicuous umbilicus **15**
Shell without umbilicus or umbilicus very small **16**

15 Shell with spire or spire reduced. Umbilicus conspicuous, with groove on
columellar region leading to it. Posterior of foot with 1 pair of metapodial
tentacles .... **Littorinidae (Genus *Lacuna*)** p. 212
Smooth, glossy shell with large body whorl and short spire. Umbilicus large.
Foot large, covers head and part of shell when animal active; without metapodial
tentacles .... **Naticidae** p. 228

16 Shell tall, narrow, with pronounced spire; with many (about 15–20) whorls **17**
Shell not as above **19**

17 Shell with up to 15 whorls; ornamentation of spiral and longitudinal ridges to give
tuberculate pattern. Aperture with short, basal extension .... **Cerithiidae** p. 211
Shell not as above **18**

18 Shell with up to 15 whorls; with pronounced, well-spaced, longitudinal ridges.
Aperture rounded .... **Epitoniidae** p. 229
Shell with up to 20 whorls; with spiral ridges. Aperture angular. Operculum with
bristles around periphery .... **Turritellidae** p. 211

19 Smooth glossy shell, of 5–6 whorls; whitish with reddish-brown streaks, with white,
domed, calcareous operculum. Up to 9 mm height .... **Phasianellidae** p. 210
Shell and operculum not as above **20**

20 Shell with pronounced spire, or spire low and flattened. Body whorl accounts for
75% or more of shell height. Shell smooth to strongly ridged with spiral ridges.
Shell height from 3 to 30 mm. Widespread and often abundant. Found at all levels
on rocky shores; some species live on soft sediments and extend into estuaries
.... **Littorinidae (Genera _Littorina_ and _Melarhaphe_)** p. 214
Shell with spire; body whorl accounts for less than 75% of shell height. Usually
up to about 6 mm shell height, never greater than 9–10 mm shell height.
Widespread and often abundant **21**

21 Shell smooth (but see _Potamopyrgus antipodarum_, p. 224); up to 6 mm in height.
Brownish-black in colour. Found on estuarine sands and muds, often in very high
densities. Also found in brackish and freshwater ditches. Not found on rocky
shores .... **Hydrobiidae** p. 221
Shell smooth to heavily ridged. Variously coloured, white, cream, orange, brown.
Usually less than about 6 mm shell height; never greater than 9–10 mm. Marine,
predominantly found on rocky shores where they may be abundant under
stones and amongst seaweeds .... **Rissoidae** p. 220

Family Haliotidae Ormers
Flattened, spiral shell with a row of rounded openings, of which 5 to 7 are open, the
rest are closed. Aperture large. Without operculum.

Haliotis tuberculata Linnaeus Ormer, abalone **Fig. 12.6**
_Flattened, spiral shell up to 90 mm in length; with row of rounded openings, 5–7 of
which are open, remainder closed. Outer surface of shell shades of red, yellow or
green; inner surface iridescent. Operculum absent._

H. tuberculata is a southern species, extending as far north as the Channel Islands. It is
found clinging to rocks, especially those covered with encrusting red seaweeds, on the
lower shore and in the sublittoral to depths of about 40 m. The ormer feeds mainly on
red seaweeds and the colour of the shell is believed to be related to the food eaten. The
sexes are separate and fertilization external. Trochophore larvae settle after a brief
free-swimming phase at a length of about 2 mm. In the Channel Islands breeding
occurs in late summer, the animals becoming sexually mature at 40 to 50 mm shell

Figure 12.6 *Haliotis tuberculata.*

length when they are two or three years old. They have a longevity of five to six years. Length of larval life is probably five to six days. The openings along the edge of the shell of *Haliotis* are the routes by which water leaves the mantle cavity. New openings form at the margin of the shell and can be seen as slit-like indentations which are surrounded as the shell grows. The oldest openings are closed. The ormer is greatly valued as food and is fished commercially in France and Spain but the industry has been beset by a number of problems, including bacterial disease and pollution. There are ongoing aquaculture projects for *Haliotis* in a number of countries.

Family Fissurellidae Slit limpets, keyhole limpets
Shell conical, with radiating ribs; a single apical aperture or a marginal slit. Without operculum.

Shell with marginal slit	*Emarginula fissura* p. 201
Shell with small, apical opening	*Diodora graeca* p. 202

Emarginula fissura (Linnaeus) (*Emarginula reticulata*) Slit limpet **Fig. 12.7**
Ribbed, conical shell; apex curved towards posterior; marginal slit at anterior. Up to 10 mm in length. Outer surface of shell white, grey or yellow; inner surface similar but nacreous. Short siphon-like structure, through which water leaves mantle cavity, extends through anterior marginal slit. Operculum absent.

E. fissura is widely distributed on rocky shores in north-west Europe but absent from the eastern English Channel and the southern North Sea. It is found on the lower shore on the undersurface of rocks and boulders and extends into the sublittoral to depths of 200 m and more. The slit limpet feeds largely on sponges but also takes some seaweeds and detritus. The sexes are separate and fertilization is external.

hardened with iron and silicate compounds, and when feeding over relatively soft rocks such as shales, zigzag impressions are left in the rock (Plate 16c,d). Feeding has been observed during high water in daylight and during emersion at night and it now appears that limpets on steep gradients are mainly active during night-time emersion while those on gently sloping rocks are active when submersed during daylight. This is a complex behaviour pattern the reasons for which are not fully understood. Limpets undergo feeding excursions, returning to the same 'home scars' from which they started (Plate 16b). This homing behaviour is achieved by the limpet following a mucous trail deposited on the rock surface and is controlled by an endogenous circatidal rhythm. The 'home scar' is a shallow depression in the rock formed by chemical action and abrasion between the limpet shell and the rock. The edge of the shell fits tightly into the depression and the homing behaviour has adaptive significance in that it reduces the rate at which water is lost from the mantle tissues during emersion.

The impact achieved by limpets and other herbivorous molluscs grazing on shore algae is well known and can be demonstrated by excluding them from marked areas on the shore. Under these conditions the sporelings of green seaweeds such as *Ulva* (p. 30) grow almost unchecked, clothing the rocks in a bright green colour (Plate 17). The common limpet is a protandrous hermaphrodite, although a few individuals remain as males. On the east coast of England, *P. vulgata* matures as male at two years of age, and then changes to female at about four years of age. Spawning occurs mainly during October to December and is believed to be induced by rough seas and onshore winds. Fertilization is external and the trochophore larvae have a pelagic life of a few days. Settlement on the shore occurs at a shell length of about 0.2 mm and the newly settled spat are usually found in rock pools or permanently damp situations. Length of life varies considerably; where growth is fast it is generally four to five years, compared with up to 15 to 17 years where conditions lead to slow growth.

Patella ulyssiponensis Gmelin (*Patella aspera, Patella athletica*) China limpet
Plate 15b

Shell conical, flattish; apex anterior: with radiating ridges; up to 60 mm in length. Outer surface whitish-grey; inner surface whitish, porcelain-like, with cream to orange apical area. Foot yellow to orange; tentacles around edge of mantle, cream. Operculum absent.

P. ulyssiponensis is distributed from southern Norway to the Atlantic coast of Spain and the Mediterranean. Around Britain it is absent from southern North Sea coasts and absent or rare on the Channel coast east of the Isle of Wight. It is found on rocks on the lower shore and in the sublittoral, also occurring in shallow rock pools on the middle shore. In exposed situations it is the commonest limpet on the lower shore. Its diet is probably similar to that of *P. vulgata* (above). Despite earlier reports that the sexes are separate throughout life, there is evidence to suggest that it is a protandrous hermaphrodite. In south-west Ireland, maturation of the gonads begins in May and June, and spawning occurs in autumn and winter, probaly induced by gales. Fertilization is external.

Patella intermedia Murray in Knapp (*Patella depressa*) Black-footed limpet
Plate 15c

Shell conical, flattened; apex anterior; with fine radiating ridges; up to 30 mm shell length. Outer surface greyish; inner surface dark with conspicuous, brown marginal rays, and creamy-yellow apical area. Foot dark grey; tentacles around edge of mantle, white, contrasting with the dark colour of the foot. Operculum absent.

P. intermedia is found on the Atlantic coast of Europe, extending northwards to south-west Britain but is not found east of the Isle of Wight. The northern limit of the species is North Wales. It favours wave-beaten, rocky shores and is usually on the middle to lower shore. The sexes are separate and on the south coast of England maturation of the gonads begins in spring and spawning, which may be related to gales and heavy wave action, occurs from July to September. Fertilization is external.

Family Trochidae Top-shells

Shell with spire; nacreous within aperture and at apex of shell when worn. Often with colourful patterning of green, red, purple lines. Aperture without siphonal canal. Umbilicus present but may be closed in some cases. Operculum rounded. In the species described here, lateral lobes of foot with 3 pairs of epipodial tentacles (see Fig. 12.5c), cf. *Calliostoma* (p. 209).

1 Single, prominent tooth on inner margin of aperture (Fig. 12.13b); apex often eroded and silvery .... ***Osilinus lineatus*** p. 209
Shell without tooth on inner margin of aperture **2**

2 Sides of whorls angular, with steps between whorls; upper surface of whorls with ridges and tubercles, tubercles especially noticeable on body whorl. Umbilicus deep .... ***Gibbula magus*** p. 208
Sides of whorls more or less smoothly convex; upper surface of whorls without tubercles. Umbilicus small or large **3**

3 Shell greyish in colour, with many narrow, red-purple lines. Umbilicus small. Apex sometimes eroded and silvery .... ***Gibbula cineraria*** p. 207
Shell greyish-green in colour, with well-spaced, diagonal purple stripes. Umbilicus large. Apex often eroded and silvery .... ***Gibbula umbilicalis*** p. 208

Gibbula cineraria (Linnaeus) Grey top-shell **Fig. 12.12a**
Shell roughly triangular in outline; up to 16 mm in height and 17 mm broad; umbilicus small. Greyish in colour with many narrow, red-purple lines. Apex sometimes eroded and silvery. Foot with 3 pairs of epipodial tentacles.

G. cineraria is widely distributed in north-west Europe and around Britain on sheltered rocky shores from the lower shore to depths of about 130 m feeding mainly on detritus and microalgae. The radular teeth of *G. cineraria*, and other top-shells, are not hardened with iron compounds (cf. patellids) and they are apparently unable to eat encrusting red seaweed or seaweeds with tough cell walls such as *Fucus* spp. (p. 63) which feature in the diet of *Patella* spp. (above). The sexes are separate, fertilization is external and it is believed that spawning occurs in response to increase in seawater temperature. In north-east England the main breeding period is from June to September. There is a free-swimming trochophore larva and laboratory studies suggest a pelagic life of eight or nine days. Longevity is believed to be up to three years.

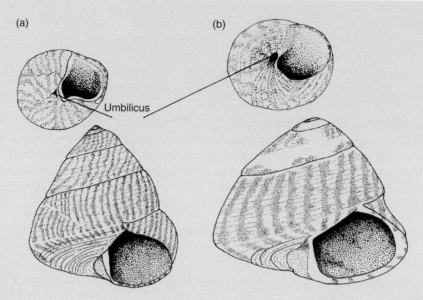

Figure 12.12 (a) *Gibbula cineraria*. (b) *Gibbula umbilicalis*.

Gibbula umbilicalis (da Costa) Flat or purple top-shell **Fig. 12.12b**

Shell roughly triangular in outline; up to 16 mm in height and 22 mm broad; umbilicus large. Greyish-green in colour, with well-spaced, diagonal purple stripes. Apex often eroded and silvery. Foot with 3 pairs of epipodial tentacles.

G. *umbilicalis* occurs on the west coast of France and south-west and west coasts of Britain as far north as Orkney, and there is some evidence to suggest that the northern limit of the species has been extended in response to warming of coastal waters. It is found on the upper and middle shore on sheltered rocky shores above the lower-shore species G. *cineraria* (above). Like G. *cineraria*, it feeds on microorganisms and detritus. The sexes are separate, fertilization is external and sexual maturity is reached in the second year at a shell height of about 8–9 mm. Breeding takes place during summer and the trochophore larvae are believed to have only a short pelagic life. Length of life has been estimated to be eight or more years.

Gibbula magus (Linnaeus) **Fig. 12.13a**

Shell broader than tall, up to 30 mm in height and 35 mm broad; upper surface of whorls with ridges and tubercles; sides of whorls angular with steps between whorls; umbilicus deep. Whitish-yellow with reddish markings. Foot with 3 pairs of epipodial tentacles.

G. *magus* is a southern species found along the west coast of Britain but is absent from the eastern English Channel and the North Sea. It is restricted to the extreme lower shore, extending into the sublittoral to depths of about 70 m on rocky substrata, as well as being found on muddy-sand. The food consists of microorganisms and detritus. The sexes are separate, fertilization is external and breeding occurs during spring and summer. The trochophore larva probably has only a short pelagic life.

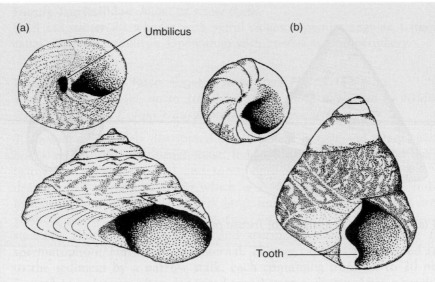

(a)   Umbilicus

(b)

Tooth

Figure 12.13 (a) *Gibbula magus.* (b) *Osilinus lineatus.*

Osilinus lineatus (da Costa) (*Monodonta lineata*) Thick top-shell **Fig. 12.13b**
Triangular shell, taller than broad, up to 30 mm in height and 25 mm broad. Umbilicus small or closed. Single prominent tooth on inner margin of aperture. Greyish-coloured shell with purplish markings; whitish and iridescent around umbilicus; apex often eroded and silvery. Foot with 3 pairs of epipodial tentacles.

O. lineatus is a southern species which suffered heavy mortalities in the cold winter of 1962–1963, particularly in populations at the northern limit of distribution in Britain. Since that time the populations have recovered and there is some evidence to suggest that the northern range of the species, which is North Wales, has been extended in response to climate change. *Osilinus* is common on south-western shores but is absent from the eastern English Channel and the North Sea. It is found on rocks and boulders from the upper to the lower shore and, as with other top-shells, its food is predominantly microorganisms and detritus. The sexes are separate, fertilization is external and breeding occurs during summer. Length of larval life is a few days and settlement occurs on the upper shore where the juvenile snails can be found on the undersurface of stones and boulders. Longevity may be 10 years or more and growth checks can be seen on the shell, particularly in specimens from the northerly part of the species range. With experience these can be used to determine the age of individual snails.

Family Calliostomatidae
The family includes a number of species of the genus *Calliostoma*, previously grouped in the family Trochidae, one of which is fairly common on the lower shore.

Calliostoma zizyphinum (Linnaeus) Painted top-shell **Fig. 12.14a**
Conical shell with straight-sided whorls; up to 30 mm in height and 30 mm broad; umbilicus not obvious in adult. Greyish-pink colour with reddish streaks; occasionally all white. Foot with 4 or 5 pairs of epipodial tentacles.

Bittium reticulatum (da Costa) Needle whelk **Fig. 12.15b**
Shell tall, narrow with 10–15 whorls; patterned with many regularly arranged tubercles; up to 10 mm in height, occasionally larger. Aperture with short basal extension. Brown in colour.

B. reticulatum is widely distributed in north-west Europe, occurring in Britain mainly in the south and west. It lives on sandy and muddy shores, especially where *Zostera* (p. 75) is present, as well as on rocks and stones. Found from the lower shore sublittorally to depths of about 250 m, it feeds on a variety of microorganisms. Breeding occurs in summer. The sexes are separate but details of fertilization are not known. Whitish eggs (up to 1000) are laid in a long, coiled ribbon approximately 24 mm long and 4–5 mm broad. After a few days the eggs hatch as free-swimming veliger larvae which are common in the plankton in spring and summer. The length of planktonic life has not been determined but after settlement the snails live for one or two years.

Family Littorinidae Genus *Lacuna* Chink shells
Shell with prominent spire, or shell flattened and spire reduced. Aperture without siphonal canal. With or without umbilicus: the species described here have an umbilicus and a groove leading to it. Posterior of foot with 1 pair of metapodial tentacles (Fig. 12.5d).

1 Shell with prominent spire, 5–6 whorls; groove leading to umbilicus smooth. Usually with broad bands of red-brown colour .... *Lacuna vincta* p. 212
   Shell with spire of 3–4 whorls; body whorl very large. Groove leading to umbilicus with longitudinal ridges **2**

2 Shell flattened, spire reduced; aperture very large. Usually green colour; white specimens present in some populations .... *Lacuna pallidula* p. 213
   Shell with low spire; aperture large. Brownish in colour, sometimes with brown bands on body whorl .... *Lacuna parva* p. 213

Lacuna vincta (Montagu) Banded chink shell **Fig. 12.16a**
Shell of 5–6 whorls; with prominent spire; up to 10 mm in height; narrow umbilicus, groove leading to it smooth. Cream or brown in colour, most specimens with red-brown bands.

L. vincta is widely distributed in north-west Europe and extends as far south as the Channel coast of France. It is most abundant in the sublittoral and is found all around Britain on semi-exposed to sheltered rocky shores, always with seaweed on which it feeds. *L. vincta* migrates onto the lower shore to breed, generally from January through to summer. The sexes are separate; following copulation and internal fertilization, ring-shaped egg masses, white, yellow or pale green in colour, are laid on seaweeds such as *Fucus serratus* (p. 63) and laminarians (p. 57) and on *Zostera* (p. 75). A single egg mass may contain more than 1000 eggs. After two to three weeks free-swimming veliger larvae emerge and these spend two or three months in the plankton before settling. Length of life is no more than one year.

Figure 12.16 (a) *Lacuna vincta* with egg masses on *Fucus serratus*. (b) *Lacuna pallidula* with egg masses on *Fucus serratus*.

Lacuna pallidula (da Costa) Chink shell **Fig. 12.16b**
Shell of 3–4 whorls, flattened; spire reduced, body whorl and aperture very large: umbilicus deep, with groove leading to it ridged longitudinally. Usually green in colour but white shells present in some populations. Females up to 8 mm in height, about twice as large as males.

L. pallidula is widely distributed in north-west Europe and extends as far south as the west coast of France. It is found all around Britain on the lower shore on semi-exposed to sheltered rocky shores, most often on *Fucus serratus* (p. 63), on which it feeds. It occurs in the sublittoral to depths of about 70 m. The sexes are separate and, following copulation and internal fertilization, round to oval egg masses are laid on the seaweed. They are white in colour and attached firmly to the weed, each containing up to 100 or so eggs. Breeding occurs during winter and spring and after eight to 14 weeks, depending on temperature, crawling young emerge from the egg mass. Length of life is no more than one year. The egg masses of *L. pallidula* resemble those of the flat periwinkles *Littorina obtusata* (p. 216) and *L. fabalis* (p. 217). Egg masses of the latter two species, however, are usually oval to kidney-shaped in outline, easily peeled from the surface of the seaweed (or in some cases the rock surface) and contain many more eggs than those of *Lacuna pallidula*.

Lacuna parva (da Costa) **Fig. 12.17**
Shell of 3 whorls, with short spire and large body whorl; up to 4 mm in height. Umbilicus deep, groove leading to it with longitudinal ridges. Shell brownish in colour; body whorl sometimes banded.

Figure 12.17 *Lacuna parva.*

L. parva is widely distributed in north-west Europe and is often found along with *L. vincta* (p. 212). Like other lacunids it is found on seaweeds on which it feeds. The sexes are separate and clear, spherical egg masses are laid on seaweeds. Longevity is one year.

Family Littorinidae Genera *Littorina* and *Melarhaphe* Periwinkles Shell with obvious spire or relatively flat (2 species). The body whorl is large, accounting for 75% or more of the shell height. Shell smooth to strongly ridged with spiral ridges. Without siphonal canal; without umbilicus. Shell colour varied, black, grey, yellow or olive. A large group of common and widespread intertidal molluscs found at all levels on rocky shores. They are often abundant on the upper levels of the shore where they are important grazers. Some species live on soft sediments and extend into estuaries.

The periwinkles show a wide range of reproductive strategies: pelagic larvae feed and develop in plankton before settlement (planktotrophy, pelagic development); embryos held in an internal brood chamber to crawl-away stage (ovoviviparity, direct development); embryos laid in gelatinous masses on seaweed and rocks, and from which juvenile snails emerge (oviparity, direct development) (see Fig. 12.18).

Figure 12.18 Life-cycle strategies in the genus *Littorina*.

For detailed information on the genus *Littorina*, reference to **Reid (1996)** is recommended.

1 Shell with distinct spire **2**
Shell flattened; spire small **4**

2 Transparent flap-like extension (the periostracum) projects beyond edge
of outer lip of shell (Fig. 12.22). Shell no more than about 9 mm in
height .... ***Melarhaphe neritoides*** p. 219
Shell without transparent flap-like extension projecting beyond outer lip **3**

3 Shell up to 30 mm in height. Outer lip joins body whorl tangentially (Fig. 12.19a).
Columellar lip white in small specimens but not obviously so in some larger
specimens. Inner surface of outer lip of small specimens with alternate light
and dark bands. Juveniles with shells less than about 7 mm in height, heavily
ridged .... ***Littorina littorea*** p. 215
Shell up to 20 mm in height, smooth to strongly ridged. Outer lip of aperture joins
body whorl at about a right angle (Fig. 12.21a,b) ... ***Littorina saxatilis*** group p. 217

4 Shell with flattened spire. Outer lip joins body whorl a little below apex. Large
aperture. Outer lip thickened in older specimens .... ***Littorina obtusata*** p. 216
Shell has little if any spire. Outer lip joins body whorl at about level of
apex (Fig. 12.20b). Outer lip thickened in older specimens. Aperture small
.... ***Littorina fabalis*** p. 217

Littorina littorea (Linnaeus) Edible periwinkle **Fig. 12.19**
*Shell with faint spiral lines on whorls but surface generally smooth; apex pointed; up to
30 mm in height. Outer lip of aperture joins body whorl tangentially. Dark grey or black*

Figure 12.19 *Littorina littorea*, (a) adult, (b) young specimen, 4 mm shell height. Note angle at which outer lip joins body whorl (*).

in colour, occasional specimens red; columella white. In small specimens inner surface of outer lip has alternate light and dark bands. Tentacles with black transverse bands.

L. littorea is widely distributed on rocky coasts in north-west Europe and all around Britain, from the upper shore into the sublittoral. It also occurs in estuaries and on mud-flats. *Littorina* grazes on microorganisms and detritus, and is also known to feed on green seaweed such as *Ulva* (p. 30), but apparently rejects *Ascophyllum nodosum* (p. 61). The sexes are separate and on the west coast of Wales the gonads develop during winter. Copulation takes place during early spring and after internal fertilization the eggs are shed into the sea in gelatinous capsules, usually containing up to three eggs, occasionally more. Egg capsule release is synchronized with the time of spring tides. The capsules are pelagic and after several days, free-swimming veliger larvae hatch which, after a pelagic life estimated to be as long as six weeks, settle on the lower shore among barnacles and in crevices, usually during June and July. Generally, shells up to about 7 mm in height are heavily ridged with a crenulate margin to the outer lip and can be confused with the young of the *Littoria saxatilis* group (p. 217). However, they can be identified by other characteristics given above, particularly the alternate light and dark banding on the inner surface of the outer lip of the shell. Maturity is generally reached at two or three years of age at a shell height of about 12 mm. Longevity is five or more years. The boring polychaete *Polydora ciliata* (p. 187) and the boring sponge *Cliona* (p. 85) often infest the shell of *L. littorea* and this is now known to weaken the shell and make the winkle more vulnerable to one of its major predators, the shore crab, *Carcinus maenas* (p. 366). The edible periwinkle has some commercial value; it is gathered by hand.

Littorina obtusata (Linnaeus) (*Littorina littoralis*) Flat periwinkle **Fig. 12.20a; Plate 20a**
Shell with flattened spire; large aperture; up to 15 mm in height. Outer lip joins body whorl a little below apex of shell; thickened in older specimens. Colour varied, usually olive-green, but yellow, brown, banded and tessellated (criss-cross patterned) varieties also common.

L. obtusata is widely distributed in north-west Europe. It occurs all around the coast of Britain and is almost always found in association with seaweed, particularly *Ascophyllum nodosum* (p. 61), *Fucus vesiculosus* (p. 64) and *Fucus serratus* (p. 63). It grazes on these large seaweeds and it has been shown that it prefers *Ascophyllum*, even though this is known to produce chemicals that deter most grazers. Shell colour is related to the degree of exposure of the shore to wave action, olive-green shells being dominant on sheltered shores and tessellated brown shells on exposed shores. These differences are believed to be maintained by visual selection of predators such as shore fishes. The sexes are separate and fertilization is internal following copulation. Whitish egg masses, usually oval or kidney-shaped in outline (Fig. 12.20a; Plate 20a) and containing up

Figure 12.20 (a) *Littorina obtusata* with egg masses on *Fucus serratus*. (b) *Littorina fabalis*. Detail of penis shown for each species.

to about 280 eggs, are laid on *Ascophyllum, Fucus vesiculosus* and *F. serratus*, and occasionally on the rock surface. They are easily confused with those of *Lacuna pallidula* (p. 213) but are more easily removed from the weed. Breeding occurs during most of the year; generally there is a peak in spring and early summer. Development within the egg mass takes up to four weeks and crawling young emerge. Sexual maturity is reached at two years of age and longevity is probably three or more years.

Littorina fabalis (Turton) (*Littorina mariae*) Flat periwinkle **Fig. 12.20b**
Shell with flattened, if any, spire; very similar to L. obtusata *(above) but flatter; up to 10 mm in height. Outer lip joins body whorl at about level of apex; thickened in older specimens giving a relatively small aperture. Colour varied, usually yellow or brown tessellated pattern.*

L. fabalis has a similar distribution to that of *L. obtusata* and is found all around the coasts of Britain, usually on *Fucus serratus* (p. 63), on which it feeds by grazing epiphytes from the surface of the weed. It tolerates more exposed situations than *L. obtusata* and occurs lower on the shore, but there is some overlap in the vertical distribution of the two species. Shell colour is varied; yellow shells are dominant on sheltered shores; criss-cross brown shells on exposed shores and, as with *L. obtusata*, these colours are believed to be maintained by visual selection of predators such as shore fishes. In some populations on sheltered shores, juvenile snails up to a diameter of 3.5 mm are often white in colour and it has been suggested that white colour at this size mimics the tube worm *Spirorbis* (p. 178), which is found attached to *F. serratus*. Shell colour changes with increase in shell height and this strategy is believed to reduce predation rates by shore fishes such as the shanny (p. 464). The sexes are separate and fertilization is internal; egg masses are laid on the fronds of seaweed, particularly *F. serratus*. They are very difficult to distinguish from those of *L. obtusata* (above) and are found throughout the year, predominantly in spring and summer. *L. fabalis* is believed to be an annual.

Some specimens of *Littorina obtusata* and *L. fabalis* can be difficult to separate. In addition to the characters given above, males can be separated on the morphology of the penis, which has a very long tip in *L. fabalis* (Fig. 12.20).

The *Littorina saxatilis* group
Littorina saxatilis, commonly known as the rough periwinkle, has a very wide geographic distribution and is characterized by extensive intraspecific variation. This has led to a history of taxonomic confusion and a literature rich in specific names and varieties; reference to the '*Littorina saxatilis* species complex' and the '*Littorina saxatilis* group' is widespread. The group was widely considered to comprise four species, namely *L. saxatilis*, *Littorina nigrolineata*, *Littorina neglecta* and *Littorina arcana*. Members of the group have different reproductive strategies but all have direct development with the opportunity for adaptation to local habitats and the establishment of ecotypes. Research using molecular techniques suggests that *L. neglecta* is not a distinct species but is an ecotype of *L. saxatilis*. This is the position adopted here and three species are described; *Littorina compressa* (= *L. nigrolineata*), *L. saxatilis* and *L. arcana*.

Littorina compressa Jeffreys (*Littorina nigrolineata*) **Fig. 12.21a**
Shell up to 20 mm in height; distinct spiral ridges on whorls separated by narrow grooves; ridges sometimes flattened. Outer lip of aperture joins body whorl at about a

Figure 12.21 (a) *Littorina compressa*. (b) *Littorina saxatilis*. Note angle at which outer lip joins body whorl (*).

right angle. Background shell colour may be white, yellow or reddish-brown; tessellated shells rare. The grooves are clearly pigmented (purplish-brown) in some populations, while in others the pigmentation may be faint or absent. When pigmentation absent, confusion with L. saxatilis *may result.*

L. compressa has a patchy distribution in north-west Europe extending from the Atlantic coast of France to northern Norway. It is widely distributed around Britain but is absent from southern North Sea coasts and the eastern English Channel. It is found on the middle to upper shore and is most common on sheltered shores, often in association with *L. saxatilis* (below), but is much less abundant than the latter species and does not extend into saltmarshes. It browses on a wide variety of microorganisms and detritus. *L. compressa* is oviparous. The sexes are separate, fertilization is internal and pinkish egg masses, each containing 100–200 eggs, are laid under stones, in crevices and other damp situations. Although egg masses are found throughout the year, there is a peak in May and June. After about four to seven weeks, depending on temperature, crawling young emerge. Length of life is up to five years.

Littorina saxatilis (Olivi) (*Littorina rudis*) Rough periwinkle **Fig. 12.21b**
Shell with pronounced spire; up to 18 mm in height, but generally smaller. Whorls with spiral ridges. Outer lip of aperture joins body whorl at about a right angle. Colour varied, yellowish-white, greenish, reddish, brown, sometimes banded, tessellated.

L. saxatilis is widely distributed in north-west Europe and around Britain on the upper reaches of shores of all grades of exposure to wave action. It is found in crevices, on stones and rocks, and with seaweeds such as *Fucus spiralis* (p. 63), occurring on mud-flats and in estuaries, in salinities down to about 8. Variation in shell colour from one population to the next is believed to be maintained, at least in part, by visually hunting predators, the different colours being camouflaged with the colour of the rock substratum. The rough periwinkle feeds on microorganisms, detritus and seaweed. The sexes are separate, fertilization is internal and the embryos develop to shelled, crawling young in a brood chamber within the body of the female (ovoviviparity). Females with embryos are found throughout most of the year, with a peak of breeding activity in late spring and early summer when the brood chamber of a single female may contain more than 100 embryos at different stages of development. Longevity is up to six years.

Small specimens of *L. saxatilis* are commonly found in the empty cases of barnacles and these include snails formerly known as **Littorina neglecta** Bean, now recognized as a barnacle-dwelling ecotype of *L. saxatilis*. They have ovoviviparous development and one of the adaptations to the barnacle habitat is the attainment of sexual maturity at a very small size; specimens of 2–3 mm shell height can be sexually mature and carry as many as 50 or more embryos in the internal brood chamber.

Littorina arcana Hannaford Ellis (not illustrated)

Shell with pronounced spire; up to 17 mm in height. Whorls with spiral ridges. Outer lip of aperture joins body whorl at about a right angle. Compared with L. saxatilis *(above), aperture large. Colour highly varied, whitish, yellow, brown to dark brown almost black; sometimes banded or tessellated.*

It is almost impossible to distinguish the shells of *L. arcana* from those of *L. saxatilis*. Both are highly variable and considerable field experience is required before differences can be detected. Study of the anatomy and reference to Hannaford Ellis (1979) is necessary for confirmation of the species.

L. arcana was first described in 1978 and for this reason, and confusion with *L. saxatilis*, records of its distribution are likely to be incomplete. It has been recorded from northern Norway to Brittany, and although it has a wide distribution around Britain it appears to be absent from southern North Sea coasts and the eastern English Channel. It has a patchy distribution on the west coast of Scotland. *L. arcana* is found from the upper middle shore to the littoral fringe on rocky shores and is tolerant of a high degree of exposure to wave action, coexisting on some shores with *Melarhaphe neritoides* (below). It is less common on boulder beaches and sheltered rocky shores and has not been recorded from saltmarshes. It browses on a wide variety of microorganisms and detritus. *L. arcana* is oviparous. The sexes are separate, fertilization is internal and gelatinous egg masses, up to 9 mm in length and containing up to 250 pinkish coloured eggs, are laid in crevices and damp situations on the upper shore. From these egg masses crawling juveniles emerge. In Britain, spawning would appear to take place over most of the year but at a reduced level in summer.

Figure 12.22 *Melarhaphe neritoides*. Note periostracum and angle at which outer lip joins body whorl (*).

Periostracum

Melarhaphe neritoides (Linnaeus) (*Littorina neritoides*) Small periwinkle Fig. 12.22

Shell with pointed apex; up to 9 mm in height. Outer lip of aperture joins body whorl tangentially. Blackish-brown in colour; transparent, flap-like extension (the periostracum), projects beyond edge of outer lip. Often with pale band on base of body whorl, running into aperture.

M. neritoides is widely distributed in north-west Europe and occurs around most of Britain, although absent from the southern North Sea. It is found in cracks and crevices on rocky shores, high in the littoral fringe, its upper limit increasing with increasing exposure to wave action. In some areas it can be found many metres above the height of the highest spring tides, where it is wetted only by spray. Microorganisms, detritus and lichens are the main food, and the snails browse over the rock surface during moist conditions. The sexes are separate, fertilization is internal and breeding generally takes place during winter and spring, although in some years many snails do not reproduce. Egg capsules, each containing one egg, are liberated fortnightly on spring tides. They are pelagic and after several days free-swimming veliger larvae emerge and spend about three weeks in the plankton before settling on the shore in cracks, crevices and empty barnacle shells. On the south coast of Ireland shell heights of less than 2 mm were recorded after the first year's growth. Longevity is five years and more but some of the larger specimens are probably considerably older.

Family Rissoidae

Small shell with spire, usually less than about 6 mm shell height and never greater than 9–10 mm. Shell smooth to heavily ridged. Variously coloured, white, cream, orange, brown. Aperture without siphonal canal. Umbilicus may be present as a narrow slit. Predominantly found on rocky shores where they are sometimes abundant under stones and among seaweed, a feature which is useful in distinguishing them from the Hydrobiidae (p. 221).

A very large family of marine snails found in littoral and sublittoral habitats. Only two species are described here. For further information the reader is referred to **Graham (1988)**.

Cingula trifasciata (J. Adams) (*Cingula cingillus*) **Fig. 12.23a**

Shell up to 4 mm in height, smooth, but with spiral lines on body whorl; yellowish-brown in colour, with conspicuous, dark brown bands. Some specimens lack the dark bands and are white. Black specimens also occur.

C. trifasciata is widely distributed in north-west Europe and around Britain, occurring in rock crevices and among barnacles on the upper shore along with *Melarhaphe neritoides* (above), small *Littorina saxatilis* (p. 217), *Otina ovata* (p. 249) and *Lasaea adansoni* (p. 266). It is also found among seaweed and under stones on the lower shore, and in the sublittoral to depths of about 20 m. *Cingula* feeds on detritus. Breeding takes place in spring and summer. The sexes are separate, fertilization is internal and rounded egg capsules are deposited in crevices. Each capsule contains up to four eggs from which crawling young emerge.

Rissoa parva (da Costa) **Fig. 12.23b,c**

Shell up to 4 mm in height; some shells smooth, others ribbed with longitudinal ridges but intermediates common; shades of brown; apex purplish; many have a brownish, comma-shaped mark on the body whorl; while useful in identification, this is absent from shells below 2 mm shell height and from many above 2 mm.

R. parva is widely distributed in north-west Europe and on British coasts, often in very high densities. It is found on the middle and lower shore and below on stones, among red seaweeds such as *Corallina officinalis* (p. 43), on *Lomentaria articulata* (p. 46) and in the holdfasts of *Laminaria digitata* (p. 57). The proportion of smooth to ribbed

Figure 12.23 (a) *Cingula trifasciata*. (b) *Rissoa parva*, apertural view and (c) lateral view.

shells varies from one locality to another. The snails feed on microorganisms and detritus. The sexes are separate, fertilization is internal and rounded egg capsules, each containing up to 50 eggs, are attached to seaweed. Veliger larvae take about three weeks to develop and probably have a long pelagic life before settling on red seaweed on the lower shore. Breeding and settlement of young apparently take place over the whole year. Length of life is up to one year, many animals living only a few months.

Family Hydrobiidae Mud snails

Shell smooth (but see the keeled form of *Potamopyrgus antipodarum*, p. 224); with spire; up to about 6 mm in height. Brownish-black in colour. Aperture without siphonal canal. With or without umbilicus but in the species described here the umbilicus is reduced to a narrow slit or closed. Found on estuarine muds and sands, often in very high densities, some species extending into brackish and fresh water. Different species of *Hydrobia* often coexist and care is needed in their separation. Not found on rocky shores.

1 Shell whorls slightly swollen, brown-yellow colour but often black with surface deposits. Some specimens have a keel and bristles on the middle of the whorls. Tentacles blue-grey, sometimes with pale line along centre. Found in fresh- or brackish-water habitats .... ***Potamopyrgus antipodarum*** p. 224
Shell whorls slightly swollen or almost straight-sided; without keel or bristles. Tentacles with or without a dark pigment patch near tip. Found in brackish to fully saline sea water **2**

2 Shell whorls almost straight-sided; apex blunt. Tentacles pale with dark pigment patch at a distance greater than its own length from tip. On mud-flats and in estuaries, often in

very high densities. During spring and summer clusters of eggs covered by sand grains are attached to many shells .... ***Hydrobia ulvae*** p. 222
Shell whorls slightly swollen. Tentacle pigmentation not as above. Usually in lagoons and ditches without direct contact with the sea **3**

3 Tentacles pale, occasionally with a faint, terminal black streak
.... ***Ventrosia ventrosa***[1] p. 224
Tentacles pale, with dark pigment patch at a distance less than its own length from tip .... ***Hydrobia acuta neglecta***[2] p. 224

Hydrobia ulvae (Pennant) Laver spire shell **Fig. 12.24a; Plate 28a,c,d**
Shell small, up to 6 mm in height; whorls almost straight-sided; apex blunt. Generally light brown to brownish-black in colour. Tentacles pale with dark pigment patch at a distance greater than its own length from tip.

H. ulvae is widely distributed in north-west Europe and around Britain on muddy-sand, mud-flats and in estuaries, often in densities as high as 20 000 individuals per square metre. It is most abundant on the middle and upper reaches of the shore in salinities ranging from 10 to 33 but can tolerate salinities as low as 1.5. When covered by the rising tide, the snail secretes a mucous raft and floats upside down in the surface film of water. The floating phase lasts for a short time during tidal cover. When the tide is out, *Hydrobia* is often seen climbing vertical objects such as the plants *Salicornea* (p. 75) and *Spartina* (p. 77) with which it is often found, and from which it launches itself on the rising tide. The food of *Hydrobia* consists mainly of silt, diatoms and fungal mycelia removed from the surface of sediment particles by the action of the radula but feeding on detritus trapped in the mucous raft also takes place when the snail is floating. Breeding occurs from spring to autumn. The sexes are separate and fertilization is internal. The female gathers sand grains under the front of the foot using the snout, and as the fertilized eggs are laid, groups of four to eight (occasionally more) are covered by sand grains and mucus to form a protective egg mass which is cemented onto solid surfaces such as the shells of neighbouring individuals. On muddy-sand habitats up to 40% of *Hydrobia* carry one or more egg masses on the shell during the breeding season whereas on mud-flat habitats the percentage is very much lower, and in some populations egg masses on the shell are difficult to find. In these situations egg masses are sometimes found on other hard surfaces such as shell fragments or the stems of *Spartina* (p. 77) but most are found free in the muddy sediment. It is believed that the differences between the two habitats reflect the fact that the mud-flat is a more stable environment than the sand-flat and there is little

[1] Some specimens of *Ventrosia ventrosa* and *Hydrobia acuta neglecta* may be difficult to separate. In these cases males of the two species can be separated on the shape of the penis, which is shown in Fig. 12.24b,c. The penis of *Hydrobia ulvae* is included for comparison (Fig. 12.24a).
[2] Recent phylogenetic research suggests that the Mediterranean species *Hydrobia acuta* comprises two subspecies, *Hydrobia acuta acuta* (Draparnaud) which is widespread in the western Mediterranean, and *Hydrobia acuta neglecta* Muus (formerly *Hydrobia neglecta*) which is found in northern Europe.

Figure 12.24 (a) *Hydrobia ulvae* with egg mass on shell. (b) *Hydrobia acuta neglecta*. (c) *Ventrosia ventrosa*. (d) *Potamopyrgus antipodarum*, specimen with keel inset. Detail of pigmentation of head and tentacles shown for each species; detail of penis shown for (a), (b) and (c). (Partly based on Muus, 1963.)

likelihood of egg masses laid freely in the sediment being washed away by the tide. At some localities peaks of spawning have been recorded in both spring and autumn while in other areas there is a single peak in summer. Within the egg mass the embryos develop into veliger larvae which emerge after 20 to 30 days. They are highly active, free-swimming larvae which can be collected in large numbers from the surface plankton during high tide. After a pelagic life of about three weeks they metamorphose and young snails settle on the shore. Sexual maturity is reached after one or two years and length of life is usually about two and a half years.

Hydrobia acuta neglecta Muus (*Hydrobia neglecta*) **Fig. 12.24b**

Shell similar to that of V. ventrosa *(below) but whorls not as rounded; apex blunt; up to 4 mm in height. Grey, brown, or black in colour. Tentacles pale, with dark pigment patch at a distance less than its own length from the tip.*

H. acuta neglecta has a patchy distribution in north-west Europe where it appears to be mostly restricted to the eastern English Channel, the coast of East Anglia, and Denmark. It is found in lagoons and ditches without direct contact with the sea, usually in salinities of 10–24 but is not abundant. It lives on sandy and muddy sediments and on *Zostera* (p. 75), feeding on diatoms and bacteria. The sexes are separate, fertilization is internal and egg capsules, similar to those of *V. ventrosa* but containing only one embryo, are laid in spring and summer on stones and empty shells. After about three weeks, crawling young emerge and the length of life is about 18 months. (see footnote 2, p. 222.)

Ventrosia ventrosa (Montagu) (*Hydrobia ventrosa*) **Fig. 12.24c**

Small shell, up to 5 mm in height; whorls rounded, apex blunt. Brownish-black in colour. Great majority have pale tentacles, occasionally there is a faint, terminal black streak.

V. ventrosa is widespread on sand and muddy deposits in north-west Europe, but much less so than *H. ulvae* (p. 222), having a patchy distribution on British coasts, where it is usually restricted to sheltered lagoons and ditches that are almost cut off from the sea. It is normally found in salinities of 6–20. Diatoms and bacteria removed from the surface of sediment particles are the main food. Breeding occurs in spring and summer. The sexes are separate, fertilization is internal and egg capsules, protected by a covering of sand grains, are laid on empty shells, stones and occasionally on the shells of neighbouring *V. ventrosa*. The capsules contain one, two or rarely three eggs and development takes about three weeks. Crawling young emerge from the capsule and length of life is about 18 months.

Potamopyrgus antipodarum (J. E. Gray) (*Potamopyrgus jenkinsi*) **Fig. 12.24d**

Brownish-yellow shell, but often black due to surface deposits; whorls slightly swollen; up to 5 mm in height. Some shells have a keel and bristles on the middle of the whorls. Tentacles blue-grey, sometimes with pale line along centre.

P. antipodarum is native to New Zealand and was introduced to Britain in the nineteenth century, being recorded in Gravesend in 1859. Since then it has spread rapidly in Europe and around Britain although it is rare in parts of mid-Wales and Scotland. It is found in fresh- and brackish-water habitats (more especially running water), with a salinity range from 0 to 20. It tends to occur in high densities and would appear to have a preference for hard water. It feeds on diatoms and detritus. Generally, all individuals in a population are female and reproduce parthenogenetically; the young undergo development in a brood chamber and leave the parent as crawling individuals. In a few populations males have been recorded, the first in 1958 from the river Thames at Sonning. Males are present in low numbers and although they may have mature sperm, their precise role in reproduction has not been determined. Hermaphroditic animals have been recorded in very low numbers. In north-west England, breeding has been recorded all year, and longevity is about seven months.

Family Aporrhaidae Pelican's foot

Shell tall, with 8 or so whorls. Outer lip drawn out into large pointed expansion to give appearance of pelican's foot. Without umbilicus. Aperture with siphonal canal.

Figure 12.25 *Aporrhais pespelecani.*

Aporrhais pespelecani (Linnaeus) Pelican's foot **Fig. 12.25**
Shell of distinctive appearance, up to 40 mm in height; whorls with longitudinal ridges and tubercles. Outer lip of adult expanded outwards to give appearance of pelican's foot. Greyish-white in colour with brown markings.

A. pespelecani is common in north-west Europe in mud, muddy-gravel and muddy-sand offshore to depths of about 180 m. Empty shells are often washed ashore. *Aporrhais* eats plant debris, feeding when buried in the sediment and when on the surface. Breeding has been recorded in spring. The sexes are separate and fertilization internal, with the eggs deposited either singly or in groups of two or three on sand grains and debris. After about two weeks, free-swimming veliger larvae emerge; they are believed to have a lengthy pelagic life. The sipunculan *Phascolion strombus* (p. 378) sometimes inhabits empty shells of *Aporrhais*.

Family Calyptraeidae Slipper limpets

Shell conical or cap-shaped with internal shelf extending across part of aperture. Without operculum.

Shell oval, cap-shaped; aperture very large; occurs in chains, large specimens at base, smaller ones above .... *Crepidula fornicata* p. 225

Shell conical, with pointed apex; not forming chains ... *Calyptraea chinensis* p. 226

Crepidula fornicata (Linnaeus) Slipper limpet **Fig. 12.26**
Oval, cap-shaped shell with much-reduced spire; surface marked with growth lines; up to 25 mm in height, 50 mm in length. Aperture very large with shelf extending about halfway across. Orange-brown, white or yellow in colour; shelf white. No operculum. Occurs in chains with large specimens at base, smaller ones above.

C. fornicata is a North American species that is believed to have entered Britain on imported oysters, the first specimens being recorded in Essex in about 1890. It is widely distributed in north-west Europe from Norway to Spain and in Britain is found on south, south-west and south-east coasts and appears to be spreading northwards. High densities of *Crepidula* on oyster beds cause much damage as the oysters are smothered. Curved chains of up to 12 animals can be found on the lower shore and sublittorally to about 10 m, often attached to stones, and the shells of oysters and mussels. They feed on small particles which are drawn into the mantle cavity in a water current and trapped on a mucous filter.

Figure 12.26 *Crepidula fornicata*, (a) lateral view, (b) apertural view.

Crepidula is a protandrous hermaphrodite; in a chain of individuals the smaller, younger animals at the top of the chain are male and the larger, older ones at the base of the chain are female. It has a long reproductive period and high fecundity. Copulation is followed by internal fertilization and the eggs are laid in stalked capsules attached to the foot or shell of the female, or to a stone or shell under the female, thereby affording them some protection. Development to the veliger stage takes three to four weeks and after a pelagic life estimated to be about five weeks, the juveniles settle under stones and shells. By the time they have reached a shell length of 3–5 mm, they migrate and take up position on an existing chain of *Crepidula* as males. Others not locating a chain settle on stones and develop as females. Breeding occurs from February/March to September; most females spawn twice a year and it has been suggested that spawning is correlated with neap tides. Length of life is about eight to nine years.

Calyptraea chinensis (Linnaeus) Chinaman's hat **Fig. 12.27**
Shell conical; apex sharp, sometimes coiled; surface smooth apart from near aperture where small tubercles occur; up to 5 mm in height, 15 mm in length. Shelf extending about halfway across aperture. External surface of shell yellowish-white in colour; shelf white. No operculum.

Figure 12.27 *Calyptraea chinensis*, (a) lateral view, (b) apertural view.

C. chinensis reaches its northern limit in Britain, where it is found on the west coast as far as Scotland, but is absent from the North Sea. It is a lower-shore and shallow sublittoral species that lives attached to shells and stones. Small particles of food are trapped on a mucous filter. It is a protandrous hermaphrodite; the smaller, younger, more active animals are male and settle temporarily on the shells of the older, larger, more sedentary females, but do not form chains, cf. *Crepidula* (above). Copulation is followed by internal fertilization. Eggs are laid in capsules that are roughly triangular in outline and attached to a stone under the front of the shell of the female. A single capsule contains up to 25 eggs. There is no veliger larva; crawling young emerge. Breeding occurs from spring to summer and length of life is three years.

Family Triviidae Cowries

Shell glossy; smooth or strongly ridged; aperture elongate, narrow; body whorl very large, may have short spire or be typical cowrie shape. Without operculum.

Shell glossy, oval in shape, with 3 dark spots; surface conspicuously marked with ridges; aperture long, narrow; typical cowrie shape .... ***Trivia monacha*** p. 227

Shell as above but without the 3 dark spots .... ***Trivia arctica*** p. 228

Note: the shell of young specimens of both species, up to a shell height of about 5–6 mm, is smooth with a wide aperture and a short spire.

Trivia monacha (da Costa) European cowrie, spotted cowrie **Fig. 12.28**

Glossy shell, oval in shape with long, narrow aperture; surface conspicuously marked with ridges; up to 13 mm aperture length. Upper surface reddish-brown with 3 brown spots; lower surface white. Mantle brightly coloured, extending to cover shell when animal is active. Operculum absent. Young shells with short spire, wide aperture and lacking the characteristic ridging of larger specimens.

(a)   (b)

Figure 12.28 *Trivia monacha*, (a) dorsal view, (b) apertural view.

T. monacha is a southern species, extending northwards to the west coast of France, and around most of Britain. It is found on the lower shore and sublittorally on rocky shores with compound ascidians such as *Botryllus schlosseri* (p. 441) and *Botrylloides leachii* (p. 442) on which it feeds. Breeding occurs during late spring and summer. The sexes are separate, fertilization is internal and flask-shaped egg capsules, each with up to 800 eggs, are deposited in holes eaten out of the ascidian by the female. The tip of the egg capsule is button-shaped and can be seen protruding from the surface of the ascidian. Free-swimming larvae emerge after a few weeks and are pelagic for a month or so.

Trivia arctica (Pulteney) **(not illustrated)**
Shell as described for T. monacha *but without the 3 dark pigment spots. Up to 10 mm aperture length.*

T. arctica is widely distributed in north-west Europe and although predominantly found in the sublittoral it is occasionally found along with *T. monacha* on the very low shore. It feeds on ascidians and its life-cycle is similar to that of *T. monacha.*

Family Velutinidae
In the species described here, the shell is internal and enclosed by mantle. Similar in appearance to sea-slugs (Opisthobranchia, p. 235).

Lamellaria perspicua (Linnaeus) **Fig. 12.29**
Shell totally enclosed by mantle; spire very small; smooth; colourless and transparent; aperture very large; up to 10 mm shell length. Animal up to 20 mm in length; colour varied, yellow, grey, lilac, sometimes with coloured flecks, white, yellow or black. Mantle covered with tubercles and thickened at edge; anteriorly forms a siphon-like extension. Tentacles smooth. May be mistaken for an opisthobranch mollusc.

Figure 12.29 *Lamellaria perspicua.*

L. perspicua is widely distributed in north-west Europe and around Britain on the lower shore and in the sublittoral, among stones and rocks where compound ascidians such as *Botryllus schlosseri* (p. 441) occur. It feeds on ascidians, breeding in spring and summer. The sexes are separate, fertilization is internal and eggs are laid in flask-shaped egg capsules deposited in holes eaten out of the ascidians by the female. Each capsule contains up to 3000 eggs. Pelagic larvae emerge after about three weeks and probably have a long pelagic life.

A second species, *Lamellaria latens* (O. F. Müller), has a similar distribution to *L. perspicua* but can be recognized by its small size, no more than 10 mm in length, and a flatter mantle surface with fewer tubercles.

Family Naticidae Necklace shells
Smooth shell with short spire. Body whorl very large. Aperture large and without siphonal canal. The foot is large and covers head and part of shell when animal is

Figure 12.30 (a) *Euspira catena*, and (b) its egg mass. (c) Shell of bivalve mollusc bored by *Euspira*.

active. The two species likely to be encountered intertidally are found on mud and sand on the very low shore; each has a distinct umbilicus.

Euspira catena (da Costa) (*Lunatia catena, Natica catena, Polinices catena*) Large necklace shell **Fig. 12.30; Plate 22b,c**
Smooth, glossy shell, up to 35 mm in height; short spire; umbilicus large and deep. Fawn to pale orange with a single row of brownish streaks or lines on upper part of whorls. Foot large; covers head and part of shell when animal is active.

E. catena extends as far north as Denmark and is found all around Britain, from the lower shore to about 125 m on sandy sediments. It preys on bivalves such as *Donax* (p. 280), entering the shell by chemical and mechanical means, the shell being softened by a chemical before being bored by the radula. Empty bivalve shells with the characteristic hole made by *Euspira* spp. (Fig. 12.30c) are often washed ashore. The sexes are separate, fertilization is internal and the eggs are laid in the spring in a coiled, greyish-brown mass having the appearance of a collar. Egg masses can be up to 170 mm in length and are often found on the shore. Only a small proportion of the eggs develop; the rest, the nurse eggs, serve as food for the developing embryos. Development is direct; juvenile snails hatch from the egg collar.

   Euspira pulchella (Risso) (*Polinices pulchellus, Natica alderi, Lunatia alderi*) lives in deeper water than *E. catena*, only occasionally being found on the lower shore. It is smaller than *E. catena* (up to 16 mm) and the body whorl has five rows of brown markings. *E. pulchella* produces egg collars from which pelagic veliger larvae hatch, and it has been estimated that off the coast of Anglesey, North Wales it reaches an age of two to three years.

Family Epltoniidae
Shell with spire, up to 15 whorls. Pronounced, well-spaced longitudinal ridges. Aperture rounded. Without siphonal canal.

Figure 12.31 *Epitonium clathrus.*

Epitonium clathrus (Linnaeus)   (*Clathrus clathrus*)
Wentletrap **Fig. 12.31**
Shell tall, narrow; many whorls with pronounced longitudinal ridges; up to 40 mm in height. Aperture rounded. Whitish to pale brown in colour, with blotches of pigment on longitudinal ridges.

E. clathrus is widely distributed in north-west Europe on sandy and muddy sediments. It is a sublittoral species which migrates onto the lower shore during spring and summer to spawn and is sometimes common on south and west coasts of Britain. There is little information regarding its feeding behaviour; it would appear to eat a range of prey including sea-anemones. Breeding occurs during summer. *Epitonium* changes sex each year; fertilization is internal and strings of polygonal egg capsules covered with sand and mud particles are laid on the shore. After about two weeks veliger larvae are released.

Family Muricidae Whelks and tingles
Stout shell with spire. Often heavily ridged. Aperture with siphonal canal: in some species this is closed over in older specimens. Outer lip often thick and sometimes with teeth. Common on rocky shores.

Shell with spiral ridges. Outer lip thin in young specimens, thicker in older shells, sometimes with teeth. Siphonal canal open ⋯⋯⋯⋯ ***Nucella lapillus*** p. 230

Shell with longitudinal and spiral ridges. Siphonal canal open in young specimens, closed in older shells ⋯⋯⋯⋯ ***Ocenebra erinaceus*** p. 232

Nucella lapillus (Linnaeus) Dog-whelk **Fig. 12.32; Plates 18, 20b**
Shell with spiral ridges; usually up to 30 mm in height, may reach 60 mm. Usually whitish-grey but grey-, brown- or yellow-banded forms also occur. Siphonal canal open, short, leads from base of aperture; outer lip of aperture may be thick and toothed.

N. lapillus is widely distributed in north-west Europe and common on the coasts of Britain. It occurs from the middle shore into the shallow sublittoral on rocky shores experiencing most degrees of exposure to wave action, but is not found on very sheltered shores. The shape of the shell and the relative size of the aperture vary with the exposure of the shore. In general, specimens from exposed shores have short, squat shells with a relatively large aperture, while those from more sheltered shores have taller shells with a relatively narrow aperture. These differences have been related both to the direct effect of exposure to wave action and to predation by shore crabs. A large aperture indicates the possession of a large foot, which enables the dog-whelk to cling to the rock surface, a useful adaptation in exposed conditions. A tall shell with a relatively narrow aperture has been shown to be an adaptation against predation by crabs, which are abundant on sheltered shores. Tall narrow shells are not easily

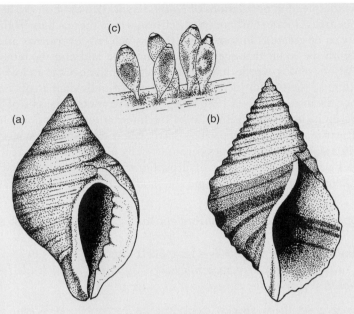

Figure 12.32 *Nucella lapillus.* (a) Specimen with thick, toothed outer lip, (b) specimen with thin outer lip, (c) egg capsules.

handled by the crab and the thick outer lip is difficult to crack open. It has also been shown that shells from sheltered shores are stronger than those from shores exposed to heavy wave action.

The dog-whelk is carnivorous; acorn barnacles and the mussel *Mytilus edulis* (p. 254) are the main prey species. The response of the edible mussel to predation by the dog-whelk is described on p. 255. Adult dog-whelks force apart the opercular plates of barnacles using the snout-like proboscis, whereas bivalves (and sometimes barnacles) are bored by a combination of mechanical and chemical means. Enzymes, particularly carbonic anhydrase, are secreted to soften the shell, which can then be bored by the radula. The prey is believed to be narcotized by secretions from the dog-whelk. The whole process of boring through a shell may take several hours and whelks are often found on the shore when the tide is out in the act of boring into a bivalve (Plate 18b). Cannibalism has also been recorded.

The sexes are separate, fertilization is internal and vase-shaped egg capsules, a few millimetres tall and usually yellow, occasionally purple, in colour are attached in clusters to rock surfaces, in crevices and under overhangs (Plate 20b). A newly deposited egg capsule may contain up to 1000 eggs but no more than a small number of these develop, the others acting as food for the embryos during the first week or so of development. After about four months crawling juveniles emerge. On the east coast of England, breeding occurs mainly in spring; in other areas it is believed to extend throughout the year. Prior to breeding, adult dog-whelks are found in non-feeding aggregations on the lower shore, where copulation takes place. Maturity is reached after about two years and individuals live for five years or more.

The dog-whelk is well known through the development of 'imposex' in response to pollution of inshore waters by tributyltin (TBT) leached from antifouling paints. Imposex is recognized as the development of a penis and vas deferens in female snails and is widespread in populations of *Nucella* in north-west Europe, being particularly evident in areas close to harbours and marinas where the leachate from antifouling paints is likely to be greatest. In some cases the vas deferens overgrows and blocks the genital duct with the result that egg capsules cannot be laid and the female is sterile. Whole populations of dog-whelks have been affected in some areas and where levels of imposex are high, a significant increase in shell size of individual snails has also been recorded. Since the introduction in 1986 of legislation to control the sale of TBT and the now total ban on its use as an antifoulant by the International Maritime Organization, levels of the pollutant in inshore waters have fallen and some recovery of whelk populations in areas where the species became locally extinct has been recorded.

Imposex affects a wide range of prosobranch molluscs, including the edible periwinkle *Littorina littorea* (p. 215) and the mud snail *Hydrobia ulvae* (p. 222).

Ocenebra erinaceus (Linnaeus) (*Oceanebra erinacea*) Sting winkle, rough tingle
Fig. 12.33a

Shell with longitudinal and spiral ridges on whorls; up to 50 mm in height. Whitish in colour with orange-brown, brownish markings. Siphonal canal leads from base of aperture: open in young shells, closed over in older shells.

O. erinaceus is a southern species extending as far north as Britain where it is commonest in the west and south-west. It is found on the lower shore on sheltered rocky shores and on muddy-gravel, extending into the sublittoral to depths of 150 m. It feeds on bivalves and at times can be an important predator of oysters, but it also eats barnacles and tube-dwelling polychaete worms. The method of boring into the prey is similar to that described for *Nucella* (above). Breeding occurs in spring and summer. The sexes are separate, fertilization is internal and yellowish egg capsules, not unlike those of *Nucella* (above) but more angular and with a conspicuous distal plug, are laid in clusters in rock crevices and on stones. They are much less common

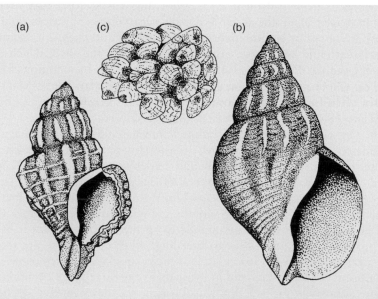

Figure 12.33 (a) *Ocenebra erinaceus*. (b) *Buccinum undatum*. (c) Egg mass of *Buccinum undatum*.

on the shore than those of *Nucella*. A single capsule may contain as many as 150 eggs (but usually far fewer), all of which normally develop into crawling, juvenile snails after about three months.

Family Buccinidae Whelks

Stout, usually large shell with spire; smooth, or ridged with spiral ridges, longitudinal ridges or both. Aperture oval, large; outer lip usually not thickened. Aperture with siphonal canal which is not closed over. A predominantly sublittoral group found on soft sediments.

Large shell, with both spiral and longitudinal ridges, the latter stop short of base of body whorl .... ***Buccinum undatum*** p. 233

Large shell, with many spiral ridges; no longitudinal ridges ....
.... ***Neptunea antiqua*** p. 234

Buccinum undatum Linnaeus Whelk, buckie **Fig. 12.33b,c; Plate 22d**
Stout shell; up to 110 mm in height. Whorls with spiral ridges and pronounced longitudinal ridges, latter stop short of base of body whorl; conspicuous ridge runs from base of siphon along columellar lip. Greyish-white, brownish-white in colour. Aperture large. Short, wide siphonal canal leads from base of aperture.

B. undatum is widely distributed in north-west Europe and Britain, from extreme low water mark into the deep sublittoral, usually on soft sediments. It is one of the largest British prosobranch molluscs and is carnivorous, feeding mainly on polychaetes and bivalves. The bivalve shell is forced open by the whelk using the edge of its own shell. The whelk is also known to use its foot to smother and kill bivalve prey such as the common cockle (p. 269) so allowing it to pull apart the shell valves. Carrion is also eaten and this is detected from some considerable distance by olfaction. The breeding period extends from about October to May, depending on locality. The sexes are separate, fertilization is internal and the eggs are deposited in capsules attached to rocks, stones and shells, often forming a substantial egg mass which can contain as many as 2000 or more egg capsules, each capsule containing up to 1000 or so eggs. Egg masses can be up to 500 mm across and contain capsules from a number of females. Only a small percentage of the eggs complete development, with the majority utilized as food by the embryos. Crawling young emerge from the capsules after several months and length of life has been estimated to be 10 years, with sexual maturity reached at a shell height of about 60 mm. Empty egg masses are often found on the strandline and are sometimes mistaken for sponges. They are known as 'sea wash balls', a name derived from the fact that the early mariners used them for washing. *Buccinum* is fished commercially using baited pots.

Tiny barnacles of the family Trypetesidae (p. 304) bore into the columella of the shells of *Buccinum* and *Neptunea* (below) occupied by the hermit crab *Pagurus bernhardus* (p. 360).

Figure 12.34 *Neptunea antiqua.*

Neptunea antiqua (Linnaeus) Red whelk **Fig. 12.34**
Stout shell, up to about 100 mm in height; whorls with many fine spiral ridges, without longitudinal ridges; conspicuous ridge runs from base of siphon along columellar lip. Cream to reddish in colour. Short siphonal canal leads from base of aperture.

N. *antiqua* is widely distributed in north-west Europe in the sublittoral, usually at depths of from 15 to 1200 m but, as with *Buccinum* (above), empty shells are often found on the shore. It is carnivorous, feeding on a wide range of invertebrate fauna. The sexes are separate, fertilization is internal and breeding occurs in spring. In the Irish Sea, males reach maturity at an age of four to five years and females at six to nine years. Egg capsules are cemented together in masses on the shells of other whelks. Development takes about six months, the young snails emerging at a shell height of about 6–12 mm. Longevity of 17 years has been estimated for *Neptunea* in the North Sea.

Family Nassariidae
Stout shell with spire, strongly ridged. Aperture with siphonal canal. Outer lip may be thickened, sometimes with teeth. In non-apertural or 'back' view the end of the siphonal canal is seen as a notch.

Stout shell, whorls straight-sided; arrangement of longitudinal and spiral ridges gives squared pattern .... ***Nassarius reticulatus*** p. 234

Stout shell, whorls well rounded; with longitudinal and spiral ridges but without squared patterning. Conspicuous brown mark in siphonal canal
.... ***Nassarius incrassatus*** p. 235

Nassarius reticulatus (Linnaeus) (*Hinia reticulata*) Netted dog-whelk **Fig. 12.35a**
Brownish shell, up to 30 mm in height; criss-crossing of longitudinal and spiral ridges gives squared pattern. Whorls straight-sided. Short siphonal canal leads from base of aperture. In non-apertural or back view, the end of the canal is seen as an indentation. Mature animals with thick outer lip with several teeth.

N. *reticulatus* is widely distributed and common in north-west Europe and around Britain. It lies just under the surface in sand and muddy-sand on the lower shore and, in particular, in the sublittoral down to about 15 m. It is also found among rocks and stones and tolerates salinities down to about 16. It scavenges on dead and decaying animal matter which it detects by following olfactory cues. Breeding occurs in spring and summer. The sexes are separate, fertilization is internal and flattened, vase-shaped egg capsules are laid in rows on *Zostera* (p. 75), on red seaweeds and sometimes on hydroids and stones. Each capsule contains up to 300 or more eggs, which hatch as free-swimming veliger larvae after about a month. The larvae have a long pelagic life of about two months. Sexual maturity has been recorded after four years and the whelks have a longevity of 10 to 15 years.

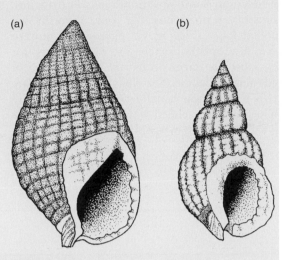

Figure 12.35 (a) *Nassarius reticulatus*.
(b) *Nassarius incrassatus*.

Nassarius incrassatus (Ström) (*Hinia incrassata*)
Thick-lipped dog-whelk **Fig. 12.35b**
Shell up to 12 mm in height, with longitudinal and spiral ridges but without squared patterning. Brownish in colour with dark bands; conspicuous brown mark in siphonal canal. Whorls well rounded. Short siphonal canal leads from base of aperture. In non-apertural or back view, the end of the canal is seen as an indentation. Mature specimens with thick outer lip and several teeth.

N. incrassatus is widely distributed in north-west Europe and occurs all around Britain. It is found in crevices and in silt on the lower shore, several specimens often occurring together, and extends into the sublittoral to depths of about 100 m. It feeds on dead and decaying animal matter. Breeding occurs in spring and summer. The sexes are separate, fertilization is internal and flattened, vase-shaped egg capsules are laid in clusters on hydroids, bryozoans and *Zostera* (p. 75), each capsule containing 50–80 eggs. After a month or so free-swimming veliger larvae emerge; they have a long planktonic life of two months or more. Longevity is believed to be up to six years.

Subclass Opisthobranchia

The opisthobranchs are among the most colourful of marine organisms. They are marine gastropods in which the shell is usually reduced and enclosed within the body, or absent. A few species, however, have an external shell and these are the primitive members of the group. The dorsal surface of those without an external shell shows a number of structures useful in identification and these are illustrated in Fig. 12.36. The anterior region can have a number of tentacle-like structures. Those on either side of the mouth are the oral tentacles, while those on the dorsal surface of the head are known as rhinophores. These vary, some being ringed and some arising from a basal cup-like sheath. Rhinophores are chemosensory. Cephalic tentacles may be present on the posterior region of the head, while propodial tentacles arise from the anterior region of the foot. Gills are often conspicuous and in some groups are in a posterior circle around the anus. Some opisthobranchs have a large number of dorsal outgrowths known as cerata which may contain extensions of the gut. The cerata produce secretions rendering the animal unpalatable to predators, an important adaptation in animals that have lost the external shell. In species that feed on sea-anemones, the nematocysts of the prey are arranged in the cerata of the opisthbranch where they continue to function, thus adding to the defensive role.

Oral tentacles
Propodial tentacles
Rhinophores
Cerata
Tubercles
Anus
Gills

Figure 12.36 Generalized opisthobranch mollusc showing features referred to in the key.

The cerata may also have a respiratory function. Some opisthobranchs burrow into soft sediments; others move over the substratum on a muscular foot, often thrown into extensive lateral lobes, the parapodial lobes (Fig. 12.42) extending onto the dorsal surface of the animal. Although a few opisthobranchs are herbivorous, for example the sea-hare, *Aplysia* (p. 242), most are carnivorous on a wide range of invertebrates, including sponges, hydroids and bryozoans. Sea-slugs and sea-hares are hermaphroditic but copulation and cross-fertilization usually occur. The eggs are generally deposited in jelly-like egg masses, often as coiled ribbons attached to the substratum, from which free-swimming veliger larvae emerge.

Key to families of opisthobranchs

The presence or absence of a shell is the first step in the identification of opisthobranchs. When present, as in *Acteon* (p. 237) and *Retusa* (p. 238), the animals must not be confused with prosobranch molluscs (p. 198). When an internal shell is present, it can be detected by careful handling. It is a delicate structure and if it is to be removed, this must be done with care.

The prosobranch mollusc *Lamellaria perspicua* (p. 228) has an internal shell and bears a superficial resemblance to the opisthobranchs with which it is easily confused. Features useful in the identification of opisthobranchs are shown in Fig. 12.36.

1 External shell present **2**
Shell not visible, may be internal or absent **3**

2 Operculum present .... **Acteonidae** p. 237
Operculum absent. Head with pair of large, posterior tentacles ....
.... **Retusidae** p. 238

3 Careful examination reveals presence of internal shell **4**
Internal shell absent **6**

4 In dorsal view body seen to be made up of 4 lobes .... **Philinidae** p. 238
Body not as above **5**

5 Parapodial lobes very large, flap-like. Pair of oral tentacles and pair
of rhinophores .... **Aplysiidae** p. 242
Body rather flattened. Pair of oral tentacles and pair of rhinophores.
Single, large gill on right side of body lying between mantle and
foot .... **Pleurobranchidae** p. 242

6 Body without a posterior, dorsal circle of gills and without cerata **7**
Body with a posterior dorsal circle of gills or with cerata **9**

7 With parapodial lobes, extending along length of body .... **Elysiidae** p. 240
Without parapodial lobes **8**

8 Three gills under posterior mantle edge on right side. Without tentacles ....
.... **Runcinidae** p. 239

Without gills under posterior mantle edge on right side. With or without
tentacles .... **Limapontiidae** p. 240

9 Body with posterior circle of gills around anus; without cerata **10**
Body without a posterior circle of gills; with cerata **12**

10 Circle of gills cannot be completely retracted, but individual gills can
retract .... **Onchidorididae** p. 244
Circle of gills can be completely retracted **11**

11 Dorsal surface with large, rounded tubercles of range of sizes, some large. Rough
to the touch .... **Dorididae** p. 245
Dorsal surface with very many, small tubercles; soft to the
touch .... **Discodorididae** p. 246

12 Rhinophores smooth, long and slender, unbranched; arise from cup-like sheaths
(Fig. 12.44) .... **Dotidae** p. 243
Rhinophores without cup-like sheaths **13**

13 Rhinophores ringed. Cerata grouped on dorsal surface .... **Facelinidae** p. 247
Rhinophores smooth; tips pale or white. Dorsal surface with transverse rows of
cerata .... **Aeolidiidae** p. 247

Family Acteonidae

Shell external, oval, glossy; pointed apex. Animal retracts fully into shell; operculum
present.

Acteon tornatilis (Linnaeus) **Fig. 12.37a; Plate 22a**

*Oval shell; up to 25 mm in height. Glossy, pinkish to brown in colour with white
bands running round shell. Single tooth present on inner lip of aperture. Body retracts
fully into shell. Head flattened, with 4 lobes. Operculum present. May be mistaken for
a prosobranch mollusc (p. 198).*

A. tornatilis is widely distributed in north-west Europe and, although occurring all
around Britain, has a patchy distribution and may be locally abundant. It is found on
sheltered sandy beaches on the lower shore and to depths of about 250 m. It burrows
in the top few centimetres of sediment using its foot and flattened head, and the
presence of the burrows is often revealed on the surface by holes 4–5 mm in diameter.
Specimens are occasionally found crawling on the sediment where they leave broad
furrows. *Acteon* is carnivorous, the diet including small polychaetes. It is hermaphroditic. In south-west Wales breeding has been recorded from April to August. Maturity
is reached at about 10 mm shell length and following copulation club-shaped/ribbon-
like egg masses are laid on the surface of the sediment. It has been estimated that one
individual can produce six or seven egg masses per year, containing as many as half a
million eggs. The egg masses are about 60 mm in length and are attached at the
narrow end to weed or stones on, or buried in, the sediment. Free-swimming veliger
larvae emerge and these are believed to have a planktonic life of one or two months.
Most individuals die after spawning at an age of about one and a half years but a few
survive to spawn a second time and may reach three years of age. *Acteon* has many

(a)

(b)

Figure 12.37 (a) *Acteon tornatilis.* (b) *Retusa obtusa.*

characters that resemble those of the prosobranchs. It is the most primitive British opisthobranch and one of the few to possess an operculum.

Family Retusidae

Shell external, delicate; may have short spire or spire flattened. Aperture very elongate, extending along most or all of shell height. Animal retracts fully into shell. Head with 2 large, posterior tentacles. Posterior margin of foot not split in 2 lobes. Without operculum.

Retusa obtusa (Montagu) Pearl bubble **Fig. 12.37b**
Delicate, whitish shell; up to 10 mm in height; aperture elongate but not as long as the main whorl of the shell; spire not always conspicuous. Retracts fully into shell. Operculum absent. Head with pair of conspicuous, posterior tentacles. Posterior margin of foot entire. May be mistaken for a prosobranch mollusc (p. 198).

R. obtusa is widely distributed in north-west Europe and all around Britain, buried in muddy-sand and mud on the shore and sublittoral, surviving salinities down to about 20. It is carnivorous, feeding on *Hydrobia ulvae* (p. 222) which it ingests whole, and on foraminiferans. *Retusa* is a protandrous hermaphrodite; copulation takes place in autumn and the eggs mature during winter and are laid in early spring. Each female produces up to four egg masses containing a total of about 50 eggs. These are found on the shells of adult snails and after three to four weeks crawling young emerge, generally from about February to May. This is followed by death of the adults which live for a little more than one year. This is an uncommon life-cycle as most opisthobranchs have free-swimming veliger larvae.

Family Philinidae

Shell internal, delicate; aperture large. Parapodial lobes conspicuous and extend onto the dorsal surface but do not meet. In dorsal view body clearly divisible into 4 regions – head, visceral mass and 2 lateral parapodial lobes.

Figure 12.38 *Philine aperta*, (a) dorsal view, (b) internal shell.

Philine aperta (Linnaeus) **Fig. 12.38**
In dorsal view body seen to be made up of 4 lobes – right and left parapodial lobes, cephalic shield (head) and posterior mantle lobe over the visceral mass. Whitish to pale yellow in colour; up to 70 mm in length. Internal shell with very wide aperture; delicate, whitish in colour.

P. aperta is widely distributed in north-west Europe. It is found on the lower shore and into the sublittoral to depths of about 500 m, burrowing in sand and muddy-sand. It preys on a wide range of invertebrates found in soft sediments, including polychaetes, small bivalves and gastropods, and foraminiferans. Acid secretion from the skin gives it some protection from predators which include fish. In Britain spawning has been recorded from spring to summer when flask-shaped egg masses are laid. Veliger larvae hatch after a few days. Longevity is believed to be three to four years.

Family Runcinidae
Shell absent. Body small, no more than 8 mm in length. Without tentacles; without parapodial lobes. Gills under posterior mantle edge on right side.

Figure 12.39 *Runcina coronata* (after Thompson, 1988).

Runcina coronata (Quatrefages) **Fig. 12.39**
Small, usually up to 6 mm in length. Brownish colour, paler areas around eyes. Without tentacles; 3 gills under mantle edge on posterior right side, visible in dorsal view when specimen relaxed. Shell absent.

R. coronata is found in rock pools on the middle shore with *Corallina* (p. 43), on which it is believed to browse. It extends into the shallow sublittoral and has been recorded on the Atlantic coast of France and west coast of Britain. It is hermaphroditic, and egg masses, each containing up to about 30 eggs, are laid in spring. There is no pelagic phase.

Family Elysiidae
Shell absent. Parapodial lobes extend onto dorsal surface. Rhinophores and propodial tentacles present.

Elysia viridis (Montagu) **Fig. 12.40**
Body up to 45 mm in length, generally much smaller. Usually green, but colour varies; shiny red, blue and green spots always present. Shell absent.

E. viridis is found throughout north-west Europe and all around Britain, especially in the south and west. It is found creeping among seaweed on the lower shore and in the shallow sublittoral. It feeds on a variety of green seaweeds including *Cladophora* (p. 32) and *Codium* (p. 33), cutting them open with the radula. Chloroplasts from the ingested algae are apparently incorporated into the digestive gland where they may continue to photosynthesize for some months. Breeding occurs from May to October. The egg masses contain up to several thousand eggs which develop into free-swimming veliger larvae after one or two weeks. The larvae metamorphose after about one month. Most individuals live for a year.

Figure 12.40 *Elysia viridis*, (a) lateral view, (b) dorsal view.

Family Limapontiidae
Shell absent. Body small, no more than 8 mm in length. Head with or without pair of tentacles, which may be ridge-like crests or finger-like. Without parapodial lobes. Eyes conspicuous, surrounded by pale areas.

(a) Head with 2 ridge-like tentacles between eyes. Found in rock
   pools .... ***Limapontia capitata*** p. 241
   or

(b) Head with 2 longish, slender tentacles. Found in rock
   pools .... ***Limapontia senestra*** p. 242
   or

(c) Head without ridge-like tentacles or long slender tentacles. Found on
   saltmarshes .... ***Limapontia depressa*** p. 241

Limapontia capitata (O. F. Müller) **Fig. 12.41a**

Small, usually up to 4 mm body length; brown or black in colour, area around eyes paler; 2 ridge-like tentacles between eyes. Shell absent.

L. capitata is widely distributed in north-west Europe and is found all around Britain, usually in rock pools on the middle to lower shore, where it feeds on seaweeds such as *Cladophora* (p. 32) and *Chaetomorpha* (p. 31). It is hermaphroditic, depositing globular or sausage-shaped masses, each containing several hundred eggs, on seaweed from spring to autumn. After one or two weeks, free-swimming veliger larvae emerge. Length of life is up to one year.

Limapontia depressa Alder & Hancock **Fig. 12.41b**

Small, rather flattened, usually up to 6 mm body length; brown-black colour but some specimens pale colour. Head without ridge-like tentacles or long slender tentacles. Eyes surrounded by pale areas. Shell absent.

Figure 12.41 (a) *Limapontia capitata.* (b) *Limapontia depressa.* (c) *Limapontia senestra.*

L. depressa is widely distributed in north-west Europe and occurs all around Britain on saltmarshes. It feeds on algae. The animals are hermaphroditic; egg masses, each containing up to 200 or so eggs, are laid from May to September. After about four weeks, free-swimming veliger larvae emerge. Length of life is believed to be one year.

Limapontia senestra (Quatrefages) **Fig. 12.41c**
Small, usually up to 6 mm body length; brown-black colour. Head with 2 longish, slender pale-coloured tentacles. Eyes surrounded by pale areas. Shell absent.

L. senestra is widely distributed in north-west Europe and all around Britain in rock pools on the lower shore, sometimes with *L. capitata* (p. 241), feeding on seaweeds such as *Cladophora* (p. 32) and *Ulva* (p. 30). It is hermaphroditic; spawn is laid from February to September. The eggs are large (400 μm) and unlike in the other species of *Limapontia* described here, development is direct.

Family Aplysiidae
Shell internal, delicate. Large, flap-like parapodial lobes. Oral tentacles and rhinophores present.

Aplysia punctata (Cuvier) Sea-hare **Fig. 12.42**
Brownish-green body, often with black spots; up to 200 mm in length; young specimens have reddish colour. Oral tentacles and rhinophores present. The flap-like parapodial lobes joined at rear. Shell internal, delicate, up to 40 mm in length. Purple secretion ejected when animal disturbed.

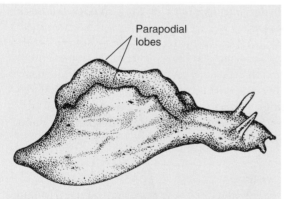

Parapodial lobes

Figure 12.42 *Aplysia punctata.*

A. punctata is widely distributed in north-west Europe and around Britain where it is essentially a shallow-water species, sometimes found on the lower shore. Contrary to some reports, *Aplysia* does not swim but moves over the surface of the substratum with a gliding motion, feeding mainly on seaweeds. It is hermaphroditic and breeding occurs mainly in spring. In some areas the animals move onto the lower shore to spawn. Large numbers of individuals form a chain and orange-pink egg masses, each containing many thousands of eggs, are laid among seaweed. After about two to three weeks, free-swimming veliger larvae emerge. Length of life is probably one year. *Aplysia* has been known since the time of Pliny, the Roman, and up to the Middle Ages was surrounded by a rich mythology related to the belief that it was poisonous.

Family Pleurobranchidae
Shell internal, delicate. Body rather flattened. Single, large gill evident on right side of body. Oral tentacles and rhinophores present.

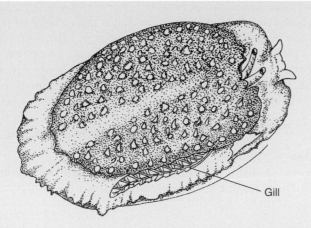

Gill

Figure 12.43 *Pleurobranchus membranaceus* (after Forbes & Hanley, 1853).

Pleurobranchus membranaceus (Montagu) **Fig. 12.43**
Body up to 120 mm in length; dorsal surface brownish, covered with tubercles. Single gill evident on right side. Oral tentacles and rhinophores present. Shell internal, thin, up to 50 mm in length.

Rhinophore

Sheath

Cerata

Figure 12.44 *Doto coronata* (after Thompson, 1988).

P. membranaceus is a shallow-water species found intertidally on rocky shores on the French coast and all around Britain. It feeds on ascidians such as *Ascidiella aspersa* (p. 438) and *Botryllus schlosseri* (p. 441) by boring into them using the radula. When the animal is disturbed, the body and foot secrete sulphuric acid which acts as a deterrent to predators. *Pleurobranchus* swims by a flapping action of the lobes of the foot and is a strong swimmer. It is hermaphroditic and breeds in spring, the egg masses containing many thousands of eggs. There is a free-swimming veliger larva. The animals probably die after spawning and have a lifespan of about one year.

Family Dotidae
Shell absent. Body with 2 rows of large, papillate cerata. Rhinophores smooth, long and slender, unbranched; arise from cup-like sheaths.

Doto coronata (Gmelin) **Fig. 12.44**
Body usually up to 12 mm in length. Pale coloured with purple-red streaks and markings. Up to 8 pairs of pale-coloured cerata. spotted red. Rhinophores emerge from smooth-rimmed, cup-like sheaths. Shell absent.

D. coronata is widely distributed in north-west Europe and has been recorded all around Britain. It is found on the lower shore and into the sublittoral to depths of 180 m, feeding on a wide range of hydroids such as *Obelia* (p. 110), *Sertularia* (p. 114) and *Dynamena* (p. 114). It is hermaphroditic; coiled bands of spawn have been recorded throughout the year, each containing many thousands of eggs, which take about 16 days to hatch as veliger larvae. There may be up to four generations per year and the length of life is believed to be up to nine months. *D. coronata* is believed to be a species complex.

Family Onchidorididae
Shell absent. Body flattened. Rhinophores ringed; not in cup-like sheaths. Circle of gills around anus; gill circle cannot be completely retracted but individual gills can retract. Mantle expansive.

Circle of up to 29 gills. Dorsal surface whitish, with brown patches and whitish tubercles .... ***Onchidoris bilamellata*** p. 244

Circle of up to 11 gills. Dorsal surface whitish with large, rounded tubercles .... ***Onchidoris muricata*** p. 245

Onchidoris bilamellata (Linnaeus) (*Onchidoris fusca*) **Fig. 12.45**
Soft body; up to 40 mm in length. Posterior circle of up to 29 gills. Dorsal surface covered with tubercles. White patterned with brown, tubercles usually white. Rhinophores ringed. Shell absent.

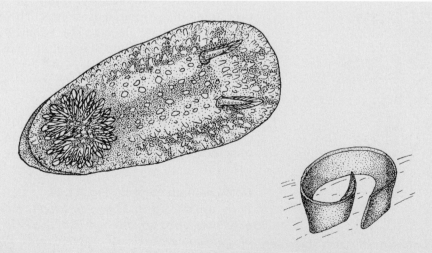

Figure 12.45 *Onchidoris bilamellata* and spawn (after Alder & Hancock, 1855).

O. bilamellata is widely distributed in north-west Europe and is common around Britain, both intertidally and in the shallow sublittoral to depths of about 20 m. It appears to feed exclusively on barnacles, especially *Semibalanus balanoides* (p. 308), forcing open the opercular plates of the barnacle and sucking out the soft parts. When disturbed, acid secretions are produced by the mantle. *Onchidoris* is hermaphroditic and breeding generally occurs from December to April, but summer and autumn spawnings have also been recorded. The spawn is laid in the shape of a flattened coil attached to the substratum by one edge and containing many thousands of eggs. *Onchidoris* aggregates at the start of the breeding season and there is evidence to suggest that it undertakes spawning migrations. As a result of this behaviour, the egg masses are sometimes abundant. After about five to six weeks, free-swimming veliger larvae emerge and these have a pelagic life of about three months before settling intertidally on the underside of boulders and rocks. The time of settlement of the young nudibranchs is believed to be closely related to the occurrence of one of their main prey items, newly settled barnacles, and it has been observed that the veligers will not metamorphose unless living barnacles are present. As the nudibranchs grow they move to the upper rock surfaces and become more conspicuous, which led to the suggestion that they migrated upshore after settling at lower beach levels. Length of life is believed to be about one year, with some individuals spawning and dying after as little as three months on the shore.

Onchidoris muricata (O. F. Müller) **Fig. 12.46**

Soft body; up to 14 mm in length. Posterior circlet of up to 11 gills. Dorsal surface covered with large, rounded tubercles. White or sometimes yellowish in colour. Rhinophores ringed. Shell absent.

Figure 12.46

Onchidoris muricata (after Alder & Hancock, 1855).

O. muricata is a northern species reaching its southern limit on the Atlantic coast of France. It is widely distributed around Britain, especially in the north, on the lower shore and in the shallow sublittoral to depths of about 15 m. It feeds on encrusting bryozoans such as *Electra pilosa* (p. 390) and *Membranipora membranacea* (p. 389), and also on *Securiflustra securifrons* (p. 391). *O. muricata* is a simultaneous hermaphrodite breeding from February to April on the east coast of England. White, gelatinous spiral ribbons, each containing thousands of eggs, are laid on rocks and stones and the larvae hatch after about two weeks. They have a pelagic life of several weeks and it appears that contact with bryozoans stimulates metamorphosis. Length of life is a year.

O. *muricata* is similar in appearance to another opisthobranch mollusc, **Adalaria proxima** (Alder & Hancock), which is sometimes common in the intertidal region and shallow sublittoral feeding on bryozoans. The tubercles on the mantle of *Adalaria* are tapering and more elongate than the rounded tubercles of O. *muricata* and it is generally yellow in colour. However, in some cases reliable separation can only be achieved by examination of the internal anatomy. Reference to **Thompson** (1988) is recommended. *Adalaria* has an annual life-cycle. Larvae hatching from the egg mass have a short pelagic life before metamorphosis.

Family Dorididae

Shell absent. Body flattened. Rhinophores ringed; not in cup-like sheaths except in one rare British genus. Circle of gills around anus. Gill circle can be completely retracted. Mantle expansive. Dorsal surface rough, with tubercles of range of sizes, some being very large.

Figure 12.47 *Doris pseudoargus* and spawn.

Doris pseudoargus Rapp (*Archidoris britannica, Archidoris tuberculata, Archidoris pseudoargus*) Sea-lemon **Fig. 12.47**
Large body; up to 120 mm in length. Dorsal surface covered with rounded tubercles of various sizes. Colour varied, usually yellowish with green, pink or brown patches, but reddish varieties also occur. Posterior circle of up to 10 whitish gills. Rhinophores ringed. Shell absent.

D. *pseudoargus* is widely distributed in north-west Europe and Britain on the lower shore and in the sublittoral. It is found under stones and in crevices on rocky shores where it feeds on sponges, especially *Halichondria panicea* (p. 84) and *Hymeniacidon perlevis* (p. 84). It is hermaphroditic, breeding in the spring. The egg mass is a long, convoluted whitish or yellowish ribbon containing thousands of eggs, attached by its edge to the rock surface. Free-swimming veliger larvae hatch after about four weeks and have a long pelagic life of, possibly, up to three months. Metamorphosis of the larvae appears to be stimulated by contact with the sponge *Halichondria*. Length of life has been shown to be about one year in animals from the Isle of Man but may be longer at other localities.

Family Discodorididae
Shell absent. Body flattened. Rhinophores ringed; not in cup-like sheaths. Circle of gills around anus. Gill circle can be completely retracted. Mantle expansive. Dorsal surface soft, with very many small tubercles.

Jorunna tomentosa (Cuvier) **Fig. 12.48**
Body up to 55 mm in length. Small tubercles of uniform size on dorsal surface. Yellowish-brown in colour with brown patches roughly in pairs along dorsal surface. Posterior circle of up to 17 whitish gills. Rhinophores ringed. Shell absent.

J. *tomentosa* is widely distributed in north-west Europe and around Britain, both intertidally and in the shallow sublittoral. It feeds on the sponge *Halichondria panicea* (p. 84). *Jorunna* is hermaphroditic and breeds during spring and summer when yellowish-coloured ribbons containing many thousands of eggs are deposited on the underside of stones on the lower shore. Free-swimming veliger larvae emerge after about three weeks.

Figure 12.48 *Jorunna tomentosa* (after Alder & Hancock, 1855).

Family Facelinidae
Shell absent. Body with groups of cerata. With the exception of two rare members of the family, the rhinophores are ringed. Oral tentacles long, propodial tentacles conspicuous.

Facelina auriculata (O. F. Müller) (*Facelina coronata*) **Fig. 12.49a**
Body up to 38 mm in length; groups of cerata on dorsal surface. Rhinophores conspicuously ringed. Body usually pinkish-white; cerata bluish, iridescent, red, brown, with white patch near tip. Shell absent.

F. auriculata is widely distributed in north-west Europe and around Britain, living among stones and rocks, on the lower shore and in the sublittoral. It feeds on hydroids, especially *Tubularia* (p. 102). It is hermaphroditic and breeding occurs during summer when convoluted egg masses are laid on the undersurface of stones. There is a free-swimming veliger larva.

Family Aeolidiidae
Shell absent. Body with transverse rows of cerata. Rhinophores smooth. Oral tentacles long, propodial tentacles conspicuous.

Aeolidia papillosa (Linnaeus) Common grey sea-slug **Fig. 12.49b**
Body up to 120 mm in length. Many cerata on dorsal surface, separated by mid-dorsal, clear whitish area. Rhinophores smooth. Usually purplish-brown, grey in colour with white tips to the cerata and rhinophores. Shell absent.

A. papillosa is widely distributed in north-west Europe and common among rocks and stones on the lower shore and shallow sublittoral. The diet includes a wide range of sea-anemones, especially *Actinia equina* (p. 91), which are detected by chemosensory means. Undischarged nematocysts from the ingested anemones are transported through the gut to the ends of the cerata where they are used by the sea-slug in defence. *Aeolidia* is hermaphroditic, breeding in spring and summer. Convoluted egg masses, usually pinkish in colour and containing thousands of eggs, are attached to stones. The larvae are planktonic. Longevity is probably about one year.

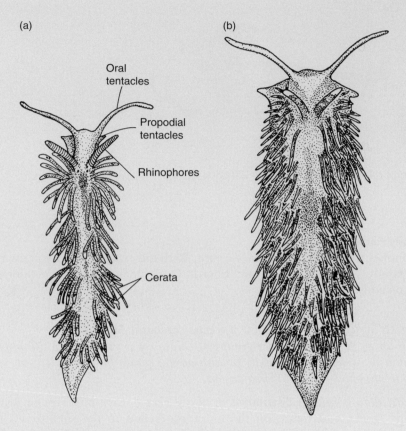

Figure 12.49 (a) *Facelina auriculata*. (b) *Aeolidia papillosa*.

Subclass Pulmonata

The best-known pulmonates are the freshwater and land snails and the slugs, but the group also includes a few marine representatives. The highly vascularized mantle cavity is adapted for air breathing and opens by a single, contractile aperture on the right side of the snail. In almost all species the operculum is absent. Pulmonates are hermaphroditic but cross-fertilization normally occurs and the eggs are laid in gelatinous capsules. Except in a very few species there is no veliger larva. In many pulmonates the shell is absent; those described in this text have a shell and may be mistaken for prosobranch molluscs (p. 198).

Auriculinella bidentata (Montagu) (*Leucophytia bidentata*) **Fig. 12.50a**
Shell thick; smooth, with pointed apex; up to 9 mm in height. Columella with 2 distinct teeth. No operculum. Creamy-white in colour. Tentacles short, flattened.

A. bidentata is widely distributed on south and west coasts of Britain, usually on rocky shores under stones and in crevices on the upper shore, but also in estuaries and on mud-flats. Its food is organic detritus. It is a protandrous hermaphrodite and

Figure 12.50 (a) *Auriculinella bidentata*. (b) *Myosotella myosotis*. (c) *Otina ovata*. ((a) After Macan, 1977.)

irregularly shaped egg masses, about 4 mm in diameter and containing up to 200 eggs, are laid in crevices in the rock. On the south coast of England, egg masses are deposited during June. There is no free-swimming larva, crawling young emerge. Longevity is probably two years.

Myosotella myosotis (Draparnaud) (*Phytia myosotis, Ovatella myosotis*) **Fig. 12.50b**
Shell thin; smooth, with pointed apex; up to 9 mm in height. Columella and outer lip each with up to 3 or more teeth. No operculum. Semi-transparent, reddish-brown, yellowish in colour. Tentacles faintly annulated; with dark streaks.

M. myosotis has a patchy distribution around Britain. It is found on the upper shore and above, in estuaries and on saltmarshes. It is hermaphroditic and egg masses containing up to 50 eggs are attached to stones and stems of plants. Crawling young emerge. Breeding probably occurs in spring.

Otina ovata (Brown) **Fig. 12.50c**
Shell with very wide aperture; about 3 mm in height. Operculum absent. Reddish-brown, purplish in colour. Shell does not enclose whole animal.

O. ovata is a high-shore species found on south and west coasts of Britain among empty barnacle shells, in crevices and with the lichen *Lichina pygmaea* (p. 73). It feeds on microorganisms and detritus. On the south coast of England breeding occurs in May and June. The animals are protandrous hermaphrodites and the egg masses, each about 5 mm in diameter and containing 20–30 eggs, are laid in crevices. Crawling young emerge.

Class Bivalvia (Pelecypoda; Lamellibranchiata)

As the names implies, bivalve molluscs have a shell of two valves. The valves have a dorsal hinge and are held together by an elastic ligament which is made of fibres that force them apart so as to open the shell. This action is opposed by the contraction of muscles that close the shell. These are the adductor muscles and they generally leave a clearly visible scar or scars on the inside of the shell (Fig. 12.51b). Also visible on the inner surface along the lower edge of each valve of the shell is a line running from one adductor muscle scar to the other. This is the pallial line, marking the position where the mantle was attached to the shell. In some but not all species, the posterior part of the pallial line is indented to give a pallial sinus showing the area occupied by the siphons (Fig. 12.51b). In bivalves such as the common cockle (p. 269), in which the siphons are small, there is no pallial sinus; however, in those species (for example the sand gaper, p. 291) that have large siphons, the pallial sinus is well developed. For many of the bivalves described in this text the position of the adductor muscle scars, the pallial line and pallial sinus is shown in the figures.

The shell ligament is internal, external or both internal and external; when external, it is often clearly visible on the outside of the shell. The internal ligament is attached to a depression known as the chondrophore. This is usually conspicuous and in a few species is developed as a spoon-shaped projection (Fig. 12.93b,c). The dorsal hinge line may have a number of interlocking teeth which prevent the two valves of the shell from slipping laterally. The teeth beneath the beak are known as cardinals, to either side of which are the anterior and posterior lateral teeth. The first-formed part of the shell is known as the beak and round it is a raised area, the umbo (Fig. 12.51b). The shell may be smooth or sculptured with grooves, lines or ribs; the ribs radiate outwards from the beaks and may bear spines. Concentric growth rings, which can be used for age determination in some species, are often clear (Fig. 12.51a). The outer layer of the shell, or periostracum, is sometimes thick and may be drawn out into spines, but it is often worn away, particularly in dead shells.

The body of the bivalve is laterally compressed and the gills, which can be very large, are used in food collection as well as respiration. Water circulates through the mantle cavity, passing over the gills where fine particles are filtered out. The water current enters and leaves by two posterior apertures, the inhalant and exhalant apertures, the exhalant being the more dorsal. In some bivalves, notably those living deep in sediment, these apertures are situated at the end of long siphons. In suspension-feeding bivalves the inhalant siphon is generally held upright and draws in a current of water containing food material such as plankton; the inhalant siphon of deposit feeders moves over the surface of the sediment, drawing in detritus. An anterior, laterally compressed foot is usually present as the organ of locomotion, but in those species that are not very mobile it is much reduced in size. Some bivalves are attached to the substratum by toughened fibres secreted by a gland in the foot and known as the byssus. The group as a whole shows adaptations to a sedentary way of life, although the scallops are able to swim by flapping the valves together. Most bivalves have separate sexes; some are hermaphroditic. Fertilization is generally external and there is usually a veliger larva in the life-cycle. This has a tiny shell of two valves and swims by means of a pair of ciliated lobes.

Bivalves are common members of the intertidal fauna. Well-known examples include mussels found on rocky shores and cockles living buried in sandy sediments.

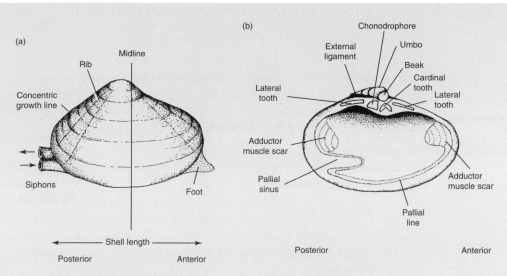

Figure 12.51 Generalized bivalve shell to illustrate features referred to in the text. (a) Right valve, external features, (b) inside of left valve. Arrows indicate direction of flow of water.

Both tend to occur in very high densities and are preyed on extensively by birds and fish. Many species occur offshore where they comprise an important part of the diet of commercial fishes and, although living specimens of such species are not likely to be found on the shore, empty shells are frequently found on the strandline. These are included in the text.

Key to families of bivalves

Shell characteristics used in the identification of bivalves are illustrated in Fig. 12.51.

1 Shell very small, covers only anterior end of long, worm-like body. Shell valves divided into regions with different sculpturing. Bores into submerged timbers; tunnels lined by calcareous tubes .... **Teredinidae** p. 295
  Shell and body not as above **2**

2 Right and left valves of shell similar (equivalve); shell almost circular, beaks central. Hinge line with many small teeth alternating with sockets (taxodont arrangement) .... **Glycymerididae** p. 253
  Teeth on hinge line not taxodont **3**

3 Shell with projections ('ears') on either side of beaks (Fig. 12.59). Shell usually with conspicuous radiating ribs. May be attached to substratum or free
  .... **Pectinidae** p. 260
  Shell without 'ears' on either side of beaks **4**

4 Shell valves markedly dissimilar (inequivalve); may be thin and delicate, or thick and rough. Valves with single adductor muscle scar; upper valve may also have additional scars (Fig. 12.63c,d). Attached to substratum **5**
Shell not as above **6**

5 Shell thin and delicate; lower valve with indented hole. Upper valve with additional scars. Attached very firmly to rocks, stones, mollusc shells .... **Anomiidae** p. 264
Shell thick, rough, with surface corrugations; shell valves without hole; without additional scars. Attached to muddy-gravel, muddy-sand, rock .... **Ostreidae** p. 258

6 Shell more or less elongate, with pronounced gape. Part of shell reflected over beaks – the umbonal reflection (Fig. 12.95b). Hinge line without teeth but has slender, finger-like projection, the apophysis. Burrowing into rock, clay, peat .... **Pholadidae** p. 293
Shell not as above **7**

7 Shell roughly triangular; beaks very much at anterior. Hinge line without teeth but may be crenulate; without chondrophore. Shell without gape ... **Mytilidae** p. 254
Shell not as above **8**

8 Hinge line without teeth. Left valve with conspicuous, spoon-shaped chondrophore projecting from below beak (Fig. 12.93b,c). Shell with posterior gape .... **Myidae** p. 290
Shell not as above **9**

9 Hinge line with triangular or semicircular chondrophore (Fig. 12.69b); cardinal teeth present, lateral teeth may be present **10**
Hinge line without chondrophore; cardinal and/or lateral teeth may be present **11**

10 Cardinal and lateral teeth present; cardinal teeth of left valve joined to give ∧-shaped ridge in front of chondrophore .... **Mactridae** p. 271
Cardinal teeth present, not joined to give ∧-shaped ridge in front of chondrophore; lateral teeth absent .... **Scrobiculariidae** p. 281

11 Hinge line of right valve with projecting cardinal tooth. Shell more or less rectangular. Beaks close to anterior. The intertidal species in this family is found boring into soft rocks and shell or nestling in rock crevices ... **Hiatellidae** p. 292
Shell not as above **12**

12 Shell long, narrow, much longer than broad; with vertical groove at anterior end; beaks, at anterior, inconspicuous. Dorsal and ventral sides of shell straight. Hinge line without lateral teeth. Shell gapes at anterior and posterior .... **Solenidae** p. 273
Shell not as above **13**

13 Shell long and narrow, much longer than broad, gaping at anterior and posterior; without vertical groove at anterior end. Dorsal and ventral sides of shell straight or curved. Beaks at anterior or just anterior of midline, inconspicuous ...**Pharidae** p. 274
Shell not as above **14**

14 Pallial sinus present (but difficult to see in the Turtoniidae, p. 289) **15**
Pallial sinus absent **19**

15 Lunule present in front of beaks (Fig. 12.83c) **16**
Shell without lunule **17**

16 Shell small, fragile; usually less than 3 mm shell length. Found in crevices, among barnacles and seaweed on rocky shores. Lunule narrow. Teeth of hinge line difficult to distinguish. Pallial line difficult to see .... **Turtoniidae** p. 289
Shell up to 75 mm or so in length. Teeth of hinge line distinct ..... **Veneridae** p. 283

17 Shell wedge-shaped. Beaks towards posterior. Margin strongly crenulate or finely serrated .... **Donacidae** p. 280
Shell not wedge-shaped; margin smooth **18**

18 Shell rounded and globular or oblong; may be truncated with raised keel running to posterior margin; beaks central or slightly anterior or slightly posterior. Hinge line with cardinal teeth; lateral teeth absent .... **Psammobiidae** p. 280
Note: this step includes *Macoma balthica* of the Tellinidae (p. 277)
Thin, flattened shell, beaks usually to the posterior. Hinge line with cardinal and lateral teeth (lateral teeth conspicuous only in right valve) ... **Tellinidae** p. 277

19 Rounded globular shell with conspicuous radiating ribs .... **Cardiidae** p. 267
Shell without radiating ribs **20**

20 Thick, heavy rounded shell, up to about 120 mm shell length. Beaks well to the anterior. Hinge line with 3 cardinal teeth in each valve. Periostracum thick and shiny .... **Arcticidae** p. 282
Shell small, delicate, no more than 10 mm shell length. Beaks towards the posterior. Hinge line with no more than 2 cardinal teeth in each valve **21**

21 Oval shell, rather plump; up to 3 mm in length. Found in crevices, empty barnacle shells, among lichens, particularly *Lichina* (p. 73) .... **Lasaeidae** p. 266
Similar to above, no more than 10 mm shell length; usually in association with echinoids (p. 392) .... **Montacutidae** p. 267

Family Glycymerididae Dog-cockles
Shell equivalve, thick rounded; beaks central. Several small teeth on either side of beaks alternating with small sockets, i.e. taxodont. Adductor muscle scars roughly same size. Pallial sinus absent.

Taxodont

Figure 12.52 *Glycymeris glycymeris*, left valve.

Glycymeris glycymeris (Linnaeus) Dog-cockle **Fig. 12.52**
Shell almost circular; up to 65 mm in length. Beaks midway between anterior and posterior ends of shell. Inner surface of shell with 6–12 teeth on either side of beak; area beneath beak without teeth. Outer surface usually patterned with brown marks; inner surface white, may be dark brown. Margin strongly crenulate.

G. glycymeris is widely distributed and common in north-west Europe and occurs all around Britain. It is found sublittorally to depths of about 70 m in all types of sediment, such as sand, gravel and mud, buried to depths of several centimetres. Dead shells are commonly washed ashore. A free-swimming veliger larva has been recorded in the plankton of the North Sea throughout the year, but mainly in late autumn. Longevity is believed to be greater than 30 years.

Family Mytilidae Mussels, crenellas
Shell roughly triangular; beaks at anterior. Attached to substratum by byssus threads. Hinge line lacks teeth and chondrophore, but may be crenulate. Periostracum obvious, may be developed into spines. Adductor muscle scars usually of different size. Pallial sinus absent.

1 Shell surface smooth; but periostracum may have spines 2
  Shell surface with radiating ribs at anterior and posterior; area between, without ribs 5

2 Beaks right at anterior 3
  Beaks not quite at anterior 4

3 Beaks rounded, not pointing downwards. Edge of mantle whitish, yellow, brown .... ***Mytilus edulis*** p. 254
  Beaks pointed, point downwards. Edge of mantle dark purple or blue .... ***Mytilus galloprovincialis*** p. 256

4 Periostracum with serrated spines on the posterior part of the shell .... ***Modiolus barbatus*** p. 257
  Periostracum without spines in adult, but developed into long, smooth spines in young animals .... ***Modiolus modiolus*** p. 257

5 Shell noticeably tumid; posterior margin somewhat elongate; 15–18 ribs anteriorly; 20–35 ribs posteriorly. Often found living in body of ascidians .... ***Modiolarca subpicta*** p. 258
  Shell not tumid; posterior margin broad and rounded; 9–12 ribs anteriorly; 30–45 ribs posteriorly .... ***Musculus discors*** p. 258

Mytilus edulis Linnaeus Common mussel, blue mussel **Fig. 12.53**
Up to 100 mm in length, occasionally larger. Beaks at anterior, rounded. Teeth absent from hinge line but up to 12 crenulations near beaks. Outer surface of shell dark blue, sometimes brownish; inner surface with pearly iridescence and dark blue edge. Edge of mantle whitish, yellow, brown.

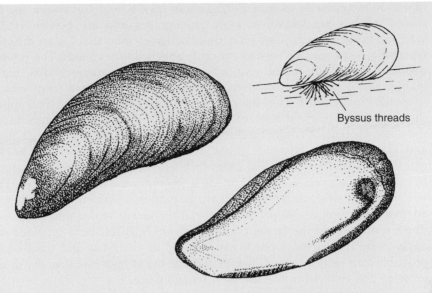

Byssus threads

Figure 12.53 *Mytilus edulis*, left valve and showing attachment by byssus threads.

M. edulis is widespread and common in north-west Europe where it is found from the middle shore to the shallow sublittoral. It is attached by byssus threads to stones, rocks and piers, and on exposed shores small mussels are often abundant. It is also found on soft sediments in estuaries and can tolerate salinities down to about 5, often forming dense beds which are fished commercially. There are fisheries at a number of localities around Britain, including the Wash, Morecambe Bay, Menai Strait, Conwy and the west coast of Scotland. At some localities mussels are farmed; banks of small overcrowded mussels are moved to more favourable areas where growth is rapid, and subsequently harvested. They are also cultivated on suspended ropes, on poles and on stakes interlaced with brushwood.

Mussels are suspension feeders. Their main predators are the dog-whelk, the shore crab, the common starfish and the oystercatcher. Despite their sessile habit and apparent vulnerability to predators, it has been shown that on dense mussel beds the mussels are able to immobilize one of their predators, the dog-whelk. The mussels close to the whelk attach large numbers of byssus threads to the shell of the whelk, eventually pulling it over. Dog-whelks feeding on mussel beds are vulnerable to attack in this way and, interestingly, the distribution of whelks in the field shows that they are not found in large numbers on mussel beds but reach their highest densities around the edge of the bed and in areas away from the highest concentration of mussels. When observed in aquarium tanks, mussels perfused with water in which natural predators such as the dog-whelk have been kept move together to form clumps. It is believed that the mussel is responding to pheromones carried in the water and the clumping behaviour reduces the risk of predation. As part of this behaviour, it has also been noted that byssus thread production is increased.

Mytilus has separate sexes, fertilization is external and around Britain there is a single peak of spawning in spring or peaks of spawning in spring and summer; the free swimming veliger larva is planktonic for up to four weeks. Primary settlement and metamorphosis and temporary attachment by the byssus occur on hydroids and filamentous seaweeds such as *Polysiphonia* (p. 51) and *Ceramium* (p. 47). Newly settled mussels are known as early plantigrades and are between 0.25 mm and 0.55 mm in shell length. They grow quickly, detach themselves, then drift to new sites where they may again become temporarily attached to seaweeds. The velum, the main organ of propulsion in the veliger larva, is lost at metamorphosis, but when the plantigrades leave the substratum they secrete a long thread from which they are suspended in the water. They hang with the ventral margin of the shell uppermost and the drag caused by the thread reduces the rate of sinking and ensures distribution by the water currents. Drifting on threads has been recorded in young of other bivalves and likened to the 'gossamer flight' of spiders. After about four weeks the late plantigrades are between 1 and 2 mm shell length. Settlement and attachment by the byssus occur on mussel beds, but some plantigrades may remain on the primary settlement site over winter. This behaviour pattern at settlement is believed to be an adaptation to reduce competition between the newly settled and adult mussels, but direct settlement of the larvae on adult mussel beds has been recorded on the west coast of Ireland. *Mytilus* can mature during its first year and the length of life is probably four or five years; however, those living at high levels on the shore are slow growing and believed to live for 17 years and more.

Careful examination of the mantle cavity of *Mytilus* may reveal the presence of the copepod **Mytilicola intestinalis** Steuer, once believed to be a damaging parasite which brought about a reduction in the mussel condition factor, with serious commercial implications. Movement of mussels from infected populations for laying in other areas was prohibited. Most recent research, however, suggests that the copepod is a commensal rather than a parasite. The pea crab, *Pinnotheres pisum* (p. 370), is also a common inhabitant of the mantle cavity of the edible mussel, and the shell of *Mytilus* is frequently bored by the polychaete worm *Polydora ciliata* (p. 187). Where heavy infestation occurs, the condition factor of the mussel is reduced and the shell is more vulnerable to predation by crabs. On commercial mussel beds where population density is high, fouling of the shell by the polychaete *Pomatoceros* (p. 175) can be a serious problem.

Mytilus galloprovincialis Lamarck Mediterranean mussel, blue mussel **Fig. 12.54**
Shell very similar to, and easily confused with, M. edulis. *Beaks at anterior end and pointed, pointing downwards. To the posterior of the beaks the ventral margin of the shell is slightly concave. Shell somewhat taller than that of* M. edulis. *Edge of mantle dark purple or blue.*

M. galloprovincialis has been recorded from the south and west coasts of Ireland, South Wales and south-west England, and extends to the coast of France and the Mediterranean. Recent research suggests that it has stronger byssal attachment to the substratum than does M. edulis and this may be reflected in its ability to withstand exposure to wave action.

The shell morphology of *Mytilus* spp. varies in response to environmental parameters and the separation of M. edulis and M. galloprovincialis is difficult. In some populations, separation based on shell morphology alone may be impossible; in others the morphological differences given above are clear

Figure 12.54 *Mytilus galloprovincialis*, left valve.

throughout all size ranges of mussel. Identification can be made more difficult by the fact that where the two species are found together, they interbreed to form hybrids with the result that populations comprising both hybrids and separate species are common.

Modiolus modiolus (Linnaeus) Horse mussel **Fig. 12.55**
Up to 200 mm in length; intertidal specimens usually smaller. Beaks not quite at anterior end of shell. Teeth absent from hinge. Outer surface of shell dark blue or purple; periostracum thick, yellow or brown and developed into long, smooth spines in young animals; inner surface of shell white. Most easily distinguished from Mytilus edulis *(p. 254) by the blunter appearance of the anterior end of the shell.*

(a)   (b)

Figure 12.55 *Modiolus modiolus*, (a) left valve, (b) young specimen, right valve.

M. modiolus is widely distributed in north-west Europe and found throughout Britain in pools and crevices on the lower shore, often with laminarians (p. 57). It is also found on soft sediments, where it lives with most of the shell buried in the sediment and attached by byssus threads which penetrate deep into the sediment. Essentially a sublittoral species, it extends to depths of about 60 m, often occurring in large densities. It is a suspension feeder. Unlike the common mussel, *M. modiolus* is not commercially important in Britain. The sexes are separate, fertilization is external and data from populations in the Irish Sea suggest that it breeds more or less continuously throughout the year in the sublittoral, while intertidal populations breed in late autumn and winter. Off the Isle of Man, *Modiolus* becomes sexually mature at three to four years of age. The horse mussel is long-lived, some specimens living for up to 48 years.

Modiolus barbatus (Linnaeus) Bearded horse mussel **Fig. 12.56a**
Up to 60 mm in length, usually smaller. Beaks not quite at anterior end of shell. Hinge line without teeth. Periostracum brownish-yellow, thick and developed into serrated spines on posterior part of shell; inner surface of shell shiny, pale blue colour.

M. barbatus is widely distributed in north-west Europe, and in Britain occurs mainly on south and west coasts. It is found among rocks and stones on the lower shore, extending into the sublittoral to depths of about 100 m.

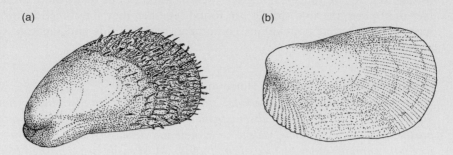

Figure 12.56 (a) *Modiolus barbatus*, left valve. (b) *Musculus discors*, left valve.

Musculus discors (Linnaeus) **Fig. 12.56b**
Up to 12 mm in length. Beaks not quite at anterior end of shell. Posterior of shell broad and rounded. Shell with 2 groups of ribs; 9–12 ribs in anterior group, 30–45 in posterior group; area between the 2 groups without ribs. Outer surface yellowish, greenish, brown in colour; inner surface pearly.

M. discors is widely distributed in north-west Europe and commonly found in Britain on the middle shore and below and into the shallow sublittoral among rocks, shells and seaweed, especially *Corallina officianalis* (p. 43). Dense populations are occasionally recorded in the shallow sublittoral.

Modiolarca subpicta (Cantraine) (*Musculus marmoratus, Modiolarca tumida*) Marbled crenella **(not illustrated)**
Up to 20 mm in length; delicate, tumid shell. Beaks not quite at anterior of shell. Posterior of shell somewhat elongate. Shell with 2 groups of ribs; 15–18 broad ribs in anterior group, 20–35 finer ribs in posterior group; area between the 2 groups without ribs. Yellowish to brown in colour with darker markings; inner surface pearly.

M. subpicta is widely distributed in north-west Europe among stones and shells and the holdfasts of laminarians but the main habitat is the tests of ascidians such as *Ascidia mentula* (p. 439) and *Ascidiella* spp. (p. 438). *Modiolarca* spawns in summer at the end of the first year of life. The sexes are separate, fertilization is external and the larvae are pelagic for about six weeks. The newly settled spat climb over the test of the ascidian and attach themselves by byssus threads close to the siphons in such a way that the test of the ascidian becomes wrapped around the mussel. The density of *Modiolarca* at settlement can be very high but generally no more than about five individuals are found on a single ascidian. Length of life is two to three years.

Family Ostreidae Oysters
Shell rough, thick, with corrugations. Beaks not obvious. Valves dissimilar; lower (left) valve convex, 'cemented' to substratum; upper (right) valve flat. Hinge line lacks teeth in adults. Single adductor muscle scar.

Ostrea edulis Linnaeus Common European oyster, flat oyster **Fig. 12.57**
Thick, rounded, but often irregular shell; up to 100 mm in length, sometimes larger. Valves dissimilar, lower (left) valve convex, upper (right) valve flat. Valves sculptured with wavy concentric lines. Attached to substratum by left valve. Outer surface usually greyish-brown; inner surface pearly white. Single adductor muscle scar white or creamish in colour. Larger specimens often bored by the sponge Cliona celata *(p. 85).*

Figure 12.57 *Ostrea edulis*, left valve.

O. *edulis* is widely distributed in north-west Europe, and although occurring all around Britain is more common in the south. It is found sublittorally to depths of about 80 m, extending into estuaries and tolerating salinities down to about 23. It often occurs in large beds on muddy-sand and muddy-gravel, and also on rocks. Like the mussel M. *edulis* (p. 254) it is commerically important and cultivated in large numbers, especially in south-east England, where Whitstable and Colchester are famous for their oysters. They are also cultivated on the French coast, especially the Bay of Arcachon, and in the Netherlands. During the nineteenth century oysters were very common in European waters but overfishing led to a drastic decline. Oysters are susceptible to a number of pests and predators, notable among which are *Crepidula* (p. 225) and *Asterias* (p. 405). They are suspension feeders.

Oysters are usually male when they first become sexually mature but after spawning change to female, thereafter regularly alternating between male and female for the rest of their lives. Breeding usually takes place during the summer months and in Britain each animal spawns twice during the breeding season, once as male and once as female. Fertilization is internal, sperm entering the bivalve in the feeding current, and the embryos remain in the oyster for one or two weeks until they become veliger larvae. The eggs are attached to the surface of the gills and as development proceeds they change colour from white through grey to black; at this time the oysters are known as 'white sick', 'grey sick' or 'black sick' and are inedible and out of season. The veligers are released and spend about two weeks in the surface waters, feeding on plankton, and by the time of metamorphosis have developed an extensible foot which enables them to move over the substratum. Once a suitable surface is found, cement is produced by the byssus gland at the base of the foot, attaching the left valve of the shell to the substratum. The larvae are attracted to surfaces already colonized by oysters. Around Britain, most oysters live for about six years but longevity is generally considered to be up to 14 years although records of them living 30 years are not uncommon.

Crassostrea gigas (Thunberg) Pacific oyster, Japanese oyster, cup oyster **Fig. 12.58**
Valves dissimilar, lower (left) deeply convex, upper (right) flat. Valves sculptured with concentric ridges, radiating ribs and grooves. Single adductor muscle scar, purple to

Adductor muscle
scar

Figure 12.58 *Crassostrea gigas*, left valve.

reddish-brown in colour. Margin of shell wavy. Outer surface white to pale brown in colour, with purple patches. Up to 180 mm shell length.

C. gigas was introduced into Europe for mariculture during the 1960s and 1970s and is now farmed commercially at a number of localities, accounting for up to 90% of British oyster cultivation. It has a faster growth rate than the flat oyster, maturing first as male and then changing sex to female, although some animals remain as male. The species is increasingly being found outside the commercial farms and although it was thought that it would not reproduce in Europe, natural spawning and recruitment have been recorded at a number of localities in England and Wales, but apparently not in Scotland.

Family Pectinidae Scallops
Shell rounded, usually with conspicuous, radiating ribs. Lies on right valve. In some species one valve is flat, the other concave. Projections ('ears') on either side of beaks. 'Ears' of equal size or anterior one the larger. May be attached by byssus or not. Hinge line lacks teeth. Single adductor muscle scar in adult.

1   Valves distorted, irregular in shape. Lower valve cemented to substratum.
    Upper valve with 60–70 ribs with spines .... ***Crassadoma pusio*** (adult) p. 261
    Valves not distorted or irregular **2**

2   Lower valve convex, upper valve flat, each valve with 15–17 broad ribs. Large,
    up to 150 mm shell length .... ***Pecten maximus*** p. 262
    Upper and lower valves convex, at least to some extent **3**

3 Anterior and posterior 'ears' of roughly same size. Each valve with 19–22
rounded ribs .... ***Aequipecten opercularis*** p. 263
Anterior 'ears' 2–3 times as long as posterior **4**

4 Shell relatively smooth or with 30 or more fine ribs, a few of which may be broad
and pronounced. Colour varied, often patterned .... **Palliolum tigerinum** p. 262
Shell with conspicuous, radiating ribs **5**

5 Shell with 25–35 ribs with blunt spines; spines most conspicuous towards
margin of shell .... ***Chlamys varia*** p. 261
Shell with 30–50 ribs, with a few spines .... ***Crassadoma pusio*** (juvenile) p. 261

See also *Pteria hirundo* (p. 264) and *Limaria hians* (p. 264)

Chlamys varia (Linnaeus) (*Mimachlamys varia*) Variegated scallop, black scallop
Fig. 12.59

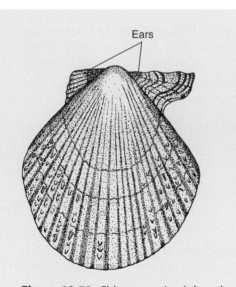

Ears

Figure 12.59 *Chlamys varia*, right valve.

Valves convex; up to 65 mm in length; 25–35 ribs with blunt spines; spines most conspicuous towards margin of shell. 'Ears' on either side of beaks asymmetrical and ribbed, anterior 2–3 times length of posterior. Attached by byssus or lying free. Extremely varied in colour; brown, red, yellow; often patterned. Internal surface of shell with single adductor muscle scar.

C. *varia* is widely distributed and common around Britain and has been recorded from Denmark to the coasts of Spain and Portugal and into the Mediterranean. It is usually found attached by the byssus to stones and shells in coarse sand and gravel, from the lower shore to depths of about 80 m. It is fished commercially in some areas, for example the Bay of Brest. Individuals usually mature first as males then undergo sex reversal one or more times during life. Breeding occurs during spring and autumn and longevity has been estimated to be nine to 10 years in populations on the west coast of Ireland. The sponge *Halichondria panicea* (p. 84) is often associated with C. *varia*. The sponge envelopes the bivalve and feeds on suspended material in the currents created by it. In return, it is believed that the scallop gains some protection from predators such as the starfish.

Crassadoma pusio (Linnaeus) (*Chlamys distorta*) Humpback scallop **Fig. 12.60a**
Valves dissimilar, up to 35 mm in length. 'Ears' on either side of beaks asymmetrical, anterior 2–3 times length of posterior. Young shells attached by byssus, older shells have the lower (right) valve cemented to the substratum, and shell shape varies with age, being oval when young and distorted when older. Young animals have 30–50 ribs on each valve, older animals 60–70 ribs with spines on upper valve; lower valve covered with encrusting organisms. Varied in colour, white, red, brown; may be patterned. Single adductor muscle scar on inside of shell.

Figure 12.60 (a) *Crassadoma pusio*, left valve. (b) *Palliolum tigerinum*, left valve.

C. pusio is widely distributed in north-west Europe and Britain from the extreme low shore into the sublittoral to depths of about 100 m. It is found on rocky shores attached to stones and shells. Most animals mature first as males, changing to female after they have spawned. In the Irish Sea, breeding occurs over a long period with a peak in spring/summer and another in autumn.

Palliolum tigerinum (O. F. Müller) (*Chlamys tigerina*) Tiger scallop **Fig. 12.60b**
Valves almost equal, small, convex, up to 25 mm in length. 'Ears' on either side of beaks asymmetrical, anterior 3 times the length of posterior. Sculpturing varied, smooth, with fine lines, or with about 30 or more ribs, a few of which may be broad and pronounced. Colour extremely varied, white, yellow, brown; may be patterned. Single adductor muscle scar on inside of shell.

P. tigerinum is widely distributed in north-west Europe and commonly found around Britain from the lower shore to depths of about 90 m among stones, gravel and sand. The sexes are separate and in the Irish Sea breeding occurs in June. The animals spawn when they are about one year old and then die.

Pecten maximus (Linnaeus) Great scallop, king scallop **Fig. 12.61**
Valves dissimilar, up to 150 mm in length; lower (right) valve convex, upper (left) valve flat, each with 15–17 broad ribs. Large, equal 'ears' on either side of beaks. Lower valve white, or whitish-brown; upper valve reddish-brown. Single adductor muscle scar on inside of shell.

P. maximus is widely distributed in north-west Europe, being recorded from Norway to the coasts of Spain and Portugal and into the Mediterranean. It is found all around Britain, usually sublittorally on sand and sandy-gravel to depths of about 110 m. Scallops are much prized as food and a number of commercial fisheries exist around Britain. Adult scallops are not attached to the substratum but lie on, or are partly buried in, the sediment. They are capable of swimming by opening and closing the valves of the shell and move through the water with the free edge of the shell leading. Scallops can also make sudden escape movements backwards if, for example, they are

Figure 12.61 *Pecten maximus*, right valve, and in sediment.

threatened by the starfish *Asterias* (p. 517), one of their main predators. Scallops are suspension feeders. They are hermaphroditic, the gametes of either sex being discharged first, and fertilization is external. The veliger larvae are planktonic for about three to four weeks and settle on a wide range of seaweeds, bryozoans and hydroids. The young scallops generally remain attached by a byssus until they are anything from 4 to 13 mm in length. The byssus is then lost and they become unattached. They breed first when two years old, and off the Isle of Man have a long breeding period with peaks in spring and autumn. The age of many individuals can be determined by annual rings laid down on the shell; longevity of up to 20 years has been recorded.

Aequipecten opercularis (Linnaeus) (*Chlamys opercularis*) Queen scallop **Fig. 12.62a**
Valves dissimilar, upper (left) convex, lower (right) only slightly convex; with 19–22 rounded ribs; up to 90 mm in length. Anterior 'ear' a little longer than posterior. Varied in colour including brown, pink, red; often patterned. Single adductor muscle scar on inside of shell.

A. opercularis is widely distributed in north-west Europe from Norway to the Mediterranean and occurs all around Britain. It is found sublittorally in sandy-gravel, often in high densities, only rarely occurring on the lower shore. Queen scallops are hermaphroditic and reach maturity after one year. In the Irish Sea the main breeding period is in autumn but spawning has been recorded in winter and spring. Gametes are discharged into the sea where fertilization takes place. Larval life is believed to be of three to four weeks duration and newly settled young are found on bryozoans, hydroids and sometimes on the kelp *Saccharina latissima* (p. 58). The young scallops are attached by a byssus but later become free and often swim by bringing the valves together sharply, the free edge of the shell leading. Longevity is five to six years. The queen scallop is fished commercially at a number of localities, especially around the Isle of Man.

Both valves of the shell of *A. opercularis* are often encrusted with a wide range of epifauna, common among which are sponges of the genus *Suberites* (p. 83). This relationship has been described as a 'protective–commensal mutualism' in which the

Figure 12.62 (a) *Aequipecten opercularis*, right valve. (b) *Pteria hirundo*, right valve. (c) *Limaria hians*, left valve.

sponges protect the scallop from predation by starfishes while the sponges are protected from one of their main predators, the sea-slug *Doris pseudoargus* (p. 246).

Two other bivalves with 'ears' on the shell, but not belonging to the family Pectinidae, are **Pteria hirundo** (Linnaeus) (Fig. 12.62b), the wing oyster, and **Limaria hians** (Gmelin) (Fig. 12.62c), the gaping file shell.

Pteria is up to 75 mm in length, the 'ears' are very large and the shell is brown on the outer surface and pearly white on the inner surface. It is found sublittorally, mainly in southern Britain, attached by the byssus in sandy-mud and gravel.

Limaria is up to 25 mm shell length, the 'ears' are small and the shell gapes anteriorly and posteriorly. It is whitish to whitish-brown in colour with about 50 radiating ribs. It is found in the south and west of Britain on the lower shore and into the sublittoral on gravelly substrata, and in the holdfasts of laminarians in a nest formed from byssus threads. A conspicuous feature of *Limaria* is the fringe of orange-red mantle tentacles which can be seen when the animal swims.

Family Anomiidae Saddle oysters
Shell thin, delicate, valves unequal. Attached firmly to solid substratum by a calcified byssus. Lower (right) valve with an indented hole through which byssus passes. This hole is formed by the shell growing around the byssus. Single adductor muscle scar; upper (left valve) also has byssal scars.

Upper valve with 3 distinct muscle scars (1 adductor, 2 byssal) on inner
surface .... ***Anomia ephippium*** p. 265

Upper valve with 2 muscle scars, just touching. Surface of muscle scars smooth.
Shell small, no more than 15 mm in shell length ..... ***Heteranomia squamula*** p. 266

Anomia ephippium Linnaeus Common saddle oyster **Fig. 12.63a,b,c**
Shell roughly circular; delicate in appearance; valves very dissimilar, up to 60 mm in length. Lower (right) valve thin and flat, with indented hole through which calcified byssus attaches shell to substratum; upper (left) valve larger, curved, irregular in shape. Inner surface of upper valve with 3 distinct muscle scars (1 adductor, 2 byssal). Lower valve with 1 adductor muscle scar. Upper valve may be encrusted with epifauna including barnacles and tube worms. Outer surface whitish in colour; inner surface white.

A. ephippium is widely distributed in north-west Europe. It is common in Britain from the middle and lower shore to depths of about 150 m, attached to seaweeds, stones, rocks and shells. The shape of the shell follows that of the object to which it is attached. *Anomia* is hermaphroditic and fertilization external. Breeding has been recorded during the summer months in populations in the Irish Sea when veliger larvae can be found in the plankton. Settlement occurs after 20 to 30 days.

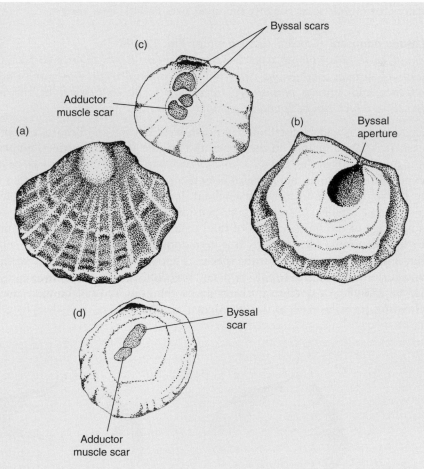

Figure 12.63 *Anomia ephippium*, (a) left valve, (b) right valve lying within left valve, (c) inner surface of left valve showing byssal and adductor muscle scars, (d) *Heteranomia squamula*, inner surface of left valve showing byssal and adductor muscle scars.

Heteranomia squamula (Linnaeus) (*Pododesmus squamula*) **Fig. 12.63d**

Shell roughly circular; delicate. Valves dissimilar, up to 15 mm in length. Lower (right) valve thin and flat, upper (left) valve larger, convex. Lower valve with oval aperture through which calcified byssus attaches shell to substratum. Inner surface of upper valve with 2 muscle scars (1 adductor, 1 byssal) just touching; surface of scars smooth. Lower valve with 1 adductor muscle scar. Upper valve white, pinkish; lower valve translucent.

H. squamula is widely distributed in north-west Europe and around Britain on the lower shore and into the sublittoral. It is found on stones and seaweed.

Two other species of saddle oyster occurring in north-west Europe are ***Pododesmus patelliformis*** (Linnaeus) and ***Pododesmus squama*** (Gmelin). They are mainly sublittoral. In both, the muscle scars on the inner surface of the upper valve of the shell are furrowed but separation of the species is difficult.

Family Lasaeidae Coin shells

Small, delicate shell; beaks towards posterior. Hinge line with poorly developed teeth. Adductor muscle scars roughly the same size. Pallial sinus absent.

Lasaea adansoni (Gmelin) (*Lasaea rubra*) **Fig. 12.64a**

Small, oval, rather plump shell, 2–3 mm in length. Beaks slightly posterior. Left valve with 1 cardinal tooth; right valve without cardinal teeth. Inner and outer surfaces whitish, with distinct pinkish-red tinge. Not to be confused with Turtonia minuta *(p. 289), which can be found in a similar habitat.*

L. adansoni is widely distributed in north-west Europe and Britain, except the southeast. It is found mainly on the upper and middle shore in crevices, empty barnacles and among tufts of the lichen *Lichina pygmaea* (p. 73), often in very large numbers on exposed rocky shores, but has also been recorded from the shingle beaches of the Fleet in Dorset on the south coast of England. It attaches by a temporary byssus, but can also crawl rapidly. *Lasaea* is a suspension feeder and highly adapted to take full advantage of the short periods of tidal cover that it experiences on the upper shore, by having a rapid filtering rate. It is hermaphroditic and believed to be self-fertilizing but it has been suggested that the species is apomictic and reproduces without meiosis and fusion of male and female gametes. Breeding occurs during the summer months. Up to 35 embryos develop in the body on the gills, and on the west coast of Wales crawling juveniles emerge during August and September at a shell length of about

Figure 12.64 (a) *Lasaea adansoni*, right valve. (b) *Tellimya ferruginosa*, right valve.

0.5 mm. There is no pelagic larval stage. Longevity is no more than three years and the animals breed in their second year.

Family Montacutidae

Small delicate shell; beaks towards posterior. Hinge line with poorly developed teeth. Adductor muscle scars roughly the same size. Pallial sinus absent. Usually commensal with echinoderms.

Tellimya ferruginosa (Montagu) (*Montacuta ferruginosa*) **Fig. 12.64b**

Delicate, oval shell, up to 9 mm in length. Beaks well to the posterior. Outer surface whitish, often with reddish-brown deposit; inner surface whitish-purple.

T. ferruginosa is common in north-west Europe and Britain, occurring in sandy deposits. It is a commensal of the sea-potato, *Echinocardium cordatum* (p. 419) and as many as 14 or more bivalves have been recorded with a single echinoderm. Adult specimens live freely in the burrow of *Echinocardium* while the young are attached to the spines of the echinoderm by byssus threads. The bivalve has also been recorded with the heart-urchins *Spatangus purpureus* (p. 419) and *Echinocardium pennatifidum* (p. 421) and the offshore species *Echinocardium flavescens* (O. F. Müller). It is a suspension feeder and presumably has an enriched food supply through its association with *Echinocardium*. *Tellimya* first matures as a female, later changing to male. Breeding occurs during the summer months and the eggs are incubated to the veliger stage on the gills of the parent. When released, the veligers are believed to have a larval life of several months.

Family Cardiidae Cockles

Rounded globular shell, with conspicuous radiating ribs. Beaks towards the anterior. Hinge line with cardinal and lateral teeth. Adductor muscle scars roughly the same size. Pallial sinus absent.

Acanthocardia echinata (Linnaeus) Prickly cockle **Fig. 12.65**

Shell stout, up to 75 mm in length; 18–22 conspicuous ribs, each bearing spines joined at the base. Beaks slightly anterior. Grooves on inner surface of shell extend long way in from edge of shell. Margin crenulate. Outer surface yellow-brown in colour; inner surface white.

Figure 12.65 *Acanthocardia echinata*, left valve.

A. echinata is widely distributed in north-west Europe and Britain. It is a sublittoral species occurring at depths of 3 m and below, living in the top few centimetres of sand, mud or gravel. Shells of dead specimens are often washed ashore. *Acanthocardia* is well known through its ability to 'leap', often as far as 200 mm, when disturbed by predators such as starfishes. It does this by rapidly extending the large foot.

    Shells of two other species, ***Acanthocardia aculeata*** (Linnaeus) and ***Acanthocardia tuberculata*** (Linnaeus), are occasionally washed ashore in south and south-west Britain. *A. tuberculata* is distinguished by the fact that the spines on the ribs of the shell are not joined at the base. The spines on the ribs of *A. aculeata*, like those of *A. echinata*, are joined at the base but these two species can be separated by examination of the hinge line; in *A. aculeata* the anterior cardinal tooth of the left valve is larger than the posterior.

Parvicardium pinnulatum (Conrad) (*Parvicardium ovale*) **Fig. 12.66a,b,c**

Small shell, up to 12 mm in length. Beaks slightly anterior. Shell with 24–26 ribs; anterior ribs with broad scales, posterior ones with small spines; ribs in between lack spines and scales. Grooves on inner surface of shell extend long way in from margin towards beak. Margin crenulate. Outer surface yellowish; inner surface white.

P. pinnulatum is widely distributed in north-west Europe and common around Britain. It is a sublittoral species occurring at depths of about 3–100 m on muddy-sand and gravel.

Parvicardium exiguum (Gmelin) Little cockle **Fig. 12.66d,e**

Small shell, up to 15 mm in length. Beaks slightly anterior. Shell with 20–22 ribs; tubercles on all ribs in young specimens, restricted mostly to anterior ribs in adults.

Figure 12.66 *Parvicardium pinnulatum*, (a) left valve, (b) anterior ribs, (c) posterior ribs. *Parvicardium exiguum*, (d) right valve, (e) anterior ribs. (After Tebble, 1976.)

Grooves on inner surface of shell present from margin to just a little way in. Margin crenulate. Outer surface whitish-yellow to brown; inner surface whitish-green.

P. exiguum is widespread in north-west Europe and around Britain in all types of sediment from the lower shore down to about 55 m. It is often found in estuaries and tolerates salinities down to about 17. Breeding takes place in spring and summer, with two peaks of recruitment recorded in north-west Spain and a single peak in Denmark.

Cerastoderma edule (Linnaeus) (*Cardium edule*) Common cockle **Fig. 12.67**
Up to 50 mm in length, but usually less. Beaks just anterior. Shell with 22–28 ribs; concentric growth lines conspicuous. Grooves on inner surface of shell present from margin to just a little way in. Margin crenulate. Outer surface off-white, yellowish or brownish; inner surface white with brownish mark near posterior adductor muscle scar.

C. edule is widely distributed in north-west Europe and around Britain on the middle and lower shore, sometimes sublittorally, in sand and muddy-sand. It usually lives in salinities between 15 and 35 and is often abundant in estuaries, as well as in sheltered bays. In areas such as Morecambe Bay, the Wash and South Wales, the cockle is fished commercially by hand-raking. Cockles are suspension feeders, found buried just below the surface of the sediment, where they are readily collected with a rake. They are easily dislodged by storms and during winter gales beds of cockles can be washed away. They have many predators, notable among which is the oystercatcher. The sexes are separate, fertilization is external and breeding takes place during spring and summer. The free-swimming veliger larvae spend some three to six weeks in the plankton before settling at a shell length of about 6 mm. Newly settled cockles have been observed to leave the sediment and drift in the water column and are dispersed further before their final settlement. While in the water column, sinking is slowed by

Figure 12.67 *Cerastoderma edule*, (a) left valve, (b) right valve, (c) in sediment. Arrows indicate direction of flow of water.

the production of byssal threads. Cockles live for up to nine years or more, but two to four years is more usual.

Cerastoderma glaucum (Poiret) (*Cardium lamarckii, Cardium glaucum*) Lagoon cockle **Fig. 12.68**
Thinnish shell, up to 50 mm in length, usually smaller. Beaks just anterior. Shell elongate posteriorly; 22–28 ribs; concentric growth lines conspicuous. Grooves on inner surface of shell extend long way in from margin towards beak. Margin crenulate at anterior, smooth at posterior. Outer and inner surfaces of shell coloured shades of brown.

Figure 12.68 *Cerastoderma glaucum*, (a) right valve, (b) left valve.

C. glaucum is widely distributed in north-west Europe and in Britain is found mainly in the south-east. It occurs predominantly in enclosed lagoons, usually permanently submerged, and has a salinity preference of 5 to 38. Its distribution is believed to be restricted by the damaging effect of wave action on the newly settled spat. It is a suspension feeder. The sexes are separate, fertilization is external and the veliger larvae have a pelagic life of from one to three weeks. Newly settled young attach temporarily by byssus threads to filamentous algae before becoming buried in the top few centimetres of sediment. Breeding occurs during late spring and early summer and individuals live for about five years.

Family Mactridae Trough shells, otter shells

Triangular to oval shell; beaks near mid-line. Below each beak is a triangular depression, the chondrophore, which houses the internal ligament. Hinge line with cardinal and lateral teeth, the laterals sometimes poorly developed. Cardinal teeth of left valve joined to give a ∧-shaped ridge in front of chondrophore. Adductor muscle scars roughly the same size. Pallial sinus present.

1 Shell elongate, gaping at anterior and posterior; thick brownish-green periostracum. Large shell, up to about 125 mm length ..... ***Lutraria lutraria*** p. 272
   Shell triangular, not gaping; without thick periostracum. Up to about 50 mm shell length **2**

2 Lateral teeth smooth .... ***Mactra stultorum*** p. 271
   Lateral teeth finely serrated .... ***Spisula solida*** p. 272

Mactra stultorum (Linnaeus) (*Mactra corallina*) Rayed trough shell **Fig. 12.69**
Smooth, triangular shell; rather thin and fragile; up to 50 mm in length. Beaks midway between anterior and posterior ends of shell. Chondrophores conspicuous; ∧-shaped ridge in front of chondrophore of left valve; lateral teeth of hinge line smooth (see Spisula, *below). Sculpturing of many very fine, concentric lines; shell feels smooth. Outer surface whitish-brown with broad, brown or yellowish rays; inner surface purplish-white. Margin smooth.*

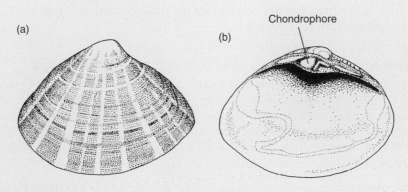

Figure 12.69 *Mactra stultorum*, (a) right valve, (b) left valve.

M. stultorum is widely distributed in north-west Europe and Britain from the very low shore to depths of about 100 m buried in clean sand. Offshore, *Mactra* oftens occurs in vast densities and empty shells are washed ashore in large numbers. It is a suspension feeder. On the south coast of England breeding occurs during spring.

Spisula solida (Linnaeus) Thick trough shell **Fig. 12.70**

Thick triangular shell up to 45 mm in length. Beaks usually mid-way between anterior and posterior ends of shell. Chondrophores conspicuous; ∧-shaped ridge in front of chondrophore of left valve small, spans about half the depth of the hinge line. Lateral teeth of hinge finely serrated (see Mactra, *above). Sculpturing of many concentric lines and several flat concentric ridges. Outer surface off-white; inner surface white. Margin smooth.*

Figure 12.70 *Spisula solida*, (a) right valve, (b) left valve.

S. solida is widely distributed in north-west Europe and Britain, living in sand from extreme low water into the sublittoral where it often occurs with *Mactra* in very high densities. It is a suspension feeder and longevity off the north Wales coast has been estimated to be up to 12 years with first maturity at about six years, whereas off the coast of southern Portugal maturity is reached in the first year.

Two other species, **Spisula elliptica** (Brown) and **Spisula subtruncata** (da Costa), commonly occur in the shallow sublittoral around Britain, often in high densities. Separation of the three species is difficult but *S. solida* is distinguished by the small ∧-shaped ridge formed by the fusion of the cardinal teeth of the left valve of the shell. In *S. elliptica* and *S. subtruncata* the ridge is relatively large and spans almost the complete depth of the hinge line.

Lutraria lutraria (Linnaeus) Common otter shell **Fig. 12.71; Plate 25**

Large, oval shell, up to 125 mm in length; shell gapes at anterior and posterior ends. Beaks slightly anterior. Chondrophores conspicuous; ∧-shaped ridge in front of chondrophore of left valve. Sculpturing of concentric lines. Outer surface white, pale orange, pale brown with thick brownish-green periostracum; inner surface white. Pallial line separate from lower margin of pallial sinus. Margin smooth. Not to be confused with Mya arenaria *(p. 291).*

L. lutraria is widely distributed in north-west Europe and Britain from the lower shore to about 90 m. It lives buried in muddy-sand and gravel at depths of up to 300 mm, maintaining contact with the surface by the very long, united siphons which have a

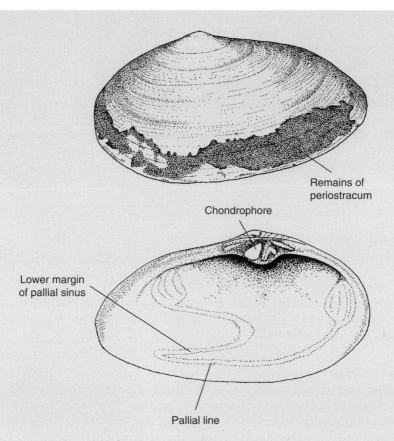

Figure 12.71 *Lutraria lutraria*, left valve. Note the pallial line and pallial sinus are separate.

tough, outer covering. The foot is small and the animal more or less sedentary, penetrating deeper into the sediment with age. It is a suspension feeder.

Two other species, ***Lutraria magna*** (da Costa) and ***Lutraria angustior*** Philippi, occur in south and south-west Britain. They are sublittoral. Shells of *L. lutraria* can be distinguished from these two species by examination of the inside of the shell. In *L. lutraria* the pallial line and the lower margin of the pallial sinus are separate while in the other species they are fused.

Family Solenidae Razor-shells
Shell long and narrow, fragile; vertical groove at anterior end. Beaks at anterior, inconspicuous. Brownish-green periostracum. Shell gaping at anterior and posterior. Hinge line with cardinal teeth; lateral teeth absent. Anterior adductor muscle scar larger than posterior. Pallial sinus present.

Solen marginatus Pulteney Grooved razor-shell **Fig. 12.72**
Fragile, elongate shell, up to 125 mm in length; straight dorsally and ventrally. Beaks small, at anterior of shell. Valves gaping at both ends; narrow, vertical groove at anterior end. Shell smooth, yellowish with pale brownish-green periostracum; inner surface white. Foot dull red.

Figure 12.72 *Solen marginatus*, (a) right valve, (b) left valve.

S. marginatus is widely distributed in north-west Europe but in Britain appears to be confined to south and west coasts. It is found in sand or muddy-sand on the lower shore and into the sublittoral.

Family Pharidae Razor-shells
Shell long, narrow and fragile; without vertical groove at anterior end (see *Solen*, above). Beaks at anterior, or just anterior of midline; inconspicuous. Shell gaping at anterior and posterior. Hinge line usually with cardinal and lateral teeth. Adductor muscle scars roughly the same size, or anterior larger than posterior. Pallial sinus present.

Razor-shells have high market value and are fished commercially at a number of localities in northern Europe using hydraulic dredges.

Ensis ensis (Linnaeus) Razor-shell **Fig. 12.73**
Narrow, fragile shell, up to 125 mm in length; dorsal and ventral margins both curved. Beaks small, at anterior of shell. Anterior and posterior ends gaping. Outer surface smooth, whitish colour with vertical and horizontal reddish-brown or purplish-brown

Figure 12.73 *Ensis ensis*, (a) right valve, (b) left valve, (c) position in sediment when feeding. Arrows indicate direction of water flow.

markings separated by diagonal line. Periostracum olive-green. Inner surface white with purple tinge. Foot pale red-brown.

E. ensis is widely distributed in north-west Europe from Norway to the Mediterranean. It is found all around Britain in sand, sometimes muddy-sand, from extreme low water into the shallow sublittoral. All species of *Ensis* have a large foot and are rapid, deep burrowers, the shell moving rapidly down into the vertical burrow when disturbed. If this happens when the animal is emersed at low water, a spout of water is ejected from the burrow indicating their presence in the sand. Their position is also indicated by keyhole-shaped openings made by the short, united siphons which extend just above the sediment surface when the animal is suspension feeding. Maturity is reached after three years and breeding probably occurs during spring. The veliger larva has a pelagic life of about a month and longevity is believed to be seven years and more.

Ensis arcuatus (Jeffreys) Razor-shell **Fig. 12.74a,b**
Fragile, elongate shell, up to 150 mm in length; dorsal margin straight, ventral margin curved. Beaks small, at anterior of shell. Valves gaping at both ends. Shell smooth; outer and inner surfaces and periostracum similar to E. ensis *(above). Foot creamy-white with brown lines.*

E. arcuatus is widely distributed in north-west Europe extending from Norway to Spain. It is found all around Britain in coarse sand and fine gravel from the lower shore to about 35 m. It lives in a coarser sediment than either *E. ensis* (above) or *E. siliqua* (p. 276). Off the coast of Spain, *E. arcuatus* spawns in winter/spring and the free-swimming veliger larva has a life of about 20 days.

Ensis directus (Conrad) (*Ensis americanus*) American jack-knife clam **Fig. 12.74c**
Dorsal margin of shell almost straight, ventral margin curved, up to 120 mm in length. Similar to that of Ensis arcuatus *(above). Reference to inside of shell is necessary to confirm identification. In* E. directus *the anterior adductor muscle scar is equal to or*

Figure 12.74 *Ensis arcuatus*, (a) right valve, (b) left valve. (c) *Ensis directus*, left valve.

only slightly longer than the external ligament, while in E. arcuatus *the scar is much longer than the ligament. In addition, the posterior adductor muscle scar is very close to the pallial sinus in* E. directus *but at least its own length from the sinus in* E. arcuatus.

E. directus was first recorded in Europe in 1979 on the North Sea coast of Germany and it has been suggested that larvae arrived there in the ballast water of ships. Since that time, the species has spread rapidly in Europe and has been found in densities in excess of 1000 per square metre in some localities. It is common on the east coast of England south of the Humber into the Wash and extends south into the English Channel where it is found on the East Sussex coast. Empty shells have been found in South Wales. In the area of the Wash it has been described as 'dominating the molluscan fauna'. *E. directus* is found from the lower shore into the sublittoral to depths of about 20 m. It lives in silty sands compared with the more sandy habitats preferred by the other species of *Ensis* and being tolerant of brackish water extends into estuaries. Spawning takes place in March/April and the larvae are pelagic for up to four weeks. Most individuals live for about four years but longevity is believed to be up to seven years.

Ensis siliqua (Linnaeus) Pod razor-shell **Fig. 12.75**
Fragile, elongate shell, up to 200 mm in length; dorsal and ventral margins straight. Beaks small, at anterior of shell. Valves gaping at both ends. Shell smooth; outer and inner surfaces and periostracum as E. ensis *(p. 274). Foot creamy-white with brown lines.*

Figure 12.75 *Ensis siliqua*, right valve.

E. siliqua is widely distributed in north-west Europe from Norway to the Mediterranean. It is found all around Britain in clean sand on the lower shore and in the shallow sublittoral. Off the east coast of Ireland, maturity is reached at an age of four years and spawning occurs from May to August. There is a free-swimming veliger larva. Longevity is up to 19 years.

Pharus legumen (Linnaeus) **Fig. 12.76**
Narrow, fragile shell, up to 120 mm in length. Beaks small and slightly anterior. Wide gape at anterior and posterior ends of shell. Sculpturing of concentric lines. Outer surface off-white to light brown with yellowish-green periostracum; inner surface white. Margin smooth. Similar in appearance to other razor-shells, but beaks are near midline of shell, not at anterior end.

Figure 12.76 *Pharus legumen*, left valve.

P. legumen is a southern species extending into south-west Britain about as far as North Wales, living in fairly clean sand on the lower shore and shallow sublittoral. It has a large, pink foot and is a rapid vertical burrower, being very difficult to catch. Unlike the other razor-shells described above, the siphons are separate throughout their length and *Pharus* is a deposit feeder. There is a free-swimming veliger larva and settlement has been recorded in summer. Length of life is up to six years. The siphons of *Pharus* are known to be cropped extensively by fishes.

Family Tellinidae Tellins
Thin, mostly flattened, relatively smooth shell. Beaks central or slightly posterior. Hinge line usually with cardinal and lateral teeth. Adductor muscle scars roughly the same size. Pallial sinus present.

1 Shell rounded, rather globular. Without lateral teeth ..... ***Macoma balthica*** p. 279
Shell oval, flattened; rather fragile. Lateral teeth present **2**

2 Shell sculpturing of concentric lines; faint diagonal lines on right valve only
.... ***Tellina fabula*** p. 278
Shell sculpturing of concentric lines; no diagonal lines .... ***Tellina tenuis*** p. 278

Tellina tenuis da Costa (*Angulus tenuis*) Thin tellin **Fig. 12.77a**
Fragile, oval shell, up to 25 mm in length; flattened; right valve slightly larger than left. Beaks a little to the posterior. Sculpturing of concentric lines. Outer surface shiny, usually pink but often orange or white; inner surface similar. Margin smooth. Not to be confused with Macoma *(p. 279).*

Figure 12.77 (a) *Tellina tennis*, right valve. Anterior profile inset. (b) *Macoma balthica*, right valve. Anterior profile inset.

T. tenuis is widely distributed in north-west Europe and Britain, where it occurs in fine, clean sand from the middle shore into the shallow sublittoral. In some areas, it is abundant. *Tellina* usually lies on the left valve a few centimetres below the surface of the sediment. The siphons are long and separate and, despite earlier reports that it was a deposit feeder, it is now suggested that it is mainly a suspension feeder. Young flatfishes often feed on the tips of the siphons when they are exposed above the surface of the sediment, but the siphons are regenerated and the bivalve usually survives. The sexes are separate and breeding occurs during summer. There is a free-swimming veliger larva and length of life is believed to be about five years but there are records of individuals living for up to 10 years

Tellina fabula Gmelin (*Fabulina fabula*) **Fig. 12.78**
Fragile, oval shell, up to 20 mm in length: flattened; right valve slightly more convex than left. Beaks a little to the posterior. Sculpturing of concentric lines on both valves; faint, diagonal lines on right valve. Outer surface shiny, whitish: inner surface similar. Margin smooth. Not to be confused with Macoma *(below).*

T. fabula is widely distributed in north-west Europe and Britain in fairly clean sand but, in contrast to *T. tenuis* (above), it is essentially a sublittoral species extending from the lower shore to about 55 m. It is found up to 100 mm deep in the sediment and lies in a more-or-less horizontal position on the left valve. It is both a suspension

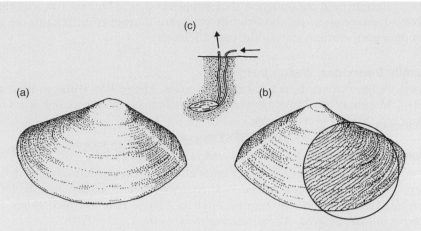

Figure 12.78 *Tellina fabula*, (a) left valve, (b) right valve, part magnified (in circle) to show diagonal striations, (c) in sediment. Arrows indicate direction of water flow.

and a deposit feeder. The sexes are separate and in the eastern North Sea breeding occurs from March to September, with those animals that breed in March breeding again in summer. There is a free-swimming veliger larva. Longevity is believed to be four or five years.

Macoma balthica (Linnaeus) Baltic tellin **Fig. 12.77b**

Rounded, rather globular shell, up to 25 mm in length; valves almost equal. Beaks midway between anterior and posterior ends of shell. Sculpturing of fine, concentric lines. Hinge line without lateral teeth. Outer surface dull, usually pinkish-purple but sometimes yellowish or white and banded; inner surface similar. Margin smooth. Not to be confused with Tellina tenuis *and* Tellina fabula *(above).*

M. balthica is widely distributed in north-west Europe and Britain, extending from Norway to the coasts of Portugal and Spain. It lives a few centimetres below the surface of sand, mud and muddy-sand and is found from the middle shore into the sublittoral. It survives in salinities down to 5 and occurs in estuaries, often with *Cerastoderma edule* (p. 269). The siphons are long and separate and although generally considered to be mainly a deposit feeder, it switches from deposit to suspension feeding when there is a high density of suspended particles in the water. During feeding, the extended siphons are frequently cropped by predators such as flatfishes. The siphons are quickly regenerated but in some situations they are so frequently cropped that the growth rate of the bivalve is inhibited. The sexes are separate and on the west coast of Wales the main breeding period is during February and March, with a second spawning in autumn in some years. There is a free-swimming veliger larva which is pelagic for up to five weeks. In some locations, after settlement the tiny bivalves leave the sediment and drift in the water column. They do this by a behaviour known as byssal drifting in which byssal threads decrease sinking rate. In this way they can disperse over considerable distances and small *Macoma* have been observed to undergo migration to higher beach levels during the summer. This is believed to reduce predation by shrimps and small crabs. Where growth of *Macoma* is fast, longevity is about three years, but in slow-growing populations specimens live for six or seven years.

Recent field data suggest that populations of *Macoma* towards the southern limit of distribution on the Spanish coast are disappearing because of increase in summer temperatures during the past 10 years.

Family Donacidae Wedge shells

Shell wedge-shaped. Beaks towards posterior. Hinge line with cardinal and lateral teeth. Margin of shell finely to strongly crenulate. Adductor muscle scars roughly the same size. Pallial sinus present.

Donax vittatus (da Costa) Banded wedge shell **Fig. 12.79**

Shell up to 38 mm in length. Beaks quite close to posterior end of shell. Sculpturing of very fine, radiating lines, absent near anterior margin. Outer surface shiny, orange, purple, yellow or brown; colours often in bands; inner surface shiny, white with purple or yellow areas. Margin strongly crenulate.

Figure 12.79 *Donax vittatus*, right valve.

D. vittatus is widely distributed in north-west Europe and Britain on exposed sandy beaches, from the lower shore to about 20 m. It lives just under the surface of the sediment and is often dislodged by pounding surf, but the presence of a large, powerful foot enables it to reburrow quickly and so reduce the dangers of predation and desiccation. It is a suspension feeder. Damage to siphons and foot through predation by flatfish is common in most populations but the bivalves generally survive. The sexes are separate and on the west coast of Scotland breeding has been recorded during summer. Where growth is rapid, *Donax* lives for two to three years but where it is slow, length of life may be up to seven years.

Family Psammobiidae Sunset shells

Shell oblong. Beaks slightly anterior or slightly posterior. Hinge line with cardinal teeth, but no laterals. Adductor muscle scars roughly the same size. Pallial sinus present.

Gari fervensis (Gmelin) Faroe sunset shell **Fig. 12.80**

Oblong shell, up to 50 mm in length; right valve slightly more convex and slightly larger than left. Beaks slightly anterior. Posterior margin truncate. Sculpturing of concentric lines and a conspicuous, diagonal posterior ridge. Outer surface pink, white or light brown with whitish rays; periostracum greenish-brown; inner surface white, purple or pink. Margin smooth.

G. fervensis is widely distributed in north-west Europe and Britain from the lower shore into the sublittoral. It is found in the top few centimetres of coarse sand and gravel. *Gari* is a suspension feeder.

Figure 12.80 *Gari fervensis*, right valve.

Gari depressa (Pennant) is occasionally found on the very low shore, particularly in the south-west. The shell is more oval than that of *G. fervensis* and the posterior margin is not truncate.

Family Scrobiculariidae

Rounded, thin shell, flattened. Beaks slightly anterior. Chondrophores conspicuous. Hinge line with cardinal teeth; no lateral teeth. Anterior adductor muscle scar slightly larger than posterior. Pallial sinus present.

Scrobicularia plana (da Costa) Peppery furrow shell **Fig. 12.81; Plate 26a,b**
Rounded, thin shell; much flattened; up to 63 mm in length. Beaks slightly anterior. Chondrophores conspicuous, semicircular. Without lateral teeth. Sculpturing of fine,

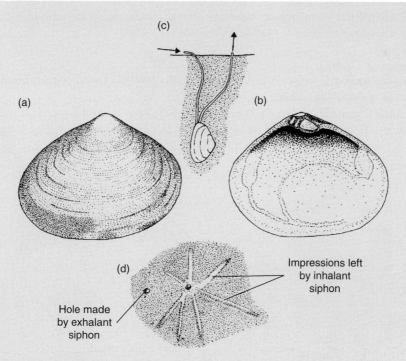

Figure 12.81 *Scrobicularia plana*, (a), (b) right valve, (c) in sediment (arrows indicate direction of water flow), (d) impressions left on surface of sediment by siphons.

concentric lines. Outer surface white, pale grey or yellowish; inner surface white. Margin smooth.

S. plana is a common intertidal bivalve in north-west Europe and occurs all around Britain, often in high densities. It is essentially estuarine and intertidal, surviving in salinities down to 10 in thick mud or muddy-sand, where it lives up to 200 mm below the surface of the sediment. It is mainly a deposit feeder and the star-shaped markings left on the surface of the sediment by the long inhalant siphon indicate the presence of *Scrobicularia*. When the siphons protrude above the sediment during feeding, they are often cropped by fishes, crabs and birds, but lost tissue is regenerated within four or five days. *Scrobicularia* is also known to suspension feed. The sexes are separate and on the coast of North Wales, breeding occurs during July and August but may be earlier in other areas. The free-swimming veliger larvae settle after a planktonic life of two or three weeks. Some specimens live for up to 18 years.

Interestingly, endocrine disruption has been recorded in populations of *Scrobicularia* from a number of estuaries in south-west England. This causes disruption of the normal pattern of development in males and an ovotestis is formed. This is believed to be similar to the now well-documented cases of endocrine disruption in male fish caused by natural and man-made substances in industrial effluents.

Family Arcticidae Cyprinas

Thick, heavy shell; beaks towards anterior. Periostracum thick and shiny. Hinge line with 3 cardinal teeth and a posterior lateral tooth. Anterior adductor muscle scar somewhat smaller than posterior. Pallial sinus absent.

Arctica islandica (Linnaeus) (*Cyprina islandica*) Iceland cyprina, ocean quahog, mahogany clam **Fig. 12.82**
Thick, heavy shell, oval to rounded; up to 120 mm in length; sculptured with concentric lines. Beaks well to the anterior. Periostracum thick, shiny and pale brown to

Figure 12.82 *Arctica islandica*, left valve.

black in colour, often peels away in dead specimens to reveal white to pale brown outer surface of shell beneath; inner surface white.

A. islandica is widely distributed in north-west Europe, extending from Iceland to Spain. It occurs all around Britain from the very low shore into the sublittoral to depths of up to 100 m. It lies buried in sandy and muddy sediments where it suspension feeds. The sexes are separate and although it has been suggested that breeding takes place throughout the year, the main breeding period is in spring and summer. Studies on the growth of *Arctica* have shown that it is slow growing and estimates of age of individuals based on annual growth-ring analysis suggest longevity of over 200 years. The oldest known specimen, with an estimated age of 405 to 410 years, was collected off Iceland in 2006. Shell analysis of such long-lived specimens is providing valuable information on the chronology of change in the oceans.

Family Veneridae Carpet shells, venus shells
Rounded to oval shell. Beaks towards anterior. Hinge line with cardinal teeth, sometimes lateral teeth. Heart-shaped region, the lunule (Fig. 12.83c), in front of beaks in all species described in this text. Adductor muscle scars roughly the same size. Pallial sinus present.

1 Inner margin of shell crenulate, at least in part **2**
Inner margin of shell without crenulations **5**

2 Shell sculpturing of 40–50 radiating ribs and many concentric grooves. Has some resemblance to a cockle .... ***Timoclea ovata*** p. 285
Shell sculpturing predominantly of concentric ridges and lines **3**

3 Shell sculpturing of concentric ridges with conspicuous raised tubercles near anterior and especially towards posterior .... ***Venus verrucosa*** p. 284
Shell without raised tubercles **4**

4 Sculpturing of concentric lines between thick, prominent, concentric ridges .... ***Clausinella fasciata*** p. 286
Sculpturing of many narrow, concentric ridges. Outer surface usually with 3 broad reddish-brown rays .... ***Chamelea gallina*** p. 284

5 Shell rounded, whitish in colour, lunule conspicuous **6**
Shell oval or elongate, lunule usually indistinct **8**

6 Shell oval, region anterior to lunule sloping, not convex. Anterior lateral tooth of left valve prominent, ridge-like .... ***Callista chione*** p. 289
Shell broadly circular, region anterior to lunule convex. Anterior lateral tooth of left valve knob-like **7**

7 Region anterior to lunule convex but relatively low .... ***Dosinia lupinus*** p. 288
Region anterior to lunule markedly convex, raised .... ***Dosinia exoleta*** p. 288

8 Shell sculpturing without radiating lines; with many concentric ridges and lines. Lunule indistinct in larger shells .... ***Tapes rhomboides*** p. 286
Shell sculpturing of radiating lines, although sometimes faint, and concentric grooves and ridges **9**

Venus verrucosa Linnaeus Warty venus **Fig. 12.83**

Stout, rounded shell, up to 63 mm in length. Beaks slightly anterior; lunule brown, conspicuous. Sculpturing of thick, concentric ridges with tubercles near anterior and especially towards posterior. Outer surface off-white or brown; inner surface white, sometimes purple or brown near posterior adductor muscle scar. Inner margin crenulate except at posterior.

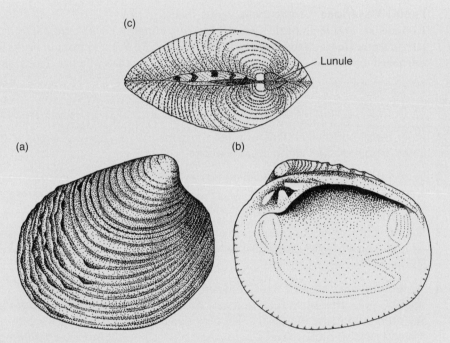

Figure 12.83 *Venus verrucosa*, (a), (b) right valve, (c) dorsal view.

V. verrucosa is a southern species which extends as far north as the west coast of Scotland. It is often found on the south and south-west coasts of Britain just below the surface in sand and fine gravel, from very low on the shore to about 100 m. It is a suspension feeder.

Chamelea gallina (Linnaeus) (*Chamelea striatula*, *Venus striatula*) Striped venus **Fig. 12.84**

Shell triangular, up to 45 mm in length. Beaks slightly anterior; lunule conspicuous. Sculpturing of many narrow, concentric ridges. Outer surface off-white, usually with 3 broad, reddish-brown rays; inner surface whitish. Inner margin crenulate towards posterior.

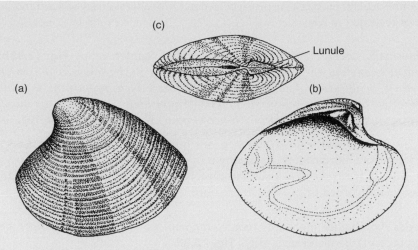

Figure 12.84 *Chamelea gallina*, (a), (b) left valve, (c) dorsal view.

C. gallina is widely distributed in north-west Europe and is commonly found in Britain buried some little way below the surface in sand and muddy-sand from the lower shore to depths of about 55 m. It is a suspension feeder. The sexes are separate, fertilization is external and on the west coast of Scotland breeding has been recorded during spring and summer. Free-swimming veliger larvae are common in the plankton during autumn. Longevity is believed to be 10 or 11 years.

Timoclea ovata (Pennant) (*Venus ovata*) Oval venus **Fig. 12.85a**
Shell thin; oval to triangular in shape; up to 20 mm in length. Beaks slightly anterior; lunule faint. Sculpturing of 40–50 radiating ribs and many concentric grooves. Outer surface off-white to light brown; inner surface whitish-orange with purple tinge. Inner margin crenulate for most of length. Bears some resemblance to a cockle.

T. ovata is widely distributed in north-west Europe and is found all around Britain. It lives just below the surface in sand, muddy-sand and gravel, at depths ranging from

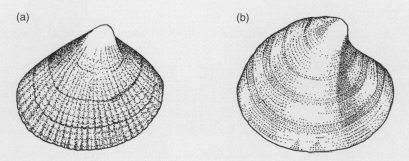

Figure 12.85 (a) *Timoclea ovata*, right valve. (b) *Clausinella fasciata*, right valve.

about 3 to 180 m. It is a suspension feeder. On the west coast of Scotland breeding has been recorded in May.

Clausinella fasciata (da Costa) (*Venus fasciata*) Banded venus **Fig. 12.85b**
Thick, triangular shell, up to 25 mm in length. Beaks towards the anterior; lunule distinct. Sculpturing of concentric lines between thick, prominent, concentric ridges. Outer surface white, yellow or pinkish-brown with darker rays or markings; inner surface white to whitish-purple. Inner margin crenulate in anterior region only.

C. fasciata is widely distributed in north-west Europe and is common all around Britain from extreme low water down to about 110 m. It lives just below the surface in sand and gravel and is a suspension feeder. On the west coast of Scotland breeding has been recorded from February to July.

Tapes rhomboides (Pennant) (*Paphia rhomboides, Venerupis rhomboides, Venus rhomboides*) Banded carpet shell **Fig. 12.86**
Elongate, oval shell, up to 60 mm in length. Beaks quite close to anterior end of shell; lunule narrow, indistinct in larger shells. Sculpturing of many concentric ridges and lines; without radiating lines, cf. Tapes decussata *(below) and* Tapes corrugata *(p. 287). Outer surface creamy-yellow with purple, brown markings, occasionally no markings; inner surface glossy and white. Margin smooth.*

T. rhomboides is widely distributed in north-west Europe and Britain from the very low shore to about 180 m in gravel and sand. It is a suspension feeder.

(a)   (b)

Figure 12.86 *Tapes rhomboides*, (a) right valve, (b) left valve.

Tapes decussata (Linnaeus) (*Tapes decussatus, Venerupis decussata*) Chequered carpet shell **Fig. 12.87**
Oval shell, up to 70 mm in length. Beaks quite close to anterior end of shell; lunule inconspicuous. Sculpturing of concentric grooves and distinct, radiating lines. Outer surface white, yellowish or pale brown, often with dark brown marks; inner surface shiny, whitish-orange, sometimes with purple marks. Margin smooth. Distinguished from T. rhomboides *(above) and* T. corrugata *(below) by the distinct radiating lines.*

T. decussata is a southern species reaching its northern limit in the south and west of Britain. It lives buried in sand or gravel on the lower shore and in the sublittoral. It is a suspension feeder. On the west coast of Scotland breeding has been recorded in July. *T. decussata* has a high commercial value.

Figure 12.87 *Tapes decussata*, left valve.

The Manila clam, ***Tapes philippinarum*** (Adams & Reeve) (*Ruditapes philip-pinarum*) was introduced into Poole Harbour for aquaculture in 1988. Despite the assumption that the species was unlikely to breed naturally in UK waters because of low temperature, successful reproduction has been reported and naturalized populations are now established. On the south coast of England breeding has been recorded from June to October when sea water temperature is about 14 °C. The species is distinguished from *T. decussata* by the distinct, dark shell patterning and coarser radiating lines, especially on the posterior part of the shell.

Tapes corrugata (Gmelin) (*Venerupis pullastra, Venerupis senegalensis*) Pullet carpet shell **Fig. 12.88**

Oval shell, up to 50 mm in length. Beaks quite close to anterior end of shell; lunule indistinct, especially in larger specimens. Sculpturing of many concentric ridges and faint, radiating lines. Outer surface off-white, cream or grey, sometimes with purple or brown markings; inner surface shiny white, occasionally with purple tinges. Margin smooth. Distinguished from T. rhomboides *(p. 286) by the sculpturing and the more angular appearance of the posterior part of the shell.*

T. corrugata is widely distributed in north-west Europe and Britain from the middle shore to depths of about 35 m. It lives in sand and gravel, often at the base of rocks, and is sometimes buried in the sediment to depths of 30–50 mm, attached to small stones or shells by the byssus. It is also found in rock crevices. *T. corrugata* is a

Figure 12.88 *Tapes corrugata*, left valve.

suspension feeder. The sexes are separate and on the west coast of Scotland breeding occurs from May to September, with free-swimming veliger larvae appearing in the plankton from June onwards. Larval life is about 30 days and individuals live for up to eight years.

Dosinia lupinus (Linnaeus) Smooth artemis **Fig. 12.89a,b,c,d**

Rounded shell, up to 38 mm in length. Beaks slightly anterior; lunule conspicuous. Region anterior to lunule convex but relatively low. Sculpturing of very many narrow, concentric ridges giving smooth appearance. Outer surface of shell off-white, yellowish or pale brown; inner surface white. Margin smooth.

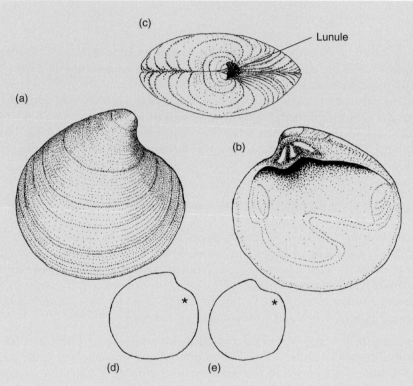

Figure 12.89 *Dosinia lupinus*, (a), (b) right valve, (c) dorsal view. (d) Profile of *D. lupinus*. (e) Profile of *Dosinia exoleta*. Note shape of shell at*.

D. lupinus is widely distributed in a variety of substrata from fine sand and muddy-sand to fine shell gravel. It is found from the lower shore to about 125 m and burrows deep into the sediment. It is a suspension feeder.

Dosinia exoleta (Linnaeus) Rayed artemis **Fig. 12.89e**

Rounded shell, up to 60 mm in length. Similar to D. lupinus *(above). Beaks slightly anterior; lunule conspicuous. Region anterior to lunule markedly convex, raised. Sculpturing of many concentric ridges. Outer surface white to light brown, with brown or pink rays; inner surface white. Margin smooth.*

D. *exoleta* is widely distributed in north-west Europe and around Britain, where it is found from the lower shore to about 70 m in gravel, burrowing deep into the sediment. It is a suspension feeder.

Callista chione (Linnaeus) Smooth venus, brown venus **Fig. 12.90**
Oval, solid shell, up to 90 mm in length. Beaks prominent, anterior to midline; lunule distinct. Left valve with single, prominent ridge-like anterior lateral tooth which fits into pit in right valve. Region anterior to lunule sloping, not convex. Sculpturing of fine, concentric lines. Outer surface of shell reddish-brown colour, often with polished appearance; inner surface white. Margin smooth.

Figure 12.90 *Callista chione*, left valve.

C. *chione* is an offshore species recorded from the Mediterranean to south-west England and North Wales. It is found in sand and sandy mud to depths of 130 m. Dead shells are frequently washed ashore and in some areas may be common. *Callista* is a suspension feeder. Maturity is reached in the second year. Spawning has been recorded over most of the year with as many as three spawning peaks. Longevity is up to 17 years. *Callista* is fished commercially off Portugal and Spain and in the Mediterranean.

Family Turtoniidae
Small, oval shell, fragile. Beaks towards anterior. Lunule narrow. Hinge line with cardinal teeth; without lateral teeth. Adductor muscle scars, pallial line and pallial sinus difficult to see. Attached by byssus.

Turtonia minuta (O. Fabricius) (*Venus minuta*) **Fig. 12.91**
Oval shell, up to about 3 mm in length. Rather plump. Beaks towards anterior. Lunule narrow and may be indistinct. Teeth of hinge line not well developed; 3 cardinal teeth but in each valve one of these is not distinct. Without lateral teeth. Pallial line may be difficult to see. Pale brown colour with pinkish tinge. Attached by byssus in crevices, among barnacles and seaweeds on rocky shores. Some shell characteristics difficult to detect. Shell size and habitat are important distinguishing features.

T. *minuta* is widely distributed in north-west Europe. It has been recorded throughout the intertidal region and is sometimes abundant on the high shore in crevices, among barnacles and seaweeds, in situations similar to those occupied by *Lasaea adansoni* (p. 266) with which it might be confused. It is a suspension feeder. Reproduction

(a)   (b)

Figure 12.91 *Turtonia minuta*, (a) left valve, (b) right valve.

occurs in summer. The sexes are separate and the eggs are fertilized in the mantle cavity of the female. The female produces gelatinous egg capsules which are attached to what have been described as modified byssus threads. Each capsule contains on average about four eggs, which develop directly into young bivalves, hatching with a shell length of about 0.3 mm.

Family Myidae Gapers

Shell valves dissimilar, right more convex than left. Beaks slightly anterior or slightly posterior. Brownish periostracum. Shell valves gape at posterior. Hinge line without teeth but in the species described in this text, a spoon-shaped chondrophore projects from the left valve; chondrophore of right valve not projecting. Anterior adductor muscle scar narrow, posterior scar rounded. Pallial sinus present.

Posterior margin of shell truncate .... ***Mya truncata*** p. 290

Posterior margin of shell slightly elongate, rounded .... ***Mya arenaria*** p. 291

Mya truncata Linnaeus Blunt gaper **Fig. 12.92**
Shell up to 75 mm in length; valves dissimilar, right more convex than left. Beaks slightly posterior. Posterior margin truncate with a large gape. Hinge of left valve with conspicuous, spoon-shaped projection, the chondrophore, to which ligament is attached. Sculpturing of concentric lines. Outer surface off-white, with brownish periostracum; inner surface white.

M. truncata is a northern species, widely distributed in north-west Europe and Britain. It lives buried at considerable depths in muddy-sand, from the middle shore to 70 m. *M. truncata* is a suspension feeder. There is a free-swimming veliger larva.

Figure 12.92 *Mya truncata*, right valve.

Mya arenaria Linnaeus Sand gaper, soft-shelled clam **Fig. 12.93; Plate 26c**
Shell up to 150 mm in length; valves dissimilar; right more convex than left. Beaks slightly anterior. Posterior part of shell somewhat elongate, rounded, with wide gape. Hinge of left valve with large, spoon-shaped projection, the chondrophore, to which ligament is attached. Sculpturing of concentric lines. Outer surface off-white, greyish or brownish, with pale brown periostracum; inner surface whitish-brown. Not to be confused with Lutraria lutraria *(p. 272).*

M. *arenaria* is a northern species widely distributed in north-west Europe and Britain but not recorded from the Isles of Scilly. It lives in muddy-sand and mud on the lower shore and into the shallow sublittoral to about 70 m but has been recorded at depths of over 190 m. It is common on mud-flats and in estuaries where it is known to feed in

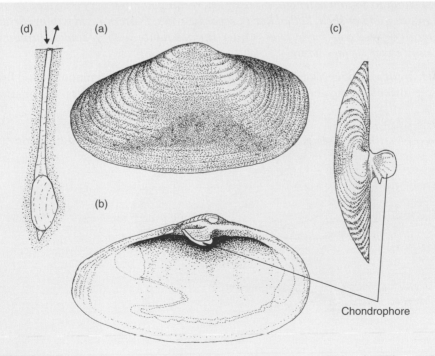

Figure 12.93 *Mya arenaria*, (a) right valve, (b) left valve, (c) dorsal view of left valve, (d) in sediment. Arrows indicate direction of water flow.

salinities down to 15, as well as surviving for some considerable time at salinities as low as 4. Adult specimens live up to 500 mm below the surface of the sediment and the siphons, joined along their length, are capable of enormous extension, making contact with the surface for suspension feeding and respiration. The presence of *Mya* on mud-flats is betrayed by a keyhole-shaped opening left on the surface of the sediment by the siphons. When contracted into the shell, the size of the siphons causes the shell valves to gape posteriorly. The foot of the adult is small and it is doubtful whether the animal can reburrow if dislodged. The sexes are separate and fertilization probably takes place in the mantle cavity, sperm entering in the inhalant current of the female. The veliger larvae are planktonic for up to three weeks. Settlement, which may be very dense in some areas, occurs on sand grains and algae, the young animals attaching by a byssus. If the substratum is unsuitable they can detach and move to another locality. Young *Mya* live in the top few centimetres of sediment and have a large foot, enabling them to move easily. With increasing age they move deeper into the sediment and byssal attachment is lost. Breeding occurs during spring and summer and veligers have a pelagic life of two to three weeks. Seasonal growth patterns can be detected on the chondrophore of the hinge line of the shell and length of life has been estimated to be 17 years. *Mya* is much prized as food, especially in America where it forms the basis of the famous clam chowder.

Family Hiatellidae Rock-borers
Shell more or less rectangular. Beaks to the anterior. Without lateral teeth. Right valve with single projecting cardinal tooth. Pallial line continuous or as separate scars. Pallial sinus present. Margin of shell smooth.

Hiatella arctica (Linnaeus) Wrinkled rock-borer **Fig. 12.94**
Shell up to 35 mm in length; more or less rectangular, but very irregular. Beaks close to anterior end of shell. Pallial line of separate scars. Posterior end gaping. Sculpturing of more or less concentric lines. Outer surface white with yellowish-brown periostracum; inner surface white.

H. arctica is widely distributed in north-west Europe. It is common around Britain from the lower shore to about 50 m and occurs either boring into soft rock, or attached by a

Projecting
cardinal
tooth

Figure 12.94 *Hiatella arctica*, right valve.

byssus and nestling in existing holes or crevices in rock or laminarian holdfasts. The siphons are long and reddish-coloured at the tips. It is a suspension feeder. The sexes are separate and on the west coast of Scotland breeding has been recorded from December to April. There is a free-swimming veliger larva and after metamorphosis the young bivalves are mobile for up to six months before selecting a more permanent habitat. The type of substratum selected determines whether the young become burrowers or nestlers; those settling on soft rock becoming burrowers, those on hard rock attach by the byssus and become nestlers. The boring action of the adults is achieved by mechanical abrasion using the valves of the shell, but the initial penetration of the rock by the young animal may involve chemical as well as mechanical means.

Family Pholadidae Piddocks

Shell elongate, oval; modified for boring into substrata such as wood, clay and rock. Beaks towards anterior of shell. Shell usually with pronounced gape. Up to 4 accessory plates in dorsal part of shell with part of the shell, known as the umbonal reflection, reflected over them. Hinge line lacks teeth but often has slender projection, the apophysis. Adductor muscle scars present, but position, size and number vary. Pallial sinus present.

1 Shell fragile, not gaping anteriorly; with 2 dorsal accessory plates ....
.... ***Barnea candida*** p. 294
Shell gapes anteriorly and posteriorly; anterior gape reveals large, sucker-like foot ... **2**

2 Shell elongate, of fragile appearance; sculpturing of concentric ridges crossed anteriorly by about 40 ribs, the anterior ones with spines .... ***Pholas dactylus*** p. 293
Shell oval, robust; sculpturing of uneven concentric ridges; crossed by vertical groove in midline .... ***Zirfaea crispata*** p. 295

Pholas dactylus Linnaeus Common piddock **Fig. 12.95; Plate 30a,c**

Elongate shell of fragile appearance, up to 75 mm in length, exceptionally 150 mm. Beaks near anterior end of shell, with part of shell, the umbonal reflection, reflected over them. Four dorsal accessory plates. Shell gapes anteriorly and posteriorly, anterior gape revealing large sucker-like foot. Hinge line with slender projection, the apophysis (the point of attachment of foot muscles) just below beak of each valve. Sculpturing of concentric ridges crossed anteriorly by about 40 ribs; spines present at anterior end. Outer surface greyish-white with yellowish periostracum; inner surface shiny white. Anterior margin of shell crenulate, remainder smooth.

P. dactylus is a southern species extending as far north as the south and south-west of Britain. It is found on the lower shore and in the shallow sublittoral boring into peat, soft rock and sometimes wood. *Pholas* is a suspension feeder. Boring is achieved solely by mechanical means; the shell valves are scraped against the rock surface and rotated to produce circular burrows which occasionally criss-cross one another. Interestingly, *Pholas* is bioluminescent, shining with a greenish-blue light while in its burrow. Breeding occurs during summer followed by settlement in autumn. There is a free-swimming veliger larva which attaches by a byssus at settlement, the byssus later being lost. Longevity is up to 14 years.

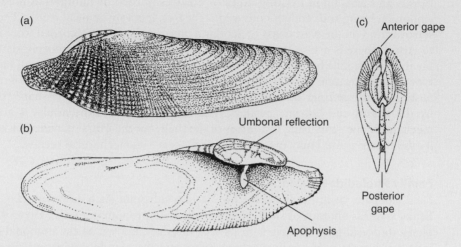

Figure 12.95 *Pholas dactylus*, (a), (b) left valve, (c) dorsal view.

Barnea candida (Linnaeus) White piddock **Fig. 12.96; Plate 30a,b**

Fragile, oval shell, up to 65 mm in length but usually smaller. Beaks near anterior end of shell. Two dorsal accessory plates. Shell gapes posteriorly. Hinge line with slender projection, the apophysis (the point of attachment of the foot muscles) just below beak of each valve. Sculpturing of concentric ridges crossed by ribs with spines, the spines well developed at anterior end of shell. Outer surface white; inner surface shiny white. Anterior margin of shell crenulate, remainder smooth.

B. *candida* is widely distributed in north-west Europe and has been recorded all around Britain. Like *Pholas* (above), it bores by mechanical means into soft substrata including

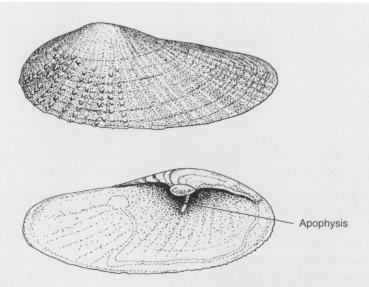

Figure 12.96 *Barnea candida*, left valve.

peat, wood and shale, from the middle shore into the shallow sublittoral, extending into estuaries in salinities down to about 20. It is a suspension feeder. The sexes are separate, fertilization is external and on the south coast of England breeding occurs in early autumn. There is a free-swimming veliger larva. Longevity is up to five years.

Zirfaea crispata (Linnaeus) Oval piddock **Fig. 12.97**

Robust, oval shell, up to 90 mm in length. Beaks near anterior end of shell, part of shell reflected over them. Single dorsal accessory plate. Hinge line with large, flat projection, the apophysis (the point of attachment of the foot muscles), just below beak of each valve. Shell gapes anteriorly and posteriorly. Sculpturing of concentric ridges, very uneven in anterior region; crossed by vertical groove in midline, most conspicuous in young specimens. Outer surface whitish with brown periostracum; inner surface white. Anterior margin crenulate, remainder smooth.

Z. crispata is widely distributed in north-west Europe and Britain. It is found from the lower shore to depths of a few metres boring into clay, shale and soft rocks. Like *Pholas* (p. 293) it burrows by mechanical means. It is a suspension feeder. The siphons are long, and close to the shell they are covered by the thick, dark-coloured periostracum. The sexes are separate, and on the north-east coast of England breeding occurs from March to October with peaks from March to May and August to October. Free-swimming veliger larvae are present in the plankton for most of the year. Animals living intertidally survive for about five years, compared with seven years for sublittoral populations.

Apophysis

Figure 12.97 *Zirfaea crispata*, left valve.

Family Teredinidae Shipworms

Shell very small, covers anterior end of long, worm-like body. Posterior end of body with pair of calcareous pallets. Bores into submerged timbers. Tunnels in timber lined by calcareous tube.

Teredo navalis Linnaeus Shipworm **Fig. 12.98a,b,c,d; Plate 31a,b**

Shell small, up to 10 mm in length. Valves divided into number of regions, each with different sculpturing. Beaks near anterior end of shell. Outer surface white, with pale brown periostracum. Shell covers only the anterior region of the long, soft body which lies in calcareous tube, up to 200 mm in length. Posterior end of body with a pair of calcareous pallets; pallets off-white, paddle-shaped with forked blade, up to 5 mm in length.

T. navalis is widely distributed in north-west Europe, where it is found boring in submerged timber and is often seen in pieces of timber washed ashore. The ravages wrought by shipworms have been known since the time of the early Greeks, but it was not until 1733 that it was established by Snellius that they were bivalve molluscs. Shipworms burrow into piles, piers, wooden ships, bridges and dykes. The only external evidence of attack is the presence of fine holes (1–2 mm across) on the surface of the timber marking the site of the original penetration, but the centre of the timber is likely to be riddled with circular tunnels up to about 20 mm across. Boring is achieved by mechanical means, the shell valves cutting into the wood, with sea water forced into the tunnelling to lubricate and cool the valves. Shipworms usually bore along the grain of the wood, neither entering the tunnel of another specimen nor leaving the wood. As they bore into the wood, all but the anterior end of the tunnel is lined by a calcareous tube secreted by the mantle, and which protects the soft body of the animal. The siphons are attached just inside the external opening to the tunnel by the pallets, and can extend a considerable distance from the

Figure 12.98 (a) Piece of timber opened to show shipworm burrows. (b) *Teredo navalis* in timber (after Yonge, 1949). (c) *Teredo navalis* removed from calcareous tube (after Eales, 1950). (d) Pallet of *Teredo navalis*. (e) Pallet of *Nototeredo norvagica*. ((d), (e) After Tebble, 1976.) Arrows indicate direction of flow of water.

entrance. When they contract they are drawn into the tunnel and the opening is closed by the pallets, enabling the animal to survive for some time when the wood is out of water. Plankton drawn in with the inhalent current and fragments of wood from the tunnelling serve as food; digestion of the wood is brought about by cellulases.

It is believed that the gonad of *T. navalis* passes through male and female phases. Fertilization is internal; sperm is discharged into the water and enters the inhalant siphon of another animal. The larvae develop in the mantle cavity and are liberated through the exhalant siphon as veligers which have been recorded in the plankton of Danish waters from March to December. Larval life is of about three to five weeks' duration and when contact is made with a suitable substratum, the larvae attach by means of a byssus and boring starts immediately. Growth rate is fast and length of life is two to three years. Many treatments have been used to protect wood from attack, the most effective being creosote.

Nototeredo norvagica (Spengler), is also common in north-west Europe. It is larger than *T. navalis* and readily separated by the shape of the pallets which are paddle-shaped but the oval blade is not forked (Fig. 12.98e). Other species of ship-worm may be found in drifting timbers around Britain.

References

CLEMAM. Check list of European marine Mollusca. (http://www.somali.asso.fr/clemam/index. clemam.html).

Polyplacophora

Jones, A. M. & Baxter, J. M. (1987). *Molluscs: Caudofoveata, Solenogastres, Polyplacophora and Scaphopoda. Keys and notes for the identification of the species.* Synopses of the British fauna (New Series), no. 37. Leiden: E. J. Brill/Dr W. Backhuys.

Gastropoda

Alder, J. & Hancock, A. (1845–55). *A monograph of the British nudibranchiate mollusca.* Parts 1–7. London: Ray Society.
Graham, A. (1988). *Molluscs: prosobranch and pyramidellid gastropods. Keys and notes for the identification of the species.* Synopses of the British fauna (New Series), no. 2, 2nd edn. Leiden: E. J. Brill/Dr W. Backhuys.
Picton, B. E. & Morrow, C. C. (1994). *A field guide to the nudibranchs of the British Isles.* London: Immel Publishing.
Reid, D. G. (1996). *Systematics and evolution of* Littorina. London: The Ray Society.
Thompson, T. E. (1976). *Biology of opisthobranch molluscs.* Vol. I. London: Ray Society.
Thompson, T. E. (1988). *Molluscs: benthic opisthobranchs (Mollusca: Gastropoda). Keys and notes for the identification of the species.* Synopses of the British fauna (New Series), no. 8, 2nd edn. Leiden: E. J. Brill/Dr W. Backhuys.
Thompson, T. E. & Brown, G. H. (1984). *Biology of opisthobranch molluscs.* Vol. II. London: Ray Society.

Bivalvia

Forbes, E. & Hanley, S. (1848–53). *A history of British Mollusca, and their shells*, Vols. I–IV. London: John Van Voorst. (Also includes Polyplacophora and Gastropoda.)
Oliver, P. G., Holmes, A. M., Killeen, I. J. & Turner, J. A. (2009). *Marine bivalve shells of the British Isles (Mollusca: Bivalvia).* Amgueddfa Cymru – National Museum Wales. (http://naturalhistory.museumwales.ac.uk/britishbivalves).
Tebble, N. (1976). *British bivalve seashells. A handbook for identification*, 2nd edn. Edinburgh: HMSO.

Nymphon spp. **Fig. 13.1a**

Body slender; up to 10 mm length. Legs very long and slender, 3 or 4 times body length. Both chelifores and palps present; chelifores with chelae, palps with 5 segments. Ovigerous legs in males and females.

The genus *Nymphon* is well known from the shores of north-west Europe and in Britain ***Nymphon gracile*** Leach is perhaps the best known. It is found under stones and seaweed on the middle and lower shore and in addition to crawling over the substratum, it is capable of swimming. It feeds on a wide range of animals, including sea-anemones, hydroids, bryozoans, small polychaetes, egg capsules of the dog-whelk, *Nucella lapillus* (p. 230), and detritus. Most individuals migrate into the sublittoral during winter when breeding occurs, returning to the intertidal zone in March/April, the males often carrying eggs. Adult size is reached in about five months and longevity is believed to be one year.

Pycnogonum litorale (Ström) (*Pycnogonum littorale*) **Fig. 13.1b**

Body stout; up to 20 mm in length. Dorsal and lateral surfaces covered with tubercles. Chelifores and palps absent; ovigerous legs, with 9 segments, in males only. Legs short and stout, about equal to body length.

P. litorale is widely distributed in north-west Europe and is the only species of the genus to be found in Britain where it lives on the lower shore and in the sublittoral, under stones and seaweed, and on sea-anemones. It crawls slowly over the substratum but does not swim. It feeds predominantly on sea-anemones and has a narrower range of food items than *Nymphon gracile* (above). The larvae feed on hydroids. Breeding occurs from spring to autumn and males and females may remain clinging together in the mating position for as long as five weeks after which the developing eggs are carried by the male for about 10 weeks. Maturity is believed to be reached after one year.

Figure 13.1 (a) *Nymphon gracile*. (b) *Pycnogonum litorale*.

Subphylum Crustacea

The crustaceans are a very large group of arthropods widely represented in the sea. They are found in the plankton, in the deep sea and on the shore and include barnacles, sand-hoppers, prawns, shrimps, lobsters and crabs. As is to be expected in a group with such diversity of habits, there is considerable variation in external morphology but the body can usually be divided into head, thorax and abdomen (see Fig. 13.7), although in some groups, for example the barnacles, the distinction is difficult to make. The body is covered by an exoskeleton which may be impregnated with calcium salts. Typically, the head has five pairs of appendages, the most anterior being the two pairs of antennae, known as antennae 1 (antennules) and antennae 2 (antennae), a feature not seen in other arthropods. Posterior to the antennae are the mandibles, maxillae 1 (maxillules) and maxillae 2 (maxillae). These appendages are used in feeding and are adapted to suit a wide variety of feeding methods including carnivorous, suspension and deposit feeding. The appendages of the thorax and abdomen vary from one group of crustaceans to the next, but the basic appendage is biramous with an inner ramus (endopod) and an outer ramus (exopod), the two rami often being very different morphologically and may serve different functions. In many crustaceans a dorsal shield or carapace is present, which to varying degrees affords protection to the body. Most crustaceans have separate sexes (but note the exception of the barnacle). After copulation the fertilized eggs may be brooded by the female, in some cases for the duration of embryonic life or, more commonly, there is at some stage in the life-cycle a pelagic larva which feeds in the plankton.

Classification of the crustacea follows Martin & Davis (2001); the subphylum is divided into five classes. Two of these, **Maxillopoda** and **Malacostraca**, are included in this text.

Class Maxillopoda

Subclass Thecostraca

Infraclass Cirripedia

Barnacles are perhaps the best known of intertidal organisms and are so common and widespread in their distribution that they have been described as ubiquitous. They are exclusively marine. On the upper reaches of most rocky shores, encrusting acorn barnacles are so numerous that they form a distinct grey-white barnacle line which marks the upper limit of the eulittoral zone (Fig. 1.4; Plate 6). Although their crustacean affinities had been noted as early as 1819, it was not until 1830 that this view was firmly established through the work of an army surgeon, Mr J. V. Thompson. Prior to Thompson's research, the sessile habit of the adult barnacle and its shell of hard, calcareous plates (superficially resembling a mollusc) had caused taxonomic

Figure 13.2 Life-cycle of an acorn barnacle (not to scale).

confusion. Thompson's great contribution was an account of the life-cycle of the barnacle in which he described a larval stage (a nauplius) showing unmistakable crustacean characters. Much of our knowledge of this group is based on the classic research of Charles Darwin.

In addition to the familiar acorn barnacles of the rocky shores, the class includes stalked barnacles and one group is highly specialized for a parasitic way of life. Stalked barnacles are occasionally washed ashore on floating debris and are believed to be among the most primitive members of the class (Plate 31c). Acorn barnacles do not have a stalk and the wall is made up of a number of plates which in some species has been reduced from the original number of eight by loss and/or fusion. The number and arrangement of these plates are important in identification. Parasitic barnacles (p. 310) are extensively modified but are recognized as cirripede crustaceans by their larval stages.

Typically, barnacles have six pairs of biramous thoracic appendages which are used, not in locomotion, but to filter suspended food particles from sea water. The common acorn barnacles of the shore are hermaphroditic and in most cases copulation and cross-fertilization take place between neighbouring individuals. After fertilization, the embryos are held within the body of the barnacle where they develop into a characteristic larval stage known as the nauplius. Nauplii are released from the adult barnacle and swim in the surface waters where they feed and grow by a series of moults, eventually developing into a second larval type, the cyprid.

This is non-feeding, surviving on fat reserves laid down during the naupliar stages. The cyprid is a unique larval stage that has evolved for the all-important role of selecting a suitable site for attachment and metamorphosis. Cyprids explore the nature of the substratum with the first pair of antennae and are attracted to rocks already colonized by adults. When a suitable site for settlement has been identified, a secretion pours from the ducts of the first antennae cementing the larva to the substratum. Metamorphosis into a juvenile barnacle then proceeds quickly, generally within 24 hours. The life-cycle of an acorn barnacle is shown in Fig. 13.2. During spring and early summer brown-coloured cyprids of *Semibalanus balanoides* (p. 308), just visible to the naked eye but clearly discernible with a hand lens, and newly metamorphosed barnacles are common on rocky shores around Britain (Plate 21).

Key to the species of barnacles likely to be found intertidally around Britain

[The key includes species that may be washed ashore attached to drifting objects.]

1 Body divisible into soft brown stalk and a head, the capitulum, with 5 calcareous plates. Generally found attached to floating objects .... **Stalked barnacles *Lepas* spp.** p. 304

Body without stalk; with rigid wall of calcareous plates capped by movable opercular plates (Fig. 13.4) .... **Acorn barnacles 2**

2 Shell wall of 4 unequal, ridged plates; opercular aperture off centre; asymmetrical in appearance (Fig. 13.5e) .... ***Verruca stroemia*** p. 305

Shell wall not of 4 unequal, ridged plates; opercular aperture in central position; symmetrical in appearance (Fig. 13.4) **3**

3 Shell wall of 4 plates .... ***Elminius modestus*** p. 308

Shell wall of 6 plates **4**

(In some specimens the plates may be fused and difficult to distinguish.)

4 Rostral plate narrow, about equal to the width of opercular aperture (Fig. 13.4b,c). Opercular aperture kite-shaped or rounded **5**

Rostral plate large, broader than width of opercular aperture (Fig. 13.4a). Opercular aperture diamond-shaped **6**

5 Opercular aperture oval or more or less circular. Junction between terga and scuta convex towards rostral plate .... ***Chthamalus stellatus*** p. 307

Opercular aperture kite-shaped, angular. Junction between terga and scuta concave towards rostral plate .... ***Chthamalus montagui*** p. 308

6 Shell base membranous, not left as a shelly plate when barnacle removed from rock .... ***Semibalanus balanoides*** p. 308
Shell base calcareous, left as a shelly plate when barnacle is removed from rock **7**

7 Shell plates of wall purplish colour or white, pinkish-white striped purple **8**
Shell plates of wall grey-white colour **9**

8 Shell plates of wall purplish colour, often ridged vertically. Apex of terga pointed; opercular aperture relatively small. Large barnacle, up to about 30 mm in height and diameter. Tissue inside opercular aperture brightly coloured, purple, pink, blue .... ***Perforatus perforatus*** p. 310
Shell plates white, pinkish-white with purplish stripes. Apex of terga not pointed. Tissue inside opercular aperture white, striped with purple-black .... ***Amphibalanus amphitrite*** p. 309

9 Shell inclined to one end when viewed in profile. Opercular aperture a broad diamond shape; upper edge of plates toothed. Tissue inside opercular aperture striped with purple and yellow or purple and white ...... ***Balanus crenatus*** p. 309
Shell not inclined. Opercular aperture an elongated diamond shape. Tissue inside opercular aperture with purple and white markings ....
.... ***Amphibalanus improvisus*** p. 310

In addition to the barnacles included in the key, two other groups, parasitic barnacles and burrowing barnacles, are also found on the shore.

Parasitic barnacles (p. 310) are extensively modified for a parasitic mode of life but are identified as cirripedes by the presence of the characteristic larval stages. They are commonly found intertidally on decapod crustaceans such as the shore crab, where they occur as a smooth yellow-brown mass on the undersurface of the abdomen of the crab (p. 366).

Burrowing barnacles bore into the shells of molluscs. Boring is achieved by chemical and mechanical means, using teeth-like structures on the mantle. The adult barnacles are tiny, being only a few millimetres in length. Members of the family **Trypetesidae** bore into the columella of shells of the whelks *Buccinum* (p. 233) and *Neptunea* (p. 234) housing the hermit crab *Pagurus bernhardus* (p. 360). The presence of the barnacle is revealed by tiny slit-like holes in the shell.

Stalked barnacles

Stalked barnacles are primitive cirripedes characterized by the possession of a stalk and a 'head' (also known as the capitulum) bearing a number of calcareous plates. The best-known examples are pelagic, attached to drifting timbers which from time to time are washed ashore. The common name of goose barnacles derives from the myth that barnacle geese hatched from barnacles growing on trees at the water's edge.

Lepas anatifera Linnaeus Goose barnacle **Fig. 13.3; Plate 31c**

Body clearly divisible into soft brown stalk and capitulum bearing 5 white plates with fine striae, the paired terga and scuta and a single carina separated by brown/black soft issue. Plates have blue sheen. Capitulum up to about 50 mm in length, overall body length 200 mm or more.

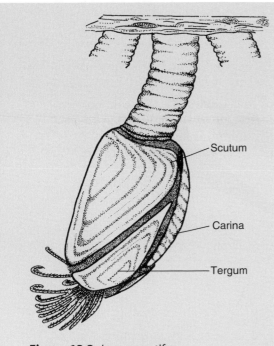

Figure 13.3 *Lepas anatifera.*

(labels: Scutum, Carina, Tergum)

There are a number of species of *Lepas, Lepas anatifera* probably being the best known from European waters. It is occasionally washed ashore on driftwood, plastics, etc. Although essentially a tropical and subtropical species, both adults and larvae have been found as far north as the Norwegian coast. It is a cross-fertilizing hermaphrodite apparently producing a succession of broods at optimum temperatures of 19–25 °C and reaching sexual maturity at these temperatures after about four to five weeks.

A second species, **Lepas hillii** Leach, is sometimes found drifting with *L. anatifera*. The species are similar but can be separated by the smooth plates of *L. hillii* and the more extensive soft tissue between the carina and scutal plates compared with *L. anatifera*. A third species, **Dosima fascicularis** (Ellis & Solander), is also found washed ashore, particularly during summer and autumn in south-west Britain and Ireland. It is readily identified by a spongy white float, secreted from the cement glands of the barnacle, to which it is attached.

Acorn barnacles

Adult acorn barnacles are sessile and characterized by a rigid wall of calcareous plates arranged in the shape of a flat-topped cone. The apex of the cone is known as the opercular aperture, and it is covered by movable plates, the terga and scuta (Fig. 13.4). When the latter are closed the animal is effectively housed within a calcareous box which affords some protection against desiccation during low water. The large basal area of acorn barnacles gives firm adhesion to the substratum, enabling them to survive in wave-beaten situations. On the rising tide, the terga and scuta open and the thoracic appendages are extended and repeatedly drawn through the water, collecting suspended food particles from the plankton. Acorn barnacles are preyed on extensively by the gastropod mollusc *Nucella lapillus* (p. 230) (Plate 18a) and shore fishes such as the shanny, *Lipophrys pholis* (p. 464). Empty barnacle cases are the habitat of a range of fauna including small specimens of periwinkles (p. 214), small bivalves such as *Lasaea adansoni* (p. 266) and the isopod *Campecopea hirsuta* (p. 326). Examination of barnacle plates with a hand lens often reveals the presence of the crustose lichen *Collemopsidium foveolatum* (p. 74) (Plate 9d).

Acorn barnacles can be identified in the field with the aid of a good-quality hand lens by the number, shape and arrangement of the plates and by the shape of the opercular aperture. The base of the shell is either soft and membranous or hard and calcareous and in the latter case a distinct white scar is left on the rock when the barnacle is removed. This is a useful character in identification. Old, worn specimens are often distorted and the individual plates obscured, making identification difficult.

Verruca stroemia (O. F. Müller) **Fig. 13.5e**

Shell wall of 4 unequal, ridged plates. Opercular aperture closed by a single tergum and a single scutum, together forming 1 movable plate. Outer wall comprises 4 plates – scutum, tergum, rostrum and carina; arrangement of plates not the same on either side

Figure 13.4 (a) *Semibalanus balanoides*. (b) *Chthamalus stellatus*. (c) *Chthamalus montagui*. (d) Typical intertidal distribution of the species, drawn from data from the west coast of Ireland. C, carinal plate; R, rostral plate; S, scutum; T, tergum.

of midline and the shell is therefore asymmetrical compared with the other species of acorn barnacles in the key all of which are symmetrical. Shell dirty white colour, may reach diameter of 10 mm, but usually much less. Shell base membranous.

V. stroemia is widely distributed in north-west Europe on the underside of stones and shells on the lower shore. It has also been recorded from laminarian holdfasts and extends into the sublittoral to a depth of about 500 m. Although self-fertilization has been recorded in isolated individuals, it is believed that cross-fertilization usually takes place. On the west coast of Scotland, fertilization takes place in winter and the main release of nauplii coincides with the spring diatom increase in the plankton. A series of smaller broods is released during summer. Cyprids settle from April to September. Sexual maturity is reached at a diameter of about 2 mm.

Chthamalus spp.

Prior to 1976 British chthamalid barnacles were all regarded as members of a single species with wide morphological variation. Since that date it has been established that there are, in fact, two species, *Chthamalus stellatus* and a 'new' species, *Chthamalus*

Figure 13.5 (a) *Elminius modestus.* (b) *Balanus crenatus.* (c) *Amphibalanus improvisus.* (d) *Perforatus perforatus.* (e) *Verruca stroemia.* (f) *Amphibalanus amphitrite.*

montagui, first described by Southward in 1976. Both species are widely distributed in north-west Europe and can be separated in the field with the aid of a hand lens. Together with *Semibalanus* (p. 308) they are the most common intertidal barnacles in Britain.

Chthamalus stellatus (Poli) **Fig. 13.4b**

Shell wall of 6, light grey plates, often difficult to distinguish because of corrosion and overgrowth of algae and lichens. Opercular aperture oval or more or less circular. Junction between terga and scuta forms curve convex towards rostral plate. Rostral plate narrow (cf. Semibalanus, p. 308). Tissue inside opercular aperture generally bright blue with black and orange marks. Diameter up to 10 mm or more. Shell base membranous.

C. stellatus is a southern species extending as far north as Shetland. It is found on western coasts of Britain extending around the north of Scotland, south to about as far as Aberdeen. Within this range it is most abundant on wave-beaten, open-coast situations exposed to the south-west while apparently being absent from many Irish Sea coasts. It is not found in the sublittoral. The species appears to favour clear, oceanic waters (cf. *C. montagui*, below). In the English Channel it extends eastwards as far as the Isle of Wight but is absent from the eastern part of the Channel and the North Sea coast. The vertical distribution of *C. stellatus* on the shore overlaps with that of *C. montagui* and *S. balanoides*; it is generally most abundant between these two (Fig. 13.4d).

Chthamalus montagui Southward **Fig. 13.4c**

Shell wall of 6, light grey or brownish plates often difficult to distinguish because of corrosion and overgrowth of algae and lichens. Opercular aperture kite-shaped and angular; junction between terga and scuta forms curve concave towards rostral plate. Rostral plate narrow (cf. Semibalanus, below). Tissue inside opercular aperture bluish with brown and black marks, usually paler than in C. stellatus (above). Similar in size to C. stellatus. Shell base membranous.

The geographical distribution of *C. montagui* is similar to that of *C. stellatus* (above) but, unlike *C. stellatus*, it has not been recorded from Shetland. It is more widely distributed on Irish Sea coasts and in contrast to *C. stellatus* apparently favours more turbid waters. It overlaps with *C. stellatus* and *Semibalanus* (below) in its vertical distribution although it is more abundant higher on the shore (Fig. 13.4d). It is not found in the sublittoral.

Chthamalus spp. breed throughout the summer and at Plymouth breeding commences when sea temperatures reach 10 °C. They are multiple brooders capable of self-fertilization, with the number of broods released per season being dependent on shore level. In the Plymouth area, one or two broods are produced by high-shore populations, increasing to four on the lower shore. Cyprids settle on the shore from July. The first brood is produced in the first year after settlement. Development of the embryos takes about three weeks.

Elminius modestus Darwin **Fig. 13.5a; Plate 20b**

Shell wall of 4, grey-white plates. Junction of terga and scuta noticeably pointed. Diameter up to 10 mm. Shell base membranous.

E. modestus is an Australasian barnacle which is believed to have become established in Britain from ship-borne populations. It was first recorded in Britain from Chichester Harbour in 1945. The rapid spread of the species since that time has been followed with interest by shore ecologists; it is now known to be established all around Britain and on southern North Sea coasts of north-west Europe and was first recorded in Ireland in 1957. It has been found from Denmark to southern Portugal. It is a cross-fertilizing hermaphrodite breeding almost continuously throughout the year. When conditions are favourable it has been known for broods to be released every 10 days. Cyprids are generally found on the shore from about May to October and the newly metamorphosed barnacles grow rapidly, reaching maturity in about eight weeks. *Elminius* apparently thrives in water of a high suspended silt content and is more tolerant of reduced salinity than *Semibalanus* (below) and *Chthamalus* spp. (above), being common in estuaries and harbours. It is also found on open coasts where it occurs on the middle shore, but it appears to be less tolerant of wave exposure than *Semibalanus* and *Chthamalus*. It extends into the sublittoral.

Semibalanus balanoides (Linnaeus) (*Balanus balanoides*) **Fig. 13.4a; Plate 21**

Shell wall of 6, grey-white plates, generally easy to distinguish. Individual specimens up to 15 mm diameter but, when crowded together, tall columnar individuals are common in which separate plates may be difficult to distinguish. Opercular aperture diamond-shaped. Rostral plate broad (cf. Chthamalus spp., p. 306). Tissue inside opercular aperture usually white, may be pinkish-white. Shell base membranous.

S. balanoides is a boreal species. It occurs extensively on British coasts, but may be absent or rare in parts of Cornwall and the Scilly Isles and the southern limit of its distribution is in France. It is found on shores of all degrees of exposure and tolerates reduction in salinity to about 20, penetrating the lower reaches of estuaries. It is mainly intertidal but extends into the sublittoral in the northern parts of its range. Together with *Chthamalus* spp. (p. 306), *Semibalanus* forms the characteristic barnacle zone seen on most rocky shores. When the two occur on the same shore, they overlap in distribution, but generally *Semibalanus* is found lower than *Chthamalus* spp. (Fig. 13.4d). *Semibalanus* is less tolerant of desiccation than *Chthamalus* spp. and this has been shown to be related to the greater permeability of its shell plates. In areas exposed to strong wave action, the upper limit of distribution of *Chthamalus* and *Semibalanus* on the shore is raised (see Fig. 1.4). On shores dominated by fucoid seaweeds, the density of barnacles is reduced, the seaweed preventing establishment of the barnacle larvae. In contrast to *Chthamalus* spp., *Semibalanus* has a single breeding season with copulation and cross-fertilization in the autumn, but the nauplii are held in the barnacle to be released into the sea from about February to May of the following year when diatom production in the plankton is high. The pelagic phase lasts some weeks, during which time the larvae feed on the rich stocks of diatoms. There are six naupliar stages, each separated by a moult. Cyprids settle on the shore during spring and early summer, depending on locality. The number of eggs produced varies according to position on the shore and age of the barnacle, and estimates as high as 8000 per individual per year have been made. Longevity depends not only on locality but also on beach level; those on the upper shore live for up to eight years compared with about two to three years for individuals on the lower shore. Maturity is generally reached after one to two years.

Balanus crenatus Bruguière **Fig. 13.5b**

Shell wall of 6, grey-white plates; up to 25 mm diameter. Shell inclined to one end when viewed in profile. Opercular aperture a broad diamond shape; upper edge of shell plates toothed. Tissue inside opercular aperture with white or yellow and purple stripes. Shell base calcareous, often left as scar on rock when barnacle removed.

B. crenatus is a northern species extending as far south as the Atlantic coast of France. It is essentially a sublittoral species, recorded at depths of up to 200 m, but is also found under stones and on shells on the lower shore, less frequently on open rock faces. It is often attached to shells that have been washed ashore. Nauplii are released from February through to about September, the cyprids settling from April onwards with a peak in April and May. Length of life is about 18 months.

Amphibalanus amphitrite (Darwin) (*Balanus amphitrite*) **Fig. 13.5f**

Shell wall of 6 white or pinkish-white plates marked with vertical purple stripes narrowing from the base; up to 25 mm diameter. Opercular aperture diamond-shaped. Tissue inside opercular aperture with 4 conspicuous purple, black bands. Shell base calcareous.

A. amphitrite is an introduced tropical species, the first records of its occurrence in Britain dating from the 1930s. It is now established at a number of localities warmed by water discharges from power stations and is widely distributed in France, Portugal, Spain and the Mediterranean. It is found both in the intertidal and sublittorally and extends into estuaries. It has been suggested that temperatures above 15 °C are needed for reproduction.

Amphibalanus improvisus (Darwin) (*Balanus improvisus*) **Fig. 13.5c**

Shell wall of 6 smooth, whitish plates; up to 15 mm diameter. Opercular aperture narrow, elongated diamond shape. Tissue inside opercular aperture with white and purple markings. Shell base calcareous.

A. improvisus has a widespread distribution in north-west Europe. It is a brackish water species rarely found in salinities above 20. Around Britain it is most common in the larger estuaries, often with *Elminius modestus*. It is capable of self-fertilization. Nauplii are found in the plankton during summer and take about two weeks to develop to the cyprid stage.

Perforatus perforatus Bruguière (*Balanus perforatus*) **Fig. 13.5d**

Shell wall of 6, purplish plates, often vertically ridged. Distinct conical shape. Opercular opening small; apex of terga pointed. Tissue inside opercular aperture marked with bright colours of purple, pink, blue. Shell sometimes up to 30 mm diameter and 30 mm tall. Shell base calcareous.

P. perforatus is a southern species extending to south-west England and South Wales where it reaches its northern limit in Cardigan Bay. It is found along the eastern English Channel and in the southern North Sea but has not been recorded from Ireland. It is found on the lower shore, sometimes being common in exposed situations. Breeding occurs during summer, the cyprids settling on the shore during August and September. The isopod crustacean *Dynamene* (p. 325) is often found in the empty cases of *P. perforatus*.

Parasitic barnacles

Sacculina carcini Thompson **Fig. 13.6**

Found on undersurface of abdomen of shore crab, Carcinus maenas *(p. 366), and other crabs of the family Portunidae, as conspicuous, yellow-brown mass which bears superficial resemblance to the crab's eggs.* Sacculina *has a smooth, apparently amorphous nature compared with granular appearance of egg mass.*

Figure 13.6 *Sacculina carcini* on the abdomen of the shore crab.

S. carcini is widely distributed and common in north-west Europe. Nauplii of the parasite are liberated into the plankton and, unlike those of other cirripedes, do not feed but survive on stored yolk (lecithotrophic development). After four or five moults, the nauplii metamorphose into cyprids of two sizes, the smaller being female and producing kentrogon larvae, while the larger are male and produce trichogon larvae. The kentrogon larvae are the infective stages and after attaching to small setae on the surface of a young *Carcinus*, cells pass from the kentrogon into the crab where they develop and spread through the tissues of the host. This is known as the soma or body of the parasite and this stage has been estimated to be anything from five to 34 months' duration. Eventually, the parasite makes its appearance on the external surface of the crab and is known as the externa. This is effectively a brood sac which is seen under the abdomen of the crab. It contains mantle tissue and ovaries and the eggs are fertilized by a trichogon larva developed from a large male cyprid. Some six weeks or so after fertilization, nauplii are produced. At Plymouth, breeding occurs throughout the year and the ratio of male to female cyprids varies seasonally. It appears that during summer the cyprids produced are mainly female, the kentrogon larvae, while the trichogon larvae predominate in the plankton during autumn and winter.

A similar life-cycle is shown by species of the genus **Peltogaster**, parasites of hermit crabs.

Class Malacostraca

Subclass Eumalacostraca

The malacostracans are the largest group of crustaceans and include the prawns, shrimps, crabs and lobsters. The generalized body form is shown in Fig. 13.7 and shows characters used in the keys. The body comprises a head, a thorax of eight segments (of which at least one is fused with the head) and an abdomen of six segments. In some groups, e.g. the decapods, the head and thorax are protected by a dorsal shield, the carapace. The appendages of the thorax are generally called pereopods and, depending on species, the first one, two or three pairs are associated with feeding; the remaining pairs are for walking and grasping. In the typical malacostracan, the first five pairs of abdominal appendages, called pleopods, are usually similar to each other. They are involved in a variety of functions, including swimming, producing currents, as gills or for carrying eggs. The last abdominal segment bears a pair of appendages known as uropods; together with the terminal telson they often form a tail fan.

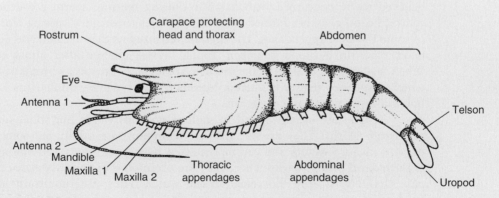

Figure 13.7 Generalized body form of a malacostracan crustacean.

Malacostracans are grouped into a number of orders which show modifications of the generalized body plan. Those orders featured in the text are recognized using the characters given below.

(a) Small (up to about 20 mm in length), shrimp-like; often transparent. Thoracic appendages biramous, lack pincers. (Mysids or opossum shrimps) .... **Order Mysida** p. 312

(b) Usually dorsoventrally flattened. Thoracic appendages similar in appearance. (Sea-lice and sea-slaters) .... **Order Isopoda** p. 317

(c) Usually laterally flattened. First 3 pairs of abdominal appendages with pleopods; last 3 pairs with uropods. (Sand-hoppers) .... **Order Amphipoda** p. 329

(d) Thorax with 5 pairs of 'walking legs'. Well-developed carapace. (Prawns, shrimps, lobsters and crabs) .... **Order Decapoda** p. 348

Order Mysida

Widely known as the opossum shrimps, the mysids are a group of small, often transparent, shrimp-like animals. A carapace is present but in most species it does not cover the last one or two thoracic segments. Antennae 1 are conspicuous, biramous appendages. The outer ramus of antenna 2 takes the form of a flattened plate known as the antennal scale (Fig. 13.8b). It is bordered along part or all of its margin with setae and is an important diagnostic character. The eyes are large and carried on stalks; the thoracic appendages are biramous and lack chelae, a feature which distinguishes them from prawns and shrimps. The first one or two pairs are generally used for feeding, while the remainder are used for a variety of functions such as catching prey and locomotion. The abdomen has six segments, the first five of which usually have small appendages, the pleopods, while the last has a pair of uropods. The inner ramus of each uropod of most mysids has a conspicuous, rounded statocyst which is believed to be concerned with balance. During mating, sperm are deposited in a ventral brood chamber formed from brood plates (oostegites) arising from the bases of some thoracic appendages. The eggs are fertilized as they are laid and development takes place in a brood chamber. Young mysids emerge after about two weeks. When feeding, the beating of the thoracic appendages creates a current of water from which suspended matter is filtered. Mysids feed extensively on detritus and diatoms but will also take small crustaceans.

Mysids are usually found in shallow coastal waters but some species occur at greater depths. They are not generally common intertidally but some are abundant in inshore waters and can be collected from rock pools and among seaweed. Some extend into estuaries where they are often seen in dense swarms, the individuals characteristically 'hovering' in the water. They are important in estuarine food webs.

Mysids are best collected by sweeping a fine-meshed net through the fringes of seaweeds around the margins of rock pools and gullies, and particularly through the creeks and channels in estuaries at low water where some species are found in surprisingly high densities.

The mysids included in this text belong to a single family, the Mysidae.

Family Mysidae

Key to species of mysids

1 Telson divided at tip **2**
   Telson not divided at tip **5**

2 Outer ramus of uropod with spines on outer margin; 5th segment of abdomen
   narrow in dorsal view and with longitudinal ridge; posterior margin of segment
   with spine .... ***Gastrosaccus spinifer*** p. 314
   Outer ramus of uropod with setae on outer margin **3**

3 Antennal scale long, fairly narrow, with single spine on outer margin about
   2/3 way along length. Eyes on long, narrow stalks, extending well beyond
   margin of carapace .... ***Schistomysis spiritus*** p. 317
   Antennal scale with single spine on outer margin close to tip of scale **4**

4 Antennal scale long and narrow, about 8 times longer than broad. Lateral margins
   of telson each with 22–28 spines .... ***Praunus flexuosus*** p. 316
   Antennal scale about 4 times longer than broad. Lateral margins of telson each
   with 15–17 spines .... ***Praunus inermis*** p. 317

5 Telson long, triangular in outline with narrow, truncated tip; short spines around
   margin .... ***Neomysis integer*** p. 316
   Telson not as above **6**

6 Telson short, bluntly rounded; with short spines around margin and single longer
   lateral spine on each side. Eye stalks very long .... ***Mesopodopsis slabberi*** p. 314
   Telson not as above **7**

7 Carapace with long rostrum, extending in front of eyes. Outer ramus of uropod
   with proximal spines and distal setae on outer margin. Telson elongate; larger
   spines around margin separated by smaller ones; 2 large spines on posterior
   margin separated by 4 small spines and 2 setae .... ***Siriella armata*** p. 313
   Carapace with short rostrum, not extending in front of eyes. Outer ramus of
   uropod with setae all around margin. Spines on margin of telson increase
   in size towards tip .... ***Leptomysis lingvura*** p. 314

Siriella armata (Milne-Edwards) **Fig. 13.8a**
*Body up to 22 mm in length, slender. Carapace with long, pointed rostrum extending
in front of eyes. Eyes on long stalks. Antenna 2 with scale lacking setae on outer
margin. Outer ramus of uropod with proximal spines and distal setae on outer
margin. Telson not divided, elongate, rounded at tip; larger spines around margin
separated by smaller ones, 2 large spines at tip separated by 4 small spines and 2 setae.*

S. armata is found all around Britain from the lower shore to depths of about 20 m.
It is often found in shoals in rock pools on the lower shore among seaweeds. Breeding
has been recorded from March to October at Plymouth.

(a)   (b)

Antennal
scale

Figure 13.8 (a) *Siriella armata*, with detail of posterior region of telson. (b) *Leptomysis lingvura*, detail of anterior region and telson. For clarity setae are shown only on the apical segment of the antennal scale (after Tattersall & Tattersall, 1951).

Leptomysis lingvura (G. O. Sars) **Fig. 13.8b**

Body translucent or reddish in colour; up to 17 mm in length. Carapace with short rostrum, not extending in front of eyes. Antenna 2 with narrow scale with setae on all margins; distal part of scale with 5 or 6 setae on each side. Outer ramus of uropod with setae all around margin. Telson more than twice as long as broad, rounded; spines on margin increase in size towards tip.

L. lingvura is widely distributed in north-west Europe and common all around Britain. It is found on the lower shore and in shallow waters, often in shoals. Breeding occurs in summer and autumn.

Gastrosaccus spinifer (Goës) (*Gastrosaccus spiniferus, Mysis spinifer*) **Fig. 13.9**

Body up to 21 mm in length; 5th segment of abdomen narrow in dorsal view, and with longitudinal ridge and spine on posterior margin. Eyes small. Antenna 2 with long scale lacking setae on outer margins. Posterior margin of carapace with fringe of spines. Outer ramus of uropod with spines on outer margin. Telson deeply divided at tip.

G. spinifer is widely distributed in north-west Europe and occurs all around Britain. It is often found in dense swarms in brackish water and among *Zostera* (p. 75) at salinities down to about 20. It has been recorded at depths of over 200 m.

Mesopodopsis slabberi (van Beneden) (*Mysis slabberi*) **Fig. 13.10a**

Body slender and colourless; up to 15 mm in length. Eyes on very long stalks. Antenna 2 with long, narrow scale, with setae on all margins. Telson short, bluntly rounded with single lateral spine on each side.

Figure 13.9 *Gastrosaccus spinifer*, lateral and dorsal views (after Tattersall & Tattersall, 1951).

Figure 13.10 (a) *Mesopodopsis slabberi*, with detail of telson. (b) *Neomysis integer*, with detail of telson. (Both after Tattersall & Tattersall, 1951.)

M. slabberi is widely distributed in north-west Europe and around Britain, but is believed to be rare on the west coast of Scotland. It tolerates very low salinities (0.5) and is found in estuaries where it is sometimes abundant. *Mesopodopsis* is omnivorous. In British waters breeding has been recorded in spring and summer.

Neomysis integer (Leach) (*Praunus integer*) **Fig. 13.10b**

Body almost transparent; up to 18 mm in length. Eyes on stalks. Antenna 2 with long, pointed scale, with setae on all margins. Telson long, triangular in outline with narrow truncated tip; short spines around margin.

N. integer is widely distributed in north-west Europe and common on the coasts of Britain, where it is the only species of the genus. It withstands salinities down to about 1 and is mainly found in estuaries where it frequently occurs in swarms. It also tolerates high salinity and is found in saltmarsh pools. Breeding has been recorded throughout the year on the west coast of Scotland, with spring, summer and autumn generations. Breeding females carry up to about 50 young in the brood chamber. Longevity is about nine months for the overwintering generation and about four months for animals spawned in spring. *Neomysis* is an omnivore, feeding on a wide variety of material including detritus, diatoms, filamentous algae and small crustaceans.

Praunus flexuosus (O. F. Müller) (*Mysis flexuosa*) Chameleon shrimp **Fig. 13.11a,b,c**

Body up to 25 mm in length; colour varied, depending on light intensity and background, from dark grey-black to colourless. Abdomen distinctly bent when seen in

Figure 13.11 *Praunus flexuosus*, (a) whole animal, (b) left antennal scale, (c) telson. *Praunus inermis*, (d) left antennal scale, (e) telson. (All after Tattersall & Tattersall, 1951.)

side view. Eyes large, on long stalks. Antenna 2 with long, narrow scale about 8 times longer than broad, lacking setae on outer margin but with small spine close to tip. Outer ramus of uropod with setae on outer margin. Telson long, divided to about 1/6 length; lateral margins with 22–28 spines.

P. flexuosus has been recorded from the coasts of Scandinavia to northern France and is common on the shores of Britain. It is found in large numbers in rock pools and among laminarians (p. 57) and *Zostera* (p. 75) and can tolerate salinities down to about 5. It has the characteristic habit of 'hovering' in the water. *Praunus* breeds in spring and summer, some of the young maturing quickly to breed in the autumn of the same year. Females mature at about 13–14 mm body length and produce up to three broods, the eggs being carried for about three weeks. Females carry up to 60 eggs in the brood chamber. Longevity is about one year. *Praunus* is omnivorous, feeding on organic debris and small crustaceans. It is probably the most abundant species of mysid in Britain but is easily confused with other species of the genus, particularly **Praunus neglectus** (G. O. Sars). The latter species tends to have a more greenish coloration.

Praunus inermis (Rathke) (*Mysis inermis*) **Fig. 13.11d,e**

Body up to 15 mm in length. Antenna 2 with scale about 4 times longer than broad; lacking setae on outer margin but with small spine on outer margin close to tip. Outer ramus of uropod with setae on outer margin. Telson divided at tip; lateral margins with 15–17 spines.

Figure 13.12 *Schistomysis spiritus*, with detail of right antennal scale (a), and telson (b) (after Tattersall & Tattersall, 1951).

P. inermis is widely distributed and common in north-west Europe among seaweed in shallow water, extending into the mouths of estuaries. It occurs all around Britain but is most common in the north, extending to Scandinavia. Breeding has been recorded from spring to autumn in British waters.

Schistomysis spiritus (Norman) (*Mysis spiritus*) Ghost shrimp **Fig. 13.12**

Body slender, transparent, glassy appearance; up to 18 mm in length. Eyes on long, narrow stalks projecting well beyond anterior margin of carapace. Antenna 2 with long, narrow scale lacking setae on outer margin but with spine about 2/3 way along length. Outer ramus of uropod with setae on outer margin. Telson divided at tip, noticeably constricted at base.

S. spiritus is widely distributed in north-west Europe, often being recorded in dense swarms. It is found in shallow water among seaweed and extends into estuaries, feeding on animal and plant detritus. *Schistomysis* breeds during summer.

Order Isopoda

Marine isopods occur from the intertidal to the deep sea. The body is usually dorsoventrally flattened and the head and first thoracic segment are fused forming a cephalothorax leaving seven, externally visible, thoracic segments. There is no cara-pace. Antennae 1 are small and antennae 2 larger and conspicuous. The eyes are not stalked. There are usually seven pairs of pereopods of similar appearance, a character from which the name of the order is derived. The pleopods are flattened plates which are respiratory and also beat to create a current of water for swimming. One or more abdominal segments are fused with the telson. Isopods show a diversity of feeding habits and include herbivores, carnivores and scavengers.

The sexes are separate and sperm are transferred to the female in capsules known as spermatophores. The eggs are held on the undersurface of the female in a brood chamber formed from overlapping oostegites. Here, the embryos develop and emerge as juveniles, closely resembling the adult but in some species having only six pairs of pereopods. In a few species, embryonic development takes place in internal pouches leading from the brood chamber. There is no pelagic larva. Many isopods are parasitic, some living ectoparasitically on fish; others parasitize crustaceans and are highly modified structurally for a parasitic mode of life. They are not dealt with in this text.

Key to families of isopods

1 With 5 pairs of pereopods of similar appearance .... **Gnathiidae** p. 318
With 7 pairs of pereopods of similar appearance **2**

2 Uropods lie under the telson (ventral) or along the sides of the telson (lateral) **3**
Uropods project in midline from the posterior margin of the telson (terminal) **6**

3 Uropods ventral, large, flattened, covering the pleopods .... **Idoteidae** p. 327
Uropods lateral. Animal typically short, compact **4**

4 Rami of uropods stick-like, not flattened; outer ramus very short and bears apical claw (Fig. 13.15). Animal wood boring .... **Limnoriidae** p. 320
Rami of uropods flattened, outer ramus not bearing an apical claw **5**

5 Abdomen with 5 clearly demarcated segments. Does not curl up tightly when disturbed .... **Cirolanidae** p. 322
Abdomen not so. Curls up tightly when disturbed .... **Sphaeromatidae** p. 323

6 Abdominal segments not demarcated. Segments of pereopods not flattened. Thoracic segments much the same size as each other. Antennae long
.... **Janiridae** p. 328
Abdominal segments clearly demarcated; animals terrestrial and in the littoral fringe .... **Ligiidae** p. 328

Family Gnathiidae
Isopods in which the first pair of pereopods (the pylopods) are modified and closely associated with the mouth parts and the last pair are absent, leaving only five typical pairs. There is marked sexual dimorphism; the male has a straight, slender body and

a pair of large mandibles which project between the eyes, while the female has a much-expanded thoracic region. There are two juvenile stages, the zuphea and the praniza.

1 pair of pereopods (pylopods) of male 5-segmented. Found in mud banks in estuaries .... *Paragnathia formica* p. 319

1 pair of pereopods (pylopods) of male, 2- or 3-segmented; anterior margin of head between mandibles gently concave but with small median lobe. Found in rock crevices, empty shells of barnacles and the holdfasts of laminarians .... *Gnathia maxillaris* p. 319

Paragnathia formica (Hesse) **Fig. 13.13; 13.14a**
Body up to 5 mm in length; male, female and juveniles differ markedly in appearance. First pereopod in the male is 5-segmented, flattened and expanded and known as a pylopod; in the juvenile it is 7-segmented, not expanded and terminates in a hooked claw.

P. formica reaches its northern limit of distribution in Britain and extends as far south as Morocco. It occurs in the muddy banks of intertidal saltmarsh creeks in sheltered estuaries tolerating salinities down to about 18. Males and females live in burrows in the sediment, a single male with as many as 25 females. The males are usually in the narrow entrance to the burrow while the females are deeper inside. Here, up to 140 embryos develop within the body of each female during the summer months and are released in the autumn. Release occurs on high tides when the burrows are submersed. After release of the young the female dies and the dead remains are removed from the burrow by the male. The newly released larva measures about 1 mm in body length and is known as a zuphea. It is an active swimmer and once it locates a fish host, typically an estuarine species such as *Pomatoschistus microps* (p. 472), it attaches itself using the hooks on the first pair of pereopods, and sucks blood. The thorax becomes much expanded and the larva, now called a praniza, with a body length of about 1.5 mm, leaves the host and returns to the sediment where it remains until the blood meal has been digested. After moulting it again becomes a zuphea and swims off to find another fish host. After three zuphea stages and three praniza stages the larva moults to become a young adult. Young females are captured by mature males using the large mandibles and taken into the burrow where breeding takes place. Males and females do not feed but survive on food reserves built up during the zupheal stages. Females live for about one year, breeding only once; males live for about two years.

Gnathia maxillaris (Montagu) **Fig. 13.14b,c**
Body up to 5 mm in length; male, female and juvenile differ markedly in appearance. First pereopod (pylopod) of the male is 2- or 3-segmented, flattened and expanded. In the juvenile it is 7-segmented, not expanded and terminates in a hooked claw. In the male the anterior margin of the head between the mandibles is gently concave with a small median lobe; anterior corners of head are square; without longitudinal ridge over eyes.

Figure 13.13 *Paragnathia formica*, (a) female, (b) male, (c) zuphea larva, (d) praniza larva.

G. *maxillaris* is common and widespread in southern Britain and Ireland where it occurs intertidally in crevices, under stones, in empty barnacle shells and in laminarian holdfasts. Up to 100 or so embryos develop within the body of the female and are released as larvae known as zupheae. They attach themselves to the body surface of fish such as *Lipophrys pholis* (p. 464), *Taurulus bubalis* (p. 455) and *Crenilabrus melops* (p. 460). After a blood meal they leave the fish host; the thoracic region is much expanded and the larvae are known as pranizae. It has been suggested that there are three praniza stages after which the larva moults to the adult. Adult G. *maxillaris* are believed not to feed, but little is known of their life.

Family Limnoriidae

Isopods that bore into timber. The body is narrow and straight-sided and the abdomen has five distinct segments. The uropods are small and lie along the sides of the telson; the outer ramus is short and bears an apical claw.

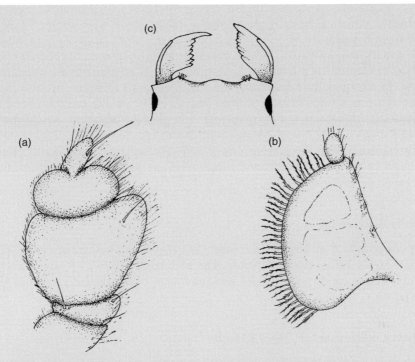

Figure 13.14 (a) Pylopod of male *Paragnathia formica*. (b) Pylopod of male *Gnathia maxillaris*. (c) Anterior of head of male *Gnathia maxillaris*. ((b), (c) After Naylor, 1972.)

Limnoria lignorum (Rathke) Gribble **Fig. 13.15**

Body up to 3.5 mm in length, rather narrow. Antennae 1 and 2 short, about equal length. Abdomen with 5 distinct segments. Uropods small, the shorter, outer ramus bearing an apical claw. Dorsal surface of telson without tubercles, marked with inverted Y-shape.

L. lignorum is a northern species extending as far as the south coast of Britain, and is well known for its habit of boring into timbers and causing extensive damage. The adult colonizes new timber, excavating a burrow using the mandibles, and feeds on fragments of wood digested by cellulolytic enzymes. Burrows are generally less than 20 mm deep and are of the same width throughout their length, communicating with the exterior via a number of pores through which a respiratory current is drawn. There is generally a male and a female in each burrow, the female producing up to 30 eggs which are incubated in a brood chamber. Breeding occurs predominantly in winter. After emergence the young establish their own burrows leading from the walls of the parent burrow. Gribbles have also been recorded swimming and crawling and at this time they presumably colonize new timbers.

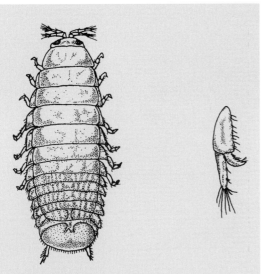

Figure 13.15 *Limnoria lignorum*, with detail of right uropod (after Sars, 1899).

There are two other British species of *Limnoria*, **Limnoria quadripunctata** Holthuis, which is usually dominant on south and west coasts of Britain, and **Limnoria tripunctata** Menzies, found only on the south coasts of England and Wales. The species are separated from *L. lignorum* by the presence of tubercles on the mid-dorsal surface of the telson. *L. quadripunctata* has four tubercles and *L. tripunctata* three. In addition, *L. tripunctata* has a row of small tubercles around the posterior margin of the telson. When the three species occur on the same timbers, they show separation of habitat – *L. tripunctata* occupies the upper, *L. quadripunctata* the middle and *L. lignorum* the lower part of the infected timber. Treatment of timber with creosote does not prevent attack and research on *L. tripunctata* suggests that bacteria in the gut of the isopod may contribute to creosote resistance by detoxification of the creosote.

Family Cirolanidae

Isopods in which all the thoracic segments except the first have conspicuous lateral expansions known as coxal plates. The abdomen has five clearly demarcated segments. The uropods are flattened and somewhat triangular and lie along the sides of the telson.

Eurydice pulchra Leach **Fig. 13.16a,b**

Thorax broader than abdomen giving oval outline; up to 8 mm in length. Large lateral eyes. Antenna 1 short, antenna 2 about 2/3 body length, with 4 segments in basal region. Abdomen with 5 distinct segments. Uropods lie along the sides of telson. Posterior margin of telson broad, convex. Coxal plates 6 and 7 extended posteriorly. Pale grey to brown colour, conspicuously marked with black chromatophores on all surfaces of body.

Figure 13.16 (a) *Eurydice pulchra*. Lateral view of (b) *Eurydice pulchra* and (c) *Eurydice affinis*. Appendages omitted from (b) and (c). Note coxal plates 6 and 7 and the distribution of chromatophores.

E. pulchra is widely distributed in north-west Europe, extending to Norway and occurs all around Britain on open coast and estuarine sandy beaches where it has been recorded in salinities down to 15 to 20. Its intertidal distribution is complicated by a progressive migration up the shore during spring tides, falling to lower levels as tidal amplitude falls to neap tides. This migration is controlled by a complex endogenous swimming rhythm in which the animal leaves the sand to swim in the water at high tide during certain phases of the tidal cycle. At such times it can be caught in large numbers in the intertidal plankton during darkness. During daytime, it can be seen swimming at the edge of the rising tide. *Eurydice* is a voracious carnivore feeding on the wide range of invertebrates associated with sandy shores and it is believed that the intertidal migration is an adaptation which extends its feeding range. The sexes are separate and mating apparently takes place during the swimming phase in spring and summer, after which embryos develop in the brood chamber for some seven to eight weeks. Maturity is reached after 12 to 20 months and the animals have a lifespan of about two years.

A second intertidal species, ***Eurydice affinis*** Hansen (Fig. 13.16c), is also found on sandy beaches on British shores. It reaches a body length of about 5 mm and the black chromatophores occur only on the dorsal surface of the body giving overall a paler appearance than *E. pulchra*. Coxal plates six and seven are not extended posteriorly. *E. affinis* has a more restricted distribution in Britain than *E. pulchra* and extends from south-west England to Morecambe Bay, and although often regarded as an open-coast species, it has been recorded in large numbers in the lower regions of estuaries on the west coast of Wales where it coexists with *E. pulchra*, tolerating salinities down to about 20. It has been recorded from the south coast of Ireland, the Netherlands and the Atlantic coast of France and, unlike *E. pulchra*, is found in the Mediterranean.

Family Sphaeromatidae

Isopods that curl up tightly when disturbed. Abdominal segments indistinct; the uropods are flattened and lie along the sides of the telson. Some genera show marked sexual dimorphism.

1 Uropod with large, paddle-shaped outer ramus; without inner ramus. May (male) or may not (female) have large backwardly directed mid-dorsal spine on 6th thoracic segment .... ***Campecopea hirsuta*** p. 326
Uropod with 2 rami **2**

2 Posterior margin of telson with a single, median indentation. May (male) or may not (female) have 2 large backwardly pointing projections on 6th thoracic segment .... ***Dynamene bidentata*** p. 325
Posterior margin of telson smooth, without any indentation **3**

3 Outer edge of outer ramus of uropod serrated; dorsal surface of telson smooth **4**
Outer edge of outer ramus of uropod not serrated or only barely so; dorsal surface of telson with tubercles either in 2 longitudinal rows or scattered **5**

4 Outer edge of outer ramus of uropod with 4–7 conspicuous serrations; long setae on pereopod 1 with many fine side branches .... ***Sphaeroma serratum*** p. 324

Outer edge of outer ramus of uropod with 4–7 poorly defined serrations; long setae on pereopod 1 smooth .... ***Lekanesphaera levii*** p. 325

5 Dorsal surface of telson with 2 longitudinal rows of tubercles ....
.... ***Lekanesphaera hookeri*** p. 325

Dorsal surface of telson with many small tubercles ..... ***Lekanesphaera rugicauda*** p. 325

Sphaeroma serratum (Fabricius) **Fig. 13.17a**

Oval-shaped body up to 11 mm in length. Antenna 1 about 1/2 length of antenna 2; antenna 2 about 1/4 body length. Abdominal segments fused and indistinct. Laterally placed uropods have 4–7 conspicuous serrations on outer edge of outer ramus. Long setae on pereopod 1 with many fine side branches. Dorsal surface of telson smooth; posterior margin without indentation. Curls up tightly when disturbed.

S. serratum is a southern species extending north to Anglesey, North Wales. It is common under stones from about the middle shore and below on south-west shores of Britain, just penetrating into estuaries. The breeding season is restricted to the summer and embryos are incubated in internal pouches opening from the brood chamber. On the coast of South Wales, the majority of young are released in August and have a lifespan of two and a half to three years, reaching maturity after about

Figure 13.17 (a) *Sphaeroma serratum*. Telson and uropods of (b) *Lekanesphaera rugicauda*, (c) *Lekanesphaera levii* and (d) *Lekanesphaera hookeri*. ((b), (c), (d) After Naylor, 1972.)

two years. In south-west Britain the commensal isopod *Jaera hopeana* (p. 328) may be found on the ventral surface of *S. serratum*.

Lekanesphaera rugicauda (Leach) (*Sphaeroma rugicauda*) **Fig. 13.17b**

Similar body shape to S. serratum *(above) but with smooth outer margin to outer ramus of uropod. Dorsal surface of telson covered with many small tubercles.*

L. rugicauda is widely distributed in north-west Europe. It withstands salinities down to about 4 and is widely distributed on the upper shore in estuaries and saltmarshes. The species is characterized by a number of different colour varieties, for example, 'yellow', 'red', 'grey', the frequency of which varies seasonally, possibly in response to environmental conditions. Breeding occurs in spring and summer. Maturity is reached in one year and the animals live for about one and a half years.

Lekanesphaera levii (Argano & Ponticelli) (*Sphaeroma monodi*) **Fig. 13.17c**

Similar body shape to S. serratum *(above), but outer margin of outer ramus of uropod with 4–7 poorly defined serrations. Long, smooth setae on pereopod 1. Dorsal surface of telson smooth.*

L. levii is a southern species extending as far north as southern Britain. It penetrates well into estuaries, usually in salinities down to 14, but is also found under stones on the open coast, particularly in areas of freshwater seepage.

Lekanesphaera hookeri (Leach) (*Sphaeroma hookeri*) **Fig. 13.17d**

Similar body shape to S. serratum *(above), but with smooth outer margin to outer ramus of uropod. Mid-dorsal surface of telson with 2 longitudinal rows of tubercles.*

L. hookeri is locally distributed around Britain and although recorded from East Anglia, appears to be absent from much of the east coast of England and Scotland. It extends northwards to southern Sweden. It is generally found in salinities ranging from 1 to 10 in the upper reaches of estuaries, but is also found in higher salinities.

Dynamene bidentata (Adams) (*Dynamene bidentatus, Naesa bidentata*) **Fig. 13.18**

Oval-shaped body up to 7 mm in length. Abdominal segments fused and indistinct. Shows pronounced sexual dimorphism; 6th thoracic segment of male with 2 large, backwardly pointing projections, absent from female. Posterior margin of telson with a single median indentation, a feature that readily distinguishes females from Sphaeroma *and* Lekanesphaera *spp. (above).*

D. bidentata has been recorded from the Mediterranean, north and west coasts of France and south and west Britain, extending northwards to south-west Scotland. In the English Channel it is found as far east as the Isle of Wight. Adult *Dynamene* are found in crevices and the empty cases of the barnacle *Perforatus perforatus* (p. 310) with the juveniles occurring among seaweeds such as *Fucus serratus* (p. 63) and *Ascophyllum nodosum* (p. 61). Embryos develop in the brood chamber of the female and on the south coast of Wales, juveniles are liberated from May to July. During summer, the juveniles live and feed on seaweed; by the autumn they have reached maturity and move to rock crevices and empty barnacle shells. Here, a single male may be found with a number of females. After fertilization, the embryos are incubated for some three to four months over winter and spring. The female produces a single brood and dies after release of the juveniles. Males, however, remain in the crevices

Figure 13.18 *Dynamene bidentata*, (a) female, (b) posterior region of male.

and empty barnacle shells to survive for a second breeding season. Mature specimens of *Dynamene* do not feed but survive on lipid food reserves, the females for about nine months and the males for up to 21 months.

Campecopea hirsuta (Montagu) **Fig. 13.19**

Oval body up to 4 mm in length. Abdominal segments fused and indistinct. Shows pronounced sexual dimorphism; large, backwardly directed, mid-dorsal spine on 6th thoracic segment of male, absent from female. In both sexes, uropod with large paddle-shaped outer ramus, but without inner ramus. Posterior margin of telson without indentation. Curls up tightly when disturbed.

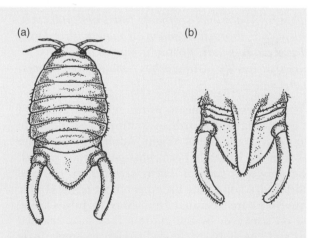

Figure 13.19 *Campecopea hirsuta*, (a) female, (b) posterior region of male.

C. hirsuta is found in south-west Britain, about as far north as Anglesey, North Wales, and on south coast of Ireland, the west coast of France and Portugal, and the Atlantic coast of Spain. It lives on the high shore in the empty cases of *Semibalanus balanoides* (p. 308), *Chthamalus* spp. (p. 306) and among *Lichina* (p. 73), and on exposed shores is often found in large numbers. It is readily collected by taking scrapings of barnacles and lichens for subsequent examination in the laboratory. The eggs are incubated in internal pouches opening from the brood chamber, and on the coast of South Wales one brood of eggs per year is produced. Females carrying embryos are recorded in summer with release of young in late summer or autumn. The lifespan is about one and a half years.

Family Idoteidae

Isopods with a more or less elongate body. The abdominal segments are indistinct. The uropods are ventral and usually have only one ramus which is large and flattened and covers the pleopods.

Idotea granulosa Rathke **Fig. 13.20a**

Elongate, somewhat oval body, narrow at posterior. Males up to 20 mm in length; females 13 mm. Antenna 2 about 1/5 body length. Abdomen with 2 distinct segments; 3rd segment only partially demarcated. Uropods ventral and cover the pleopods. Sides of telson slightly concave; posterior margin with rounded median extension; without a mid-dorsal keel. Colour varied, includes shades of brown, green or red; some specimens have longitudinal white markings.

I. granulosa is widespread in north-west Europe, where it is found on a variety of intertidal seaweeds. It is a northern species extending south to the coast of Normandy and in Britain is probably the commonest intertidal species of *Idotea*. Adults occur mainly on the large brown seaweeds such as *Ascophyllum* (p. 61) and *Fucus* (p. 63) (but avoid laminarians), while young specimens are found on *Polysiphonia* (p. 51) and *Cladophora* (p. 32). *I. granulosa* is omnivorous and the colour of the animal is dependent on the seaweed on which it feeds, the gut contents being visible through the semi-transparent body wall. Breeding females have been found throughout the year with a peak in early summer on the Isle of Man; breeding in winter has been recorded off the south-east coast of Ireland. Longevity varies from a few months to about one year.

Several other species of *Idotea* occur in British waters. **Idotea pelagica** Leach often outnumbers *I. granulosa* on exposed shores where it lives among barnacles and fucoid seaweeds. **Idotea chelipes** (Pallas) is found in estuaries, and withstands salinities down to about 8. Other species, such as **Idotea balthica** (Pallas), normally found

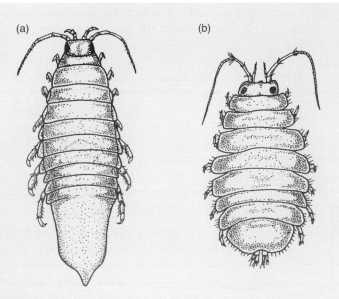

(a)   (b)

Figure 13.20 (a) *Idotea granulosa*. (b) *Jaera albifrons*. (Both after Sars, 1899.)

in the sublittoral, are occasionally washed ashore in large numbers on decaying seaweed; separation of the species is based on the shape of the telson and habitat.

Family Janiridae

Isopods generally with an oval body. The abdominal segments are not demarcated. The uropods are narrow and stick-like and project from the posterior margin of the telson. The anterior pleopods are modified and cover the other pleopods in the female, or function as a copulatory organ in the male.

Jaera spp. Fig. 13.20b

Oval-shaped body fringed with setae; margins of thorax deeply notched. Females up to 6 mm in length, usually larger than males. Antenna 1 short, antenna 2 about 1/2 length of body. Abdominal segments not visible dorsally. Telson large, with smooth, non-serrated margin. Uropods very small; project from indentation in posterior margin of telson.

The genus *Jaera* is widely distributed in north-west Europe and Britain and includes a number of species. Generally speaking, they are common in moist situations on the middle regions of the shore. Some species withstand reduced salinity and are found under stones where freshwater streams run across the beach and in estuaries. The species overlap in distribution but each has its own ecological niche and there appears to be little evidence of hybridization. One species, **Jaera hopeana** Costa, is no more than 2.5 mm in length and lives ectocommensally on the ventral surface of *Sphaeroma serratum* (p. 324). It is a southern species found in south-west Britain, very occasionally free-living but always where *S. serratum* is found.

Family Ligiidae

Terrestrial and littoral fringe isopods. Antenna 1 inconspicuous, antenna 2 large with at least 10 small distal segments. Abdominal segments clearly demarcated. Uropods large, each with two long rami, project from the posterior margin of the telson.

Ligia oceanica (Linnaeus) Sea-slater
Fig. 13.21

Oval-shaped body more than twice as long as broad; up to 30 mm in length. Antenna 1 inconspicuous, antenna 2 about 2/3 body length. Abdominal segments clearly demarcated. Uropods large, each with 2 long processes which project from posterior margin of telson. Various shades and intensities of grey, olive.

L. oceanica is widely distributed in north-west Europe. It is the only species of the genus in Britain and is common in crevices and under stones on the upper shore. It is omnivorous, emerging at night to feed on detritus, decaying seaweed and encrusting diatoms. Breeding occurs in spring and summer. On the east coast

Figure 13.21 *Ligia oceanica.*

of England maturity has been recorded in the first year, although most animals do not breed until they are two years of age. The majority of individuals breed only once, and length of life is two and a half to three years.

Order Amphipoda

Amphipods have many characters in common with the isopods (p. 317), such as the lack of a carapace. As a general rule, however, the body of the amphipod is laterally flattened and one important distinguishing feature is the fact that the first three segments of the abdomen typically each bear a pair of flattened pleopods, while the last three have narrow, backwardly directed uropods. There are seven externally visible thoracic segments and the appendages of the first two are known as gnathopods and are used in feeding and grasping. The remaining five segments of the thorax bear pereopods. In many amphipods, the terminal claw of the gnathopods closes against the preceding segment, a condition known as subchelate (Fig. 13.22c) and this is an important character in identification. Other characteristics useful in identification include the form of the coxal plates (lateral expansions along the sides of the thorax), the presence or absence of an accessory flagellum on antenna 1, and the shape of the telson (Fig. 13.25). The range of feeding habits shown by amphipods includes suspension feeding, scavenging and detritus feeding. The sexes are generally separate and some species show precopula behaviour, during which the male holds the female using the gnathopods, and carries her for some days before mating. Sperm are deposited in the brood chamber of the female, where external fertilization takes place, and where the eggs develop to the juvenile stage. The brood chamber is formed of brood plates (oostegites) arising from the bases of some thoracic appendages (Fig. 13.27a).

Amphipods are generally benthic in habit and many burrow into sand and mud. Others are found among shingle, under stones and seaweed, and in the holdfasts of laminarians. Characteristically, they move away quickly on their sides when disturbed. Amphipods swim using the pleopods and some are caught in high densities in the intertidal plankton, especially at night. Over 100 species have been recorded from British shores; only some of the most common are included here. Reference to **Lincoln (1979)** is recommended for further information.

Key to families of amphipods

Amphipod taxonomy has been the subject of a number of reviews including those of Costello, Holmes, McGrath & Myers (1989) and Myers & Lowry (2003). Identification to families can be difficult and often relies on the combination of a number of characteristics. These are shown in Fig. 13.25. In a small number of cases it has proved most convenient to key to the level of genus.

1 Body rounded; with pleopods and uropods. Eyes very large, bulbous. Pelagic or found on the undersurface of jellyfishes .... **Hyperiidae** p. 347
   Not as above **2**

2 Body very elongate and narrow; abdomen tiny. Some pereopods absent or reduced in size; pleopods and uropods absent. Found clinging to seaweeds, hydroids, sponges; move by characteristic looping movements .... **Caprellidae** p. 346
Body usually, but not always, laterally compressed; with pleopods and uropods. Found buried in sandy and muddy sediments, in timbers, among shingle, under stones and seaweed **3**

3 Antenna 2 stout and stumpy in appearance; uropods dissimilar, uropod 3 very large; found in timbers .... **Cheluridae** p. 344
Not so **4**

4 Body stout, broad, solid; of very distinctive appearance (Fig. 13.28). Pereopods 5, 6, 7 very much expanded with many setae and spines; without terminal claw .... **Haustoriidae** p. 338
Not so **5**

5 Antenna 1 with very large, rectangular basal segment; remaining segments arising at right angles (Fig. 13.27). Basal segments of pereopods 5, 6, 7 flattened and expanded .... **Pontoporeiidae (Genus *Bathyporeia*)** p. 336
Not so **6**

6 Head with small rostrum; antenna 1 with accessory flagellum; pereopod 5 with horizonal rows of stout spines, quite different in appearance from other pereopods .... **Urothoidae** p. 338
Not so **7**

7 Head with large, downwardly pointing rostrum; eyes large, extend towards dorsal surface of head. Antennae shortish; antenna 1 without accessory flagellum. Pereopod 7 elongate and narrow, much longer than preceding pereopods .... **Oedicerotidae** p. 339
Not so **8**

8 Body dorsoventrally flattened; coxal plates small, not touching one another. Eyes small. Antenna 2 large and stout. Pereopod 7 narrow, longer than preceding pereopods .... **Corophiidae (Genus *Corophium*)** p. 343
Not so **9**

9 Antenna 1 much shorter than antenna 2, without accessory flagellum. Uropod 3 uniramous, small. Amphipods which are capable of jumping considerable distances ... **10**
Not so **11**

10 Antenna 1 shorter than large basal segments of antenna 2. Common on the strandline .... **Talitridae** p. 331
Antenna 1 longer than large basal segments of antenna 2 but less than total length of antenna 2 .... **Hyalidae** p. 333

11 Last 3 segments of abdomen with conspicuous clusters of spines and setae on dorsal surface .... **Gammaridae** p. 334
Last 3 segments of abdomen without conspicuous clusters of spines and setae, or at most with only a few spines. Dorsal surface of some segments of body may be toothed ... **12**

12 Antenna 1 considerably longer than antenna 2; antenna 1 with accessory flagellum. Eyes small. Dorsal surface of abdomen usually toothed. Gnathopod 2 larger than gnathopod 1. Telson bilobed .... **Melitidae** p. 335
Not so **13**

13 Antenna 1 shorter than antenna 2, without accessory flagellum. Abdomen with only 5 visible segments. Abdominal segment 4 with 2 dorsal teeth, the posterior one the larger. Dorsal surface of other abdominal segments may have tooth-like posterior projections. Telson bilobed .... **Dexaminidae Genus *Atylus*** p. 341
Not so **14**

14 Uropod 3 large, biramous; rami elongate, flattened, fringed with setae and spines. Gnathopods 1 and 2 small, of similar size. Telson not divided, large ....
.... **Eusiridae (Genus *Apherusa*)** p. 340
Uropod 3 usually biramous; rami short, shorter than basal region; outer ramus with terminal hooks and/or small teeth. Gnathopods 1 and 2 well developed, gnathopod 2 larger than gnathopod 1 .... **15**

15 Basal segment 3 of antenna 1 much smaller than basal segment 2 ....
.... **Ampithoidae** p. 341
Basal segment 3 of antenna 1 nearly as long as basal segment 2. Antennae usually fringed with long setae .... **Ischyroceridae** p. 345

Family Talitridae Sand-hoppers

Stout-bodied amphipods; laterally compressed. Head without a rostrum; antenna 1 shorter than the large basal segments of antenna 2; antenna 1 without an accessory flagellum. Antenna 2 longish; basal joints stout in male. Gnathopods subchelate or not; gnathopod 2 sometimes large in males. Uropod 3 small, uniramous. Telson usually divided only at tip.

Generally found among debris and decaying seaweed on the strandline; known as sand-hoppers because of their ability to jump considerable distances.

Small terminal segments of antenna 2 with teeth on inner margin, giving a rough, spiky appearance. Gnathopod 2 of male small, not subchelate. Uropod 3 terminating in long spine .... ***Talitrus saltator*** p. 332

Small terminal segments of antenna 2 without teeth on inner margin. Gnathopod 2 of male large and subchelate .... ***Orchestia*** **spp.** p. 331

Orchestia spp. Sand-hopper **Fig. 13.22**

Greenish-brown body up to 22 mm in length. Eyes black. Antenna 1 small, shorter than large basal segments of antenna 2; antenna 2 about 1/3 body length. Small terminal segments of antenna 2 without teeth on inner margin. Gnathopod 1 small and subchelate; gnathopod 2 small in female, large and subchelate in male.

The genus *Orchestia* is widely distributed in north-west Europe, with *Orchestia gammarellus* (Pallas) being the commonest species. It is found under stones and rotting seaweed, particularly on shingle shores on the upper-shore/terrestrial fringe, and is also found in estuaries. It is an omnivorous scavenger and is particularly active at night. Although reproduction has been recorded as continuous on the coast of

Figure 13.22 *Orchestia gammarellus*, (a) female, (b) detail of gnathopod 1, male, (c) detail of gnathopod 2, male (after Sars, 1895).

Portugal, around Britain breeding has been recorded from April to about September at a number of localities and there is some evidence to suggest that young released in early summer will mature by the autumn of the same year. Longevity is estimated to be about 18 months. When disturbed, *Orchestia* jumps considerable distances.

Talitrus saltator (Montagu) Sand-hopper **Fig. 13.23a,b**

Greyish-green body up to 25 mm in length. Eyes black. Antenna 1 small, much shorter than large basal segments of antenna 2. Small terminal segments of antenna 2 with teeth on inner margin giving a rough, spiky appearance, especially in male. Gnathopods 1 and 2 small, not subchelate. Uropod 3 with long, slender spine on posterior margin.

T. saltator is the only species of the genus found in north-west Europe. It lives on the upper-shore/terrestrial fringe and is widely distributed on sandy beaches around Britain where it is associated with decaying seaweed. It is an omnivorous scavenger, usually spending the daytime buried in the sediment, emerging at night to feed on detritus on the strandline. Although essentially nocturnal, *Talitrus* has been shown to be capable of navigating by means of a Sun compass. If for some reason the animals are transported from the upper shore, they can return by orientating to the angle of the Sun and other features. The common name sand-hopper (which includes *Orchestia*, above) is associated with the animal's ability to jump up to 300 mm by rapid backwards extension of the abdomen and telson. On Irish Sea coasts *T. saltator* breeds during the summer months with peak activity from May to August. Females are believed to produce more than one brood and the incubation period varies from eight to 20 days, depending on temperature. Longevity is between 18 months and two years.

Talorchestia deshayesi (Audouin) is another sand-hopper occurring amongst shingle and strandline debris around Britain. It is similar to *Talitrus* in that the small terminal segments of antenna 2 are toothed but differs in that gnathopod 2 of the male is very large and pincer-like (Fig. 13.23c), and the posterior margin of uropod 3 does not bear a long slender spine.

Figure 13.23 (a) *Talitrus saltator*, with (b) detail of gnathopod 2, male. (c) *Talorchestia deshayesi*, gnathopod 2, male.

Family Hyalidae

Stout-bodied amphipods, laterally compressed. Head without a rostrum; antennae 1 and 2 short; antenna 1 longer than large basal segments of antenna 2, but less than total length of antenna 2. Antenna 1 without an accessory flagellum. Gnathopods 1 and 2 subchelate, more or less the same size in female; in male, gnathopod 2 much larger than gnathopod 1. Uropod 3 small and uniramous. Telson bilobed.

Live intertidally among seaweed and in crevices; able to jump considerable distances.

Hyale spp. Fig. 13.24

Brownish-green, usually up to about 8 mm in length. Eyes black or reddish colour. Antenna 1 shorter than antenna 2, but longer than the large basal segments of antenna 2 (cf. Orchestia *(p. 331) and* Talitrus *(p. 332)). Gnathopods 1 and 2 subchelate. Gnathopod 2 of male much larger than gnathopod 1. Gnathopods 1 and 2 about same size in female.*

The genus *Hyale* is widely distributed in north-west Europe. Some four species have been recorded from Britain, of which **Hyale prevosti** (Milne-Edwards) (*Hyale prevostii, Hyale nilssonii*) is widely distributed and probably the best known. Found among seaweeds such as *Fucus* and *Ascophyllum* on rocky shores, it occurs from the upper shore into the sublittoral and also penetrates into estuaries. It is able to jump considerable distances. Specimens on the high shore have been shown to exhibit

(a)

(b)

Figure 13.24 *Hyale prevosti*, (a) female, (b) detail of gnathopod 2, male.

a preference for feeding on the fertile tips of the brown seaweed *Pelvetia canaliculata* (p. 64). Breeding has been recorded on the west coast of Scotland during spring and summer; the eggs are green coloured.

Family Gammaridae

Mostly stout-bodied amphipods; laterally compressed. Dorsal surface smooth, not toothed; last three segments of abdomen with clusters of spines and setae on the dorsal surface. Head with small rostrum; eyes mostly large and kidney-shaped. Antennae 1 and 2 longish, and of similar length; antenna 1 with conspicuous accessory flagellum. Gnathopods 1 and 2 subchelate; well developed, much the same size. Uropods biramous, rami of uropod 3 subequal, inner ramus shorter than outer and may be very small; with many marginal spines and setae. Telson bilobed.

Amphipods *par excellence*. A large family, common in marine and freshwater habitats. Representatives are found in all salinity concentrations, typically under stones and among seaweed and in shingle; seen wriggling on their sides when disturbed.

The systematics of 'gammarid' amphipods continues to be the subject of much change (Costa, Henzler, Lunt, Whiteley & Rock, 2009) and species described below under the genus *Gammarus* may be referred to in other texts under the names given in parentheses.

Gammarus spp.   Fig. 13.25

Body usually brownish-green; up to 33 mm in length. Eyes rounded or kidney-shaped. Antenna 1 with accessory flagellum; antenna 2 of similar length to antenna 1. Gnathopods 1 and 2 subchelate. Last 3 segments of abdomen have clusters of spines or spines and setae on posterior, dorsal margins giving spiny appearance. Rami of 3rd pair of uropods subequal, inner ramus shorter than outer, and may be very small.

Many species of *Gammarus* have been described from marine and freshwater habitats. They are mostly intertidal, extending from fully marine through brackish-water to freshwater environments and individual species are often used as indicators of salinity.

Gammarus locusta (Linnaeus) is widespread and often abundant in Britain and north-west Europe where it is found in damp situations under stones and seaweed on the middle and lower shore, surviving in salinities down to about 4. It feeds mainly on

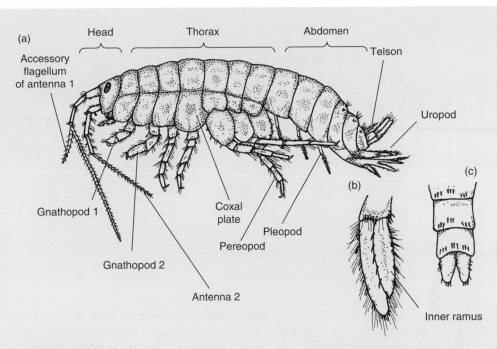

Figure 13.25 *Gammarus locusta* (a), with (b) detail of uropod 3 and (c) dorsal surface of posterior of abdomen (uropods omitted).

detritus and seaweed. The male carries the female for a number of days before fertilization, and females with embryos in the brood chamber are found throughout most of spring and summer.

Two other species of *Gammarus* are widely distributed and common around Britain. ***Gammarus obtusatus*** Dahl (*Eulimnogammarus obtusatus, Marinogammarus obtusatus*) is common and widespread among seaweed and in gravel and shingle, and ***Gammarus marinus*** Leach (*Echinogammarus marinus, Chaetogammarus marinus, Marinogammarus marinus*) is found among seaweed and under stones on the upper and middle shore. In both these species, the inner ramus of the third uropod is less than one-third of the length of the outer ramus, compared with *G. locusta* in which the inner ramus is more than one-third of the length of the outer (Fig. 13.25b).

Echinogammarus planicrurus (Reid) (*Pectenogammarus planicrurus*) is found in mobile gravel and shingle deposits on open coast beaches. Few species are found in such hostile environments and although *E. planicrurus* has a restricted distribution in north-west Europe, in some localities it is an abundant and characteristic species of this habitat. In this species the inner ramus of the third uropod is less than one-third of the length of the outer ramus. The distinctive habitat of *E. planicrurus* is a useful indicator of species identification.

Family Melitidae

Mostly slender-bodied amphipods; laterally compressed. Dorsal surface of thorax smooth; abdomen usually toothed dorsally; last three segments sometimes with a few spines on dorsal surface, but without distinct clusters of spines and setae

Figure 13.26 *Melita palmata*, (a) female, (b) detail of gnathopod 2, male.

(cf. Gammaridae, p. 334). Head without a rostrum; eyes small; antennae 1 and 2 longish, antenna 1 usually considerably longer than antenna 2; antenna 1 with accessory flagellum. Gnathopods 1 and 2 subchelate, gnathopod 1 small, gnathopod 2 large, much larger in male. Uropod 3 biramous. Telson bilobed.

Commonly found among stones on sandy and muddy sediments, and among seaweed.

Melita spp. **Fig. 13.26**

Body rather slender, up to about 16 mm in length. Colour varied, reddish-brown, grey, whitish. Eyes rather small, rounded or oval. Antenna 1 with accessory flagellum; antenna 1 longer than antenna 2. Gnathopods 1 and 2 subchelate; gnathopod 2 much larger than gnathopod 1 in male with segment next to claw triangular in shape. Dorsal surface of abdomen usually, but not always, toothed. Last 3 segments of abdomen may have a few spines but without clusters of spines and setae. Outer ramus of 3rd pair of uropods elongate, longer than basal region and with marginal spines; inner ramus very small.

The genus *Melita* is widely distributed in north-west Europe and of the intertidal species, **Melita palmata** (Montagu) is the most widespread. It is found among stones on sandy and muddy sediments.

Family Pontoporeiidae
Genus *Bathyporeia*

Slender-bodied amphipods, laterally compressed. Head without rostrum. Antenna 1 shorter than antenna 2. Antenna 1 with very large basal segment; with accessory flagellum. Gnathopods 1 and 2 not strongly developed. Pereopods 3–7 with small terminal claw. Basal segments of pereopods 5, 6, 7 flattened and expanded. Uropod 3 biramous. Telson bilobed.

The only genus of the family found in Britain. Found in clean, sandy sediments; common and widespread and extends into estuaries.

Figure 13.27 *Bathyporeia pilosa*, (a) female with eggs in brood chamber, (b) head of male.

Bathyporeia spp. Fig. 13.27

Body up to 8 mm in length; semi-transparent to white, with varying degrees of red pigment associated with abdomen. Dorsal surface of abdominal segment 4 notched, with setae or setae and spines. Eyes red. Antenna 1 with accessory flagellum and shorter than antenna 2. Basal segment of antenna 1 very large, more or less rectangular; other segments smaller and arise at right angles to it, a feature known as geniculate and characteristic of the genus. Antenna 2 about twice the length of antenna 1 in females but equal to, and often exceeding, body length in mature males.

The genus *Bathyporeia* is well represented around Britain and eight species have been described. Morphologically they are very similar. **Bathyporeia pilosa** Lindström is one of the commonest species, often reaching densities of several thousand per square metre. It is widely distributed in north-west Europe on the upper shore of sheltered sandy beaches and is also common in estuaries, tolerating salinities down to about 5. It is an epistrate feeder, individual sand grains being rotated by the mouthparts and organic matter removed. Although *Bathyporeia* lives buried in sandy sediments, it leaves the protection of the sand to swim in the surface waters during spring tides; large numbers can be taken in intertidal plankton hauls, particularly during darkness. On the west coast of Wales, breeding has been recorded throughout the year, with peaks of reproductive activity in spring and autumn, while in some localities a single peak of breeding has been recorded in summer. Males and females pair while swimming but there is no prolonged precopula behaviour. During the breeding season, gravid females are readily identified by the presence of blue eggs in the brood chamber, each female producing a succession of broods. In summer, development takes about 14 days and is synchronized with spring tides. Longevity is about one to one and a half years.

Other species of *Bathyporeia* may be common on sandy shores, at some localities showing a zonation of species down the shore. **Bathyporeia pelagica** (Bate) is widely distributed around Britain and has been recorded from the Netherlands and the Channel coast of France. It is found from about mean tide level into the sublittoral on open coasts in clean, sandy deposits and, although unable to tolerate such low salinities as *B. pilosa*, it is often found at the mouths of estuaries. It has a similar life-cycle to *B. pilosa*, with a peak of breeding activity during the summer. **Bathyporeia sarsi** Watkin is common on the coast of Wales, south-west England and the North Sea at about the level of mean high water of neap tides and below. Where the species coexist, it tends to be distributed between *B. pilosa* and *B. pelagica*. **Bathyporeia elegans** Watkin and **Bathyporeia guilliamsoniana** (Bate) are widely distributed on the lower shore and at some localities may be common. Both are found from about the level of mean low water neap tides into the sublittoral.

Family Haustoriidae

Solid, stout-bodied amphipods of very distinctive appearance. Segments of pereopods 5, 6, 7, except the most distal one, flattened and very much expanded, with many setae and spines. Head with small rostrum. Antennae 1 and 2 short; antenna 1 with accessory flagellum. Gnathopods 1 and 2 not strongly developed. Pereopods 3–7 without terminal claw. Uropod 3 biramous. Telson bilobed.

Haustorius arenarius (Slabber) **Fig. 13.28**

Body broad, stout, up to 13 mm in length; whitish colour. Eyes small, difficult to see. Antenna 1 slightly shorter than antenna 2, both relatively short. Pereopods 3–7 without a terminal claw; pairs 5, 6, 7 very much flattened. Sexes difficult to distinguish, but brood plates present in mature female. Animal of unmistakable and distinctive appearance.

Figure 13.28 *Haustorius arenarius.*

H. arenarius is widespread in north-west Europe and is the only species of the genus found in Britain. It is commonly recorded in clean, sandy deposits, into which it burrows very quickly when disturbed, and is predominantly found on open coasts and at the mouths of estuaries, being most abundant on the lower shore. It is primarily a suspension feeder. When *Haustorius* leaves the sediment, it swims with the ventral surface uppermost and mating is believed to take place while the animals are swimming. Little is known of the breeding patterns of populations in Britain, but females with eggs are found in the summer months and longevity is believed to be two, possibly three, years.

Family Urothoidae

Solid, compact-bodied amphipods; laterally compressed. Head with small rostrum. Antenna 1 short; antenna 2 usually short in female, elongate in male. Antenna 1 with

(a)

(b)

Figure 13.29 *Urothoe brevicornis*, (a) female, (b) head of male.

accessory flagellum. Gnathopods 1 and 2 not strongly developed. Pereopods 3–7 with terminal claw. Pereopod 5 with horizontal rows of stout spines. Basal segments of pereopods 6 and 7 flattened and expanded. Uropod 3 biramous. Telson usually bilobed.

There is a single genus in Britain, found in clean, sandy sediments. It is common and widespread.

Urothoe spp. Fig. 13.29

Body mostly yellowish-white, with superficial resemblance to Bathyporeia *(p. 336) but without geniculate antenna 1; up to 8 mm in length. Head with very small rostrum. Eyes large in male and, in some species, they almost merge on dorsal surface of head. Antenna 1 with accessory flagellum. Antennae 1 and 2 shortish in female but antenna 2 nearly as long as body in male. Pereopod 5 spiny, noticeably different in appearance from other pereopods.*

The genus *Urothoe* is widely distributed in clean, sandy sediments, some species extending into deep water. Of the intertidal species **Urothoe brevicornis** Bate is probably the best known. It is found from the middle shore to depths of 40 m and has also been recorded from the Netherlands to the north-west coast of Spain. It reproduces in summer and autumn and, although little is known of its life-cycle, it probably lives for one to two years.

Family Oedicerotidae

Somewhat slender-bodied amphipods, slightly laterally compressed. Head usually with a large downwardly pointing rostrum; eyes large, extending onto dorsal surface of head. Antennae 1 and 2 shortish; antenna 1 without an accessory flagellum. Gnathopods usually subchelate, sometimes gnathopod 2 with thin, elongate terminal segments. Pereopod 7 elongate and narrow, much longer than preceding pereopods. Uropod 3 biramous. Telson not divided.

Live buried in clean, sandy deposits, widespread and common.

Figure 13.30 *Pontocrates altamarinus*, (a) female, with (b) detail of gnathopod 2.

Pontocrates spp. **Fig. 13.30**

Whitish body up to 7 mm in length. Head with conspicuous downwardly pointing rostrum. Eyes large, rounded. Antenna 1 slightly shorter than antenna 2 in female. Antenna 2 of male long, at least equal to body length in mature specimens. Gnatho-pod 1 well developed, subchelate; gnathopod 2 thin, elongate, with narrow process extending beyond the lower edge of the terminal claw. Pereopod 7 very long and narrow.

The genus *Pontocrates* is widely distributed in north-west Europe. Two species, **Pontocrates arenarius** (Bate) and **Pontocrates altamarinus** (Bate & Westwood), are commonly recorded from Britain. They are found from the middle shore into the sublittoral on clean, sandy beaches, and often occur in high densities. In populations of *P. arenarius* from Irish Sea coasts, breeding has been recorded throughout the year. A third species, **Pontocrates arcticus** G. O. Sars, found in the shallow sublittoral, has its southern limit of distribution on the west coast of Scotland.

Family Eusiridae
Genus *Apherusa*

Stout-bodied amphipods. Laterally compressed; dorsal surface sometimes with tooth-like projections. Head with rostrum; eyes small or large. Antenna 1 shorter than antenna 2; antenna 1 without accessory flagellum. Gnathopods 1 and 2 subchelate, small; of similar size. Uropod 3 biramous; rami elongate, flattened, fringed with setae and spines. Telson large, not divided.

Live among seaweed; widespread and common.

Apherusa jurinei (Milne-Edwards) **Fig. 13.31**

Colour varied, pink, reddish, yellowish-brown; may be banded; sometimes with white patches. Up to 8 mm in length. Dorsal surface of body smooth. Eyes large,

Figure 13.31 *Apherusa jurinei.*

kidney-shaped. Antenna 1 slightly shorter than antenna 2; without an accessory flagellum. Gnathopods 1 and 2 subchelate, small.

A. jurinei is widely distributed in north-west Europe and Britain and is found among seaweed on the lower shore and in the shallow sublittoral.

Other species of *Apherusa* have been recorded from among seaweed on the lower shore and may be locally common. These can generally be distinguished from *A. jurinei* by the presence of tooth-like, dorsal projections on the posterior margins of the first two abdominal segments.

Family Dexaminidae
Genus *Atylus*

Stout-bodied amphipods, laterally compressed; dorsal surface of at least some segments of abdomen usually with tooth-like projections. Abdomen with only five visible segments. Antenna 1 without accessory flagellum. Gnathopods 1 and 2 small, subchelate. Uropod 2 shorter than uropod 1. Uropods biramous. Telson bilobed.

On lower shore in sandy deposits or among seaweed.

Atylus swammerdami (Milne-Edwards) (*Nototropis swammerdami*) **Fig. 13.32**
Body whitish in colour with brown patches; up to 10 mm in length. Head with small rostrum. Eyes large, kidney-shaped. Two dorsal teeth on abdominal segment 4, the posterior one the larger; other thoracic and abdominal segments smooth dorsally. Second pair of uropods shorter than 1st.

A. swammerdami is widely distributed in north-west Europe and Britain. It is found mainly in sandy deposits on the lower shore and sublittorally.

Family Ampithoidae

Slender-bodied amphipods, slightly compressed laterally. Head without rostrum; eyes small. Antenna moderately long; antennae 1 and 2 same length or antenna 2 shorter; basal segment 3 of antenna 1 much smaller than basal segment 2. Antenna 1 without accessory flagellum or flagellum very small. Gnathopods 1 and 2 subchelate; gnathopod

Figure 13.32 *Atylus swammerdami.*

2 the larger, more so in male. Uropod 3 biramous, rami shorter than basal region; inner ramus with terminal setae, outer ramus with one to three large hooked spines. Telson not divided; sometimes with hooks.

Live in tubes of algae and detritus among seaweed and stones.

Ampithoe rubricata (Montagu) **Fig. 13.33**

Body reddish, brown, green, sometimes mottled with small, white spots; up to 20 mm in length. Eyes small, red. Antenna 1 about equal to or slightly longer than antenna 2. Gnathopods 1 and 2 large, subchelate; gnathopod 2 larger than 1, more so in males. Outer ramus of uropod 3 with 2 hooked spines and a group of setae. Telson without large hooked spines; posterior margin convex.

A. rubricata is widely distributed, often commonly occurring on the shores of north-west Europe and Britain. It lives in tubes which it builds from algae and detritus, and attaches to stones and seaweed on the middle and lower shore. Most tubes house only one animal but a few contain a male and a female and it is believed that fertilization occurs within the tube. Breeding occurs during the summer. *Ampithoe* feeds mainly on seaweed.

Figure 13.33 *Ampithoe rubricata.*

Family Corophiidae
Genus *Corophium*

Narrow, elongate body, not laterally compressed but dorsoventrally flattened; coxal plates small, not touching one another. Last three abdominal segments often indistinct. Head with or without a rostrum; eyes small. Antennae well developed; antenna 1 narrow, shorter than or the same length as antenna 2; antenna 2 large and stout, especially so in male. Antenna 1 without accessory flagellum. Gnathopods 1 and 2 small, densely fringed with setae; gnathopod 2 with middle two segments joined along their length. Pereopod 7 much longer than preceding pereopods. Uropod 3 uniramous. Telson small, not divided.

The genus *Corophium* includes many species that live in burrows in soft sediments, or construct tubes among hydroids, sponges and laminarian holdfasts. They are found in marine, brackish and fresh water and may be locally abundant. Two species characteristic of intertidal mud and sand deposits are described here.

Outer margin of basal region of uropod 1 with 10–12 stout spines; inner margin with 3–4 slender spines (Fig. 13.34b) .... ***Corophium volutator*** p. 343

Outer margin of basal region of uropod 1 with 11–15 spines and 1–5 long, proximal setae; inner margin with a single spine (Fig. 13.34c) .... ***Corophium arenarium*** p. 344

Corophium volutator (Pallas) **Fig. 13.34a,b,d**

Body not laterally compressed; up to 10 mm in length. Head with very small pointed rostrum. Antenna 1 without accessory flagellum, shorter than antenna 2 which is

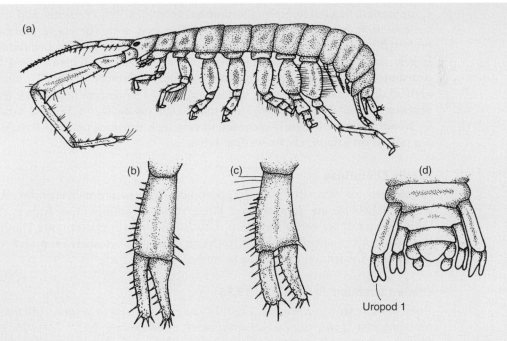

Figure 13.34 (a) *Corophium volutator*, male. (b) *Corophium volutator*, uropod 1. (c) *Corophium arenarium*, uropod 1. (d) Posterior region of abdomen of *Corophium arenarium/volutator*, setae omitted.

large and stout, particularly in males, and characteristically extended in front of animal. Basal segment 3 of antenna 1 much shorter than basal segment 1. Gnathopods 1 and 2 small, with many setae. Coxal plates small and separate. Last pair of pereopods much longer than preceding pairs. Last 3 abdominal segments distinct. Outer margin of basal segments of uropod 1 with 10–12 stout spines; inner margin with 3–4 slender spines.

C. volutator is widely distributed in north-west Europe and Britain. It lives in U-shaped burrows in muddy, estuarine sediments, tolerating salinities down to about 2, and when present in high densities the apertures of the burrows are clearly visible on the surface of the sediment. *Corophium* is a selective deposit feeder but also feeds by filtering particles from the respiratory current drawn through the burrow by the action of the pleopods. On the west coast of Wales, breeding takes place from April to October and development is synchronized with spring tides. Although the animals are known to swim on spring tides, mating takes place in the burrow. As the tide recedes, adult males crawl over the surface of the moist sediment in search of burrows occupied by mature females which they enter if the female is receptive. Laboratory experiments have shown that males are attracted by chemical signals (pheromones) released by the females. Embryos are carried in the brood chamber of the female for about 14 days when the juveniles emerge at the time of the next spring tide. Each female produces a number of broods.

Corophium arenarium Crawford **Fig. 13.34c,d**
Body not laterally compressed; up to 7 mm in length. Closely resembles C. volutator (above) but outer margin of basal region of uropod 1 has 11–15 spines and 1–5 long setae; inner margin with single spine.

C. arenarium is similar in its distribution, ecological requirements and life-cycle to *C. volutator*, but lives in muddy-sand, compared with the more glutinous muds preferred by the latter species. Records of its distribution around Britain are patchy, largely as a result of misidentification. Nevertheless, it is believed to be widely distributed, having a slightly narrower salinity tolerance than *C. volutator*. The vertical distribution of the two species appears to be determined by the presence or absence of a suitable sediment rather than height on the shore. Both *C. volutator* and *C. arenarium* are frequently recorded in very high densities on mud-flats, where they are preyed on extensively by wading birds.

Family Cheluridae
Amphipods of very distinctive appearance; flattened dorsoventrally. Antenna 1 shorter than antenna 2; antenna 2 large, stout, of stumpy appearance. Antenna 1 with very small accessory flagellum. Gnathopods small; gnathopods 1 and 2 similar size. Uropods modified, and very different from one another, uropod 3 very large. Telson not divided. Found boring in timber.

Chelura terebrans Philippi **Fig. 13.35**
Body not laterally compressed. Light brown; up to 6 mm in length. Antenna 1 shorter than antenna 2; antenna 2 stout and densely covered with setae. Posterior margins of body segments with many setae giving overall 'hairy' appearance. Third abdominal segment extended backwards into curved mid-dorsal projection, larger in males than females. Uropods 1, 2 and 3 dissimilar, 2nd and 3rd pairs much modified and showing pronounced sexual dimorphism.

Figure 13.35 *Chelura terebrans*, (a) female, (b) posterior region of male (after Sars, 1895).

The genus *Chelura* is represented in north-west Europe by a widely distributed single species, *C. terebrans*. It is found in timbers that have been attacked by *Limnoria* (p. 321), and populations of *Chelura* will not survive any length of time if *Limnoria* is not present. It feeds predominantly on its own faecal pellets but also ingests wood particles and the faeces of *Limnoria*.

Family Ischyroceridae

Amphipods somewhat flattened dorsoventrally. Head usually without a rostrum. Antennae well developed, usually densely fringed with setae; basal segment 3 of antenna 1 nearly as long as basal segment 2; antenna 1 with or without small accessory flagellum. Gnathopods well developed; gnathopod 2 larger than gnathopod 1 and in male usually very well developed with a distinctive claw. Uropod 3 biramous or uniramous, rami much shorter than basal region; without terminal setae; with hooked terminal spine or small teeth or both. Telson usually not divided; without terminal hooks.

Live in tubes among hydroids, seaweeds, mooring ropes and buoys.

Jassa spp.   Fig. 13.36

Body slender; up to 13 mm in length. Colour varied, often greyish with brownish and reddish patches. Antenna 1 about 2/3 length of antenna 2 and with small accessory flagellum. Antenna 2 large, particularly in male. Gnathopod 2 larger than gnathopod 1, terminating in characteristic claw which differs in size and form between sexes. Uropod 3 biramous; outer ramus with hooked spine and 1–3 small teeth.

The genus *Jassa* has a worldwide distribution and in some localities occurs in high densities. It is widely distributed in north-west Europe. ***Jassa falcata*** (Montagu) is one of the most familiar species but its polymorphism has caused taxonomic problems and reference to Conlan (1990) is recommended. Although recorded from rock pools, typically it is found attached to mooring ropes, buoys, ships' hulls, hydroids and seaweeds. Males and females live in separate tubes constructed of pieces of debris cemented together, the tubes often forming dense mats, particularly in warm water discharge pipes from power stations and in some situations densities as high as 500 000 per square metre have been recorded. They are suspension feeders. Reproduction occurs throughout the year, females producing a succession of broods. There is a marked sexual dimorphism between males and females, the males acquiring

Figure 13.36 *Jassa falcata*, (a) female, (b) gnathopod 2, male (after Sars, 1895).

a conspicuous 'thumb' on the second gnathopod (Fig. 13.36b). Mature females remain in the burrow but on reaching sexual maturity, males leave the burrow in search of receptive females. They enter the burrow of a female where mating takes place and the embryos develop in the brood pouch of the female. Each female can have several broods.

Family Caprellidae

Characteristic body form. Body elongate and narrow, abdomen tiny. Some pereopods absent or reduced in size, pleopods and uropods absent. Two or three pairs of gills. Found clinging to seaweeds, hydroids, sponges. Move by characteristic looping movements.

Caprella spp. Ghost shrimp **Fig. 13.37**

Body very elongate and narrow; up to 20 mm in length; reddish-brown. Antenna 1 longer than antenna 2. Gnathopod 2 often larger in male than female and differing in shape. Female less elongate than male and carries brood chamber on thoracic segments 3 and 4. In both sexes these segments lack pereopods. Abdomen minute.

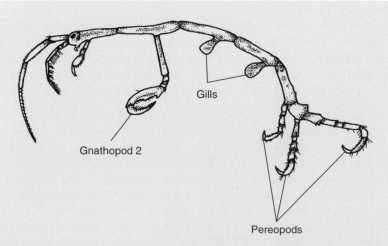

Figure 13.37 *Caprella linearis*, male (after Sars, 1895).

Caprella linearis Linnaeus is probably the best-known caprellid in British waters. It is widely distributed on the lower shore among seaweeds, hydroids and sponges, where it is found clinging by means of the well-developed posterior, thoracic appendages. It crawls in a characteristic looping movement achieved by alternately grasping and releasing the substratum with anterior and posterior appendages. It is a suspension feeder. Breeding probably occurs during autumn and winter; longevity is believed to be about one year.

An invasive species of *Caprella*, ***Caprella mutica*** Schurin, was first recorded in north-west Europe from the Netherlands in 1994 and from Britain in 2000. A native of north-east Asia, it has spread rapidly in the northern hemisphere and is now recorded from the English Channel, the North Sea, the west coast of Ireland, the Irish Sea and the west coast of Scotland. It is found on aquaculture units and oil rigs, and in ports and marinas on structures such as cages, buoys and ropes but has also been recorded from hydroids, ascidians and drifting *Sargassum* (p. 67). It is believed to have been introduced in ballast water and during transport of oysters. *C. mutica* is orange-red in colour and is a large species compared with others in the genus, reaching a body length of up to 40 mm. It tolerates salinities down to about 20 and is a very successful invasive species with females producing successive broods of over 100 juveniles. In some areas it is now out-competing *C. linearis*.

Family Hyperiidae
Body rounded, bulbous; abdomen with pleopods and uropods; eyes very large and conspicuous. Pelagic or found on the undersurface of jellyfishes.

Hyperia galba (Montagu) **Fig. 13.38**
Dorsal margin of body markedly curved, less so in male; up to 14 mm in length. Eyes very large, bulbous, greenish in colour. Antennae 1 and 2 very small in female; long and whip-like in male and reaching to at least 1/2 body length.

H. galba is widely distributed in north-west Europe and extends south to about the English Channel. It is widely distributed in Britain and can often be found in the cavities on the undersurface of jellyfishes, particularly *Rhizostoma* (p. 124), and occasionally on the ctenophore *Beroe* (p. 127), where it clings by means of the last

(a)　　　　　　　　　(b)

Figure 13.38 *Hyperia galba*, (a) female, (b) anterior region of male.

three pairs of thoracic appendages. Its diet includes the tissues of the host, including the gonads of mature jellyfishes, as well as prey captured by the host. In Danish waters breeding has been recorded from July to December. Mature females carry large numbers of embryos; when newly released, the juveniles are unable to swim and remain within the protection of the jellyfish. After a few moults, the pleopods and uropods are fully developed and the juveniles are capable of swimming away to seek a new host. While in association with the jellyfish, the body of the amphipod is almost transparent, but when free-swimming it is a reddish-brown colour.

Order Decapoda

The decapods are perhaps the best-known crustaceans and include the commercially important prawns, shrimps, lobsters, crayfish and crabs. The anterior three pairs of thoracic appendages, the maxillipeds, are used in feeding, while the other five pairs are the pereopods, or 'walking legs'. The first pair of pereopods, the chelipeds, often terminate in large pincers or chelae. The head and thorax are protected by a well-developed carapace which extends laterally to protect the gills. As a rule the first five abdominal segments each have a pair of pleopods, which are sometimes known as swimmerets and, as the name implies, are in some groups used in locomotion. In females the eggs are held on the pleopods until hatching. In prawns, shrimps and lobsters, the uropods form a tail fan with the telson, while in the true crabs the abdomen is reduced in size and folded under the thorax.

Most decapods have separate sexes and elaborate courtship behaviour often takes place. In many cases the female undergoes a moult, the precopulatory moult, just before mating. In prawns, shrimps and lobsters fertilization is external, the eggs being fertilized as they leave the female, but in the true crabs fertilization is internal, sperm being transferred to the spermathecae of the female. The fertilized eggs are attached by sticky secretions to fine setae on the pleopods of the female where they may be held for several months. During this time the female is referred to as 'berried'. The eggs hatch as pelagic larvae.

Key to the different groups of decapods

The decapods are a large, diverse order of crustaceans. A useful first step in identification is separation on the basis of habit. The prawns and shrimps, which have a swimming habit, are sometimes referred to in older texts as the Natantia, while the lobsters and crabs, of characteristic walking habit, are referred to as the Reptantia. Classification of the order is complex but the species described here can be subdivided into a number of readily identifiable groupings using the characters given below.

1 Exoskeleton relatively thin and light; not heavily calcified. Abdomen extended. Swimming habit .... **Prawns and shrimps** p. 349
Exoskeleton generally hard; usually heavily calcified. Abdomen extended or folded under thorax. Walking habit. Some occupy empty gastropod shells **2**

2 Abdomen extended, terminates in a well-developed tail fan; or abdomen extended, asymmetrical, soft and twisted **3**
Abdomen reduced and folded, either tightly or loosely, under thorax; symmetrical **4**

3   Abdomen extended, terminates in a well-developed tail fan. First 3 pairs of
pereopods with chelae, those of the 1st pair very large .... **Lobsters** p. 355

Abdomen extended; asymmetrical, soft and twisted. Animal lives in empty
gastropod shell .... **Hermit crabs** p. 359

4   Fifth pair of pereopods much the same size as preceding 3 pairs. Abdomen small,
tightly folded under thorax .... **True crabs** p. 363

Fifth pair of pereopods very much smaller than preceding pairs. Abdomen tightly
or loosely folded under thorax **5**

5   Abdomen much reduced; tightly folded under thorax. Carapace rounded
.... **Porcelain crabs** p. 358

Abdomen slightly reduced; folded loosely under thorax. Carapace longer than
broad .... **Squat lobsters** p. 357

In addition to the groups described above, the order also includes stone crabs of deep water and the burrowing shrimps found in burrows at low water and into the sublittoral. Neither of these groups is included.

Prawns and shrimps

The words 'prawn' and 'shrimp' lack precise definition but popular usage has led to the larger species being referred to as 'prawns' and the smaller as 'shrimps'. They are distinguished from other decapods by their swimming habit. The exoskeleton is relatively thin and light and the abdomen is extended. The second antennae each bear a plate-like extension which gives stability during swimming. The carapace is often extended between the eyes to form a spine or rostrum, particularly in those animals referred to as prawns. The pleopods are the organs of propulsion. Most prawns and shrimps have separate sexes and mature after one year. Prior to copulation the female undergoes a precopulatory moult and takes on what has been described as 'breeding dress' in preparation for copulation and carrying the fertilized eggs. At copulation spermatophores are placed under the abdomen of the female by the male and soon afterwards the eggs are laid and fertilized as they pass over the spermatophores. The eggs are then attached to setae on the first four pairs of pleopods where they are carried by the female for varying periods of time. Several thousand eggs are carried in a single brood by some species and these hatch as planktonic larvae known as zoeae (Fig. 13.39b) which feed in the plankton before moulting into a post-larval juvenile stage. Prawns and shrimps are omnivorous; they are preyed on extensively by fishes and birds.

Key to families of prawns and shrimps

1   First pair of pereopods subchelate (Fig. 13.43a); rostrum reduced to a spine or
very short .... **Crangonidae** p. 354

First pair of pereopods with or without chelae **2**

2 First pair of pereopods with chelae. Pereopod 2 without two or more small segments next to chela .... **Palaemonidae** p. 350

Pereopod 2 with two or more small segments next to chela **3**

3 Pereopods 1 and 2 clearly chelate; chela of pereopod 1 not greatly enlarged .... **Hippolytidae** p. 352

Pereopod 1 with tiny chela (seen under magnification) or chela absent; pereopod 2 with tiny chela .... **Pandalidae** p. 353

Family Palaemonidae

Rostrum conspicuous (in specimens included here) with teeth on dorsal and ventral margins. Pereopods 1 and 2 terminate in chelae, chela of pereopod 2 being the larger.

1 Rostrum curves upwards; teeth on dorsal margin do not extend to tip of rostrum but are restricted to basal 2/3 .... *Palaemon serratus* p. 351

Rostrum straight, or only slightly curved upwards; teeth all along dorsal margin **2**

2 Rostrum with 4–6 dorsal teeth, one of which is posterior to the orbit; 2 ventral teeth .... *Palaemonetes varians* p. 352

Rostrum with 7–9 dorsal teeth, 3 (sometimes 2) of which are posterior to the orbit; 3 (occasionally 2 or 4) ventral teeth .... *Palaemon elegans* p. 350

Palaemon elegans Rathke (*Leander squilla*) European glass prawn, prawn
Fig. 13.39a,b
Semi-transparent, often with brownish markings. Up to 60 mm in length. Rostrum more or less straight with 7–9 teeth along dorsal edge, 3 (sometimes 2) of which are posterior to the orbit; 3 (occasionally 2 or 4) teeth along ventral edge of rostrum. Tip of rostrum usually divided into two. Antenna 1 triramous. First and 2nd pereopods terminate in chelae, those of the 2nd pair being the larger. Lateral margin of telson with 2 pairs of spines.

P. elegans is widely distributed around Britain and is found in rock pools from the middle shore into the shallow sublittoral but is known to tolerate salinities down to 6 and has been recorded from lagoons. Large *Palaemon* migrate into deeper water during winter leaving the smaller animals in pools on the high shore. This migration is most marked in northern Britain and there is evidence to suggest that it is a response to low temperature, smaller animals having better osmoregulatory powers than larger. Breeding has been recorded from April to September and the eggs are carried for about six weeks. Some females produce two broods a year and longevity is two to two and a half years. *Palaemon* feeds on a wide range of organisms, including algae and crustaceans.

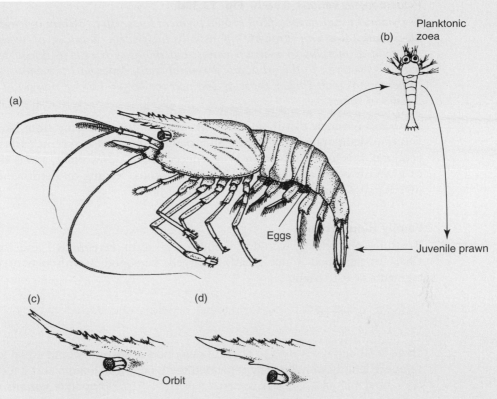

Figure 13.39 (a) *Palaemon elegans*, berried female, with (b) planktonic zoea stage of life-cycle (not to scale). (c) *Palaemon serratus*, rostrum, (d) *Palaemonetes varians*, rostrum.

Palaemon serratus (Pennant) (*Leander serratus*) Common prawn **Fig. 13.39c**
Similar to P. elegans *(above). Pinkish-brown, marked with reddish lines and spots. Up to 100 mm in length, generally about 60 mm. Rostrum curves upwards, usually with tip divided into two; 6 or 7 teeth along dorsal edge of rostrum restricted to basal 2/3 of length: 2 of dorsal teeth posterior to orbit. Four or 5 teeth along ventral edge of rostrum. Antenna 1 triramous. First and 2nd pairs of pereopods terminate in chelae, those of 2nd pair the larger. Lateral margin of telson with 2 pairs of spines.*

P. serratus extends as far north as Denmark, occurring commonly in Britain on south and west coasts but less so on the east coast. It is found among seaweed in rock pools on the lower shore, where it can often be caught in large numbers on the rising tide using a hand net. It occurs offshore to depths of about 40 m and is fished commercially by trawl and baited pots on the south and west coasts of Britain. Breeding occurs from about November to June; a single female can carry as many as 4000 eggs which are brooded for up to four months. Young settle from the plankton in July/August and reproduce the following February. Longevity is about three to four years. Most of the population moves offshore in winter and returns in spring. *P. serratus* is omnivorous.

Some individuals of this species and of *P. elegans* (above) are infected by a parasitic isopod, ***Bopyrus squillarum*** Latreille, which gives rise to pronounced swellings on the sides of the carapace, a condition known as 'face ache'.

Palaemonetes varians (Leach) **Fig. 13.39d**

Body almost transparent; faint brown pigment lines. Up to 50 mm in length, generally 20–30 mm. Rostrum straight, tip usually undivided. Number of teeth on dorsal and ventral margins of rostrum varies, but majority have 4–6 dorsal teeth, one of which is posterior to the orbit, and 2 ventral teeth. Antenna 1 triramous. First 2 pairs of pereopods with chelae, those of 2nd pair the larger. Lateral margin of telson with 2 pairs of spines.

P. varians is widely distributed in north-west Europe extending from Norway to the western Mediterranean. It occurs all around Britain but is rare in parts of northern Scotland. It is found only in areas of reduced salinity, down to 1, and is characteristic of brackish water and estuaries. It is omnivorous, feeding on detritus, diatoms and invertebrates. Breeding occurs from April to July.

Family Hippolytidae

Rostrum conspicuous, with dorsal and ventral teeth. Pereopods 1 and 2 with chelae. Chela of pereopod 1 not greatly enlarged. Pereopod 2 with two or more small segments next to chela.

Three small segments next to chela of pereopod 2. Rostrum about equal to carapace length; with 2 widely separated teeth on dorsal margin and 2 teeth on ventral margin .... *Hippolyte varians* p. 353

Seven small segments next to chela of pereopod 2. Rostrum short; with 2–5 teeth on dorsal margin .... *Eualus pusiolus* p. 352

Figure 13.40 *Eualus pusiolus.*

Eualus pusiolus (Krøyer) **Fig. 13.40**

Semi-transparent, flecked with various colours. Up to 28 mm in length. Rostrum short; straight or slightly curved downwards, tip almost always undivided; 2–5 teeth along dorsal margin. Antenna 1 short, biramous. Carapace with 1 spine beneath eye. First and 2nd pair of pereopods terminate in chelae; 1st pair shorter than others; 7 small segments next to chela of pereopod 2. Lateral margins of telson with 4 or 5 pairs of spines.

E. pusiolus is widely distributed in north-west Europe and Britain where it is commoner in the north. It occurs under stones and among seaweed (especially laminarians) on the lower shore and into the sublittoral to depths of about 500 m. Breeding occurs throughout the year.

Hippolyte varians Leach Chameleon prawn **Fig. 13.41**

Green, red, brown in colour, but sometimes transparent with red, yellow markings; up to 32 mm in length, usually less. Rostrum nearly equal to carapace length; more or less straight with 2 widely separated teeth on dorsal margin; 2 teeth on ventral margin; tip undivided. Antenna 1 short, biramous. Carapace with single spine above eye, 2 spines below eye. First pair of pereopods shorter than others; 1st and 2nd pairs with chelae; 3 small segments next to chela of pereopod 2. Lateral margins of telson with 2 pairs of spines.

Figure 13.41 *Hippolyte varians.*

H. varians is widely distributed in north-west Europe but is commoner in the south than the north of Britain. It occurs on the lower shore in rock pools and under seaweed, extending to depths of 50 m. Breeding occurs throughout the year.

Family Pandalidae

Rostrum conspicuous, in some cases longer than carapace and with marked upward curve; with dorsal and ventral teeth. Pereopod 1 with tiny chela or chela absent; pereopod 2 with tiny chela and with two or more, sometimes many more, small segments next to chela.

Pandalus montagui Leach Aesop prawn, pink shrimp **Fig. 13.42**

Semi-transparent, pink colour; generally 40–50 mm in length; may be much longer. Rostrum long, curves markedly upwards, tip divided into two; 10–12 teeth on basal half of dorsal edge of rostrum, 4 of which usually posterior to the orbit; 5 or 6 teeth on ventral edge of rostrum. Carapace with spine just beneath eye. Antenna 1 biramous; antenna 2 longer than body and marked with light and dark bands. First pair of pereopods with tiny chelae; 2nd pair of pereopods of unequal size, left larger than right. Right pereopod 2 with 20–22 small segments next to chela; left pereopod 2 with 50–65 small segments next to chela. Lateral margin of telson with 5–7 pairs of spines.

P. montagui is a northern species extending as far south as the English Channel. It occurs all around the coast of Britain extending to depths of 100 m or more but is occasionally found in rock pools on the lower shore. Although some individuals are

Figure 13.42 *Pandalus montagui.*

found in relatively shallow water throughout the greater part of the year, offshore migration takes place in most areas in October/November, followed by a migration into shallow water, including the mouths of estuaries, in the spring. In the sand and mud in shallow waters off the Wash, the south-east coast, Morecambe Bay and the Solway Firth, densities are high enough to support commercial exploitation. About 50% of a population are protandrous hermaphrodites becoming female at an age of about 13–16 months; the remainder are female and stay as such throughout life. Individuals mature and breed in the first year of life and in the North Sea eggs are laid from November to February and hatch in April/May; young settle in July/August. Each female lays a single batch of eggs in a breeding season and length of life is usually three to four years. *Pandalus* feeds mainly on hydroids, small crustaceans and polychaetes.

Family Crangonidae
Rostrum reduced to a small spine (as in species described here) or very short. Pereopod 1 more heavily built than others; subchelate.

Crangon crangon (Linnaeus) (*Crangon vulgaris*) Common shrimp, brown shrimp
Fig. 13.43
Brownish in colour, flecked: generally 30–50 mm in length, but up to 90 mm. Rostrum reduced to small spine. Antenna 1 biramous; antenna 2 about 3/4 body length. First pair of pereopods more heavily built than others, subchelate; 2nd pair reach to 3/4 the distance along large penultimate segment of pereopod 1, with tiny chelae. Telson with 2 pairs of small, lateral spines; dorsal surface of 6th abdominal segment smooth, not grooved.

C. crangon is very common in European waters, buried in sand from the middle shore to depths of 150 m. It extends into estuaries where it can be caught in large numbers using a push net on the rising tide. It is also found in the gullies and creeks left after the

Figure 13.43 *Crangon crangon* (a), with dorsal view of anterior of head (b).

tide has receded and these are now believed to be important nursery areas for *Crangon*. It is fished commercially in some areas, for example Morecambe Bay. The sexes are separate, although there are suggestions that the species is a protandrous hermaphrodite. Maturity is reached after one to two years and the breeding period differs according to locality. In the Bristol Channel breeding has been recorded in spring and summer, while in other localities three breeding periods have been recorded. Eggs are carried for three to four weeks during summer and 10 weeks in spring, with larval life lasting approximately five weeks. Average longevity is three years and it has been estimated that a female spawning three times a year will produce over 30 000 eggs. *Crangon* is an important predator of the invertebrate communities of sand deposits in shallow water, such as polychaete worms, crustaceans and molluscs, including *Macoma balthica* (p. 279) in which the siphons are often heavily cropped. *Crangon* also feeds extensively on the young-of-the-year flatfishes such as plaice and is itself preyed on by a wide range of fishes.

A second species, **Crangon allmanni** Kinahan, is found offshore over sandy and muddy sediments. It is very similar to *C. crangon* but can be distinguished by the presence of longitudinal grooves on the dorsal surface of the 6th abdominal segment.

Lobsters
Family Nephropidae
Exoskeleton heavily calcified. Abdomen extended and ends in a well-developed tail fan. The first three pairs of pereopods have chelae, the first pair large.

Homarus gammarus (Linnaeus) European lobster, common lobster **Fig. 13.44**
Body up to 400 mm or more in length; specimens found intertidally generally much smaller. Dark blue in colour. First pair of pereopods very large, held forward, terminate in very large chelae; unequal in size, one used for cutting, the other for crushing

Figure 13.44 *Homarus gammarus.*

prey; 2nd and 3rd pereopods with small chelae. Uropods and telson form prominent tail fan.

H. gammarus is widely distributed on European coasts where it frequents holes and gullies on the lower shore and into the sublittoral to depths of about 60 m. It is a valuable commercial species and is fished using pots baited with pieces of rotting fish. Lobsters mature at about five or six years of age. The sexes are separate and during the breeding season the female moults into the soft-shelled stage in which she is receptive to the male. During copulation, the male attaches spermatophores to the ventral surface of the abdomen of the female and as the eggs are laid they are fertilized and attached to the pleopods. Females carrying eggs, which are bright blue in colour, are known as 'berried' and it has been estimated that a single female can carry up to 1 million eggs. Mating occurs in summer, after which the eggs are held on the pleopods of the female for about nine months. The zoea larvae are planktonic for about three weeks and then settle on the sea bed. Although little is known of the behaviour of very young lobsters, aquarium observations suggest that they seek refuge from predators in the interstitial spaces of coarse boulder and cobble deposits which they rarely leave. They also live in fine mud, in which they construct complex burrows. Spawning takes place once a year, declining to about once every two years in later life and individuals can be long-lived, surviving to 15 years and more. They are scavengers.

Other crustaceans, also commonly known as lobsters, are *Nephrops norvegicus* (Linnaeus) and *Palinurus elephas* (Fabricius) both of which are extensively fished commercially. The former, variously referred to as the Norway lobster, the Dublin Bay prawn and langoustine, is the well-known scampi. It is in the same family as *Hommarus*, reaching up to 250 mm in length and orange in colour, but lives in soft sediments from depths of about 14 m into deep water. *Palinurus elephas*, the spiny lobster or crawfish, is in the family Palinuridae. It is about the same size as the common lobster and is found sublittorally on rocky ground, from depths of about 10 m. It is red-brown in colour and, unlike in the common lobster, the first three pairs of pereopods lack chelae.

Squat lobsters
Family Galatheidae
Abdomen slightly reduced and loosely folded under thorax. Carapace longer than broad. Rostrum pointed, extends beyond eyes. Fifth pair of pereopods much smaller than preceding four pairs; first pair of pereopods with chelae.

Figure 13.45 *Galathea squamifera.*

Galathea squamifera Leach **Fig. 13.45**
Greenish-brown colour, sometimes with red markings; body up to 60 mm in length. Pointed rostrum broad and triangular in shape with 4 tooth-like projections on either side, posterior pair smallest. First pair of pereopods longer than body, with well-developed chelae covered in scales and with spines on outer margin; basal segments with spines on inner margins; 5th pair of pereopods small and folded along side of body.

G. squamifera is the most common of the intertidal squat lobsters and is widely distributed in north-west Europe. It is a suspension feeder living under stones on the lower shore and extending to depths of about 70 m. In the English Channel breeding occurs in winter and spring.

Other species of *Galathea* are predominantly sublittoral but occasionally are found under stones on the lower shore. They are identified to the genus by the broad, triangular rostrum which is sharply pointed and has spines on the lateral margins.

Porcelain crabs
Family Porcellanidae

Abdomen much reduced, tightly folded under thorax. Carapace rounded. Porcelain crabs move sideways and have general appearance of true crabs (p. 363) but can be distinguished by the much reduced fifth pair of pereopods which are folded under the carapace, and the long antenna 2.

1st pair of pereopods long and broad, with well-developed chelae. Pereopods, edge of carapace and outer edge of chelae with many long setae .... ***Porcellana platycheles*** p. 358

1st pair of pereopods long and slender. Pereopods, edge of carapace and outer edge of chelae with very few setae .... ***Pisidia longicornis*** p. 359

Porcellana platycheles (Pennant) Broad-clawed porcelain crab, hairy porcelain crab
Fig. 13.46a

Rounded carapace about 16 mm in length; brownish colour. Antenna 2 long. First pair of pereopods long, broad, flattened; have well-developed chelae; 5th pair of pereopods much reduced. Abdomen folded under thorax. Pereopods, edge of carapace and outer edge of chelae with many conspicuous, long setae.

P. platycheles is widely distributed, and is believed to reach its northern limit in Shetland although there is a report of its presence in Norway. It is common under stones on the middle and lower shore, particularly those among mud and gravel; although essentially an intertidal species, it is found from the shallow sublittoral to

(a)   (b)

Figure 13.46 (a) *Porcellana platycheles*. (b) *Pisidia longicornis*.

depths of about 30 m. It is a suspension feeder. In the English Channel and South Wales, egg-bearing females have been recorded from March/April to August. Maturity is reached at a carapace width of about 4 mm and females carry eggs during their second year on the shore.

Pisidia longicornis (Linnaeus) (*Porcellana longicornis*) Long-clawed porcelain crab
Fig. 13.46b

Similar in appearance to P. platycheles *(above), but smaller, up to 10 mm carapace length; body shiny, not heavily clothed with setae. First pair of pereopods long and slender.*

P. longicornis is widely distributed in north-west Europe from Scandinavia to the Mediterranean. It is common under stones at low water and in laminarian hold-fasts, extending into the sublittoral to depths of about 90 m where it is often in very high densities. Egg-bearing females have been recorded in spring and summer. Maturity is reached at a carapace width of about 3.5 mm and females carry eggs during their second year on the shore, possibly surviving to breed two or three times. Development of the egg to the zoea larva takes about a month. It is a suspension feeder.

Hermit crabs

Hermit crabs live in the empty shells of a variety of prosobranch molluscs. The abdomen is soft and twisted, allowing it to fit into the coiled shell. The crab is held securely by the specially adapted left uropod which is hook-like and grips the inside of the shell (Fig. 13.48), while the fourth and fifth pereopods are held against the wall of the shell to give further purchase. The pleopods have been lost in males but are retained on the left side of the body in females and are used to carry the eggs. The eggs hatch as zoea larvae which after a period of growth in the plankton moult into megalopa larvae and these resemble a juvenile hermit crab. The first pair of pereopods is generally large and equipped with chelae. The fourth and fifth pairs are usually small. Only the anterior part of the carapace is calcified and this is often referred to as the hard carapace or shield (Fig. 13.48). When disturbed, hermit crabs retreat completely into the gastropod shell. When the crab becomes too big for the shell, it changes to another. A new shell is not occupied until it has been thoroughly tested for weight and size, and in choosing a new shell the crab goes through a complex ritual of behaviour which can be observed in aquarium tanks if the crabs are kept with a supply of empty shells. Empty shells on the shore are clearly a valuable resource.

Two families of hermit crabs are found intertidally in British waters. These can generally be identified by the size of the chelae.

Chelae about equal in size, or left chela larger than right ...... **Diogenidae** p. 360
Right chela larger than left .... **Paguridae** p. 360

Family Diogenidae
Diogenes pugilator (Roux) **Fig. 13.47**

Bluish in colour. Hard carapace up to about 5.5 mm in length; surface with a few groups of setae. Antenna 2 with many setae. First pair of pereopods with chelae, the right with many setae. Left chela larger than right and each with white tip.

D. *pugilator* reaches its northern limit in Anglesey, North Wales. It has been recorded on the south and west coasts of Ireland, the western English Channel and the southern North Sea. It is found from extreme low water to about 40 m in clean, sandy sediments. It is a scavenger but also suspension feeds using long setae on the antennae. In south-west Britain it is locally abundant during some years and the crabs are occasionally washed ashore in large numbers on the edge of the incoming tide when they quickly anchor the body in the sediment with the large chela and then bury into the sand using the walking legs. Maturity is reached at about 1–2 mm hard carapace length and egg-bearing females are found from June to September. At some localities females carry two broods of eggs per year.

Hard carapace

Figure 13.47 *Diogenes pugilator*, anterior of body.

Family Paguridae

1 Hard carapace with almost straight lateral margins; roughly triangular in shape; smooth surface .... ***Anapagurus hyndmanni*** p. 362
Hard carapace with convex lateral margins **2**

2 Anterior (rostral region) of carapace pointed. Right chela with 2 longitudinal rows of large tubercles plus many smaller tubercles. Often with the sea-anemone *Calliactis parasitica* (p. 96) fixed to the shell .... ***Pagurus bernhardus*** p. 360

Anterior (rostral region) of carapace rounded. Right chela with fine tubercles. Almost always found with the sea-anemone *Adamsia carciniopados* (p. 96) wrapped around the shell .... ***Pagurus prideaux*** p. 362

Pagurus bernhardus (Linnaeus) (*Eupagurus bernhardus*) Common hermit crab
Fig. 13.48

Carapace red, yellow or brownish colour, lateral margins of hard carapace convex; rostral region pointed. Hard carapace up to 20 mm in length, intertidal specimens usually about 12 mm. Well-developed chelae on 1st pair of pereopods, right larger than left and with many tubercles. Red line down centre of right chela, to each

Hard
carapace

Left uropod

Figure 13.48 *Pagurus bernhardus*, removed from prosobranch shell.

*side of which is row of large tubercles. Terminal segment of 2nd and 3rd pairs of
pereopods sharply pointed.*

P. bernhardus is widely distributed in north-west Europe and may be common on the
lower shore to depths of about 20 m but has been recorded to depths of about 400 m.
Small specimens are abundant on most British shores during spring and summer, their
frequent occurrence in the shells of *Littorina littorea* (p. 215), *Littorina obtusata*
(p. 216) and *Nucella lapillus* (p. 230) no doubt reflecting the availability of these
shells. In deeper waters offshore, shells of other species are occupied. Intertidal
populations breed from November to May while egg-bearing females have been
recorded throughout the year in sublittoral populations, with a peak in January and
February. Small females probably carry a single brood per year, but larger specimens

are believed to produce multiple broods. The eggs are carried for up to two months and the pelagic larval stage lasts several weeks. Most individuals reach maturity in the first year at about 4 mm carapace length and although longevity is difficult to ascertain, the largest specimens are believed to be at least three years old. The hermit crab is a scavenging omnivore but also filters small particles from the water. Shells occupied by *P. bernhardus* are often colonized by a variety of animals such as barnacles, polychaete worms and sponges. Familiar associations are those with *Hydractinia echinata* (p. 106) and *Neanthes fucata* (p. 161), but by far the best known is that between the crab and the sea-anemone *Calliactis parasitica* (p. 96). A parasitic barnacle of the genus *Peltogaster* (p. 311) is sometimes found under the abdomen of the crab, and appears as a smooth, yellowish mass. It should not be confused with the crab's eggs.

Pagurus prideaux Leach (*Pagurus prideauxi*) **Fig. 13.49b**

Reddish-brown, hard carapace up to about 15 mm in length, lateral margins of hard carapace convex; rostral region rounded. Right chela with fine tubercles; larger than left. Second and 3rd pairs of pereopods grooved longitudinally, terminal segments sharply pointed.

P. prideaux reaches its northern limit in southern Norway but is apparently absent from Denmark and the southern North Sea. It is very occasionally found on the shore but is more common offshore and has been recorded from depths of over 300 m. This hermit crab is almost always found in association with the sea-anemone *Adamsia carciniopados* (p. 96). Apparently, the sea-anemone cannot reach sexual maturity without the crab, although the crab can live without the anemone, at least under aquarium conditions. Breeding females have been found throughout the year and most produce two broods a year.

Anapagurus hyndmanni (Bell) **Fig. 13.49a**

Hard carapace up to about 5 mm in length; carapace roughly triangular in shape, with straight lateral margins, surface smooth. Well-developed chelae on 1st pair of

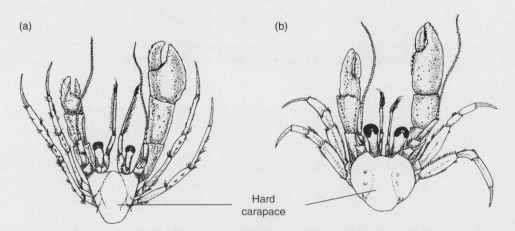

Figure 13.49 (a) *Anapagurus hyndmanni*, anterior of body. (b) *Pagurus prideaux*, anterior of body (modified after Ingle, 1992).

pereopods; right larger than left and without, or with very few, setae. Right chela usually whitish in colour; much wider than other segments of that appendage.

A. hyndmanni extends as far north as Shetland and is widely distributed around Britain but is apparently absent from southern North Sea coasts. It occurs on the lower shore and extends into the sublittoral to depths of about 35 m on a wide variety of substata, including coarse sand, gravel and shell deposits. On the west coast of Ireland, breeding has been recorded from March to September and most females carry two broods of eggs a year.

True crabs

In true crabs the abdomen is tightly folded under the carapace. The first pair of pereopods is armed with chelae; the fifth pair is similar in size to the preceding three pairs. The sexes are separate and the abdomen of the female is usually broader than that of the male. Only the first two pairs of pleopods are present in males and these are modified for the transference of sperm during copulation. Sperm are stored in the spermathecae of the female and fertilization is internal. The eggs are carried on the pleopods of the female for a varying length of time before hatching as planktonic larvae known as zoeae. These undergo several moults, the exact number depending on species, before developing into a second larval type, the megalopa (Fig. 13.53b,c). The megalopa is planktonic and moults into a benthic, juvenile crab.

Soft crabs, known as peelers, are occasionally found on the shore, the soft, pliable exoskeleton indicating that the crab has recently moulted. The exoskeleton requires a few days to harden and while soft the crab seeks shelter under rocks and stones to avoid predators.

Key to families of true crabs

The measurements of carapace length and width are shown in Fig. 13.50a.

1 Carapace longer than wide; with or without bifid rostrum between eyes **2**
  Carapace not longer than wide; without bifid rostrum **3**

2 Carapace with pronounced, bifid rostrum; roughly triangular in shape; broader towards posterior. Antenna 2 not densely fringed with setae. May be spider-like in appearance .... **Spider crabs** p. 371
  Carapace with 3, sometimes 4, sharp teeth on lateral margins; 2 teeth between eyes. Antenna 2 longer than width of carapace, densely fringed with setae
  .... **Corystidae** p. 364

3 Carapace rectangular, almost square .... **Varunidae** p. 370
  Not as above **4**

4 Carapace rounded, smooth; region between eyes not lobed or toothed. Small crab living as a commensal in mantle cavity of bivalve molluscs, particularly the common mussel *Mytilus edulis* (p. 254) .... **Pinnotheridae** p. 369
  Carapace wider than long, not as above **5**

5 Terminal segment of pereopod 5 flattened and usually fringed with setae; often segment is very much expanded and used in swimming. From 5 to 9 sharp teeth on anterolateral margins of carapace .... **Portunidae** p. 365
   Terminal segment of pereopod 5 not flattened **6**

6 Region of carapace between eyes with 3 lobes. Carapace much wider than long, with up to 10 rounded lobes on anterior and lateral margins, giving pie-crust effect .... **Cancridae** p. 368
   Carapace not as above **7**

7 Anterolateral margins of carapace with 5 well-defined, pointed teeth
   .... **Pilumnidae** p. 369
   Anterolateral margins of carapace with 5 blunt lobes .... **Xanthidae** p. 368

Family Corystidae   Masked crabs

Carapace longer than broad, with two teeth between the eyes and three or four teeth on lateral margins. Antenna 2 longer than carapace width, densely fringed with setae.

Corystes cassivelaunus (Pennant) Sand crab, masked crab **Fig. 13.50**
Reddish-brown to yellow carapace; up to 40 mm in length, 25 mm across; 2 teeth between eyes and 3, sometimes 4, teeth on lateral margins of carapace. Antenna 2 longer than carapace width, heavily clothed with setae. First pair of pereopods with chelae and about twice length of carapace in male, equal to carapace length in female.

C. cassivelaunus is the only member of the genus. It is widely distributed on sandy beaches from low water of spring tides to depths of over 100 m. Large numbers of

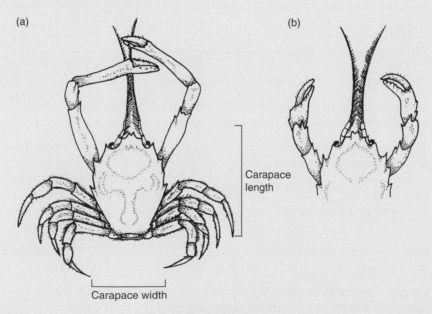

Figure 13.50 *Corystes cassivelaunus*, (a) male, (b) female.

dead and dying specimens are often found on the strandline after storms. It extends as far north as the west coast of Sweden although absent from much of the north of Scotland. If exposed and placed on the surface of the sand the crab will burrow quickly by digging with pereopods 2 to 5, going into the sediment backwards. Once in the sediment the long, second pair of antennae make contact with the surface and are held parallel to one another so that the setae interlock and form a tube, down which an inhalant current of water is drawn for respiration. Egg-bearing females have been recorded in June and July and the eggs are carried for about 10 months. Females breed successively for a number of years. *Corystes* feeds on a wide range of burrowing invertebrates, particularly bivalve molluscs, polychaete worms and amphipods.

Family Portunidae Swimming crabs

Carapace wider than long. Terminal segment of pereopod 5 flattened and usually fringed with setae; often segment is very much expanded and used in swimming. Usually five to nine sharp teeth on anterolateral margins of carapace.

A large family of crabs. In addition to the common species described here, others are found in the shallow sublittoral and occasionally on the lower shore. Reference to **Ingle (1996)** is recommended for further study.

1 Pereopod 5 with flat, but narrow, pointed terminal segment (Fig. 13.53), slightly wider than terminal segments of pereopods 2–4. Carapace with 5 sharp teeth on anterolateral margins; 3 small, rounded teeth between eyes. Widely distributed, common .... ***Carcinus maenas*** p. 366

Pereopod 5 with broad, flat terminal segment **2**

2 Seven to 10 unequal-sized teeth between eyes, central pair often the largest. Eyes red; joints between segments of pereopods bluish with red markings .... ***Necora puber*** p. 365

Three similar-sized, rounded teeth between eyes. Dorsal surface of carapace with short transverse rows of small tubercles with setae ..... ***Liocarcinus depurator*** p. 366

Necora puber (Linnaeus) (*Liocarcinus puber, Portunus puber, Macropipus puber*) Velvet swimming crab, fiddler crab **Fig. 13.51**
Reddish-brown carapace; broader than long; up to 90 mm across; dorsal surface covered with fine setae. Joints between segments of pereopods bluish, with red markings. Eyes red; anterior margin of carapace with 7–10 small teeth, central pair largest; 5 sharp teeth on anterolateral margins of carapace. First pair of pereopods with well-developed chelae; 5th pair with broad, flattened, terminal segment.

N. puber is widely distributed in north-west Europe. Small specimens are common at low water on rocky shores, particularly in the south and west of Britain, while larger specimens are found to depths of 80 m. It is a powerful swimmer and an aggressive crab. Maturity is reached at a carapace width of about 40 mm and on British shores egg-bearing females are found throughout most of the year. *Necora* feeds largely on

Figure 13.51 *Necora puber.*

Figure 13.52 *Liocarcinus depurator.*

brown seaweeds such as the laminarians and *Fucus,* and although crustaceans and molluscs are important food items, they are less so in intertidal populations. The diet of juveniles is mostly crustaceans such as barnacles and small crabs. The species is fished commercially using baited pots. It is a relatively small fishery in British waters but the catch has increased since the early 1980s.

Liocarcinus depurator (Linnaeus) Swimming crab, harbour crab, cleanser crab **Fig. 13.52**

Carapace broader than long, up to 40 mm across; dorsal surface with short transverse rows of small tubercles with setae. Reddish-brown colour; terminal segment of last pair of pereopods violet in larger crabs. Three similar-sized rounded lobes between eyes. Anterolateral margins of carapace with 5 pointed teeth. First pair of pereopods with well-developed chelae: 5th pair with broad, flat, terminal segment.

L. depurator is widely distributed in north-west Europe extending from Norway to the Mediterranean and in some localities is abundant. It is found on mud and sandy mud from the lower shore to depths of about 280 m, feeding on a wide range of invertebrates and fishes. On the west coast of Scotland, breeding has been recorded from January to June with females having two broods a year.

Carcinus maenas (Linnaeus) Shore crab, green crab **Fig. 13.53**

Carapace broader than long; up to 80 mm across. Dark green, drab colour to orange-red; young forms often conspicuously marked with white patches. Three small, rounded lobes or teeth between eyes, sometimes difficult to detect, particularly in young. Anterolateral margins of carapace with 5 pointed teeth. First pair of pereopods with well-developed chelae. Terminal segment of 5th pair of pereopods flat, but pointed and narrow.

C. maenas is one of the commonest crabs on the shores of north-west Europe and is a recent colonizer of Australasia, South Africa and North America. It is tolerant of a wide range of temperature and desiccation and is found on all types of shore from high water into the subittoral to depths of about 60 m. It extends into estuaries, the adults

Juvenile crab

Figure 13.53 (a) *Carcinus maenas*, with stages in life-cycle: (b) zoea larva, (c) megalopa larva (not to scale).

surviving salinities down to about 4. *Carcinus* shows a wide range of carapace colours from dark green to orange and red and it has been suggested that the green crabs are actively growing and moulting while the orange-red specimens are in an extended intermoult. Red crabs are generally most common in the sublittoral, the green dominating on the shore and on saltmarshes, and this has led to the suggestion that colour reflects differing abilities to withstand low salinity. Maturity is generally reached after about one year at a carapace width of 25–30 mm (males) and 15–31 mm (females). The male seeks out a female crab about to moult and carries her under his body for a few days. When the female moults, copulation takes place. The female then burrows in sand and forms a large cavity beneath her. While in this position the eggs are laid and attached to the pleopods. Females carrying eggs are found on the shore throughout the year and in south-west Britain there is a peak of breeding activity in summer. Although adult crabs survive low salinities, the eggs will not develop at salinities less than 20. The eggs are held by the female for several months and it has been estimated that as many as 185 000 eggs can be carried at any one time. After a zoeal larval life of two to three months in the plankton megalopae settle on the bottom. Maximum size is reached after about four years. Very small crabs are commonly found on mud-flats and among *Spartina* (p. 77). The shore crab feeds on a wide variety of invertebrates including polychaete worms, molluscs and crustaceans; young crabs feed on barnacles and small molluscs. Predators include fishes and gulls.

The abdomen of female specimens of *Carcinus* is broader than that of the male and has seven segments compared with five visible segments in the male. Male crabs parasitized by *Sacculina carcini* (p. 310) show a broadening of the abdomen and may be mistaken for females.

It has been suggested that variations in the morphology of Atlantic and Mediterranean populations of *C. maenas* reflect separate species, that as described above and **Carcinus aestuarii** Nardo from the Mediterranean. It would seem that the position has not been resolved but the latest research suggests that the morphological variation recorded between these populations has both environmental and genetic components.

Pirimela denticulata (Montagu) (Family Pirimelidae) may be confused with *Carcinus maenas*. It is found on the lower shore, reaching a carapace width of about

20 mm. It is not as abundant as *Carcinus* and can be recognized by the presence of an anterior projection between the eyes, which has three sharp points, and the last segment of pereopod 5 which is not flattened.

Family Cancridae

Carapace much broader than long; margins with many rounded lobes giving characteristic appearance. Region of carapace between eyes with three blunt lobes.

Cancer pagurus Linnaeus Edible crab **Fig. 13.54**

Carapace much broader than long; may be more than 200 mm across; reddish-brown in colour. Carapace has 3 small lobes between eyes and about 10 larger, rounded lobes on anterolateral margins giving what is often referred to as a 'pie-crust' effect. First pair of pereopods with very large chelae, with black tips; 2nd to 5th pairs with many setae.

Figure 13.54 *Cancer pagurus.*

C. pagurus is widely distributed in north-west Europe in pools and gullies from the lower shore to depths of about 90 m. The larger specimens are usually found offshore where they are fished commercially using pots baited with pieces of fish. Most crabs reach maturity at about 110 mm (male) and 127 mm (female) carapace width when they are about three to five years old. Prior to copulation, the male holds the female and has been observed to help her cast the exoskeleton during the precopulatory moult. Immediately after moulting, copulation occurs and the eggs are laid some months later. The female seeks a sandy substratum where she excavates a small depression which she occupies during egg laying. Breeding takes place in winter and the eggs are carried for seven to eight months, hatching in spring and summer. The edible crab lives for up to 20 years and it has been estimated that large females carry more than 2 million eggs at any one time. The larvae are planktonic for 23–30 days. Adults take a wide variety of food and like *C. maenas* are important predators of molluscs.

Family Xanthidae

Carapace broader than long; region between eyes broad, with two broad lobes. First pair of pereopods with large broad chelae. Fifth pair of pereopods shorter than first pair. Anterolateral margins of carapace with five blunt lobes.

Xantho hydrophilus (Herbst) (*Xantho incisus*) Montagu's crab **Fig. 13.55b**

Carapace much broader than long; up to 70 mm across. Yellowish or reddish-brown; chelae black. Anterolateral margins of carapace with 5 blunt lobes. First pair of pereopods with large chelae. Margins of pereopods 2–5 with few simple setae.

X. hydrophilus is common under stones on the lower shore in the south and west of Britain, where it is sometimes abundant. It extends into the sublittoral to depths of about 35 m and reaches the northern limit of its distribution on the west coast of Scotland. Egg-bearing females are found from March to July and the larvae are

Figure 13.55 (a) *Pilumnus hirtellus*, with club-shaped setae inset. (b) *Xantho hydrophilus*.

present in the plankton for most of the summer. *Xantho* is a herbivore feeding on a variety of algae.

A second species, **Xantho pilipes** A. Milne-Edwards, is found on the lower shore of sandy and stony beaches. It is most readily separated from X. *hydrophilus* by dense fringes of setae (not club-shaped) on the lateral and posterior margins of the carapace and the second to fifth pairs of pereopods.

Family Pilumnidae

Carapace broader than long; region between eyes broad, with two broad lobes, sometimes finely serrated. First pair of pereopods with large broad chelae. Fifth pair of pereopods shorter than first pair. Anterolateral margins of carapace with five well-developed, pointed teeth.

Pilumnus hirtellus (Linnaeus) Hairy crab **Fig. 13.55a**

Carapace broader than long; up to 30 mm across; reddish-brown colour; very finely serrated between eyes; anterolateral margins with 5 teeth, anterior 2 small. First pair of pereopods with chelae, large and dissimilar in size, the smaller with rough tuberculate surface. Pereopod 1 larger in male than female. Pereopods 2–5 banded purple and white. Carapace and appendages covered with long, club-shaped setae.

P. hirtellus is widely distributed in north-west Europe and in Britain is most frequent in the south and west. It is found under stones and among laminarian holdfasts from the lower shore to depths of about 75 m. Silt accumulates among the setae on the body making the crab difficult to detect. Egg-bearing females are found from April to August. *Pilumnus* is an omnivore, its diet including small mussels.

Family Pinnotheridae

Carapace rounded, smooth; often soft. Region between eyes not lobed or toothed. Eyes small, widely separated. Antennae very small. Small crabs living as commensals in the mantle cavity of bivalve molluscs, particularly the edible mussel, *Mytilus edulis* (p. 254).

Figure 13.56 *Pinnotheres pisum*, (a) male, (b) female.

Pinnotheres pisum (Linnaeus) Pea crab **Fig. 13.56**

Carapace rounded and smooth; up to 6 mm across in male, 13 mm in female. Cara-
pace of male, but not female, projects in front of orbits. Eyes small, particularly in
female. Male yellowish in colour with dark patches; female semi-translucent, with
pale colour of internal organs showing through. In some specimens carapace hard, in
others soft (see below).

P. pisum is widely distributed in north-west Europe, where it can be found living in the
mantle cavity of a range of bivalve molluscs, including the edible mussel, *Mytilus*
(p. 254). The life-cycle usually involves two hosts. The larvae are planktonic. The
newly metamorphosed crab has a hard cuticle and swims with the aid of setae on some
of the pereopods. It invades the bivalve *Spisula solida* (p. 272) where it undergoes a
series of moults, during which time the cuticle is soft and the pereopods lack setae.
Eventually, a hard-shelled stage is formed, developing long setae on the third and fourth
pairs of pereopods. The crab then leaves *Spisula*, mates and finds the second host, which
is most often *Mytilus edulis* (p. 254), but can sometimes be *Modiolus modiolus* (p. 257)
or *Tapes decussata* (p. 286). In the second host the crabs have a soft shell and it is here
that the female breeds. In some populations of *Pinnotheres* newly metamorphosed
crabs are found in *Mytilus* and it has been suggested that a change of host does not
always take place. In south-west England, the main breeding period lasts from May to
August and females are capable of producing several broods. The male crabs apparently
pass through a number of hard and soft stages. The crab feeds on strings of food
collected by the bivalve and may cause erosion of the gill surfaces of its host.

Family Varunidae

Carapace rectangular, almost square. First pair of pereopods equal. Eyes conspicuous.
A family of essentially warm water species that includes the Chinese mitten crab, an
invasive species in Britain and Europe.

Eriocheir sinensis H. Milne Edwards Chinese mitten crab **Fig. 13.57**

Carapace rectangular, may be almost square; up to 80 mm across, light brown-olive in
colour. Carapace with 4 obvious anterolateral teeth, area between eyes with 4 points.
First pair of pereopods with well developed chelae, clothed with dense setae giving
very characteristic 'hairy' appearance; chelae with white tips suggestive of 'mittens'.
Pereopods 3 and 4 about twice width of carapace.

Figure 13.57 *Eriocheir sinensis*.

E. sinensis, a native of China and Korea, is an invasive species first recorded in Europe in 1912 in the river Aller, a tributary of the river Weser, Germany. It is believed to have arrived in ships' ballast water and has since spread widely being recorded in 'pest proportions' in a number of localities. It lives in fresh water migrating to estuaries to breed. It was first recorded in Britain in 1935 from the river Thames and has since been recorded from a number of river systems including the Medway, Teign, Humber and Tyne. *Eriocheir* feeds on a wide variety of detritus, algae and invertebrates and can cause serious damage to freshwater and brackish-water ecosystems by burrowing into the soft sediment and causing bank erosion. It also competes with indigenous species, sometimes resulting in their loss from the area. After three to five years in fresh water, the crabs migrate downstream into the estuary where they mature and breed in brackish water. The fertilized eggs are carried for some months by the female and on hatching the larvae develop in the estuary and the juveniles eventually migrate upstream, sometimes for substantial distances. Both sexes are believed to die after reproducing.

Spider crabs

Carapace longer than broad, roughly triangular in shape, broader towards posterior; usually with bifid rostrum between eyes. First pair of pereopods usually shorter than other pereopods. Pereopods 2–5 sometimes long and slender and spider-like.

The spider crabs described here belong to the families **Inachidae**, **Majidae** and **Oregoniidae**. Morphological characters are not given for the separate families; the species of spider crab commonly found and described here can be separated on the gross morphological characters used in the key below.

1 Rostrum slender, bifid, the 2 parts touching along length, not curving upwards or downwards. Eyes not retractile. Pereopods 2–5 long, slender, spidery
.... (Family Inachidae) ***Macropodia rostrata*** p. 374

Not as above **2**

2 Rostrum bifid, the 2 parts touching at tips   .....(Family Oregoniidae) *Hyas araneus* p. 373
Rostrum not as above **3**

3 Rostrum short, bifid, the 2 parts well separated. Four tubercles in a row across
anterior part of carapace. Small, delicate, up to 30 mm carapace width
.... (Family Inachidae) *Inachus dorsettensis* p. 373
Rostrum strong, bifid, the 2 parts diverging or parallel. Carapace with many sharp
spines, particularly on the anterolateral margins. Large crab, up to 150 mm across
carapace   .... (Family Majidae) *Maja brachydactyla* p. 372

Maja brachydactyla Balss (*Maja squinado*) Common spider crab **Fig. 13.58**
*Carapace longer than broad; narrows anteriorly; up to 150 mm across; the largest of
British spider crabs. Reddish-brown colour; tip of terminal segment of pereopods 2–5
black. Rostrum strong, bifid, the 2 parts usually diverging, or nearly parallel in larger
specimens. Eyes retract into orbit. Many sharp spines over carapace, those on anterolateral margins well developed. First pair of pereopods with small chelae; 5th pair of pereopods relatively short. Carapace and pereopods 2–4 with hooked setae. Often camouflaged with seaweed attached to carapace.*

Figure 13.58 *Maja brachydactyla.*

M. *brachydactyla* is widespread in north-west Europe and reaches its northern limit in the south of Britain. There are records of its presence in the southern North Sea and on the coasts of the Netherlands. It is sometimes found under rocks and seaweed on the lower shore and in some years is abundant on a wide range of substrata in the south and west of Britain, extending into the sublittoral to depths of about 75 m. Off the south coast of England, copulation takes place in July and August and the eggs are laid about six months later. Egg-bearing females have been recorded throughout spring and summer at a range of localities. The eggs are carried for up to nine months, each female producing a single brood. The common spider crab has the interesting behaviour pattern of forming heaps or mounds of individuals in relatively shallow water during late summer and autumn. These can be over 600 mm high and 900 mm in diameter and contain as many as 100 crabs which remain together for some months. Presumably the heaps offer protection to the soft-shelled, recently moulted crabs which are found in the centre. The behaviour may also be important as a means of bringing together the sexes for reproduction. *Maja* is omnivorous, feeding on a wide range of both seaweeds and animals such as *Ulva* (p. 30), *Corallina* (p. 43), bryozoans and hydroids. It is heavily exploited commercially.

The common spider crab shows wide morphological variation over its geographic range and detailed morphological studies on populations from the Atlantic and the Mediterranean suggest that two species should be recognized: the Mediterranean *Maja squinado* (Herbst) and the Atlantic *Maja brachydactyla* as described above.

(a)

(b)

Plate 1   (a) Exposed rocky shore. The eulittoral is dominated by acorn barnacles and limpets with patches of mussels and scattered fucoid seaweeds. Cardigan Bay. (b) Extremely sheltered rocky shore. The eulittoral is dominated by the knotted wrack, *Ascophyllum nodosum*, above which is the channelled wrack, *Pelvetia canaliculata*. Patches of yellow and white lichens show in the littoral fringe. Outer Hebrides.

Plate 2   **(a)** Encrusting red seaweeds, *Lithophyllum* and *Lithothamnion* commonly found on bedrock and in rock pools on the middle and lower shore, together with the bushy *Corallina officinalis*. Pembrokeshire. **(b)** Encrusting red seaweed, *Hildenbrandia rubra*. Cardigan Bay. **(c)** Sea lettuce, *Ulva lactuca*. Cardigan Bay.

Plate 3   **(a)** Channelled wrack, *Pelvetia canaliculata*. Co. Clare, Ireland. **(b)** The brown seaweed, *Fucus ceranoides*, characteristic of areas of reduced salinity. Dyfi estuary, W. Wales. **(c)** Thong weed, *Himanthalia elongata*. Long reproductive fronds grow from the vegetative 'buttons'. Co. Clare, Ireland. **(d)** Vegetative 'button' of *Himanthalia* showing early growth of reproductive frond in centre.

(a)

(b)

(c)

Plate 4   (a) Oyster thief, *Colpomenia peregrina*. Co. Clare, Ireland. (b) Knotted wrack, *Ascophyllum nodosum* with the epiphytic red seaweed, *Polysiphonia* sp. Cardigan Bay. (c) Stunted growths of *Ascophyllum* found on exposed rocky shores. Cardigan Bay.

(a)

(b)

Plate 5   **(a)** Moderately sheltered rocky shore. The conspicuous green seaweed, *Ulva lactuca* is growing on patches of *Sabellaria* reef. Cardigan Bay. **(b)** Sublittoral fringe marked by the kelp, *Laminaria digitata*. Pembrokeshire.

(a)

(b)

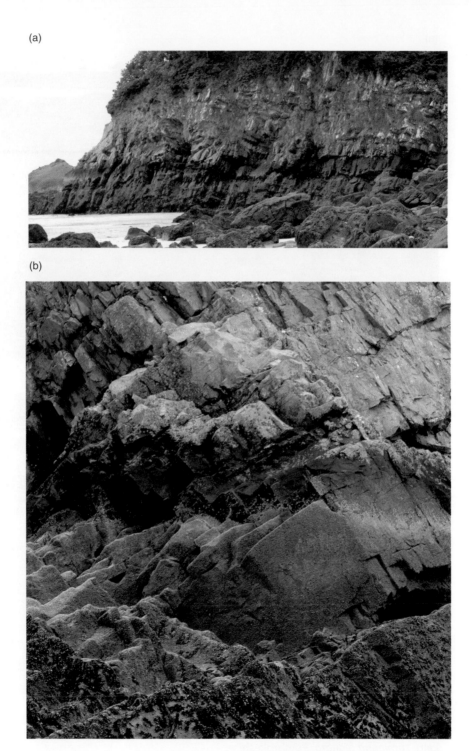

Plate 6   **(a)** Zonation of the littoral fringe and upper eulittoral. A band of the black lichen, *Verrucaria maura* is conspicuous. Pembrokeshire. **(b)** Detail of the zonation in **(a)** showing orange lichens, *Caloplaca* and *Xanthoria*; black lichen, *Verrucaria maura*; acorn barnacles; mussels.

Plate 7   Rocky shore lichens. **(a)** *Ramalina siliquosa*. **(b)** *Caloplaca marina*. **(c)** *Caloplaca thallincola*. Pembrokeshire.

Plate 8   Rocky shore lichens. **(a)** *Tephromela atra*. **(b)** *Ochrolechia parella*. **(c)** *Lichina pygmaea*. Pembrokeshire.

Plate 9   Rocky shore lichens. **(a)** *Xanthoria parietina.* **(b)** *Anaptychia runcinata* (J. P. Savidge).
(c) *Verrucaria mucosa.* **(d)** *Collemopsidium foveolatum* on the acorn barnacle *Chthamalus
montagui* (A. Pugh). Pembrokeshire.

Plate 10 **(a)** Breadcrumb sponge, *Halichondria panicea*. Co. Clare, Ireland. **(b)** Encrusting sponge, *Ophlitaspongia papilla*. Cardigan Bay.

Plate 11   **(a)** Purse sponge, *Grantia compressa*. Co. Clare Ireland. **(b)** Stone bored by the sponge, *Cliona celata*. Cardigan Bay. **(c)** Encrusting sponge, *Hymeniacidon perlevis*. Cardigan Bay.

(a)

(b)

Plate 12   **(a)** Beadlet anemone, *Actinia equina*. Note the prominent acrorhagi in the partially contracted specimen (R. A. Moore). **(b)** Strawberry anemone, *Actinia fragacea* (R. A. Moore). Cardigan Bay.

(a)

(b)

Plate 13   Dahlia anemone, *Urticina felina*, relaxed **(a)** and contracted **(b)** specimens (R. A. Moore). Cardigan Bay.

(a)

(b)

Plate 14   **(a)** Snakelocks anemone, *Anemonia viridis*. Cardigan Bay. **(b)** Plumose anemone, *Metridium senile*. Anglesey.

Plate 15 **(a)** Common limpet, *Patella vulgata*. **(b)** China limpet, *Patella ulyssiponensis*. **(c)** Black-footed limpet, *Patella intermedia*. **(d)** Blue-rayed limpet, *Ansates pellucida*, on the stipe of *Laminaria digitata*. Pembrokeshire.

(a)   (b)

(c)   (d)

Plate 16   **(a)** Common limpet, *Patella vulgata*. Cardigan Bay. **(b)** Limpet 'homing scar'. Cardigan Bay. **(c)** Limpet grazing track left on surface of rock by the action of the radula. (By kind permission of J. R. Lewis.) **(d)** Scanning electron micrograph of the radula of *Patella vulgata* showing rows of teeth.

(a)

(b)

Plate 17   (a) Rocky shore in Trevone, Cornwall, photographed in the summer of 1955. (b) The same
shore photographed in the summer of 1967 following inundation with crude oil from the
Torrey Canyon oil spill in March 1967 and subsequent spraying with detergent to remove
the oil. This resulted in the destruction of limpets, top-shells and other herbivores allowing
the green seaweeds *Ulva* spp. to grow almost unchecked until the populations of herbivores
recovered. (By kind permission of the Marine Biological Association of the United Kingdom.)

(a)

(b)

Plate 18   **(a)** Dog-whelks, *Nucella lapillus* preying on acorn barnacles. **(b)** *Nucella lapillus* disturbed while drilling through the shell of the common mussel, *Mytilus edulis*. Cardigan Bay.

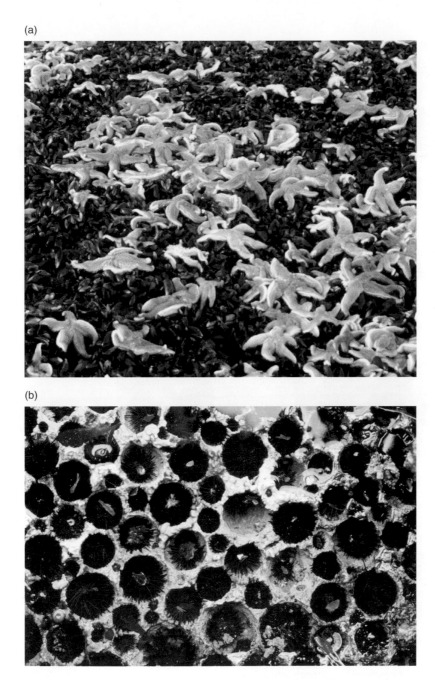

(a)

(b)

Plate 19 (a) Common starfish, *Asterias rubens* preying on *Mytilus edulis* (W. Cook). Morecambe Bay. (b) Sea-urchin, *Paracentrotus lividus*. Each urchin occupies a cavity excavated in the rock. Co. Clare, Ireland.

Plate 20   (a) Flat periwinkle, *Littorina obtusata* and egg mass laid on *Fucus serratus*. Cardigan Bay. (b) Dog-whelk, *Nucella lapillus* and egg capsules. The barnacles on the whelk are *Elminius modestus*. Pembrokeshire.

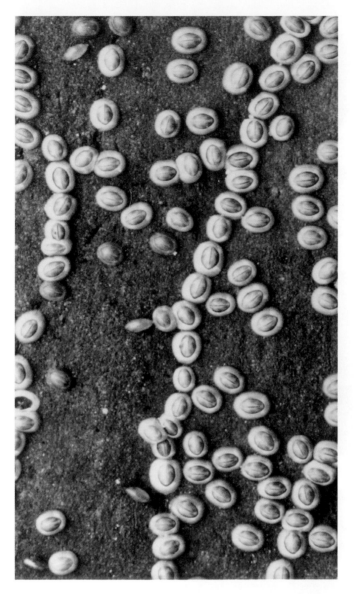

Plate 21   Newly settled cyprids of *Semibalanus balanoides* together with recently metamorphosed barnacles (J. Morgan). Cardigan Bay.

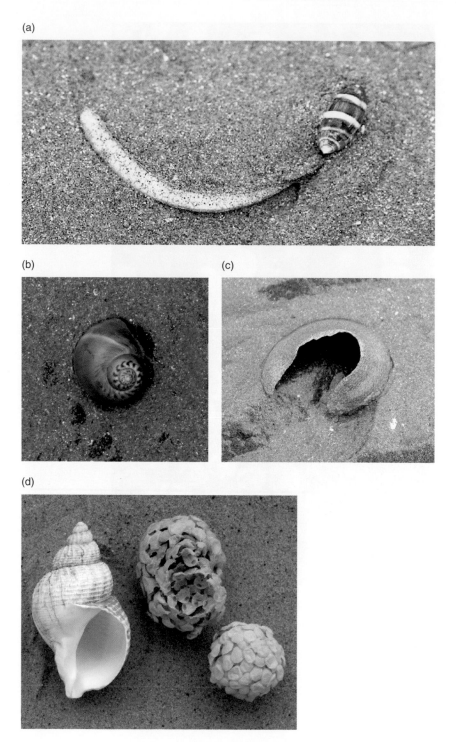

Plate 22 (a) Opisthobranch mollusc, *Acteon tornatilis* with egg mass being deposited in the sand (Gwen Jenkins). (b) Large necklace shell, *Euspira catena*. (c) Egg collar of *Euspira catena*. (d) Whelk, *Buccinum undatum* with egg masses. Pembrokeshire.

Plate 23   **(a)** *Sabellaria* reef at low water. **(b)** Tubes of *Sabellaria alveolata*. **(c)** Anterior end of tube of the sand mason, *Lanice conchilega* protruding from sediment. Cardigan Bay.

Plate 24 (a) Opening on surface of sediment indicating the presence of the sea-potato, *Echinocardium cordatum*. (b) Sediment removed to show *Echinocardium in situ*. Note the path of the respiratory channel opening on the surface in (a). The 'black layer' shows clearly a few centimetres below the surface. (c) Depression and cast marking the presence of the lugworm, *Arenicola marina*. (d) Anterior end of tube of the peacock worm, *Sabella pavonina* protruding from the sediment at low water. Pembrokeshire.

(a)

(b)

Plate 25   **(a)** Inhalant and exhalant siphons of **(b)** the common otter shell, *Lutraria lutraria* which lives up to 300 mm deep in the sediment. The siphons can be seen at the surface when undisturbed (Gwen Jenkins). Pembrokeshire.

Plate 26   **(a)** Impressions left on the surface of the mud by the inhalant siphons during feeding indicate the presence of the peppery furrow shell, *Scrobicularia plana*. **(b)** Sediment removed to show *Scrobicularia in situ*. Note the channels left by the separate inhalant and exhalant siphons. The foot shows clearly. **(c)** Sand gaper, *Mya arenaria* removed from the sediment alongside a characteristic hole left on the surface of the mud by the siphons. The posterior gape of the shell and the projecting siphons show clearly. Dyfi estuary, W. Wales.

(a)

(b)

(c)

Plate 27   **(a)** Islands of the common cord-grass, *Spartina anglica*, early stages in the development of saltmarsh. **(b)** Common glasswort *Salicornia europaea* and *Spartina anglica* at the saltmarsh edge. **(c)** Channels draining dense stands of *Spartina anglica*. Dyfi estuary, W. Wales.

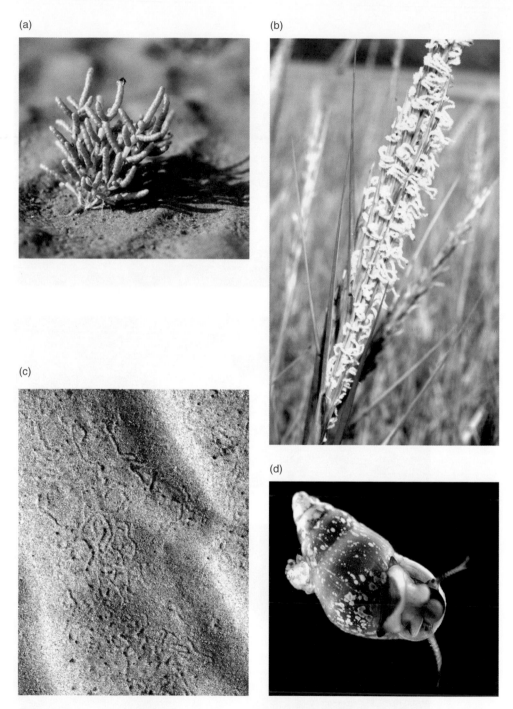

Plate 28   **(a)** *Salicornia europaea* (and the mud snail, *Hydrobia ulvae*). **(b)** Flowering head of *Spartina anglica*. **(c)** Trails left on the surface of the mud by *Hydrobia ulvae* seen when the tide retreats. **(d)** *Hydrobia ulvae*, with egg mass on the shell, floating on the tide (Gwen Jenkins). Dyfi estuary, W. Wales.

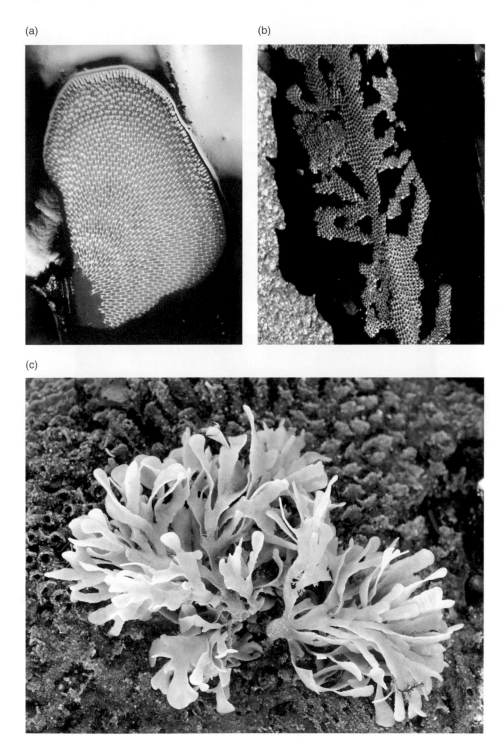

(a)   (b)

(c)

Plate 29   **(a)** Sea-mat, *Membranipora membranacea* on *Laminaria* sp. **(b)** Hairy sea-mat, *Electra pilosa* on *Fucus serratus*. Note the irregular margin. **(c)** Hornwrack, *Flustra foliacea* washed ashore. Cardigan Bay.

Plate 30 (a) Piddock holes. (b) White piddock, *Barnea candida*. (c) Common piddock, *Pholas dactylus*. Cardigan Bay.

Plate 31   (a) Timber bored by shipworm (Gwen Jenkins). (b) *Teredo navalis*. Calcareous tubes *in situ*. (c) Goose barnacles, *Lepas anatifera* on timber washed ashore. Cardigan Bay.

(a)

(b)

Plate 32   **(a)** Jack sail-by-the-wind, *Velella velella*, washed ashore (Gwen Jenkins). **(b)** The strandline on an exposed sandy shore is often a rich collecting ground, particularly after a storm. Cardigan Bay.

Hyas araneus (Linnaeus) Spider crab **Fig. 13.59**

Reddish-brown carapace, triangular narrow anteriorly, broad and rounded posteriorly; generally 30–50 mm across; tubercles on dorsal surface. Rostrum bifid, the 2 parts meeting at tips. Eyes not completely concealed when withdrawn. First pair of pereopods thick, with chelae, others long and thin. Carapace often covered with encrusting invertebrates.

Figure 13.59 *Hyas araneus.*

H. araneus is widely distributed in north-west Europe, reaching its southern limit on the south and west coasts of Britain. It is found on the lower shore among rocks, seaweed and on sand, extending to depths of about 80 m. Egg-bearing females have been recorded throughout the year at a number of localities in Britain. It is omnivorous and includes a large proportion of seaweed in its diet.

A second species, *Hyas coarctatus* Leach, also reaches its southern limit around Britain. It is found on the very low shore and is common in the sublittoral. It is similar in appearance to *H. araneus* but can be separated from that species by the lyre-shaped carapace compared with the more rounded, triangular carapace of *H. araneus*.

Inachus dorsettensis (Pennant) Scorpion spider crab **Fig. 13.60**

Reddish-brown carapace, triangular in outline; up to 30 mm across; 4 tubercles lie in a row across anterior part of carapace, other tubercles posterior to them. Short, bifid rostrum, the 2 parts well separated. Eyes retractile. First pair of pereopods with chelae, larger in male than female; 2nd pair very long, longer and stouter than remaining pairs. Carapace often covered with encrusting organisms.

I. dorsettensis is widely distributed in north-west Europe extending just north of Britain. It is occasionally found among rocks and seaweed on the lower shore and also on muddy-sand extending to depths of about 180 m. Egg-bearing females have been found throughout the year on the west coast of Britain and aquarium observations suggest that the eggs take about 15 weeks to develop to pelagic larvae. It has been estimated that each female carries three batches of eggs. *Inachus* is omnivorous.

Figure 13.60 *Inachus dorsettensis*, female.

Figure 13.61 *Macropodia rostrata*, male.

Macropodia rostrata (Linnaeus) Long-legged spider crab **Fig. 13.61**

Yellow-brown to red carapace; triangular in shape, up to about 15 mm across; with 8 or so tubercles on surface. Bifid rostrum, the 2 parts touching along their length; less than 1/2 length of large basal segments of antenna 2; more or less straight, not curving markedly upwards or downwards. Long eye stalks, clearly visible, without prominent tubercle on anterior margin; eyes not retractile. First pair of pereopods with chelae, larger in male than female; 2nd to 5th pairs slender and spider-like, with 4th and 5th pairs having curved terminal segments.

M. rostrata is widely distributed in north-west Europe and is the commonest spider crab in Britain. It is found under seaweed on the lower shore and on muddy-sand and sand, extending to depths of about 120 m. Egg-bearing females have been found throughout the year. Aquarium observations suggest that the eggs are carried for about 14 weeks and it has been estimated that each female carries up to three broods of eggs per year. *Macropodia* is omnivorous, feeding mainly on seaweed and crustaceans.

Subphylum Hexapoda

Class Insecta

The body of the insect is divided into head, thorax and abdomen, although these are not always clear. The head has a single pair of antennae and there are three pairs of legs (and usually two pairs of wings) on the thorax. Insects are essentially terrestrial but some do extend onto the shore. Two genera are of particular interest to the shore ecologist.

Figure 13.62 (a) *Anurida maritima*. (b) *Petrobius maritimus*.

Anurida maritima (Guérin) (*Lipura maritima*) **Fig. 13.62a**

Body roughly cigar-shaped; dull, dark blue; about 3 mm in length. Antennae short. Thorax has 3 pairs of short legs; no wings.

A. maritima is widely distributed in north-west Europe, where it is found on the upper shore. Characteristically, it is seen floating on the surface film of sheltered rock pools, often in dense aggregations. A film of air saturated with water is trapped around the cuticle, reducing desiccation and allowing the insect to extend its distribution onto the high shore. It has also been recorded from saltmarshes. When the tide comes in, *A. maritima* takes shelter in crevices in the rock and among seaweeds. Breeding occurs in summer. The male leaves a spermatophore on the rock surface and this is picked up by the female. Groups of pale yellow-coloured eggs arc laid in August in cracks and crevices in the rock where they overwinter, becoming a dark orange colour and hatching the following spring. The adults die before winter and longevity is believed to be less than one year. *Anurida* feeds on a variety of invertebrates and has been described as one of the most important scavengers on the shore.

Petrobius spp. Bristle tail **Fig. 13.62b**

Grey coloured, body about 10 mm in length. Head bears pair of very long antennae, and pair of palps. Thorax has 3 pairs of legs, no wings. Abdomen with 3 conspicuous bristles at posterior end; central bristle very long.

Two species of the genus are widespread in north-west Europe. These are **Petrobius maritimus** (Leach) and **Petrobius brevistylis** Carpenter, while a third species, **Petrobius lohmanderi** Agrell, has been recorded from the east coast of Sweden. Where *P. maritimus* and *P. brevistylis* coexist there is some separation of the species. *Petrobius* spp. occur higher on the shore than *Anurida* (above) and are widely distributed in the littoral fringe, extending into the terrestrial environment. Shiny black eggs are laid in crevices in groups of up to 100 in October/November and hatch in May/June of the following year. *Petrobius* is active at night, feeding on microorganisms and lichens; longevity is believed to be between one and two years.

References

Cheliceriformes

King, P. E. (1974). *British sea spiders. Arthropoda: Pycnogonida. Keys and notes for the identification of the species*. Synopses of the British fauna (New Series), no. 5. London: Academic Press.

King, P. E. (1986). Sea spiders: a revised key to the adults of littoral Pycnogonida in the British Isles. *Field Studies*, 6, 493–516.

Crustacea: Cirripedia

Southward, A. J. (2008). *Barnacles. Keys and notes for the identification of British species*. Synopses of the British fauna (New Series), no 57. Shrewsbury: Field Studies Council.

Crustacea: Mysida

Makings, P. (1977). A guide to the British coastal Mysidacea. *Field Studies*, 4, 575–95.

Tattersall, W. M. & Tattersall, O. S. (1951). *The British Mysidacea*. London: Ray Society.

Crustacea: Isopoda

Naylor, E. (1972). *British marine isopods. Keys and notes for the identification of the species*. Synopses of the British fauna (New Series), no. 3. London: Academic Press.

Sars, G. O. (1896–9). *An account of the Crustacea of Norway*, Vol. 2. Isopoda. Christiania: Cammermeyers.

Crustacea: Amphipoda

Lincoln, R. J. (1979). *British marine Amphipoda: Gammaridea*. London: British Museum (Natural History).

Sars, G. O. (1895). *An account of the Crustacea of Norway*, Vol. 1. Amphipoda. Christiania: Cammermeyers.

Crustacea: Decapoda

Crothers, J. & Crothers, M. (1983). A key to the crabs and crab-like animals of British inshore waters. *Field Studies*, 5, 753–806.

Ingle, R. W. (1992). *Hermit crabs of the north-eastern Atlantic Ocean and Mediterranean Sea. An illustrated key*. London: Chapman & Hall.

Ingle, R. W. (1996). *Shallow-water crabs. Keys and notes for identification of the species*. Synopses of the British fauna (New Series), no. 25, 2nd edn. Shrewsbury: Field Studies Council.

Ingle, R. W. & Christiansen, M. E. (2004). *Lobsters, mud shrimps and anomuran crabs. Keys and notes for the identification of the species*. Synopses of the British fauna (New Series), no. 55. Shrewsbury: Field Studies Council.

Smaldon, G. (1993). *Coastal shrimps and prawns. Keys and notes for identification of the species*. Synopses of the British fauna (New Series), no. 15, 2nd edn. Revised and enlarged by L. B. Holthuis & C. H. J. M. Fransen. Shrewsbury: Field Studies Council.

Sipuncula

The phylum Sipuncula comprises worm-like, marine animals occurring both intertidally and sublittorally. They are found in burrows in sandy and muddy sediments, in the empty tubes of polychaetes, the empty shells of molluscs, and in rock crevices. Sipunculans have an unsegmented body in which the anterior part, or introvert, is narrow and can be withdrawn, and the posterior part, the trunk, which is wider and more or less cylindrical. In relaxed specimens the mouth is seen at the anterior end of the introvert surrounded completely, or in the dorsal aspect, by lobes or tentacles. The anus is at the anterior end of the trunk in a mid-dorsal position. Sipunculans are mainly deposit feeders, feeding on mud, sand and detritus but filter feeding, using a crown of modified tentacles, has been described in one group. Almost all species have separate sexes with external fertilization, but see *Nephasoma minutum* (p. 377). There is a free-swimming larva and the length of pelagic life varies from a few days to a month or so.

Identification to species can be difficult and often requires examination of internal anatomy. The species described here can be recognized on external morphology and habitat.

Golfingia elongata (Keferstein) **Fig. 14.1a**
Slender body up to 150 mm in length. Anterior end with up to 36 tentacles (usually 24–28) encircling mouth. Anterior region of introvert with rings of hooks. Yellow-brown in colour.

G. *elongata* is widely distributed around Britain and north-west Europe, extending northwards to Denmark. It lives in vertical burrows up to 500 mm below the surface in muddy-sand on the lower shore and sublittorally to depths of 400 m. The burrows, and those of *Golfingia vulgaris* (below), are sometimes occupied by the bivalve mollusc **Kurtiella bidentata** (Montagu) (*Mysella bidentata*). Breeding has been recorded in July and August at Plymouth. The larva has a pelagic life of about one week.

A second species, **Golfingia vulgaris** (de Blainville), is widespread in north-west Europe in mud, sand and gravel from the lower shore to depths of about 2000 m. It is similar in length to *G. elongata* (150–200 mm), but is distinguished by the irregularly arranged hooks on the introvert (Fig. 14.1b). It breeds in summer.

Nephasoma minutum (Keferstein) (*Nephasoma minuta, Golfingia minuta*) **Fig. 14.1c**
Small, up to 15 mm in length and about 1.5 mm wide. Anterior end with short tentacles (lobes) encircling mouth. Anterior region of introvert with irregularly arranged hooks, particularly in small specimens, but may be few or absent in larger specimens.

N. *minutum* is widespread in north-west Europe being recorded from Sweden to the Atlantic coast of France. It lives in sandy and muddy deposits on the middle shore and

Figure 14.1 (a) *Golfingia elongata*. (b) *Golfingia vulgaris*, introvert showing irregularly arranged
hooks (after Gibbs, 2001). (c) *Nephasoma minutum*, introvert (after Gibbs, 2001).
(d) *Phascolion strombus* in empty shell of *Turritella* (after Cuénot, in Grassé, 1959).
(e) *Sipunculus nudus* (after Gibbs, 2001).

is also found in rock crevices and among the reefs of *Sabellaria* (p. 169). It extends
into the sublittoral to depths of about 50 m. Unlike other sipunculans, *Nephasoma* is a
simultaneous hermaphrodite and on the south coast of England breeding has been
recorded from November to January. The embryos undergo direct development in the
burrow of the parent where they hatch as juvenile sipunculans. There is no pelagic
stage in the life-cycle.

Phascolion strombus (Montagu) (*Phascolion strombi*) **Fig. 14.1d**
*Elongate body up to 50 mm in length. Anterior end with up to 50 tentacles encircling
mouth. Anterior region of introvert with many hooks, irregularly arranged. Body
covered with papillae. Usually found in empty prosobranch shells; also lives freely in
sediment. Greyish coloured.*

P. strombus usually lives in mud and sand accumulated in empty shells, especially
those of *Aporrhais* (p. 225) and *Turritella* (p. 211). It is widely distributed and
sometimes common in north-west Europe from the shallow sublittoral to depths of
over 3000 m and shells containing *Phascolion* are often washed ashore. When *Phascolion* is present, a small round hole can be seen in the plug of sediment in the mouth

of the shell, and from this the introvert of the sipunculid projects. Breeding probably occurs during most of the year; the larva has a pelagic life of about one week.

Sipunculus nudus Linnaeus **Fig. 14.1e**
Cylindrical body up to 350 mm in length. Anterior end with convoluted lobes surrounding mouth. Introvert with many papillae over entire length. Trunk marked with rectangular pattern. Whitish, greyish in colour.

S. nudus is a southern species occurring as far north as the west coast of Ireland and the southern North Sea. It is occasionally found in sandy sediments on the lower shore and sublittorally, usually to about 30 m but there are records of the species from depths of 700 m and more. Breeding has been recorded during summer and the larva has a pelagic life of about one month.

Reference

Gibbs, P. E. (2001). *Sipunculans. Keys and notes for the identification of British species.* Synopses of the British fauna (New Series), no. 12, revised edn. Shrewsbury: Field Studies Council.

Echiura

Members of the phylum Echiura are marine, worm-like animals found burrowing in mud and sand, under rocks and in crevices, both on the shore and sublittorally. The body is divided into two parts, an anterior proboscis with a longitudinal ventral groove and a posterior, cylindrical or sac-like trunk, which has two large, ventral setae near its anterior end. The mouth is at the posterior end of the proboscis. The species vary in length from a few millimetres to a metre or more in those species in which the proboscis is long and highly extensible. Echiurans feed on detritus. In most, the animal remains in the burrow and the proboscis is extended over the surface of the sediment and detritus trapped in mucus secreted by the proboscis. The mucous string is carried along the ventral groove to the mouth. In some species, a mucous net is secreted by the proboscis at the entrance to the burrow and the animal filter feeds. The sexes are separate and fertilization is usually external; there is a free-swimming trocophore larva.

Echiurus echiurus (Pallas) **Fig. 15.1a**
Trunk cylindrical, up to 200 mm in length. Anterior end with large proboscis with conspicuous ventral groove. Trunk covered with small papillae; 2 anterior, ventral setae; 2 rings of setae at posterior. Proboscis orange with brown stripes, trunk yellowish-grey.

E. echiurus is widely distributed in north-west Europe where it is found in burrows in sand and mud on the lower shore and sublittoral. Breeding occurs during winter.

Thalassema thalassemum (Pallas) (*Thalassema neptuni*) **Fig. 15.1b**
Trunk sac-like in shape, pointed posteriorly; up to 70 mm in length. Anterior end with long, highly extensible proboscis with conspicuous ventral groove; may be 200 mm in length when extended. Trunk with many papillae on surface, especially on posterior region; 2 anterior, ventral setae. Colour varied, blue, grey, yellow, pink or orange, with white line on mid-ventral surface.

T. thalassemum is described as a southern species found as far north as south-west Britain, where it is sometimes common on the lower shore but it has been recorded off south-west Scotland. It lives in cracks and crevices in rocks and among stones but has also been found sublittorally burrowing in soft sediments. It is a deposit feeder using the extensible proboscis to collect particles from the surface of the sediment. Breeding occurs during summer.

Figure 15.1 (a) *Echiurus echiurus* (after Greef, 1879). (b) *Thalassema thalassemum* (after Rietsch, 1886).

Reference

Stephen, A. C. & Edmonds, S. J. (1972). *The phyla Sipuncula and Echivra*. London: British Museum (Natural History).

Bryozoa (Ectoprocta)

Bryozoans are small, colonial animals found mainly in marine environments although a few species are present in fresh water. They are almost all sessile and occur in a variety of shapes and sizes on substrata such as seaweeds, rocks, stones and shells, both intertidally and sublittorally, sometimes in high densities. Each individual in the colony is known as a zooid and is usually no more than a millimetre in length. The zooids are surrounded by a cuticle which is often strengthened by calcareous material. One of the most characteristic features of bryozoans is the lophophore, a group of ciliated tentacles which encircles the mouth and collects microorganisms such as bacteria and diatoms on which the animals feed. Ciliated tentacles distinguish bryozoans from hydroids, with which they may be confused. In hydroids the tentacles lack cilia. The lophophore projects through an aperture in the cuticle and in some species this is closed by an operculum or by contraction of muscles. The alimentary canal is U-shaped with the anus outside the lophophore. Other groups of invertebrates have a lophophore and these include the phylum Phoronida (p. 394) and the phylum Brachiopoda, marine benthic invertebrates in which the body is enclosed by two shell valves. In some texts the bryozoans, the phoronids and the brachiopods are grouped together as the lophophorates, but see p. 394.

Most bryozoans exhibit polymorphism and, in addition to the feeding zooids, there are zooids adapted for anchoring the colony to the substratum and others for brooding developing embryos. The two most modified types of zooid are the avicularium and the vibraculum. In the avicularia, which are somewhat like a bird's head in appearance (see Fig. 16.7a), the operculum is in the form of a beak. Avicularia pick up and remove larvae and other invertebrates which settle on, and tend to smother the bryozoan. In the vibracula, the operculum is in the form of a seta which is moved over the surface of the colony keeping it free of debris. The role of avicularia and vibracula has been likened to that of the pedicellariae of echinoderms (see p. 396).

Bryozoan colonies are hermaphroditic but cross-fertilization is believed to occur. In some species, testes and ovaries develop in the same zooid, in others there are separate male and female zooids within a colony. A few species have a free-swimming planktonic larva with a triangular, bivalve shell. This is known as a cyphonautes larva and has a pelagic life of several weeks. In most species, however, the eggs are brooded in special chambers in the colony and the larvae have only a short free-swimming life. The larvae of bryozoans are very selective in their choice of substratum and some species will settle only on a specific area of a seaweed. The first zooid formed after settlement is known as an ancestrula and growth of the colony occurs by budding. Length of life varies from a few months to 10 or more years. The main predators of bryozoans are sea-slugs and to a lesser extent, sea-spiders, sea-urchins and fishes.

Classification of the phylum is based on **Hayward & Ryland (1999)**.

Phylum BRYOZOA (ECTOPROCTA)

Class Phylactolaemata	Freshwater bryozoans.
Class Stenolaemata	**Marine bryozoans, cylindrical zooids; cuticle calcified. Includes many fossil species.**
Order Cyclostomatida	**Aperture of zooid rounded, terminal, closed by muscular contraction. Operculum absent. Colony usually upright or cushion-like.**
Class Gymnolaemata	**Nearly all are marine; zooids vase-shaped and upright, or flattened and box-like; cuticle sometimes calcified. Aperture closed by operculum or muscular contraction.**
Order Ctenostomatida	**Cuticle not calcified. Zooids vase-shaped or flat. Aperture closed by muscular contraction. Colony upright or as a soft encrustation.**
Order Cheilostomatida	**Cuticle calcified to varying degrees. Zooids box-like. Aperture closed by operculum. Colony upright or as a firm encrustation.**

Key to species of bryozoans

Bryozoans are widely distributed on the shore attached to a variety of solid substrata and may be common in some situations. The form of the colonies is highly varied, from encrusting to upright, to creeping. Some are calcified and hard to the touch or partly calcified; others are gelatinous. These features are important in identification and are used in conjunction with other details as revealed by examination under a stereo-microscope. For further detail reference should be made to the four volumes covering bryozoans in the *Synopses of the British Fauna* (see p. 393).

1 Colony upright, sometimes creeping; attached to seaweed, stones, shells **2**
Colony encrusting; as a gelatinous mass, a glassy, shiny patch or a lacework; on seaweed, stones, shells **8**

2 Colony upright, a brownish-lobed gelatinous mass; may be much branched. Surface smooth, without papillae. Zooids appear as tiny blunt areas on the surface. Very low shore, but usually sublittoral; often washed ashore .... ***Alcyonidium diaphanum*** p. 386
Colony upright and tufted, sometimes creeping, ***or*** upright, lobed, with many papillae ***or*** an upright, much-divided flattened frond **3**

3 Colony upright, fleshy, may be lobed; up to 100 mm or more long; sometimes attached to red seaweeds. Surface of colony with many small papillae; when out of water matt, not shiny .... ***Alcyonidium hirsutum*** p. 387
Colony not so **4**

4 Colony upright and tufted, sometimes creeping **5**
Colony an upright, much-divided, flattened frond. Common on the strandline; frequently misidentified as a 'dried seaweed' **7**

5 Colony upright, tufted with branches curving inwards, jointed; 5–7 zooids between each joint. Zooids calcified and cylindrical with rounded aperture, without spines .... ***Crisia eburnea*** p. 385
Colony not so **6**

6 Colony upright, tufted, sometimes creeping; zooids large, semi-transparent and vase-shaped, not calcified, arranged in groups along the branches but not spirally arranged. Free end of zooids squarish. Without avicularia .... ***Bowerbankia imbricata*** p. 388
Colony upright, tufted, branches spirally arranged; zooids predominantly in 2 rows; rectangular and box-like with 1 spine at each corner of distal end. Avicularia large, rounded with short, sharply bent beak .... ***Bugula turbinata*** p. 392

7 Frond bluntly rounded at ends. Zooids rectangular, with short spines. Pale greyish-brown in colour. Common on the strandline .... ***Flustra foliacea*** p. 391
Frond narrow, truncate at ends. Zooids rectangular, without spines. Pale yellow in colour. On the strandline .... ***Securiflustra securifrons*** p. 391

8 Colony a purplish-brown, coarse-textured mass, usually on *Fucus serratus,*
Mastocarpus stellatus and *Chondrus crispus*. Zooids rectangular, with spines
.... ***Flustrellidra hispida*** p. 388
Colony not so **9**

9 Colony a gelatinous mass, *or* a brittle, shiny, rounded, glassy patch; zooids
without spines **10**
Colony a whitish/silvery lacework; zooids with spines **12**

10 Colony a shiny, glassy patch, especially in depressions on frond of *Saccharina*
latissima .... ***Celleporella hyalina*** p. 392
Colony a gelatinous mass, on seaweeds, mostly *Fucus serratus*; also on stones
and shells **11**

11 (a) Colony a thin, transparent, gelatinous mass, nearly always on seaweeds.
Surface of colony smooth, without papillae .... ***Alcyonidium polyoum*** p. 387
or
(b) Colony smooth, an opaque gelatinous mass on stones, shells; also on *Fucus*
serratus. Zooid walls white .... ***Alcyonidium gelatinosum*** p. 388
or
(c) Colony a thick, brownish gelatinous mass on *Fucus serratus, Chondrus*
crispus and other red seaweeds. Surface of colony with many small
papillae .... ***Alcyonidium hirsutum*** p. 387
or
(d) Colony smooth, almost transparent when small, thicker and greyish-white
when larger. Surface of larger colonies with regularly spaced dark spots. On
mussel shells and stipe of *Laminaria hyperborea* .... ***Alcyonidium mytili*** p. 388

12 Colony whitish, mostly even-edged; usually on fronds of *Laminaria*, often covering
large areas. Zooids rectangular, with blunt spine at each corner. Front of zooids
without pores .... ***Membranipora membranacea*** p. 389
Colony silver-grey, usually uneven in outline; on seaweeds, stones and shells.
Zooids oval, surrounded by 4–12 spines, 1 of which is often long. Front of zooids
with numerous pores .... ***Electra pilosa*** p. 390

Class Stenolaemata

Order Cyclostomatida

Crisia eburnea (Linnaeus) Fig. 16.1
*Colony upright, tufted, jointed. Branches curving inwards. Zooids calcified, alternate;
cylindrical, without spines; usually 5–7 between each joint. Eight ciliated tentacles.
White in colour, joints yellow or pale brown. Up to 20 mm in height.*

Figure 16.1 *Crisia eburnea.*

C. eburnea is widely distributed in north-west Europe and Britain on the lower shore and into the sublittoral attached to red seaweeds, hydroids, the bryozoan *Flustra foliacea* (p. 391), stones and shells. It is often found washed up on the shore. On Irish Sea coasts, embryos can be found in the colonies from April to August. After fertilization each egg divides by a process known as polyembryony into a number of embryos which develop in large, swollen zooids known as gonozooids. On release, the larvae swim rapidly and settlement and metamorphosis occur within minutes.

Class Gymnolaemata

Order Ctenostomatida

Genus *Alcyonidium*
The genus *Alcyonidium* has been the subject of much taxonomic confusion and misidentification and, as a result, species designations in the literature are not consistent and in many cases unreliable. The taxonomy of the group and information on the different species given here is based on **Ryland & Porter (2006)**.

Alcyonidium diaphanum (Hudson) (*Alcyonidium gelatinosum, Alcyonium gelatinosum*) **Fig. 16.2a**
Colony form very varied, generally a lobed gelatinous mass, may be extensively branched. Surface smooth. Always upright in form, not encrusting; attached to rocks, stones and shells. Zooids appear as tiny, blunt areas on the surface; 14–16 ciliated tentacles. Colour varied, often yellowish-brown, but also brown, grey, dark red. Up to 300 mm in height.

A. diaphanum is probably widely distributed in north-west Europe. It is found all around Britain and abundant in the southern North Sea. It is essentially sublittoral,

Figure 16.2 (a) *Alcyonidium diaphanum* (detail after Hayward, 1985). (b) *Alcyonidium polyoum* on *Fucus serratus*.

extending to depths of about 100 m and although small colonies have been recorded in pools on the lower shore, they are rare. It is frequently washed ashore, often in large quantities in the autumn. It is a brooding species and in the Irish Sea embryos have been recorded from November to February with release of larvae in March. Interestingly, contact with *A. diaphanum* can cause an allergic dermatitis, commonly referred to by fishermen as 'Dogger Bank itch'.

Alcyonidium polyoum (Hassall) (*Alcyonidium gelatinosum*) **Fig. 16.2b**
Colony in the form of thin, encrusting, gelatinous mass; nearly always on Fucus serratus *(p. 63), but occasionally on other seaweeds including* Ascophyllum nodosum *(p. 61) and* Chondrus crispus *(p. 42); rarely on rock, stones and shells. Surface smooth, lacking papillae; 18–21 ciliated tentacles. Transparent at growing edge but darker-coloured in older parts of colony.*

A. polyoum has been recorded from a number of localities on the west coast of Britain and in Ireland. Timing of breeding is dependent on locality and has been recorded from spring to autumn when the zooids contain up to six or seven buff-coloured embryos. The larvae have a brief pelagic life.

Alcyonidium hirsutum (Fleming) **Fig. 16.3**
Colony in the form of a thick encrusting, gelatinous mass, usually on Fucus serratus *(p. 63), but also on red seaweeds such as* Chondrus crispus *(p. 42) and* Mastocarpus stellatus *(p. 42). More rarely it occurs as erect lobes up to 100 mm or more in length. Surface covered with many conspicuous, small papillae; 14–19 ciliated tentacles. Colour brown or pale brown. Surface matt, not shiny, when out of water.*

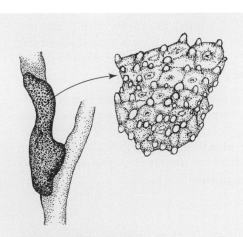

Figure 16.3 *Alcyonidium hirsutum* on stipe of *Fucus* sp.

A. hirsutum is widely distributed in north-west Europe and around Britain. In Irish Sea populations, embryos develop within the zooids during autumn and winter and each brooding zooid contains four to eight whitish-coloured embryos arranged in a circle. Larvae are released into the sea. Colonies are believed to live for one or two years.

Alcyonidium gelatinosum (Linnaeus) (*Alcyonidium mytili, Alcyonidium reticulum*) **(not illustrated)**

Colony smooth, in the form of an encrusting, gelatinous mass on stones, mollusc shells and Fucus serratus *(p. 63). 15–18 ciliated tentacles. Young colonies almost transparent, becoming opaque when older. Walls of zooids white, conspicuous. Very similar to, and often mistaken for* A. polyoum *(p. 387), but the latter has 18–21 tentacles.*

A. gelatinosum has been recorded from a wide range of localities around Britain and Ireland but the taxonomic confusion surrounding this species means that many records of its distribution published prior to **Ryland & Porter** (2006) may be unreliable. It is found intertidally and in the shallow sublittoral. Breeding has been recorded from autumn to spring when zooids contain three or four embryos which are white, pinkish or red in colour, compared with the buff-coloured embryos of *A. polyoum*. Larvae are released into the sea.

Alcyonidium mytili Dalyell **(not illustrated)**

Colony a gelatinous mass, smooth, almost transparent when small, thicker and greyish-white when larger. 15–17 ciliated tentacles. Surface of larger colonies with regularly spaced dark spots, these are not seen in A. gelatinosum *or* A. polyoum. *Found on* Mytilus edulis *(p. 254) and shells of* Ostrea edulis *(p. 259), also stones and the stipe of* Laminaria hyperborea *(p. 57).*

A. mytili appears to have a wide distribution around Britain and Ireland but its distribution is patchy. It is found from the lower shore into the sublittoral. Breeding occurs in autumn and winter; eggs are discharged into the sea where they develop into cyphonautes larvae.

Flustrellidra hispida (O. Fabricius) **Fig. 16.4a**

Colony in the form of thick mass, usually on Fucus serratus *(p. 63), but also on other seaweeds including* Mastocarpus stellatus *(p. 42),* Chondrus crispus *(p. 42) and the holdfasts of laminarians, but not usually on rocks and stones. Spines present round the individual zooids (a condition known as hispid), giving coarse texture; 28–40 ciliated tentacles. Colour purplish-brown, reddish-brown.*

F. hispida is widely distributed in north-west Europe and Britain, where it is predominantly intertidal. Embryos develop within the zooids, each brooding zooid containing as many as eight, yellowish-coloured embryos. Breeding occurs during summer but there is variation between localities and on the Menai Strait, North Wales settlement has been recorded from April to November. Larvae are similar to cyphonautes larvae in having a bivalve shell. They have a short pelagic life.

Bowerbankia imbricata (Adams) **Fig. 16.4b**

Colony usually upright, growing in tufts; sometimes creeping; branches usually thick. Zooids large, 1 mm in height; vase-shaped, arising in groups along branches, but not spirally arranged. Free end of zooids squarish; 10 ciliated tentacles. Brownish-yellow in colour. Up to 70 mm in height.

Figure 16.4 (a) *Flustrellidra hispida* on stipe of *Fucus* sp. (b) *Bowerbankia imbricata*.

B. imbricata is widely distributed in north-west Europe and Britain where it occurs from the middle shore into the shallow sublittoral, usually on *Ascophyllum nodosum* (p. 61), *Fucus vesiculosus* (p. 64), *Fucus serratus* (p. 63), other byrozoans such as *Flustra foliacea* (p. 391) and stones. The zooids are hermaphoditic; sperm are released through a tiny pore at the tip of the tentacles and enter a neighbouring zooid in the feeding current. Only one embryo is brooded within a zooid, the embryos giving a yellowish tinge to the colony. After release, the larvae have only a brief pelagic phase. Breeding occurs during summer.

Order Cheilostomatida

Membranipora membranacea (Linnaeus) Sea-mat **Fig. 16.5a; Plate 29a**
Colony in the form of an encrusting, mostly even-edged, lacework, usually on fronds of laminarians (p. 57), but occasionally on Fucus serratus *(p. 63) and other seaweeds. May cover very large areas. Zooids rectangular in shape with short, blunt spine at each corner. In some colonies there are tall columns, known as towers, projecting from the front of the zooids. About 17 ciliated tentacles. Colour whitish.*

M. membranacea is one of the best-known bryozoans. It is widely distributed in north-west Europe and Britain on the lower shore and shallow sublittoral, breeding during spring and early summer. The zooids are hermaphroditic; sperm are released into the sea through pores at the tip of the tentacles and are drawn in feeding currents into a neighbouring zooid where fertilization takes place. The fertilized eggs are released into the sea and develop into cyphonautes larvae. Although *Membranipora* is a simultaneous hermaphrodite and sperm and eggs are mature at the same time giving the potential for self-fertilization, the release of sperm and their subsequent capture by a neighbouring individual is believed to promote cross-fertilization. The cyphonautes larvae have a planktonic life of several weeks and in the Irish Sea they are

Figure 16.5   (a) *Membranipora membranacea* on *Fucus serratus*. (b) *Electra pilosa* on shell and (c) on *Fucus serratus*.

found in the plankton from February to November. *Membranipora* lives for several years. The colonies are flexible and able to withstand flexure of the seaweed frond. They grow towards the base of the frond, so ensuring that not all the colony is lost when the laminarian is damaged and eroded at the tip.

Electra pilosa (Linnaeus) Hairy sea-mat **Fig. 16.5b,c; Plate 29b**

Colony in the form of an encrusting mat, often uneven in outline; occasionally upright. Found on variety of seaweeds, including Fucus serratus *(p. 63), but also on laminarians (p. 57),* Mastocarpus stellatus *(p. 42) and* Palmaria palmata *(p. 45); also on stones, shells, hydroids and often on the shells of the whelk* Buccinum *(p. 233) inhabited by hermit crabs. Zooids more or less oval, surrounded by 4–12 spines, one of which (the median proximal) is often long and conspicuous; 11–15 ciliated tentacles. Front of zooids with numerous pores. Colour silvery grey.*

E. pilosa is widely distributed in north-west Europe and Britain, where it is one of the most common, if not the commonest, species of bryozoan. It is found from the middle shore into the sublittoral, breeding during August and September. Sperm are released into the sea through pores at the tip of the tentacles and it is believed they are drawn into a neighbouring zooid where fertilization occurs. Cyphonautes larvae have been recorded in the plankton during all months of the year but they settle mainly in the autumn. Drifting, spherical colonies of *Electra* up to 70 mm across are occasionally washed ashore on the coasts of the North Sea. They are made up of branches of *Electra* attached to an object such as a stone or shell fragment.

Two other species, ***Electra crustulenta*** (Pallas) and ***Electra monostachys*** (Busk), occur around Britain but are found only in estuarine areas. In these species, the proximal medial spine is short and the front of the zooid is without pores.

Figure 16.6 (a) *Flustra foliacea*. (b) *Securiflustra securifrons*.

Flustra foliacea (Linnaeus) Hornwrack **Fig. 16.6a; Plates 29c, 32b**

Colony upright; much divided, flattened fronds, bluntly rounded at ends. Zooids more or less rectangular with 4–5 short spines at one end; 13–14 ciliated tentacles. Greyish-brown in colour, dead specimens paler. Up to 200 mm in height, generally smaller.

F. foliacea is widely distributed in north-west Europe and Britain, growing sublittorally attached to rocks, shells and gravel. It smells strongly of lemons, although the smell has been likened to that of violets, roses or fish. Large quantities are often washed ashore after storms and may be misidentified as 'dried seaweed'. Breeding occurs during autumn and winter, the colonies containing separate male and female zooids. Pale orange-coloured embryos develop in an outgrowth of the zooid known as an ovicell, and are present from October to February in colonies on Irish Sea coasts. The larvae are released in early spring and have a short pelagic life. During the first year after settlement, the colony is flat and encrusting and only assumes the upright form in its second year. Colonies of *Flustra* only grow in spring and summer, growth ceasing for the rest of the year, and, as a result, annual growth lines can be detected on the surface of the frond enabling an estimate of age to be made. Longevity is up to 12 years. Other byrozoans such as *Crisia eburnea* (p. 385), hydroids and sedentary polychaetes are often found living on *Flustra*.

Securiflustra securifrons (Pallas) **Fig. 16.6b**

Colony upright; similar to Flustra foliacea *(above) but fronds narrower and truncate at ends. Zooids rectangular, lacking spines. Pale yellow in colour. Up to 150 mm in height.*

S. securifrons is a northern species widespread in north-west Europe and occurring all around Britain but is most abundant in the north. It is sublittoral, attached to rocks, stones and shells, and is often washed ashore in large quantities. Off the north-east

Figure 16.7 (a) *Bugula turbinata* (after Hincks, 1880). (b) *Celleporella hyalina* on *Saccharina latissima*.

coast of England, breeding has been recorded in November, March and April. The developing embryos are pinkish-orange in colour.

Bugula turbinata Alder **Fig. 16.7a**

Colony upright, tufted, firm to the touch. Branches arranged spirally on central axis. Zooids in 2 rows at base of branches, 3 or more distally. Zooids rectangular with 1 short spine at each corner of distal end; 13 ciliated tentacles. Avicularia (p. 382) squat and rounded, with short, sharply bent beaks; conspicuous among zooids. Orange-brown in colour. Up to 60 mm in height.

B. *turbinata* is a southern species extending northwards to south-west Scotland. It is found on the lower shore and in the shallow sublittoral under rocky overhangs and on stones and shells. Breeding has been recorded during August and September. The embryos are yellowish in colour and develop in outgrowths of the zooids known as ovicells. The larvae have only a brief pelagic life. Several species of *Bugula* are found around Britain; *B. turbinata* is probably the most common intertidally.

Celleporella hyalina (Linnaeus) **Fig. 16.7b**

Colonies in rounded patches; rather brittle; on seaweed, especially in the depressions along the fronds of Saccharina latissima *(p. 58). Zooids more or less oval with rounded aperture; without spines; 12 ciliated tentacles. Shiny and glassy in appearance when young, becoming opaque, whitish. Colonies up to 20 mm diameter.*

C. hyalina is widely distributed in north-west Europe and Britain where it occurs on the lower shore and into the sublittoral down to 55 m. There are separate male and female zooids within a colony and although *Celleporella* is a simultaneous hermaphrodite, there is little evidence of self-fertilization. Fertilization is internal and yellowish-coloured embryos develop in outgrowths of the female zooids known as ovicells which have a number of pores. Development takes three to four weeks and the larvae are released almost always during daylight. They have a brief pelagic life of up to some four hours before settling and metamorphosing. In Irish Sea populations, breeding occurs during most of the year. Length of life is believed to be only three to four months. *Celleporella hyalina* has been described as a species complex.

References

Hayward, P. J. (1985). *Ctenostome bryozoans. Keys and notes for the identification of the species*. Synopses of the British fauna (New Series), no. 33. Leiden: E. J. Brill/Dr W. Backhuys.

Hayward, P. J. & Ryland, J. S. (1985). *Cyclostome bryozoans. Keys and notes for the identification of the species*. Synopses of the British fauna (New Series), no. 34. Leiden: E. J. Brill/Dr W. Backhuys.

Hayward, P. J. & Ryland, J. S. (1998). *Cheilostomatous Bryozoa. Part 1. Aeteoidea – Cribrilinoidea. Notes for the identification of British species*. Synopses of the British fauna (New Series), no. 10, 2nd edn. Shrewsbury: Field Studies Council. (Includes all cheilostomatous bryozoans described in this text, with the exception of *Celleporella hyalina*.)

Hayward, P. J. & Ryland, J. S. (1999). *Cheilostomatous Bryozoa. Part 2. Hippothooidea – Celleporoidea. Notes for the identification of British species*. Synopses of the British fauna (New Series), no. 14, 2nd edn. Shrewsbury: Field Studies Council. (Includes *Celleporella hyalina*.)

Hincks, T. (1880). *A history of the British marine Polyzoa*, 2 Vols. London: John Van Voorst.

Ryland, J. S. & Porter, J. S. (2006). The identification, distribution and biology of encrusting species of *Alcyonidium* (Bryozoa: Ctenostomatida) around the coasts of Ireland. *Biology and Environment: Proceedings of the Royal Irish Academy Series B*, **106**, 19–33.

Phoronida

The phylum Phoronida comprises small, worm-like marine animals, usually no more than 200 mm in length found on the shore and sublittorally to about 400 m. Phoronids inhabit chitinous tubes to which sediment particles are often attached, and are found in sandy and muddy sediments, encrusting or burrowing into rocks and shells. The body is narrow, elongate and unsegmented, the anterior end being characterized by a lophophore, a group of ciliated tentacles surrounding the mouth. The tentacles are hollow and clothed with cilia and increase in number as the animal grows. They are used in suspension feeding, the cilia drawing in plankton and detritus which are trapped in mucus. As in the Bryozoa, the alimentary canal is U-shaped and the anus opens outside the lophophore. Either the sexes are separate or the animals are hermaphroditic, but in both cases internal cross-fertilization occurs. The eggs are either brooded in the parental tube or on the lophophore and have only a short pelagic life. In some cases the eggs are liberated directly into the sea. and develop as larvae, known as actinotrochs, with a pelagic life of several weeks. Asexual reproduction by budding also occurs. The animals have considerable powers of regeneration and can regenerate the anterior end in as little as two or three days.

There is continuing debate on the classification of phoronids. In some texts the phoronids, bryozoans and brachiopods are grouped together as the lophophorates (p. 382). However, recent molecular phylogenetic analysis suggests that phoronids should be in a subphylum of the phylum Brachiopoda.

A single genus, containing several species, has been recorded from north-west Europe; separation of the species is difficult, sometimes requiring preparation of material for sectioning and examination under a microscope. Two species are likely to be encountered on the shore.

Phoronis hippocrepia Wright **Fig. 17.1**

Body long and narrow; up to 100 mm in length and 1.5 mm wide; horseshoe-shaped lophophore with up to 150 tentacles of 2–3 mm in length. Green-grey, or yellowish in colour.

P. hippocrepia has been recorded from a few widely separated localities in north-west Europe and Britain on the lower shore and sublittorally to depths of 10 m or more. It sometimes occurs in high densities and is found either encrusting rocks and stones or burrowing into calcareous rock, shells and encrusting red seaweeds. The species is hermaphroditic and breeds throughout the year. The fertilized eggs are brooded on the lophophore and develop into free-swimming larvae.

Figure 17.1 *Phoronis hippocrepia* (after Dyster, 1859).

A second species, **Phoronis ovalis** Wright, is also found intertidally. It is much smaller than *P. hippocrepia*, usually being about 6 mm in length, occasionally up to 15 mm, and is no more than 0.3 mm in width. The lophophore is oval-shaped and has up to 28 small tentacles, each about 1 mm in length. It is found from the lower shore into the sublittoral, often in extremely high densities, burrowing in empty mollusc shells such as those of *Arctica* (p. 282) and *Ostrea* (p. 259), and into barnacle shells and limestone rocks. Larval development takes place in the tube of the worm and the larva has only a short pelagic life. *P. ovalis* also reproduces asexually by budding and transverse fission.

Reference

Emig, C. C. (1979). *British and other phoronids. Keys and notes for the identification of the species*. Synopses of the British fauna (New Series), no. 13. London: Academic Press.

Echinodermata

The echinoderms include starfishes, brittle-stars, sea-urchins, sea-cucumbers, feather-stars and sea-lilies and are one of the most distinct and well-known groups of shore animals. All are marine, rarely being recorded from brackish water, and are found from the seashore to the deepest waters. The surface of the body is often brightly coloured and is generally spiny or warty. Echinoderms are characterized by a unique five-rayed or pentamerous radial symmetry very clearly displayed in the starfishes and brittle-stars and usually readily detected in the globular sea-urchins and the elongate sea-cucumbers. Calcareous plates are present in the body wall. In starfishes these are held together by connective tissue rendering the body wall flexible; in sea-urchins the plates are fused to give a rigid test, and in sea-cucumbers they are isolated within the tissue. In all echinoderms, a delicate epidermis overlies the skeletal plates and the skeleton is therefore an endoskeleton. Spines are characteristic of the phylum, and in asteroids and echinoids specialized structures known as pedicellariae (see Figs. 18.7, 18.12) are also found on the body surface. These are pincer-like in appearance, sometimes on long stalks, and equipped with adductor muscles enabling the jaws to be snapped shut. Some are equipped with poisonous glands. They are used both in defence and as a protection against smothering by particles of detritus and small organisms such as settling larvae. They are important in identification.

Echinoderms are characterized by a unique water vascular system made up of fluid-filled canals projecting to the exterior as rows of blind ending tube-feet (podia), usually found in grooves or ambulacra running across the surface of the body. These thin-walled projections are alternately extended and contracted by changes in hydrostatic pressure and are used in a variety of functions such as respiration, feeding and locomotion. The canals of the water vascular system make contact with the exterior at the madreporite (Fig. 18.7a), a calcareous plate perforated by very fine canals and believed to be the site at which a balance is achieved between the hydrostatic pressure in the water vascular system and the exterior. In starfishes and sea-urchins it is often seen as a conspicuous button or plate-like structure on the upper surface. In the brittle-stars it is on the lower or oral surface while in the majority of holothurians, the madreporite opens not on the surface but into the body cavity. The crinoids lack a madreporite. Small, thin-walled protuberances on the upper surfaces of some echinoderms are sites of gaseous exchange.

Echinoderms show a wide range of feeding habits including scavenging, grazing, suspension and deposit feeding. Some are carnivorous. Most species have separate sexes and external fertilization and although many have free-swimming pelagic larvae, some brood the young.

Phylum ECHINODERMATA

Class Crinoidea	The sea-lilies and feather-stars. Central disc or calyx with 10 feathery arms. Mouth and anus on upper surface of disc. May be attached to substratum by a stalk, or unattached.

Class Asteroidea	The starfishes. Five or more arms radiate from a central disc. Arms thick, narrowing gradually from central disc.

Class Ophiuroidea	The brittle-stars. Slender, long arms. Distinct, rounded central disc, sharply demarcated from arms.

Class Echinoidea	The sea-urchins and heart-urchins. Body globular, oval or flattened; covered with spines. Rigid, calcareous test. No arms.

Class Holothurioidea	The sea-cucumbers. Elongate body, cucumber or sausage-like in shape; lacking arms. Branched tentacles (modified tube-feet) around mouth.

Class Crinoidea

The crinoids are primitive echinoderms attached to the substratum permanently by a stalk (the sea-lilies) or attached only during early development (the feather-stars). The body is made up of a cup-shaped central disc or calyx with both mouth and anus on the upper surface, the anus often on a raised cone-like structure. The disc has five arms but these divide into two close to the disc to give the appearance of 10 arms. In some tropical species, further subdivision gives the appearance of many more arms. In the feather-stars, a number of cirri arise from the undersurface of the disc. These are prehensile and allow temporary attachment to the substratum.

The arms are clothed with side branches or pinnules giving them a feathery appearance. These pinnules have tube-feet arranged in groups of three. Crinoids suspension feed by extending the arms and pinnules; suspended particles of food are trapped on the mucus-covered tube-feet and transported along ciliated grooves to the central mouth. Feather-stars are found clinging to rocks, hydroids and seaweed by the prehensile cirri on the undersurface of the disc. They crawl over the surface on their arms and swim by moving the arms up and down through the water. Crinoids are dioecious; the gametes develop in the pinnules at the base of the arms and are shed by rupture of the walls of the pinnules. In some species the young are brooded, but in the majority there is a free-swimming larva known as the doliolaria or vitellaria. This is non-feeding and has lecithotrophic development.

Around Britain, the stalked sea-lilies are found in deep water and are not included here. One family of feather-stars, the Antedonidae, is found in shallow waters and representatives may occur on the very low shore.

Antedon bifida (Pennant) Rosy feather-star **Fig. 18.1**
Without stalk. Central disc; 5 pairs of feathery arms, up to 100 mm long. Up to 25, rarely 30, short cirri on undersurface of disc. Colour varied, red, pink, orange, yellow; sometimes banded.

Figure 18.1   *Antedon bifida.*

A. *bifida* is found from the shallow sublittoral to depths of 450 m but is occasionally seen at the level of the lowest spring tides. It is free-moving but is most often found clinging by means of the cirri on the undersurface of the disc to the walls of gullies and on seaweeds, particularly the stipes of laminarians, in situations where there is fast current flow. It is widely distributed in north-west Europe and is found around most of Britain, except the southern part of the east coast, mainly in shallow water and often in densities of over 1000 per square metre. *Antedon* feeds on fine particles in suspension and takes a wide range of detritus, plankton and inorganic particles. The food particles are trapped in mucus on the tube-feet which are of three sizes and arranged on the pinnules in groups of three. The tube-feet are wiped against rows of cilia and the food particles transferred to ciliated grooves leading to the mouth. The sexes are separate. Mature individuals can be recognized by swollen genital papillae at the base of the arms. Eggs escape through splits which appear in the pinnule walls, and adhere in groups to the external wall of the pinnule where fertilization takes place. The embryos are held on the pinnules in a mucous net during which time the female holds its arms close together in what has been described as brooding behaviour, allowing spawned females to be readily identified. The embryos hatch as free-swimming larvae which, after a short pelagic phase, attach to the substratum and develop a short stalk. At this stage they are known as pentacrinoid larvae. The pentacrinoids eventually detach and by this time the small, prehensile cirri have developed on the undersurface of the disc. Sexual maturity is reached in the second year, and in populations in the English Channel eggs and early larvae are brooded during May to July. The corkwing wrasse, *Crenilabrus melops* (p. 460), has been observed to feed extensively on the genital pinnules.

A second species, **Antedon petasus** (Düben & Koren), has a more northerly distribution around the British Isles than *A. bifida* and is sometimes confused with the latter species. It is larger, with arms up to 130 mm long, and has 50 or so cirri on the undersurface of the disc.

Class Asteroidea

The asteroids or starfishes are readily identified by the presence of arms which grade into the central disc. In some cases the arms are very short, and in some the animal has a flattened, star-shaped appearance. The undersurface has a central mouth and rows of tube-feet running along the length of each arm. The anus is usually on the upper surface but is often very difficult to see. Spines are developed to varying degrees and in some asteroids they are modified as pedicellariae (p. 396). These can often be seen under a hand lens, but in some cases it is necessary to take scrapings from the body surface and examine under a microscope. The tips of some spines end in a brush- or comb-like arrangement and these are known as paxillae (Fig. 18.5). The presence or absence of pedicellariae and paxillae can be important in identification. The tube-feet are used in locomotion and, in some species, for feeding, and are often equipped with terminal suckers. In burrowing forms the tube-feet may be pointed. Starfishes are widely distributed in the sea, with some species moving sluggishly over the bottom, while others are adapted to burrowing in soft sediments. Many are carnivorous and feed on a wide range of invertebrates, sometimes by everting thin-walled lobes of the stomach through the mouth to engulf the prey. Cannibalism has been recorded for a number of asteroids. Some species are suspension feeders and extend the arms to trap suspended material in a mucous coating.

A few asteroids are hermaphroditic, but most have separate sexes. The gametes are shed into the sea where external fertilization takes place with the development of a planktonic larva known as a bipinnaria. This feeds in the plankton (planktotrophic) and has a free-swimming life of up to several weeks. Eventually it develops short arms, tipped with adhesive cells and is known as a brachiolaria larva. The brachiolaria attaches temporarily to the substratum and metamorphoses into a tiny starfish. In some asteroids either the bipinnaria or brachiolaria stage is omitted; others brood their eggs and lack a free-swimming larval stage altogether. Asteroids are capable of regenerating lost arms.

Key to species of asteroids

1 Marginal plates (see Fig. 18.2) along arms, large and conspicuous on upper
   and/or lower surfaces. Tube-feet without suckers **2**
   Marginal plates of arms inconspicuous **3**

2 Both upper and lower marginal plates of arms conspicuous;
   5 arms .... ***Astropecten irregularis*** p. 400
   Lower marginal plates of arms conspicuous but upper plates replaced by
   brush-like spines (paxillae); 7 arms .... ***Luidia ciliaris*** p. 401

3 More than 5 arms (usually 9–12) **4**
   With 5 arms (rarely 4 or 6) **5**

4 Eight to 14 (usually 10–12) arms. Brush-like spines (paxillae) (see Fig. 18.5a) along
   margins of arms in single, conspicuous row .... ***Crossaster papposus*** p. 403
   Seven to 13 (usually 9–10) arms. Brush-like spines (paxillae) along margins of
   arms inconspicuous, in double row .... ***Solaster endeca*** p. 405

5 Very short, blunt arms; flattened pentagonal shape, but upper surface of disc
   rather rounded; usually olive-green to brown in colour; up to 50 mm across **6**
   Arms not short; characteristic starfish appearance **7**

6 Upper surface of central disc with dark, star-shaped area. Up to 15 mm
   across .... ***Asterina phylactica*** p. 403
   Upper surface of central disc not marked with star-shaped area. Up to 50 mm
   across .... ***Asterina gibbosa*** p. 402

7 Upper surface not conspicuously spiny. Arms rather stiff, cylindrical in section.
   No pedicellariae .... ***Henricia* spp.** p. 403
   Upper surface of arms with longitudinal rows of spines. Pedicellariae present **8**

8 Upper surface with short spines; row of whitish spines along midline of arms.
   Arms broad at base, tapering .... ***Asterias rubens*** p. 405
   Upper surface very spiny, with conspicuous rows of spines on upper and lateral
   margins of arms. Pedicellariae grouped around base of spines. Arms long,
   narrow, tapering gradually .... ***Marthasterias glacialis*** p. 407

Astropecten irregularis (Pennant) **Fig. 18.2**

Flattened body, stiff; 5 arms with conspicuous marginal plates on upper and lower
surfaces with spines arising from lower marginal plates; up to 120 mm across, rarely
up to 200 mm. Pointed tube-feet without suckers. Upper surface varied in colour,
pink, yellow, orange, brown, sometimes with purple marks; undersurface pale.

A. irregularis is a sublittoral species, widely distributed in north-west Europe and
found all around Britain in sand and gravel down to 1000 m. It lies buried in the

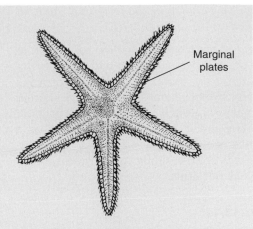

Figure 18.2 *Astropecten irregularis.*

sediment and contact with the surface is maintained by the tips of the arms. Specimens are often washed ashore, particularly after storms. It migrates into deeper water during winter possibly to avoid sediment disturbance during storms. It is carnivorous and uses the tube-feet to push food into the mouth. It feeds mainly on young bivalves, particularly *Spisula subtruncata* (p. 272), often digging down into the sediment to locate prey, and is believed to be an important predator of benthic animals, including brittle-stars, polychaete worms and crustaceans. The sexes are separate and breeding takes place during spring and summer. There is a bipinnaria larva but no brachiolaria in the life-cycle.

Luidia ciliaris (Philippi) **Fig. 18.3**
Body with 7 arms; usually up to 400 mm across but occasionally 600 mm. Marginal plates on undersurface of arms conspicuous and bear spines; upper marginal

Figure 18.3 *Luidia ciliaris.*

plates replaced by brush-like spines known as paxillae. Tube-feet long, without suckers. Upper surface red, lower surface paler.

L. ciliaris is widely distributed in north-west Europe from the shallow sublittoral to depths of 400 m and is common around Britain. It is occasionally found on the lower shore, particularly on south and west coasts but is not found off southern North Sea coasts. *Luidia* burrows into sandy and gravelly sediments and feeds mainly on other echinoderms. It is a fragile species and specimens are often found with one or more regenerating arms. Off the south coast of England it breeds in summer; it has a bipinnaria larva.

A second species, ***Luidia sarsi*** Düben & Koren, is similar to *L. ciliaris*, but is found in deeper water in muddy deposits. It has five arms.

Asterina gibbosa (Pennant) Cushion-star **Fig. 18.4a**
Body with 5 (rarely 4 or 6) stubby arms; up to 60 mm across. Ventral surface flat, dorsal surface rather rounded. Colour uniform but varied, olive-green, brown, orange.

(a)

(b)

Figure 18.4   (a) *Asterina gibbosa*. (b) *Asterina phylactica*.

A. gibbosa is a southern species reaching the northern limit of its distribution on west and south-west coasts of Britain. There are only scattered records of the species north of the Isle of Man and it has not been recorded from North Sea coasts. Found under boulders and stones on the lower shore on sheltered and semi-exposed shores, and also in rock pools, it extends into the sublittoral to depths of about 100 m. Specimens from the lower shore and sublittoral are generally pale orange or grey in colour, compared with olive-green for specimens from rock pools higher on the shore. *Asterina* is an omnivorous scavenger feeding on microorganisms, decaying seaweed and dead invertebrates by everting thin-walled lobes of its stomach through the mouth. It is a protandrous hermaphrodite and in south-west Britain matures at a diameter of about 10 mm when two years old, first as a male, then changes to female when about four years old, at a diameter of about 20 mm. In May, females lay as many as 1000 orange-coloured eggs in crevices and on the undersurface of rocks, and after two to three weeks these hatch as tiny starfishes. They reach a diameter of about 5 mm in the first year and live for about seven years, producing between three and seven broods.

Asterina phylactica Emson & Crump **Fig. 18.4b**

Body with 5 (rarely 4 or 6) stubby arms; up to 15 mm across. Ventral surface flat, dorsal surface rather rounded. Colour usually dark olive-green; very similar to A. gibbosa *(above) but with dark star-shaped area in centre of disc.*

A. phylactica was first described from British waters in 1979 and has now been recorded from a number of sites on the south and west coasts of Britain, where it appears to favour rock pools with a rich growth of seaweed. It has also been recorded from the sublittoral. Small specimens of up to about 4.5 mm diameter (one year old) cannot reliably be separated from *A. gibbosa*. Above this size the colour differences are established. *A. phylactica* is an omnivorous scavenger. The species is a simultaneous hermaphrodite. Sexual maturity is reached at a diameter of 5–6 mm, at an age of two years, and in May aggregations of up to 10 animals can be found in crevices on the shore where the eggs are laid. The adults stay together and, unlike *A. gibbosa*, protect the developing embryos for about three weeks until they hatch as tiny starfishes. Aggregation during egg laying is believed to ensure cross-fertilization, but the animals are capable of self-fertilization. Longevity is up to four years and up to three broods are produced.

Crossaster papposus (Linnaeus) (*Solaster papposus*) Common sun-star **Fig. 18.5a**

Body with 8–14 (usually 10–12), short, thick arms; up to about 300 mm across but usually 100–200 mm. Conspicuous brush-like spines (paxillae) in single row along margins of arms. Dorsal surface covered with spines and thinly distributed paxillae. Reddish-purple in colour; arms may be banded, giving overall pattern of white rings on body. Undersurface whitish.

C. papposus is a northern species with a circum-polar distribution. It is widely distributed in north-west Europe and common around the coasts of Britain, reaching its southern limit in the English Channel. It is occasionally found on rocks and stones on the lower shore and has been recorded in the sublittoral in sand and gravel to depths of over 200 m. The common sun-star feeds on a wide range of food items, including other echinoderms, particularly the common starfish *Asterias rubens* (p. 405). It reproduces in spring; the sexes are separate and eggs and sperm are shed into the water where fertilization takes place. The eggs are large and yolky and the larvae do not feed.

Henricia spp. **Fig. 18.5b**

Body with 5 rounded, rather stiff arms, cylindrical in section; usually up to 120 mm across, occasionally much larger. No pedicellariae. Dorsal surface usually dark purple, red, occasionally yellowish; ventral surface pale.

Prior to the 1980s, all British specimens of the genus *Henricia* were referred to as *Henricia sanguinolenta*. Following a review by Madsen (1987) it has been suggested that there may be four species around Britain. These are difficult to separate and microscopic examination of the calcareous plates of the endoskeleton is required.

Figure 18.5 (a) *Crossaster papposus*, with detail of paxilla. (b) *Henricia* sp.

Henricia oculata (Pennant) has been recorded from a wide range of localities around Britain, particularly the English Channel and west coast, and is believed to be the commonest British species. It is orange, yellow or red in colour and found on hard substrata such as stones, shell and gravel, occasionally on the lower shore and to depths of 50 m or more. *H. oculata* is a suspension feeder, extending the arms in the water to trap suspended particles in mucus. It also feeds on sponges, hydroids and detritus by stomach eversion, and scavenging on dead fish has been recorded. The sexes are separate and breeding occurs in spring, the gametes being liberated directly into the sea. The larva is believed to be pelagic.

Henricia sanguinolenta (O. F. Müller) is also recorded from Britain, particularly the north-east coast, and Orkney and Shetland. It tends to be blood-red in colour. In this species the female broods the eggs under the central disc until they hatch. There is no larval stage.

Solaster endeca (Linnaeus) Purple sun-star **Fig. 18.6**

Body with 7–13 (usually 9–10) tapering arms; up to 400 mm across. Brush-like spines (paxillae), inconspicuous, in double row along margin of arms. Upper surface very hard and rough with somewhat inconspicuous spines and small paxillae. Cream-coloured or shades of violet, purple, orange in colour.

S. endeca is a northern species, widely distributed in north-west Europe and absent from English Channel coasts and the southern North Sea. It is found from the lower shore to depths of about 450 m. It feeds on other echinoderms. Development is direct; there is no larval stage in the life-cycle.

Figure 18.6   *Solaster endeca.*

Asterias rubens Linnaeus Common starfish   **Fig. 18.7a; Plate 19a**

Body with 5 arms (occasionally 4 or 6), broad at base, tapering; usually up to 150 mm across but much larger specimens have been recorded. Arms round in section. Upper surface rough with row of whitish spines along midline of arms; pedicellariae present, especially on lateral walls of arms. Colour varied, upper surface from yellowish-brown to red, sometimes violet. Lower surface pale with conspicuous rows of tube-feet.

A. rubens is widely distributed in north-west Europe and abundant in many areas. It is commonly found on British shores, especially among mussels and barnacles, extending into the sublittoral to depths of about 400 m. In many areas it shows an onshore migration during spring. Although infrequently recorded from estuaries, laboratory studies suggest that North Sea populations tolerate salinities down to about 23. The tube-feet of *Asterias* have terminal suckers enabling it to attach to

Figure 18.7 (a) *Asterias rubens*, with detail of pedicellariae. (b) *Marthasterias glacialis*, with detail of a single spine to show clusters of pedicellariae at base.

the substratum, so that when the animal is moving across a flat rock surface it is essentially stepping on its tube-feet. *Asterias* preys on a wide range of food items including molluscs, polychaete worms and other echinoderms. Small crustaceans are occasionally caught on the suction discs of the tube-feet, but the common starfish preys particularly on bivalve molluscs. The valves of the shell are forced open by the tube-feet, the tips of which attach to the bivalve shell by suction and pull the valves apart. As soon as a tiny gap, believed to be as small as 0.1 mm, is established between the valves of the shell, the starfish everts its stomach lobes through its mouth and these pass into the bivalve and commence digestion. Dense aggregations of starfishes are often seen feeding on the shore: one such aggregation on the west coast of England, measuring 1.5 km long and 15 m wide, was estimated to have cleared about

50 hectares (about 3500 to 4500 tonnes) of young of the common mussel in four months (Plate 19a). Starfishes are also important predators of oysters.

The sexes of *Asterias* are separate and fertilization external and it has been estimated that a starfish of 140 mm diameter spawns as many as 2.5 million eggs. Breeding takes place in spring and summer and bipinnaria larvae are often abundant in summer plankton samples. The bipinnaria is followed by a brachiolaria larva and larval life may be as long as three months. On the south coast of Britain, sexual maturity has been recorded in the second year, when the starfish has a diameter of about 50 mm.

Marthasterias glacialis (Linnaeus) Spiny starfish **Fig. 18.7b**
Body with 5 arms, long, narrow; tapering gradually; up to 700 mm across but usually 250–300 mm. Upper surface of arms very rough, with conspicuous spines; spines also present on lateral margins. Many pedicellariae grouped round base of spines. Colour varied, yellowish, reddish, bluish-green.

M. glacialis is widely distributed in north-west Europe; around Britain it is restricted to west and south-west coasts, where it is essentially a sublittoral species extending to depths of about 180 m, but occasional specimens are found at extreme low water on rocky shores. It feeds on a wide range of organisms, including seaweed, crustaceans, molluscs and other echinoderms and has been described as a voracious carnivore. The sexes are separate, fertilization is external and breeding aggregations have been recorded in summer. It has been suggested that spawning takes place in shallow water. Both bipinnaria and brachiolaria larval stages are present in the life-cycle and there is a long pelagic life, estimated to be some four months.

Class Ophiuroidea

The ophiuroids, or brittle-stars, are well represented on the lower shore where they are found in crevices, under stones and among seaweed. They extend into the sublittoral to the deep sea and are often found on gravelly bottoms or burrowing into soft sediments. In both situations they may be present in very high densities. Most are relatively small and have five long, slender arms and a distinct, rounded central disc, a character which is important in the separation of ophiuroids from asteroids. The arms of ophiuroids are made up of articulating calcareous ossicles which generally allow considerable twisting in the lateral plane. These ossicles, also known as vertebrae or segments, are each encased in a regular arrangement of four plates, known as the arm plates. There is a dorsal plate, a ventral plate and two lateral plates, the latter carrying spines known as the arm spines. The dorsal surface of the disc is also covered with plates and these are of varying size. At the base of each arm there is a pair of plates known as the radial shields (Fig. 18.8a). These are sometimes large and conspicuous. Plates known as arm combs may also be present at the base of each arm (Fig. 18.10a). The undersurface of the disc has a complex arrangement of plates and papillae and of these the plates surrounding the mouth, known as the oral shields, may be conspicuous (Fig. 18.10b). The mouth lies centrally on the undersurface of the disc; there is no anus. The tube-feet are without suckers and are used in feeding but not locomotion. Pedicellariae are lacking. Ophiuroids are very active and when disturbed move quickly in characteristic fashion, the arms pushing against the substratum. They have a wide range of feeding habits including scavenging, deposit and suspension feeding. In most species the sexes are separate and fertilization external, leading to the development of a pelagic larva, the ophiopluteus. In a few, however, the embryos develop in a thin-walled sac within the body of the female until they emerge as juveniles.

Key to species of ophiuroids

Ophiuroids can be difficult to identify to species. In most cases detailed examination of skeletal plates on the disc and the arms using a microscope or hand lens is required.

1 Arms very long, up to 15 times the diameter of the disc ....
.... ***Acrocnida brachiata*** p. 410
Arms up to about 8 times the diameter of the disc **2**

2 Spines along margins of arms lie almost flat against arms. Disc covered by distinct scales. Conspicuous oral shields on undersurface (Fig. 18.10b) **3**
Spines along margins of arms stand out, set at right angles to arm **4**

3 Arm combs (Fig. 18.10a) at base of each arm, each with about 30 papillae. Ventral plates at base of arm separated by grooves (Fig. 18.10b) ...... ***Ophiura ophiura*** p. 412
Arm combs at base of each arm, each with fewer than 20 papillae. Ventral plates at base of arm not separated by grooves .... ***Ophiura albida*** p. 413

4 Upper surface of disc with pair of conspicuous radial shields at base of each arm (Fig. 18.8a) **5**
Radial shields obscured **6**

5 Disc spiny. Radial shields large, up to 2/3 radius of disc; triangular in shape and separate along length .... ***Ophiothrix fragilis*** p. 408
Disc covered with small scales. Radial shields conspicuous, joined for almost entire length .... ***Amphipholis squamata*** p. 410

6 Disc covered with scales and small granules. Dorsal plates of arms oval, ringed by small plates .... ***Ophiopholis aculeata*** p. 410
Disc completely covered with fine granules, smooth. Dorsal plates of arm not ringed by small plates .... ***Ophiocomina nigra*** p. 410

Ophiothrix fragilis (Abildgaard) Common brittle-star **Fig. 18.8a**
Central disc up to 20 mm diameter, spiny; specimens found on shore generally much smaller. Arms about 5 times diameter of disc, slender, with many serrated spines. Arms very fragile. Upper surface of central disc with 5-rayed pattern of spines; radial shields

Figure 18.8 (a) *Ophiothrix fragilis,* whole animal inset. (b) *Ophiocomina nigra.* (c) *Ophiopholis aculeata.*

large, triangular in shape; up to 2/3 radius of disc; separate along length and without spines. Very varied in colour; red, white, brown, sometimes banded.

O. fragilis is widely distributed in north-west Europe and often abundant around British coasts, extending to depths of about 150 m. Sublittorally, it forms dense aggregations, in which densities as high as 10 000 individuals per square metre have been recorded. It is from studies on such populations that most of our knowledge of the biology of this species has been gained. Intertidally it is found in crevices and under stones and seaweed. *Ophiothrix* is a suspension feeder; the arms are raised above the substratum and the tube-feet extended to filter suspended material from the water. Food, consisting of detritus and diatoms, passes along the arms to the mouth. Scavenging on decaying material has also been recorded. The sexes are separate and mature males and females have been recorded at disc diameters of over 3 and 5 mm, respectively. In the Irish Sea, breeding occurs in summer. Electron microscope studies on the ossicles of the arms reveal banding patterns which are believed to result from annual deposition of calcareous material, allowing age to be determined. It is suggested that *O. fragilis* is a long-lived species, living for 10 years or more. There is a planktonic larval stage, the ophiopluteus, and in dense, offshore aggregations larvae settle on the adults. It has been suggested that they feed semi-parasitically for a short while before migrating to other areas. Newly settled brittle stars have also been recorded in large numbers on sponges where they tend to aggregate around the ostia

(p. 79), presumably feeding on the suspended material in the inhalant currents drawn into the sponge. *Ophiothrix* is often preyed on extensively by starfishes.

Ophiocomina nigra (Abildgaard) **Fig. 18.8b**

Central disc up to 30 mm diameter. Arms about 5 times diameter of disc, with many long, smooth spines. Upper surface of central disc covered with many fine granules obscuring radial shields. Colour usually uniform black, grey or brown; may be pink or whitish.

O. nigra is widely distributed in north-west Europe and around Britain but has a patchy distribution on North Sea coasts, where it is absent south of Northumberland. It lives on the lower shore under stones and seaweed, extending into the sublittoral to depths of about 400 m. Dense offshore aggregations have been recorded but, despite the high density of animals, individuals are spaced out, barely touching one another. *Ophiocomina* shows a range of feeding habits including deposit feeding, herbivorous browsing, carrion feeding and suspension feeding in which fine material is trapped in a mucous net and carried to the mouth by the tube-feet. *Ophiocomina* breeds in summer; the sexes are separate and fertilization external, with the development of a pelagic ophiopluteus larva.

Ophiopholis aculeata (Linnaeus) Crevice brittle-star **Fig. 18.8c**

Disc up to 20 mm diameter; covered with scales and small granules obscuring radial shields. Arms about 4 times diameter of disc. Dorsal arm plates oval-shaped, each surrounded by a circle of much smaller plates; 6–7 spines on either side of each arm segment. Colour red, purple; arms characteristically banded with darker colour.

O. aculeata reaches the southern limit of its distribution in Britain, where it is common in the north but much less so in the south and the English Channel. It is found on the lower shore under stones and particularly in crevices with the arms often protruding. It has been recorded sublittorally to depths of about 200 m. There is an ophiopluteus larva in the life-cycle.

Acrocnida brachiata (Montagu) (*Amphiura brachiata*) **Fig. 18.9a**

Disc up to 13 mm diameter. Arms very long, slender, up to 15 times diameter of disc; with short spines. Upper surface of central disc covered with small scales; radial shields separate, conspicuous. Arm plates with many short spines, the number decreasing from base to tip of arm. Brownish-grey in colour.

A. brachiata is widely distributed on the shores of north-west Europe, extending sublittorally to depths of about 40 m. In Britain it appears to be most frequently found in the south and west, buried in sandy sediments, often in dense aggregations, with the tips of the arms extending above the surface. *Acrocnida* is a suspension feeder. It spawns during summer. Development is not known; it has been suggested that it has a brief pelagic phase.

A bivalve mollusc, **Kurtiella bidentata** (Montagu) (*Mysella bidentata*), which reaches a shell length of about 4 mm, and the young stages of a polychaete worm of the genus **Malmgreniella** are sometimes commensal with this brittle-star.

Amphipholis squamata (Delle Chiaje) **Fig. 18.9b**

Small, central disc up to about 5 mm diameter. Arms about 4 times diameter of disc, with short spines. Upper surface of central disc with small scales; radial shields

Figure 18.9 (a) *Acrocnida brachiata*. (b) *Amphipholis squamata*.

conspicuous, joined for almost entire length. Grey, white, blue-grey in colour, shows a range of colour patterns.

A. squamata is widely distributed in north-west Europe and around Britain, extending into the sublittoral to depths of about 250 m. It is found under stones, in sand, gravel and rock pools, often associated with *Corallina* (p. 43) and bryozoans. *A. squamata* is a deposit feeder using the tube-feet around the mouth to take a wide range of plant and animal material; it also suspension feeds, raising the arms above the sediment to trap fine material on mucus-coated spines. Food is transported to the mouth by the tube-feet. It is a bioluminescent species. *Amphipholis* is hermaphroditic and probably self-fertilizing. There is no pelagic larval stage; young are brooded by the parent. Maturity is believed to be reached at a disc diameter of about 1.5 mm, when the brittle-stars are in the first year of growth. Juveniles are released from the brood pouch during the summer at disc diameters of 0.8–1.2 mm. Studies on a rock pool population in Devon suggest a longevity of 12–18 months, but in the Firth of Forth some adults survive a second winter to reproduce in the following spring. It has been suggested that the different colour morphs are reproductively isolated sibling species; molecular studies suggest that *A. squamata* is a species complex.

Ophiura ophiura (Linnaeus) (*Ophiura texturata*) **Fig. 18.10a,b**
Disc up to about 35 mm diameter. Arms about 4 times diameter of disc. Upper surface of disc with coarse scales. Arm spines lie almost flat against arms. Radial shields large, and lie close to pair of smaller plates (the arm combs) which have comb-like edge with about 30 papillae along the margin. Oral shields, on undersurface of disc, large. Grooves between ventral plates at the base of the arms. Dorsal surface of disc red/ brown, mottled, ventral side paler; radial shields red/brown.

O. ophiura is widely distributed in north-west Europe and common all around Britain in sand and muddy-sand. It extends from the lower shore to depths of about 200 m and in the shallow sublittoral has been recorded in densities as high as 700 per square metre. It is a suspension feeder taking a wide range of food. Breeding occurs during summer. The sexes are separate and fertilization external, and there is a pelagic ophiopluteus larva in the life-cycle. Newly metamorphosed juveniles have a disc

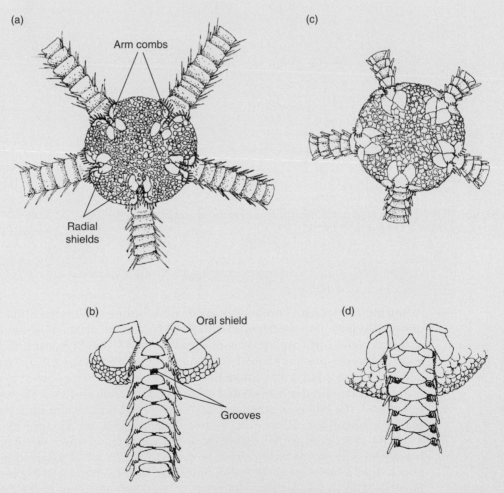

Figure 18.10 *Ophiura ophiura*, (a) upper surface of disc, (b) undersurface of disc. *Ophiura albida*, (c) upper surface of disc, (d) undersurface of disc. (Based on Mortensen, 1927.)

diameter of 0.2–0.3 mm, and studies on the microscopic growth bands on the arm ossicles suggest a longevity of five or more years.

Ophiura albida Forbes **Fig. 18.10c,d**

Disc up to 15 mm diameter. Arms about 4 times diameter of disc. Upper surface of disc with coarse scales. Arm spines lie almost flat against arms. Radial shields lie close to pair of smaller plates (the arm combs) which have comb-like edge with fewer than 20 papillae along the margin. Oral shields on undersurface of disc conspicuous. Ventral plates at base of arms not separated by grooves. Dorsal surface of disc red/ brown; radial shields often whitish.

O. albida is widely distributed in north-west Europe. It is common all around Britain in soft sediments from the lower shore into the sublittoral to depths of over 200 m. It feeds on microorganisms, organic debris and carrion. It breeds in summer; the sexes are separate, fertilization is external and there is a pelagic ophiopluteus larva in the life-cycle. It has been suggested that longevity is up to three years.

Class Echinoidea

The echinoids or urchins are characterized by a hard, calcareous test, which is globular, heart-shaped or flattened. There are no arms and the test is covered in movable spines. Urchins with a globular test are known as regular echinoids and in these forms, for example the common sea-urchin *Echinus* (p. 415), the spines are generally long and well developed. Those with an oval or heart-shaped, or much-flattened test are known as irregular echinoids and include the heart-urchins and sand dollars. These generally have small spines. In the regular echinoids the mouth lies on the undersurface of the test and the anus is on the upper surface. In irregular urchins the mouth is on the undersurface and the anus either on the posterior edge of the test or on the undersurface. In some echinoids the mouth is equipped with a complex feeding apparatus, the Aristotle's lantern, which includes five hard, calcified teeth capable of tearing seaweed and removing encrusting organisms from the rock surface. These teeth can generally be seen protruding from the mouth. The body of an echinoid has five double rows of tube-feet which are often clearly seen on the external surface, and in living specimens many of the tube-feet can be seen to be long and highly extensible. On the dry, denuded test, the position of the tube-feet is indicated by the five rows of ambulacral plates which have pairs of tiny pores (Fig. 18.12e) through which the tube-feet emerge in life. Among the spines and tube-feet are large numbers of pedicellariae consisting of a long stalk bearing a head region. Different types of pedicellariae are shown in Fig. 18.12b,c,d. The stalk is supported by a delicate calcareous rod which in some cases stops short of the head leaving a flexible region, the neck; in others, the skeletal rod runs up the base of the head. The head is made up of usually three delicate calcareous jaws, and in the globiferous pedicellariae the dentition of these is important in identification. Globiferous pedicellariae secrete a toxin.

Regular echinoids move slowly using the tube-feet and the spines, the former pulling and the latter pushing the animal along. Some species are adapted for boring into soft rocks such as limestone. Irregular echinoids are generally found burrowing into soft sediments and some of the spines have flattened ends and are adapted for burrowing. Other small, club-shaped spines are ciliated and create a flow of water through the burrow. They lie in definite tracks which in some species form conspicuous rings (fascioles).

Many echinoids graze on plant and animal material and a number, including the common sea-urchin *Echinus* (p. 415), remove encrusting organisms by means of the Aristotle's lantern. Some of the irregular urchins or heart-urchins lack an Aristotle's lantern and use modified tube-feet to feed on particles of detritus. As echinoids grow, individual skeletal plates increase in size by deposition of calcite round their margins. In some species, for example *Echinus esculentus* (p. 415) and *Paracentrotus lividus* (p. 416), growth rings can be detected as contrasting light and dark bands in individual plates which have been carefully ground down with fine sand paper and examined under the microscope. Echinoids have separate sexes. Fertilization is external and the majority have a free-swimming pelagic larva known as an echinopluteus. In some, the eggs are brooded among the spines or in depressions on the surface of the test until they hatch as juveniles.

Key to species of echinoids

Although many echinoids can be identified by the shape and coloration of the test, the nature and colour of the spines and the habit or habitat, in some cases further criteria are needed to confirm identification. The shape of the jaws of the *globiferous* pedicellariae and the arrangement of the tube-feet are two important characteristics. Pedicellariae have stalks measuring up to about 4 mm in length and are conspicuous when the test is seen under a stereo-microscope. They can be removed with fine forceps without serious damage to the echinoid and should be carefully examined to ensure that it is the globiferous pedicellariae that have been removed (see Fig. 18.12c); they should then be compared with the drawings given for the particular species. In order to determine the arrangement of the tube-feet, the spines have to be removed and the test cleaned with a bleach solution. This reveals the arrangement of the tiny pores in the plates (ambulacral plates) of the ambulacral region of the test (Fig. 18.12e) through which the tube-feet project in life. In the following key, the shape and coloration of the test and colour of the spines are given first and these are followed, where appropriate, by the nature of the globiferous pedicellariae and arrangement of tube-feet, to be used for confirmation as required.

1 Test rounded, radially symmetrical – regular echinoids **2**
   Test oval, heart-shaped – irregular echinoids **5**

2 Test somewhat flattened, greenish-brown in colour. Spines green, red or violet, often with white tips. Restricted to Shetland and the east coast of Britain. Jaws of globiferous pedicellariae without teeth below terminal tooth ....
.... ***Strongylocentrotus droebachiensis*** p. 418
   Coloration of test and spines not as above. Jaws of globiferous pedicellariae with 1 or more teeth below terminal tooth **3**

3 Test flattened, greenish in colour; with long violet- or olive-coloured spines. Bores into rock. Found on west coast of Ireland, rarely south-west England. In denuded test, each ambulacral plate has 5 or 6 pairs of pores ...... ***Paracentrotus lividus*** p. 416

Test coloration and habitat not as above. In denuded test, each ambulacral plate has 3 pairs of pores **4**

4 Test slightly flattened; overall greenish colour. Tips of spines with violet tinge. Jaws of globiferous pedicellariae with row of teeth below terminal tooth .... ***Psammechinus miliaris*** p. 415

Test may be slightly flattened at poles, but overall globular; reddish colour. Jaws of globiferous pedicellariae with 1 tooth below terminal tooth .... ***Echinus esculentus*** p. 415

5 Feeding apparatus (Aristotle's lantern) present. Mouth and anus on lower surface, mouth central. Very small, no more than 15 mm in length, distinctive .... ***Echinocyamus pusillus*** p. 419

Aristotle's lantern not present. Mouth on undersurface, towards anterior; anus on posterior edge of test. Large, up to 120 mm in length **6**

6 Violet colour. Spines mostly directed backwards; long curved spines on upper surface paler than other spines, may be white .... ***Spatangus purpureus*** p. 419

Yellow-brown colour; close-set spines directed backwards **7**

7 Conspicuous anterior furrow on test. In profile, highest point of test towards posterior .... ***Echinocardium cordatum*** p. 419

No anterior furrow on test. In profile, highest point of test towards centre .... ***Echinocardium pennatifidum*** p. 421

Psammechinus miliaris (P. L. S. Müller) **Fig. 18.11**

Test up to 50 mm diameter; somewhat flattened; greenish in colour. Spines green with purple tips. Aristotle's lantern present. Jaws of globiferous pedicellariae with row of teeth below terminal tooth. In denuded test, each ambulacral plate has 3 pairs of pores.

P. miliaris is widely distributed in north-west Europe and occurs all around Britain under stones and in crevices on the lower shore. Often found with seaweeds, it may be camouflaged by pieces of weed trapped among the spines. *Psammechinus* is omnivorous and consumes a wide variety of seaweed and animal material including barnacles, bivalve molluscs and ascidians. The sexes are separate, fertilization is external and breeding occurs during spring and summer. The echinopluteus larva is pelagic for about four weeks before metamorphosing and settling on the shore at 0.5–1 mm diameter. Sexual maturity is reached after one or sometimes two years when the urchins are generally between 6 and 10 mm diameter. Longevity is believed to be up to 10 years.

Echinus esculentus Linnaeus Common sea-urchin **Fig. 18.12**

Test almost globular; usually 100–120 mm diameter, may be larger. Aristotle's lantern present. Pinkish-red colour, also shades of green, yellow or purple. Spines rather short; often coloured red and white with lilac tips. Jaws of globiferous pedicellariae with 1 tooth below terminal tooth. Denuded test with white tubercles; each ambulacral plate with 3 pairs of pores.

Figure 18.11 *Psammechinus miliaris*, (a) lateral view of test, (b) detail of globiferous pedicellaria. (Note teeth on jaws*.)

E. esculentus is widely distributed in north-west Europe. Although sometimes found on the lower shore, highest densities are recorded offshore and estimates of as many as 2000 individuals per hectare have been made for populations in the English Channel. It is generally found to depths of about 40 m but it has been recorded at depths of 100 m and more. The common sea-urchin shows considerable variation in the shape of the test. In some areas specimens from shallow water have a relatively flat test, while those from deeper water are taller. Coloration is also varied; in some specimens the spines are slender, in others they are more robust. *Echinus* browses on seaweed and encrusting invertebrates using the sharp teeth of the Aristotle's lantern and is capable of clearing large areas of rock. The sexes are separate and breeding occurs in spring when male and female gametes are shed into the sea where fertilization takes place. It has been estimated that large females contain as many as 20 million eggs. In some countries sea-urchin roe is considered a delicacy. The fertilized eggs develop into echinopluteus larvae which are planktonic for seven to eight weeks before metamorphosing into tiny urchins about 1 mm in diameter. Rate of growth depends on locality and it has been calculated that urchins can reach a diameter of 40 mm in the first year. Studies on the west coast of Scotland suggest that urchins are slow growing and may survive for more than 16 years.

Paracentrotus lividus (Lamarck) **Fig. 18.13a,b,c,d; Plate 19b**

Test somewhat flattened; up to 70 mm diameter. Aristotle's lantern present. Brownish-green in colour; closely covered with long spines. Spines long, sharply pointed; purple, brown or green in colour. Jaws of globiferous pedicellariae with 1 tooth below terminal tooth. In denuded test, each ambulacral plate has 5 or 6 pairs of pores.

P. lividus is found as far north as the Channel Islands and the west coast of Ireland. There are few records of its presence in south-west England and it is on the west coast of Ireland that it is widespread, being abundant in some areas such as the shores of

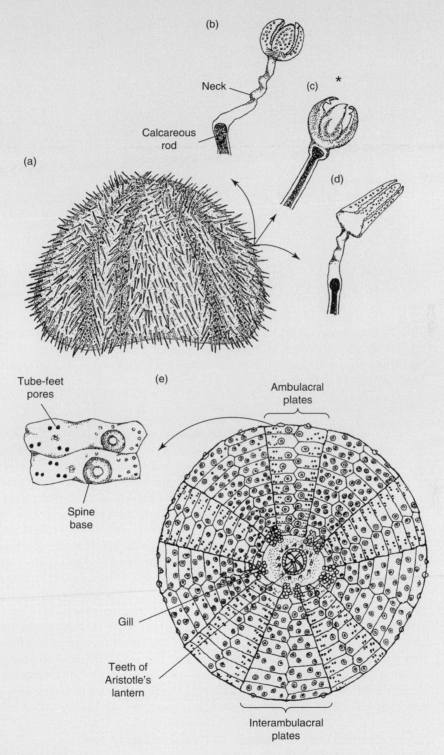

Figure 18.12 *Echinus esculentus*, (a) lateral view of test, with detail of (b) ophiocephalous, (c) globiferous (note teeth on jaws*) and (d) tridentate pedicellariae. (e) Oral view of denuded test with detail of ambulacral plates showing spine bases and tube-feet pores.

Figure 18.13 *Paracentrotus lividus*, (a) lateral view of test, with detail of (b) ambulacral plates, (c) globiferous pedicellaria. (d) Specimens in characteristic depressions in rock. (e) Globiferous pedicellaria of *Strongylocentrotus droebachiensis*. (Note arrangement of teeth on jaws of pedicellariae*).

County Clare. It has some commercial importance, particularly in the Mediterranean where the roe is considered a delicacy and on some shores there are clear signs of over-exploitation. It is found on the lower shore in rock pools and typically bores into rock, often limestone, by means of its spines, and the teeth of the Aristotle's lantern. Each urchin fits tightly into its cavity and thus derives some protection from wave action and from desiccation during low tide. Some are believed never to leave the cavity. They are found in the sublittoral to depths of 30 m. *Paracentrotus* actively covers the test with a wide range of materials, including pieces of bivalve shell, fragments of seaweed and the broken tests of urchins. This has been described as 'masking' and it has been suggested that it protects the urchin from damage by floating material and may also protect it from the harmful effects of ultraviolet radiation. *Paracentrotus* grazes on seaweed, detritus and sporelings, keeping in check, and in some situations eliminating, the growth of seaweeds. It has been suggested that an important source of food, particularly in very dense aggregations, may be the fragments of seaweed brought in by the tide and trapped on the spines of the urchin. The sexes are separate and fertilization external with the development of an echinopluteus larva. On the west coast of Ireland, breeding has been recorded in January–March and August–September, individuals reaching a diameter of 35–50 mm in about four years and living for six to nine years.

Strongylocentrotus droebachiensis (O. F. Müller) **Fig. 18.13e**

Test somewhat flattened; up to 80 mm diameter. Aristotle's lantern present. Greenish-brown. Spines green, red or violet, often with white tips. Wide variation in colour. Jaws of globiferous pedicellariae without teeth below terminal tooth; with spicules on sides of jaws. In denuded test, each ambulacral plate has 5 or 6 pairs of pores.

S. droebachiensis is widespread and often abundant in north-west Europe. While very similar to *Paracentrotus lividus* (above), it is, however, only found in Scotland and on the north-east coast of England and the northern North Sea. It is found on the lower shore where it often bores into rock and extends well into the sublittoral to depths of about 200 m. *Strongylocentrotus* grazes on the rock surface and feeds on seaweeds and encrusting invertebrates. The sexes are separate, fertilization is external and breeding occurs in spring. There is an echinopluteus larva.

Strongylocentrotus pallidus (G. O. Sars) is found around Orkney and Shetland. Like *S. droebachiensis* it has wide morphological variation and the species can only be separated on the basis of the detail of the ambulacral plates and globiferous pedicellariae. Detailed morphometric studies suggest that the two are distinct species.

Echinocyamus pusillus (O. F. Müller) Green urchin, pea-urchin **(not illustrated)**

A distinctive, very small echinoid. Oval, much-flattened test with short, closely set spines; up to 15 mm in length; mouth and anus on lower surface, mouth central; Aristotle's lantern present. Greenish-grey in colour, becoming brighter green when damaged. In denuded test the ambulacral plates form a petal-like arrangement.

E. pusillus is widely distributed in north-west Europe and found all around Britain in coarse sand or gravel from extreme low water into the subittoral to depths of about 200 m. It feeds on a wide variety of small organisms and detritus. Breeding occurs in summer. The sexes are separate and fertilization external with the development of an echinopluteus larva.

Spatangus purpureus O. F. Müller Purple heart-urchin **Fig. 18.14**

Heart-shaped test broad and low; up to 120 mm in length; covered with large number of shortish spines, mostly directed backwards. Mouth on undersurface towards anterior. Conspicuous anterior furrow on test. Violet in colour; long, curved spines on upper surface paler or white in colour. In denuded test the ambulacral plates form a petal-like arrangement.

S. purpureus has a wide but patchy distribution in north-west Europe. It is found all around Britain buried in coarse sand and gravel on the lower shore, extending into the sublittoral to depths of about 900 m. It is sometimes locally common. The bivalve *Montacuta substriata* (Montagu) is often commensal with *Spatangus*. The sexes are separate, fertilization is external and breeding occurs in summer. There is an echinopluteus larva.

Echinocardium cordatum (Pennant) Sea-potato, heart-urchin **Fig. 18.15; Plates 24a,b, 32b**

Heart-shaped test, usually 40–50 mm in length, but up to 90 mm; covered with large number of closely set spines, mostly directed backwards. Mouth on undersurface towards anterior. Conspicuous anterior furrow on test. In profile, highest point of test towards posterior. Yellow-brown colour. In denuded test the ambulacral plates form a petal-like arrangement.

E. cordatum is widely distributed in north-west Europe and occurs all around the coasts of Britain, extending to depths of over 200 m. The denuded tests of dead specimens are frequently washed ashore, particularly after storms. Living specimens are found on the lower shore buried to depths of about 150–300 mm in sandy

Figure 18.14 *Spatangus purpureus*, (a) dorsal view of test, (b) denuded test showing petal-like arrangement of ambulacral plates. ((b) After Forbes, 1841.)

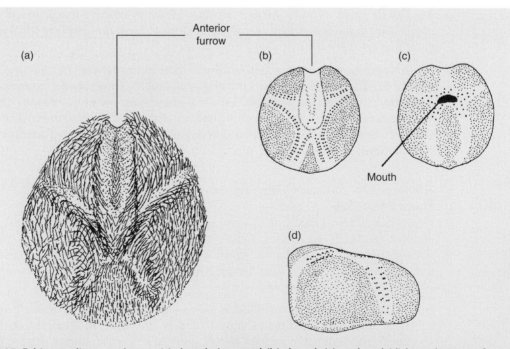

Figure 18.15 *Echinocardium cordatum*, (a) dorsal view, and (b) dorsal, (c) oral and (d) lateral views of denuded test.

sediments. Contact with the surface is maintained by a respiratory channel that leaves an opening on the surface of the sediment and reveals the position of the urchin. *Echinocardium* is a deposit feeder. Tube-feet pick up particles of sand and detritus from in front of the mouth and, in addition, long, highly extensible tube-feet extend up the respiratory funnel and sometimes pick up particles of sediment and detritus. The sexes are separate and fertilization external, with the development of an echinopluteus larva. Breeding occurs in summer with individuals from a littoral population on the north-east coast of England breeding for the first time when three years old. On the west coast of Scotland, breeding has been recorded at the end of the second year. Longevity is believed to be about 10 years. The bivalve *Tellimya ferruginosa* (p. 267) is often commensal with *Echinocardium*.

Genetic studies suggest that *Echinocardium cordatum* is a species complex.

Echinocardium pennatifidum Norman (not illustrated)

Heart-shaped test, up to 70 mm in length; covered with large number of closely set spines, mostly directed backwards. Mouth on undersurface towards anterior. No anterior furrow on test; in profile, highest point of test is central, cf. E. cordatum (above). Yellow-brown colour. In denuded test the ambulacral plates form a petal-like arrangement.

E. pennatifidum is widely distributed in north-west Europe and around Britain appears to be absent from the south-east coast. Although generally not common, it is found on the lower shore in large numbers in some areas, such as the Channel Islands. It extends to depths of 250 m.

Class Holothurioidea

The holothurians or sea-cucumbers are elongate, often worm-like echinoderms. They are without arms. The mouth is anterior, the anus posterior and the animal lies on one side, the ventral or trivium, which is often marked by three, longitudinal rows of tube-feet. The dorsal surface, or bivium, is often marked by two rows of tube-feet. In some holothurians, the tube-feet are scattered over the surface of the body, while in others they are absent. When relaxed, the mouth is seen to be surrounded by a ring of retractile, branched tentacles which are modified tube-feet. The skin is generally thick and tough. When dried and smoked it is known as 'trepang' or 'bêche-de-mer', which is regarded as a delicacy in China. The calcareous endoskeleton is much reduced, generally consisting of microscopic, calcareous plates which in some species, such as *Leptosynapta* (p. 425) and *Labidoplax* (p. 426), are in the shape of wheels and anchors (Fig. 18.20d). These plates are important in species identification. Holothurians are found buried in mud and sand, sometimes in U-shaped burrows, or attached to stones and rock. Although not generally common on the shore, they are sometimes abundant offshore, extending into the deep sea. They are either suspension feeders, using mucus-coated tentacles to trap suspended particles, or deposit feeders in which the tentacles push bottom sediments into the mouth. The sexes are generally separate, although hermaphroditic species are known. Many have a pelagic larva known as an auricularia followed by a doliolaria larva which metamorphoses into a juvenile sea-cucumber. Either the auricularia or the doliolaria may be omitted from the life-cycle. In some holothurians, the eggs are brooded, either in pouches on the outside of the body or internally.

Key to species of holothurians

Holothurians are generally recognized by the shape of the body, the number and shape of the tentacles, and the nature and distribution of the tube-feet. Specimens in which the tentacles are retracted can be relaxed by immersing in sea water containing 7% magnesium chloride. Where tube-feet are absent, the presence of calcareous spicules in the body wall confirms that the specimen is a holothurian and, if necessary, these can be isolated for examination under the microscope by digesting a small piece of skin in bleach.

1 Tube-feet well developed **2**
Without tube-feet. Worm-like body, may be tacky to the touch because of the presence of calcareous spicules projecting through skin **5**

2 Twenty short, stumpy, branched tentacles. Ventral tube-feet well developed, in 3, sometimes 4, longitudinal rows; reduced to conical papillae on dorsal surface. Up to 250 mm in length .... ***Holothuria forskali*** p. 422
Tentacles long, tree-like; usually 10 in number, 8 long, 2 short. Tube-feet in 5 longitudinal rows, single or double; some tube-feet may be scattered over surface ... **3**

3 Tube-feet in irregular, single rows. Skin thick, smooth, white in colour. No more than 40 mm in length .... ***Ocnus lacteus*** p. 424
Tube-feet in 5 rows, rows may be single or double **4**

4 Skin tough, wrinkled. Tube-feet in 5 double, occasionally multiple, rows .... ***Aslia lefevrei*** p. 424
Skin relatively thin and smooth. Tube-feet in 5 rows, 3 ventral rows double, with cylindrical tube-feet; 2 dorsal rows irregular with conical, papilla-like tube-feet .... ***Pawsonia saxicola*** p. 423

5 Twelve much-branched tentacles, the branches increasing in length to tip .... ***Leptosynapta inhaerens*** p. 425
Twelve tentacles, each with 2 pairs of branches at tip .... ***Labidoplax digitata*** p. 426

Holothuria forskali Delle Chiaje Cotton spinner **Fig. 18.16**
Cylindrical body up to 250 mm in length, flattened on ventral side. Skin soft and thick. Ventral tube-feet well developed, in 3 longitudinal rows; conical papillae on dorsal surface; 20 short stumpy, branched tentacles. Dorsal surface usually dark brown, black; ventral surface paler; tentacles yellowish.

Figure 18.16 *Holothuria forskali.*

H. *forskali* is found on the Atlantic coast of France and south and west coasts of Britain. It is occasionally found at extreme low water among rocks and stones, but is more common in soft sublittoral sediments and has been recorded at 300 m. When roughly handled, it discharges a mass of sticky threads, the Cuvieran organs or cotton glands, from which it gets its common name. On discharge, the threads disorientate and entangle would-be predators. An annual reproductive cycle has been described and in a sublittoral population off Brittany, spawning occurs in April. The sexes are separate and fertilization external, and there is a pelagic auricularia larva. Breeding occurs in summer.

Pawsonia saxicola (Brady & Robertson) (*Cucumaria saxicola*) Sea-gherkin **Fig. 18.17**
Cylindrical body up to about 150 mm in length. Skin relatively thin and smooth. Tube-feet in 5 rows; 3 ventral rows most distinct and double, with cylindrical tube-feet; 2 dorsal rows irregular, single, with conical, papilla-like tube-feet; 10 tentacles, tree-like; 8 long, 2 short. Body white, grey, sometimes blackish; tentacles blackish. Darkens on exposure to light.

Figure 18.17 *Pawsonia saxicola.*

P. saxicola occurs on the south coast of England, the south and west coast of Ireland and the Atlantic coast of France. It lives in crevices, under stones and on mussel beds, from the lower shore to 50 m. It is often found with *Aslia* (below) with which it may be confused, but *Pawsonia* has smoother skin and a different arrangement of tube-feet. It is a suspension feeder. Breeding is believed to occur during spring. There is a pelagic larva.

Aslia lefevrei (Barrois) (*Cucumaria normani*) **Fig. 18.18**

Cylindrical body up to about 180 mm in length. Skin tough, wrinkled. Tube-feet in 5 double, occasionally multiple, rows; 10 tentacles, tree-like; 8 long, 2 short. Body brownish colour, tentacles blackish; darkens on exposure to light.

Figure 18.18 *Aslia lefevrei.*

A. lefevrei occurs on the Atlantic coast of France, the south and west coasts of England, the west of Ireland and the west coast of Scotland to Orkney. It lives in crevices and under stones, and has been recorded in dense aggregations of more than 70 per square metre. It extends from the lower shore into the sublittoral to depths of about 30 m. *Aslia* is a suspension feeder. Suspended material is trapped on extended tentacles, equipped with adhesive papillae, and the tentacles are then pushed into the mouth where the food is removed. On the west coast of Ireland, breeding takes place in early spring. The sexes are separate and fertilization external. There is a pelagic larva and, following metamorphosis, a size of 30–35 mm body length is achieved in the first year, with growth of up to 80 mm after two and a half years. A body length of 180 mm has been recorded for specimens living in the sublittoral, and it has been suggested that longevity may be more than 10 years.

Ocnus lacteus (Forbes & Goodsir) (*Cucumaria lactea*) **Fig. 18.19**

Cylindrical body up to 40 mm in length, usually smaller. Skin thick, smooth. Tube-feet in irregular, single rows; 10 tentacles, tree-like; 8 long, 2 short. Body white in colour, tentacles yellowish.

O. lacteus is widely distributed in north-west Europe from Iceland to the Atlantic coast of France. In Britain it is apparently absent from southern North Sea coasts. It is found on the lower shore on stones and shells, often among hydroids and *Corallina* (p. 43), extending sublittorally to about 100 m. It is a suspension feeder.

Figure 18.19 *Ocnus lacteus.*

Leptosynapta inhaerens (O. F. Müller) **Fig. 18.20a**

Elongate, worm-like body, generally up to 150 mm in length, may reach 300 mm. Skin soft, semi-transparent; adhesive to touch due to presence of tiny, anchor-shaped calcareous plates which project through skin. Without tube-feet; 12 (sometimes 11 or 13) much-branched tentacles, the branches increasing in length to the tip. Body pale-pink to greyish-white in colour.

Figure 18.20 (a) *Leptosynapta inhaerens*, tentacle (after Eales, 1950). (b) *Labidoplax digitata*. (c) *Labidoplax digitata*, tentacle. (d) *Labidoplax digitata*, calcareous plates from body wall.

L. inhaerens is widely distributed in north-west Europe. Around Britain it is found on the west coast as far north as the Shetlands, and on the north-east coast of Scotland. It lives buried in sand and mud from the lower shore to depths of about 50 m. There is some confusion in the literature about breeding in this species, but there are reports of it breeding in spring with the development of a doliolaria larva.

Labidoplax digitata (Montagu) **Fig. 18.20b,c,d**
Body elongate, worm-like, usually 150 mm in length, may be up to 300 mm. Skin soft, adhesive to touch due to presence of tiny, anchor-shaped calcareous plates which project through skin. Without tube-feet; 12 tentacles, each with 2 pairs of branches at tip. Pink or brownish in colour.

L. digitata is found on the Atlantic coast of France and south-west and western coasts of Britain in muddy-sand from the lower shore sublittorally to depths of about 70 m. Breeding occurs in summer. The sexes are separate and fertilization external with the development of an auricularia larva.

References

Clark, A. M. & Downey, M. E. (1992). *Starfishes of the Atlantic*. London: Chapman & Hall.

Forbes, E. (1841). *A history of British starfishes, and other animals of the class Echinodermata*. London: John Van Voorst.

Mortensen, T. (1927). *Handbook of the echinoderms of the British Isles*. London: Oxford University Press. (Reprinted 1977. Leiden: E. J. Brill.)

Picton, B. E. (1993). *A field guide to the shallow-water echinoderms of the British Isles*. London: Immel Publishing.

Southward, E. C. & Campbell, A. C. (2006). Echinoderms. *Keys and notes for the identification of British species*. Synopses of the British fauna (New Series), no. 56. Shrewsbury Field Studies Council.

Hemichordata

The hemichordates are marine animals with the body divided into three regions: proboscis, collar and trunk. Some are sessile and colonial and live in deep water, but others are worm-like and found intertidally burrowing in soft sediments.

Phylum HEMICHORDATA

Class Enteropneusta	**The acorn worms. Solitary, unsegmented, worm-like hemichordates, often burrowing in mud and sand. Also found under rocks and stones. The body is clearly divided into a proboscis, a short collar and a long trunk.**
Class Pterobranchia	Sessile hemichordates, no more than 12 mm in length, usually found in deep water. Colonial. Tube-dwelling.
Class Planctosphaer-oidea	Spherical body, jelly-like; pelagic. Affinities poorly understood, may be the larval stage of an enteropneust.

Class Enteropneusta

The enteropneusts have a soft, unsegmented, worm-like body divided into three regions: a conical, ciliated proboscis, a short, cylindrical collar and a long trunk. The mouth lies on the ventral side between the proboscis and collar and the anus is terminal. The anterior region of the trunk has pairs of gill openings on the dorsal surface. Some species live in U-shaped burrows in sandy and muddy deposits and ingest large quantities of sediment. Others are found under rocks and stones where they suspension feed using the cilia on the proboscis to create feeding currents. The sexes are separate, fertilization is external and the characteristic pelagic larval stage, the tornaria, resembles the larva of some echinoderms. Some species show direct development and asexual reproduction by fragmentation has been recorded.

A small number of species is found in north-west Europe. In Britain they have a patchy distribution, occurring mostly on west and south coasts. They are readily separated from other worm-like animals by the division of the body into three distinct regions. Enteropneusts are fragile and must be handled with care if fragmentation is to be avoided. They are not common intertidally; a single species is described here.

Figure 19.1 *Saccoglossus ruber* (after Brambell & Cole, 1939).

Saccoglossus ruber Tattersall (*Saccoglossus cambrensis*) **Fig. 19.1**
Proboscis long, up to 30 mm; pink to pale orange in colour; collar longer than broad, reddish in colour; trunk long and coiled; fragile, with 50–95 pairs of gill openings; orange, yellow in colour, with small red spots. Whole animal up to 200 mm in length, generally smaller.

S. *ruber* has been recorded from a number of localities on the west coasts of Ireland, Scotland and Wales. It lives in U-shaped burrows on the lower shore in sand and muddy-sand. Coiled faecal casts are left on the surface of the sediment and the burrows extend to depths of about 200 mm. Breeding probably takes place in spring and summer.

Chordata

The chordates include such diverse forms as the sea-squirts found attached to rocks and stones on the lower shore, and the fishes, amphibians, reptiles, birds and mammals. Although there is great diversity of structure within the phylum, chordates are united by four features present in at least some stage in the life-cycle. These are a notochord, a hollow dorsal nerve cord, pharyngeal clefts and a post-anal tail containing extensions of the nerve cord and notochord. The notochord is a dorsal, flexible rod which has a skeletal function and may persist throughout life. It is often completely or partly replaced by cartilaginous or bony material to form a backbone. The nerve cord is dorsal to the notochord and often expanded anteriorly to form a brain, but in some groups, such as the sea-squirts, it is, like the notochord, lost in the adult. The pharyngeal clefts (sometimes known as gill slits), lead from the cavity of the pharynx to the exterior and have a variety of functions: in aquatic chordates they may be used in both feeding and respiration, but are generally reduced or lost in the adult terrestrial chordate. The post-anal tail persists in many aquatic vertebrates such as the fishes as a muscular, flexible structure which is important in locomotion.

There is considerable debate concerning the classification of the phylum; the scheme followed here is based on Nelson (2006).

Phylum CHORDATA

Subphylum Urochordata (Tunicata)	The tunicates. All are marine. Larvae, but not usually adults, possess chordate characteristics.

Class Ascidiacea	Sessile tunicates, the sea-squirts.

Class Thaliacea	Translucent, planktonic tunicates. Includes the salps.

Class Appendicularia (Larvacea)	Very small, planktonic tunicates in which the adults show some larval characteristics.

Subphylum Cephalochordata (Acrania)	The lancelets (amphioxus). Free-swimming, fish-like chordates.

Subphylum Craniata	**Chordates possessing a head. With the exception of the class Myxini all have a vertebral column made up of a large number of separate units, the vertebrae. Show wide diversity of body form and habit.**

Class Myxini	The hagfishes. Eel-like, primitive craniates lacking jaws and paired fins. No vertebrae; skeleton of cartilage. Marine.

Class Petromyzontida	The lampreys. Eel-like primitive vertebrates lacking jaws and paired fins. Skeleton of cartilage. Freshwater and coastal, moving into freshwater to breed.

Class Chondrichthyes	Skates, rays and sharks. Fishes with a cartilaginous skeleton.

Class Actinopterygii	**The ray-finned fishes. Bony skeleton, body usually covered by scales. Gills open into a space covered by a flattened plate the operculum.**

Class Sarcopterygii	The coelacanths, lungfishes and tetrapods.

Subphylum Urochordata

Class Ascidiacea

Ascidians are widely distributed on the lower shore and are characteristically found in crevices and attached to rocks, stones and seaweeds. They also occur on soft sediments and some species are adapted to living in the tiny spaces between sand grains. The body is surrounded by a jelly-like coating, or test, which varies considerably in thickness. It may be hard or soft and is sometimes encrusted with sand and fragments of shell. Ascidians show a wide range of colours as the test is sometimes pigmented; however, it is often translucent, the colours of the internal organs being visible. Some species are solitary, others colonial, and the generalized body structure is shown in Fig. 20.1.

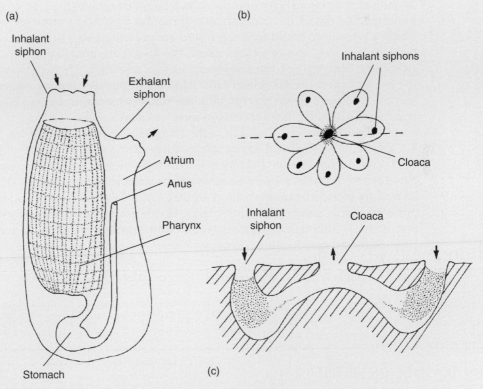

Figure 20.1   (a) Generalized body form of a solitary ascidian based on *Ciona*. (b) Surface view of compound ascidian such as *Botryllus* showing arrangement of zooids round a common cloaca. (c) Diagrammatic representation of a compound ascidian seen in section along dotted line in (b). Arrows indicate flow of water.

In the solitary species, the free end of the body has an inhalant or oral siphon through which water enters and an exhalant or atrial siphon through which water leaves the body. The water current is maintained by the beating of cilia. Water entering the body passes through the pharynx, the walls of which are perforated by rows of small slits or stigmata, and into a space surrounding the pharynx, the atrium, before leaving through the exhalant siphon. Ascidians are suspension feeders and as the water passes through the pharynx, suspended detritus and plankton are trapped in mucus which is then carried by cilia to the oesophagus. When the animal is disturbed, rapid contractions of longitudinal muscles in the body wall lead to the expulsion of a jet of water and it is from this that the group derives its common name.

The basic organization of colonial ascidians is essentially as described above. In some, the individuals or zooids are united at the base by a stolon. In others, the colony forms flat encrusting growths in which the zooids are embedded in a common gelatinous test. Here, the exhalant siphons of several zooids lead into a common chamber, the cloaca, opening to the exterior via a common exhalant opening (Fig. 20.1b,c). These forms are known as compound ascidians.

Ascidians are generally hermaphroditic with self-fertilization, although some rely on cross-fertilization. In most solitary forms eggs and sperm are liberated into the sea where fertilization and development to the larval stage take place, while in compound forms the eggs are generally held in the atrium until the larvae are fully developed. The typical larva is known as a tadpole larva and, as the name implies, it has a flexible tail as the main organ of propulsion. It is the structure of the tail, in which a notochord and dorsal nerve cord are present, which betrays the relationship between ascidians and other chordates. The larvae are non-feeding and have a free-swimming life varying from a few minutes to a few days. The selection of a suitable site for settlement is their primary role and they often settle among adults of the same species, leading to dense aggregations. Asexual reproduction by budding is common and reproduction by fission occurs in some colonial species.

Ascidians are distributed from the intertidal region to the deep sea. Some species are abundant on sheltered shores while others thrive in wave-swept gullies, on rock faces and on laminarian holdfasts. Encrusting colonial forms are often strikingly patterned with bright colours. Some ascidians are tolerant of salinities down to about 25 and are found in estuaries; others are well known as fouling organisms, often being present in large numbers on ships' hulls, floating buoys and quays and aquaculture installations. Although it is not easy to determine the age of ascidians, particularly the colonial forms, length of life is generally one or two years. Ascidians have a number of predators, including gastropod molluscs, echinoderms and fishes, but the body wall of some species is highly acidic rendering them unpalatable to predators.

Key to species of ascidians

Although some ascidians can be identified on the basis of external morphology, accurate identification of others is difficult and requires dissection followed by examination of the detail of the pharynx and the position of the gut and gonads. The following key to the common intertidal species is based on growth form and distinctive morphological features although it should be borne in mind that growth form and colour are variable, and some individuals are highly contractile and this might render the characters difficult to see. Dissection is not required.

1   Individuals or zooids occur singly or aggregated in clusters; sometimes attached
    to one another at base or along sides; may be joined at base by stolon **2**
    Zooids embedded in common test; growth form encrusting or as upright lobes
    often encrusted with sand **11**

2   Zooids transparent, very delicate; joined at base by stolon; zooids with white ring
    round pharynx; no more than 20 mm in height .... ***Clavelina lepadiformis*** p. 434
    Not as above **3**

3   Individuals may occur singly or aggregated in dense clusters or sheet-like
    aggregations, attached at base or along sides. Firm to the touch; usually no
    more than 10–20 mm in height; reddish-brown colour, without adhering sand **4**
    Individuals occur singly; attached to substratum at base or along one side.
    Cylindrical or rounded in shape; may have adhering sand grains; up to 120 mm
    in height **5**

4   Zooids in sheet-like aggregations; joined at base and along sides; siphons relatively
    inconspicuous. Up to 10 mm in height .... ***Distomus variolosus*** p. 441
    Individuals may occur singly or aggregated in dense, rather bumpy clusters; body
    shape varies from flattened and rounded to elongate and cylindrical; siphons
    conspicuous. Up to 20 mm in height .... ***Dendrodoa grossularia*** p. 440

5   Test very soft to the touch, gelatinous, translucent; highly contractile. Attached
    at base; siphons conspicuous, with red spots .... ***Ciona intestinalis*** p. 437
    Test to varying degrees rough; may have projections, papillae or fibrils; sometimes
    with adhering sand grains **6**

6   Body rounded; test thin, with fibrils and adhering sand; siphons at free end;
    attached to substratum at base, or unattached .... ***Molgula* spp.** p. 442
    Not as above **7**

7   Body club-shaped, up to 120 mm in height; with narrow stalk about equal to
    length of body. Test tough, folded .... ***Styela clava*** p. 440
    Body more or less cylindrical, attached to substratum at base or along one side **8**

8   Inhalant siphon at free end; exhalant siphon about 1/2 to 2/3 of way down body **9**
    Inhalant siphon at free end; exhalant siphon no more than 1/3 of way down body **10**

9   Test thick, uneven, mostly attached by part of one side; translucent.
    Siphons small; inhalant at free end, exhalant 1/2 to 2/3 of way down body ....
    .... ***Ascidia mentula*** p. 439
    Test attached by most of one side; thin and rough, often with attached
    shell. Inhalant siphon terminal, exhalant about 2/3 of way down body
    .... ***Ascidia conchilega*** p. 439

10 Test stiff and rough to touch, with small papillae. Attached at base. Inhalant siphon
    at free end; exhalant about 1/3 of way down body .... ***Ascidiella aspersa*** p. 438
    Test stiff, reddish papillae at base of siphons, otherwise smooth. Inhalant
    siphon terminal, exhalant nearly so. May be attached by most of one side ....
    .... ***Ascidiella scabra*** p. 438

11 Colonies rounded, club-shaped or flat-topped, up to 50 mm in height; may be
    encrusted with sand **12**

Colonies encrusting on rocks and seaweed; may be brightly coloured and
gelatinous or hard; bluish, purplish, marked with dark lines **13**

12 (a) Club-shaped lobes with distinct narrow stalk; stalk red in colour and encrusted
with sand. Rounded head without sand. Each zooid with 4 orange-red spots.
Inhalant siphons of zooids with 8 lobes .... ***Morchellium argus*** p. 435
or
(b) Colonies flat-topped lobes or club-shaped with narrow stalks; joined at
base by creeping stolon; test semi-transparent with conspicuous exhalant
openings; inhalant siphons of zooids with 8 lobes. Not encrusted with sand
.... ***Aplidium turbinatum*** p. 436
or
(c) Colonies of rounded lobes, sometimes flat-topped, narrowing towards base;
forming irregular masses usually encrusted with sand; each lobe with 1 or
more conspicuous exhalant openings; inhalant siphons of zooids with 6 lobes
.... ***Polyclinum aurantium*** p. 435
or
(d) Colony stalked, club-shaped or broad-based, flat-topped. Exhalant
openings large; inhalant openings with 6 lobes with white pigment ....
.... ***Aplidium proliferum*** p. 436

13 Colony thin, encrusting, sheet-like or thrown into small lobes; hard. Greyish,
bluish, purplish in colour; marked with dark lines .... ***Didemnum* spp.** p. 436
Colony generally thinnish, gelatinous; individual zooids arranged around
exhalant openings in star-shaped or linear patterns; often brightly coloured **14**

14 Zooids arranged in star-shaped patterns with common exhalant opening
in centre .... ***Botryllus schlosseri*** p. 441
Zooids arranged around elongate exhalant opening ...... ***Botrylloides leachii*** p. 442

Figure 20.2 *Clavelina lepadiformis.*

Clavelina lepadiformis (O. F. Müller) Light bulb
ascidian **Fig. 20.2**
*Colonial; groups of very delicate zooids joined at base
by stolon; up to 20 mm in height. Pharyngeal region
short. Siphons close together. Zooids transparent with
pale yellow, pink and brown lines; white ring around
pharynx.*

C. lepadiformis is widely distributed in north-west
Europe and occurs all around Britain. In some areas it
is common on the lower shore attached to seaweed,
rocks and stones, extending into the sublittoral to depths
of about 50 m. It is tolerant of salinities as low as 14. On
the west coast of Scotland breeding has been recorded
during summer. The eggs are fertilized in the atrium,
where development to the tadpole larval stage takes
place. After release the larvae are free-swimming for
only about three hours before settling to establish new
colonies. Asexual reproduction by budding also takes

place. Towards the end of summer, many of the adult zooids regress and the colony survives as winter buds from which new zooids develop in spring.

Polyclinum aurantium Milne-Edwards **Fig. 20.3a**
Colonial; zooids embedded in common test and forming rounded lobes, each narrowing towards the base and often with a flat top. Up to 30 mm across and 15 mm in height. Zooids irregularly arranged round conspicuous exhalant openings. Inhalant siphons with 6 lobes. Yellow-brown colour, usually encrusted with sand.

P. aurantium is widely distributed in north-west Europe. It is found on the lower shore in gullies and attached to rocks and seaweed, particularly in exposed areas, extending to depths of about 100 m. On the west coast of Scotland breeding occurs in spring and summer. The eggs are fertilized in the atrium where development to the tadpole larva takes place. After hatching, the larvae have a short free-swimming life of about two or three hours before settling. Asexual reproduction by budding also takes place, usually during winter.

Morchellium argus (Milne-Edwards) (*Aplidium argus*) **Fig. 20.3b**
Colonial; zooids embedded in common test forming club-shaped lobes with distinct head and stalk. Inhalant siphons with 8 lobes. Stalk up to 40 mm in height, encrusted

Figure 20.3 (a) *Polyclinum aurantium.* (b) *Morchellium argus* (after Millar, 1970). (c) *Aplidium turbinatum.* (d) *Aplidium proliferum.* ((a), (c), (d) After Alder & Hancock, 1912.)

with sand; red in colour. Head up to 30 mm across, not encrusted with sand; yellowish colour, each zooid with 4 orange-red spots around base of inhalant siphon.

M. *argus* has been recorded from the west coast of France and south and west Britain, extending as far north as the west coast of Scotland. It is found on the lower shore attached to rocks and stones, often on the underside of rocky overhangs in wave-swept situations where it is sometimes abundant. Breeding occurs in summer. Fertilized embryos are held in the atrium where they develop to tadpole larvae with a brief free-swimming life of two or three hours. Asexual reproduction by budding generally takes place during late summer and autumn.

Aplidium turbinatum Savigny (*Sidnyum turbinatum*) **Fig. 20.3c**
Colonial; zooids embedded in common test to form rounded, sometimes club-shaped lobes, generally flat-topped, narrowing at base; 10–20 mm in height. Lobes joined at base by creeping stolon; exhalant openings conspicuous, surrounded by 6–12 zooids. Inhalant siphons with 8 lobes. Test semi-transparent; zooids with white markings, occasional red spots. Not encrusted with sand.

A. *turbinatum* is widely distributed in north-west Europe and Britain where it is most common in the south and west. It is found on rocks, stones and seaweed from the lower shore to depths of about 200 m. On the west coast of Scotland breeding has been recorded during spring and summer. The fertilized eggs are held in the atrium and released as tadpole larvae.

Aplidium proliferum (Milne-Edwards) **Fig. 20.3d**
Colonial; zooids embedded in common test. Two growth forms recognized: club-shaped lobes up to 50 mm in height, and broad-based, flat-topped colonies 5–10 mm thick and up to 60 mm across. Zooids irregularly arranged round, large exhalant openings. Inhalant siphons of zooids with 6 lobes, distinctly white pigmented. Some colonies yellowish-white in colour, others pink.

A. *proliferum* is widely distributed in north-west Europe. It is sometimes common on the west coast of Britain but is absent from North Sea coasts. It is found on rocks, stones and seaweed, extending from the lower shore into the shallow sublittoral.

    Some authors regard the two growth forms as separate species, the club-shaped colonies being A. *proliferum* and the flat-topped colonies *Aplidium nordmanni* (Milne-Edwards).

Didemnum spp. **Fig. 20.4a**
Colonial; thin, encrusting, sheet-like or thrown into lobes; surface hard, smooth, often with dark lines; individual zooids small, difficult to distinguish, irregularly arranged around a common exhalant opening. Siphonal openings inconspicuous in contracted specimens. Colour varied, white, grey, bluish, purple.

The genus *Didemnum* is widely distributed around Britain, particularly on south and west coasts, and in some situations may be very common on the lower shore encrusting stones, rock, laminarian stipes and a wide range of seaweeds. It extends into the sublittoral. The different species are difficult to identify and microscopic examination is required. In some cases the species can only be separated on the basis of the morphology of the ascidian tadpole larva.

    The genus has attracted the attention of marine ecologists worldwide over the past decade as one species is a very successful, and unwelcome, invader. Its

Figure 20.4   (a) *Didemnum* sp. (b) *Ciona intestinalis*.

morphology varies widely from surface encrustations to long, rope-like colonies and in many reports the invading species has been referred to as *Didemnum* sp. Recent morphological and genetic studies, however, suggest that it is **Didemnum vexillum** Kott, a fast-growing species that rapidly covers hard substrata, smothering the indigenous fauna with the potential of disrupting local ecosystems. It is also found on aquaculture installations where there is the danger that shellfish are smothered. *D. vexillum* has been recorded worldwide from the intertidal to depths of 65 m.

Ciona intestinalis (Linnaeus) **Fig. 20.4b**

Solitary, attached at base; body elongate with soft gelatinous, translucent test. Longitudinal bands of muscle in body wall. Up to 120 mm in height. Siphons large, conspicuous with delicate yellow margins; inhalant at free end, with 8 lobes with reddish spots between them; exhalant siphon just posterior to inhalant, 6-lobed, with reddish spots. Pharyngeal region extensive. Yellow-greenish colour, sometimes orange. Ciona *is highly contractile and in contracted specimens some of the characters are difficult to see.*

C. intestinalis is widely distributed in north-west Europe and although generally not common, in some situations it may be abundant reaching densities of up to 5000 per square metre. It is found on the lower shore in dense groups attached to the underside of stones, rocks and seaweed, and also commonly occurs in harbours attached to piers, and on ships' hulls and buoys. It extends into the sublittoral to depths of over 500 m. *Ciona* is tolerant of reduced salinity and breeding populations have been recorded at salinities of 11. It is a protandrous hermaphrodite. Spawning normally occurs just before dawn, the gametes being shed into the sea, where cross-fertilization takes place. In Britain breeding generally takes place in spring and summer but breeding throughout the year has been recorded at some localities. The larvae are free-swimming for up to 36 hours. Two generations per year have been recorded in Scandinavia and Britain, with this reduced to a single generation at some localities. On the west coast of Scotland, *Ciona* reaches sexual maturity at a body height of about 25–30 mm and longevity is one and a half years. It does not undergo asexual reproduction by budding.

Ascidiella aspersa (O.F. Müller) **Fig. 20.5a**

Solitary, attached at base; body more-or-less oval-shaped; test with small papillae, stiff and rough to touch; with attached detritus. Up to 100 mm in height. Siphons obvious; inhalant at free end, sometimes frilled; exhalant about 1/3 of way down body. Greyish-black or brown in colour.

A. aspersa is widely distributed in north-west Europe and Britain where it appears to be most common on south and west coasts. It is found on the lower shore and sublittorally to 80 m attached to stones, seaweed and pilings. It is tolerant of salinities down to 18 and is often common in estuaries. *A. aspersa* is a protandrous hermaphrodite. Gametes are discharged into the sea where fertilization and development of the tadpole larvae take place. Larvae settle during summer. On the west coast of Scotland, sexual maturity has been recorded at heights ranging from 25 to 50 mm and breeding takes place in the summer following settlement. On British coasts, longevity is one to one and a half years compared with two to three years in Norwegian waters. The small bivalve *Modiolarca subpicta* (p. 258) is occasionally found as a commensal in the test.

Figure 20.5   (a) *Ascidiella aspersa*. (b) *Ascidiella scabra*.

Ascidiella scabra (O.F. Müller) **Fig. 20.5b**

Solitary, may be attached by most of one side; body oval-shaped; test stiff, with reddish papillae at base of siphons. Up to 50 mm in height. Siphons obvious, inhalant at free end, exhalant nearly so. Semi-transparent with some reddish markings.

A. scabra is widely distributed in north-west Europe and Britain. It lives on the lower shore attached to seaweeds, rocks and shells, extending to depths of about 300 m. *A. scabra* is a protandrous hermaphrodite. The gametes are discharged into the sea where fertilization and development of the tadpole larvae take place. Breeding occurs during summer; on the west coast of Sweden, one generation a year has been recorded. Longevity is about one year.

Ascidia mentula O. F. Müller **Fig. 20.6a**
Solitary, mostly attached by part of one side; body oblong; test thick, uneven. Up to 100 mm in height, may be more. Siphons small; inhalant at free end, with 8 whitish lobes; exhalant 1/2 to 2/3 of way down body. Test translucent, pinkish in colour, sometimes greyish.

A. mentula is widely distributed in north-west Europe and around Britain where it is most commonly found on south and west coasts, extending sublittorally to depths of about 200 m. Intertidally it occurs on the lower shore, particularly in shaded gullies and crevices on rocky shores, sometimes attached to laminarian holdfasts. It is also found attached to shells and stones on sand and mud. It survives in salinities down to 20. At a number of localities breeding has been recorded throughout the year. Fertilization is external and the tadpole larva has a short pelagic life. A longevity of seven years has been recorded for a population on the west coast of Sweden. The small bivalve *Modiolarca subpicta* (p. 258) is sometimes found as a commensal in the test.

Figure 20.6 (a) *Ascidia mentula*. (b) *Ascidia conchilega* (after Alder & Hancock, 1905).

Ascidia conchilega Moeller **Fig. 20.6b**
Solitary, body flattened, oblong; attached by most of one side; test thin and rough; may be encrusted with sand and shell fragments. Up to 60 mm in length, generally smaller. Inhalant siphon at free end, with 8 red spots; exhalant siphon about 2/3 of way along body, with 6 red spots. Test translucent, greenish colour.

Figure 20.7 *Styela clava.*

A. conchilega is widely distributed in north-west Europe and Britain on stones, shells and, occasionally, seaweeds. It is found on the lower shore and extends into deep water. Fertilization is external. There is a tadpole larva in the life-cycle.

Styela clava Herdman (*Styela mammiculata*) Leathery sea-squirt **Fig. 20.7**

Solitary, attached by narrow stalk widening out to cylindrical body up to 120 mm in height. Siphons close together at free end, each marked with 4 brown stripes. Test tough and leathery, much folded. Brown, reddish in colour.

S. clava is an introduced species first recorded in Britain from Plymouth in 1953. It is believed to have been introduced from the north-western Pacific on the hulls of ships and has since spread along the west coast to Scotland and along the English Channel. It has also been recorded from Ireland and is now widely distributed in north-west Europe where it is believed to have been spread through the transport of juvenile oysters. It is most often found in sheltered, fully marine sites, although it can tolerate salinities down to about 20. *S. clava* is fast growing and densities as high as 1000 individuals per square metre have been recorded. Reproduction only occurs at temperatures above about 15 °C.

Dendrodoa grossularia (van Beneden) **Fig. 20.8a**

Solitary or aggregated in dense, bumpy clusters. Firm to the touch; shape varied, flattened and rounded in solitary individuals; usually upright, elongate and cylindrical in aggregated individuals. Siphons conspicuous, noticeably extended upwards from free end, apertures rounded. Up to 20 mm in height. Reddish-brown colour. Easily confused with Distomus variolosus *(below).*

D. grossularia is widely distributed in north-west Europe and Britain, occurring on the lower shore and extending sublittorally to depths of about 600 m. It is found on a variety of substrata such as rocks, shells and seaweeds. Solitary individuals have a wide but patchy distribution in Britain and are found on most coasts, while aggregates are especially common in the south and west. Clusters of *Dendrodoa* have a superficial resemblance to *Distomus variolosus* (below). Dense groups of individuals result

(a)   (b)

Figure 20.8   (a) *Dendrodoa grossularia.* (b) *Distomus variolosus.*

from the gregarious settlement of larvae, young often settling on older individuals. Asexual reproduction by budding has not been recorded in *Dendrodoa* (unlike *Distomus variolosus*, below). On the west coast of Scotland breeding occurs throughout summer and autumn, while spring and autumn peaks in breeding have been recorded in a sublittoral population off the south-east coast of Britain. Fertilization takes place in the atrium where the eggs are held and released as tadpole larvae. Newly settled young grow to a height of about 3–3.5 mm before their first winter, reaching maximum size the following year. Longevity is one and half to two years.

Distomus variolosus Gaertner **Fig. 20.8b**

Colonial, encrusting in sheet-like aggregations. Zooids firm to the touch, joined at base and along sides. Oval to cylindrical in shape; up to 10 mm in height. Reddish-brown colour, often sheen-like appearance. The siphons are less conspicuous than in Dendrodoa grossularia *(above) but the 2 species are easily confused.*

D. variolosus is found on the lower shore and shallow sublittoral attached to rocks, stones and hydroids, and on laminarian holdfasts and stipes. It has been recorded on south coasts of Britain and the west of Ireland, but at more northerly localities it may have been confused with *Dendrodoa grossularia* (above) and confirmation of records is needed. Dense aggregations of *Distomus* result from asexual budding but the colony does not form the thick clumps and masses so characteristic of *Dendrodoa*.

Botryllus schlosseri (Pallas) Star ascidian **Fig. 20.9a**

Colonial; zooids embedded in common test to form colonies which are very varied in form, usually thin encrusting sheets, but thick, or lobed when living in deep or calm water. Test gelatinous. Individual zooids arranged in groups of 3–12 round exhalant opening to give star-shaped patterns. Colonies vary in size from a few millimetres to 150 mm and more across. Colour varied, includes shades of blue,

Figure 20.9 (a) *Botryllus schlosseri*. (b) *Botrylloides leachii*, both on rock.

yellow and brown; zooids often contrasting in colour with test. In some locations a particular colour morph may be abundant.

B. schlosseri is widely distributed and sometimes abundant in north-west Europe. It is found all around the coast of Britain on stones and seaweeds on the lower shore and into the shallow sublittoral. Surviving in salinities down to about 25, it is found in harbours and estuaries. On the west coast of Scotland breeding occurs during spring and summer. Eggs are fertilized in the atrium where they develop to tadpole larvae which escape through the common exhalant openings. The free-swimming larvae survive for up to about 16 hours before settling to form new colonies. In Britain there are one or two generations each year. Asexual reproduction by budding also occurs and appears to take place at any time of year. Longevity is one to one and a half years. *Botryllus* and *Botrylloides* (below) are preyed on by cowries (*Trivia* spp., p. 227), which in addition to feeding on the ascidians also lay their egg capsules in holes bitten out of the test.

Botrylloides leachii (Savigny) (*Botryllus leachi*) **Fig. 20.9b**

Colonial; zooids embedded in common test to form thin, encrusting colonies. Test gelatinous. Individual zooids arranged round elongate, exhalant opening, and not in star-shaped patterns (see Botryllus, *above). Colour varied but generally yellow, orange or grey.*

B. leachii has a similar distribution to that of *Botryllus schlosseri* (above) and is found on rocks and seaweed on the lower shore and in the shallow sublittoral. Colony form and colour are less varied than in *Botryllus*. On the west coast of Norway, colonies have been recorded in salinities as low as 16. Breeding occurs during summer and the tadpole larvae escape through the common exhalant openings. Asexual reproduction by budding also takes place. Longevity is about two years.

Molgula spp.

Several species of *Molgula* are found on the lower shore and in the shallow sublittoral. They occur individually or in groups, attached to stones, shells and seaweed, but some are unattached. The test bears adhering sand and shell to varying degrees and in some species only the siphons are free of adhering material. The species are difficult to separate.

Molgula manhattensis (De Kay) **Fig. 20.10**

Solitary, body rounded, attached at base; siphons long, both at free end; inhalant 6-lobed, exhalant 4-lobed. Test relatively thin, with fine fibrils and often with encrusting sand grains. Up to 30 mm across. Greenish-grey in colour.

M. manhattensis is widely distributed in north-west Europe and in Britain appears to be most common in the south and west. It is found from the lower shore to depths of about 90 m attached to stones, shells and seaweeds. It is also found on sand and survives salinities down to about 23. Fertilization is external, leading to the development of a tadpole larva. On the south coast of England breeding has been recorded throughout the year.

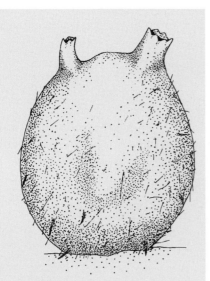

Figure 20.10 *Molgula manhattensis* (after Millar, 1970).

Subphylum Craniata

Class Actinopterygii

The actinopterygians or ray-finned fishes are frequently referred to as the bony fishes (Fig. 20.11) or Osteichthyes to distinguish them from the cartilaginous skates, rays and sharks. Also known as teleosts, they are found throughout the world's oceans from the intertidal region to the greatest depths. They have wide diversity of body form and habit and the class includes a wide range of species fished commercially such as the cod, haddock and plaice. As the name implies, actinopterygians have fins comprising a number of rays often displayed in a fan-like fashion. They usually have a bony (ossified) skeleton and typically are covered by bony scales, although the latter are often absent or reduced in size in shore fishes because of the danger of them being rubbed off or damaged, exposing sites of possible infection. When present, the number of scales counted in a row along the lateral line is sometimes important in identification. In some species, growth rings on the scales can be 'read' to give the age of the fish. The mouth is terminal and the body has a number of fins, supported by fin rays. Dorsal, anal and tail (caudal) fins are single, the pectorals and pelvics paired. Pectoral fins are usually found on the sides of the body just behind the head while the position of the pelvic fins varies. They have a ventrolateral position and in the more

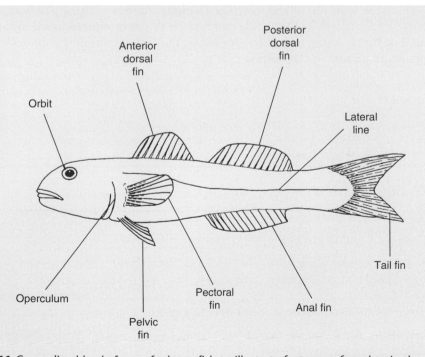

Figure 20.11 Generalized body form of a bony fish to illustrate features referred to in the text.

primitive bony fishes are situated close to the anus while in more advanced forms they lie in a much more anterior position, either slightly anterior or posterior to the pectoral fins. The fin rays of the primitive actinopterygians are soft, while in more advanced forms stiff spines are often present on dorsal, anal and pelvic fins. The arrangement of the fins and the number of rays and spines are important in identification. A well-developed lateral line system, sensitive to vibrations and changes in pressure, is often present, but in shore fishes it is generally reduced, presumably as a result of them living in wave-beaten situations where the effectiveness of such a system would be limited. In some groups, for example the gobies, the lateral line system is absent and sensory pores are found on the head. Most have a swim bladder, an outgrowth of the anterior region of the alimentary canal, which functions as a hydrostatic organ. In some shore fishes it is reduced or absent resulting in negative buoyancy, a further adaptation to living in turbulent, inshore waters. The gill openings are covered by a bony plate, the operculum, which is free along its posterior margin to allow water to leave after entering through the mouth and passing over the gills. Body shape varies widely.

The majority of fishes have separate sexes with external fertilization and the development of pelagic larvae which feed in the plankton. In some species there is a complex behaviour pattern associated with breeding whereby the eggs are laid in nests where the young develop. These nests are often guarded by one or both parents.

The term 'shore fishes', as used here, includes species typical of rock pools and damp crevices, such as the blennies and gobies. They are truly intertidal and highly adapted to the sharply fluctuating conditions found there. Also included are fishes which, although more typical of the shallow sublittoral, migrate in with the tide and frequently take refuge in crevices, under stones and among seaweed during low water. Other fishes, not included in the text, feed over the intertidal region during high water and retreat with the falling tide. For example, flatfish such as plaice and flounders move into estuaries with the tide and feed on the invertebrates living in the sand and mud deposits. Along with the shore fishes, they are important predators of intertidal plant and animal communities.

Key to families of fish

Characters referred to in the key are shown in Fig. 20.11.

1   Body very slender, elongate; divided into rings along length. Head extended
    into tube-like snout. Dorsal fin conspicuous, other fins reduced or absent
    .... **Syngnathidae** p. 451
    Body not divided into rings. Head not extended into snout **2**

2   Body long, slender, slimy. Dorsal fin continuous with anal fin. Scales tiny or
    absent; no pelvic fins. The eels **3**
    Not as above **4**

3 Anterior margin of dorsal fin some way behind pectoral fins. Lower jaw longer than upper .... **Anguillidae** p. 446
Anterior margin of dorsal fin just behind pectoral fins. Upper jaw longer than lower .... **Congridae** p. 447

4 Body elongate with forked tail fin. Dorsal fin long. No pelvic fins. Scales small .... **Ammodytidae** p. 466
Pelvic fins present. Sometimes reduced to spines or fused to form crescent-shaped disc (Fig. 20.34c) or wholly or partly modified to form conspicuous sucker (Fig. 20.23) **5**

5 Pelvic fins modified to form sucker or joined to form crescent-shaped disc **6**
Pelvic fins not as above **9**

6 Pelvic fins joined to form crescent-shaped disc, enabling fish to attach loosely to substratum; scales present; eyes large, protruding, dorsolateral in position. Two dorsal fins .... **Gobiidae** p. 469
Pelvic fins modified wholly or partly to form a conspicuous sucker **7**

7 Body of distinctive shape, rounded; with rows of bony tubercles along sides .... **lumpsuckers – Cyclopteridae** p. 456
Body without bony tubercles along sides; no scales **8**

8 Skin loose, prickly to touch, 1 long dorsal fin **sea-snails – Cyclopteridae** .... p. 456
Skin smooth. Head flattened, almost triangular in shape. Single dorsal fin, short, close to tail fin .... **Gobiesocidae** p. 449

9 Pelvic fins reduced to 1 long spine and a small ray, situated posterior to the base of the pectoral fins. Conspicuous row of spines anterior to dorsal fin. Without scales, but may have bony plates along sides of body. .... **Gasterosteidae** p. 450
Not as above **10**

10 Pelvic fins very small, reduced to small spines just below pectoral fins. Body long and slender, with row of black spots along base of long, slender dorsal fin .... **Pholidae** p. 463
Not as above **11**

11 Lips thick and pronounced. Body deep and laterally compressed. A single dorsal fin .... **Labridae** p. 459
Not as above **12**

12 Two dorsal fins, the anterior one spiny but not black in colour. Spines just in front of operculum. Without scales .... **Cottidae** p. 454
Not as above **13**

13 Two dorsal fins, anterior dorsal fin spiny, black in colour. Large spine on operculum. Mouth sloping sharply downwards. Scales present .... **Trachinidae** p. 463
Not as above **14**

14 Two dorsal fins, separate; anterior dorsal of 4 or more spines. Large scales. Found in coastal waters; young fish (up to c. 50 mm) in rock pools and estuarine creeks during summer **15**
Not as above **16**

15 Two dorsal fins, anterior with 8–10 spines, posterior with 1 spine and 11–14 rays .... **Moronidae** p. 457

Two dorsal fins, anterior with 4 spines, posterior with 1 spine and 8 rays. Thick upper lip .... **Mugilidae** p. 458

16 One or 2 dorsal fins; anterior dorsal may be reduced to 1 long ray with row of much smaller rays in groove behind. Fins not spiny, soft. Head usually with 1 or more barbels .... **Lotidae** p. 447
Single dorsal fin, long **17**

17 Dorsal fin continuous with tail and anal fin .... **Zoarcidae** p. 462
Tail fin separate from dorsal and anal fins. Dorsal fin divided into anterior and posterior portions by a depression .... **Blenniidae** p. 464

Family Anguillidae Eels

Body long and narrow, slimy. Dorsal fin continuous with anal fin, anterior margin some way behind pectoral fins. No pelvic fins. Eyes small. Lower jaw longer than upper. Scales tiny and embedded in the skin.

Anguilla anguilla (Linnaeus) Common eel **Fig. 20.12a,b**

Body elongate, skin slimy. Dorsal fin long, anterior margin well behind pectoral fins; continuous with tail and anal fin. Margin of pectoral fin rounded. Without pelvic fins. Lower jaw longer than upper, protuberant (cf. Conger conger, below). Eyes round, small. Scales tiny, embedded in the skin. Colour varied, dorsal surface brown, black to olive-green; ventral surface yellowish. Some specimens silvery. Up to about 1 m in length.

A. anguilla is widely distributed in the rivers and estuaries of north-west Europe and is frequently common on the shore. Part of the life-cycle is spent in fresh water and part in the sea. Males and females are believed to spend up to 12 and 20 years, respectively, in fresh water and at a length of 400–500 mm they take on a silvery colour and begin their migration to the Sargasso Sea where they spawn and die. Although it was believed that the leaf-like larvae (leptocephali) drifted in the plankton of the Atlantic

(a)

(b)

(c)

Figure 20.12 (a), (b) *Anguilla anguilla*. (c) *Conger conger*.

for two to three years before reaching European coasts, it is now understood that the larvae swim actively on a journey which takes about one year. At the end of the migration the leptocephali metamorphose into transparent young eels or 'glass eels'. These darken in colour and in British waters ascend freshwater streams in large numbers in late winter and spring. At this stage they are known as elvers, and are about 50 mm in length. Both elvers and mature eels are fished commercially. Research indicates a significant reduction in the number of eels in European waters over the past 20 years.

Family Congridae Conger eels
Body long and narrow, slimy. Dorsal fin continuous with anal fin, anterior margin just behind pectoral fins. No pelvic fins. Upper jaw longer than lower. Eyes oval-shaped. No scales.

Conger conger (Linnaeus) Conger eel **Fig. 20.12c**
Body elongate, skin slimy. Dorsal fin long, anterior margin just behind pectoral fins; continuous with tail and anal fin. Margin of pectoral fin pointed. Without pelvic fins. Upper jaw longer than lower (cf. Anguilla anguilla, *above). Eyes oval. Without scales. Dorsal surface grey-brown, ventral surface paler. Up to about 2 m in length.*

C. conger is widely distributed in north-west Europe where it is found in rock crevices on the lower shore and in deep water. It is common on south and west coasts of Britain but much less so on the east coast. It feeds on a wide variety of fishes and crustaceans. Maturity is reached at anything between five and 15 years and conger eels of 20 years of age have been recorded. Mature eels migrate to the Atlantic as far south as the Azores where they spawn. The larvae (leptocephali) have a long larval life, drifting from the spawning grounds, to be found on the shores of north-west Europe one or two years later. They do not enter fresh water.

Family Lotidae Rocklings
Fins soft, without spines. One or two dorsal fins; one anal fin. Anterior dorsal fin may be reduced to a single, prominent ray behind which is a row of very short rays. Pelvic fins anterior to pectoral fins. Head usually with one or more barbels. Scales very small. Includes common shore fishes, the rocklings.

Two dorsal fins, the anterior a single prominent ray behind which is a row of short rays in a groove. Head with 3 barbels, 1 on lower jaw and 1 close to each anterior nostril .... ***Gaidropsarus mediterraneus*** p. 447

Two dorsal fins, the anterior a single prominent ray behind which is a row of short rays in a groove. Head with 5 barbels, 1 on lower jaw, 2 on upper jaw and 1 close to each anterior nostril .... ***Ciliata mustela*** p. 448

Gaidropsarus mediterraneus (Linnaeus) Shore rockling **Fig. 20.13a**
Rather slender body. Two dorsal fins, anterior one a single prominent ray behind which is a row of very short rays lying in a groove. Posterior dorsal and anal fins long;

Figure 20.13 (a) *Gaidropsarus mediterraneus.* (b) *Ciliata mustela* (after Day, 1884).

15–17 rays in pectoral fin. Head with 3 barbels. Mouth extends to, or just beyond, posterior margin of orbit. Body covered with tiny scales. Dorsal surface uniform in colour, brown, reddish-brown or almost black; ventral surface paler. Length usually up to 250 mm.

G. *mediterraneus* is common on the north-west and north coasts of France and the south and west coasts of Britain, extending north to Orkney and southern Norway. It has been recorded sublittorally to 27 m but is mainly intertidal in pools and among seaweed on rocky shores. It feeds on a wide range of invertebrates and small fish. The head barbels are believed to be used to locate food and, contrary to earlier opinion, the groove and hair-like rays of the anterior dorsal fin apparently play little, if any, role in food detection. In South Wales maturity is reached in the second or third year. Breeding occurs offshore in summer and both the eggs and larvae are pelagic. Young are found in rock pools from late September.

Gaidropsarus vulgaris (Cloquet), the three-bearded rockling, is widely distributed in north-west Europe. It is similar to G. *mediterraneus* in having three barbels on the head and in the arrangement of the fins, but the pectoral fins have 20–22 fin rays. It is mainly salmon-pink in colour with darker patches. It is sublittoral, generally over rocky grounds, usually to depths of about 30 m but it has been recorded from deeper water.

Ciliata mustela (Linnaeus) Five-bearded rockling **Fig. 20.13b**
Rather slender body. Two dorsal fins, anterior one a single, prominent ray behind which is a row of very short rays lying in a groove. Posterior dorsal and anal fins long. Head small, with 5 barbels. Mouth extends just beyond posterior margin of orbit. Body covered with tiny scales. Dorsal surface brown, black or reddish; ventral surface paler. Length up to 250 mm.

C. mustela is widely distributed in north-west Europe and occurs all around Britain where it is often common intertidally in rock pools and among seaweed. It is also found over sandy and muddy shores and extends to depths of 20 m. It feeds on a variety of invertebrates and small fishes which are believed to be located by the head barbels. Breeding has been recorded in spring and summer, and in South Wales maturity is reached in the second year. Spawning occurs offshore but specimens with ripe ovaries have been recorded intertidally, suggesting that some spawning takes place on the lower shore. Both eggs and larvae are pelagic and young are found inshore in summer.

Family Gobiesocidae Clingfishes

Head flattened, almost triangular in shape. Single dorsal fin, short, close to tail fin. Pelvic fins modified, in part, to form a conspicuous sucker on the ventral surface. No scales.

Lepadogaster purpurea (Bonnaterre) (*Lepadogaster lepadogaster*) Shore clingfish, Cornish sucker **Fig. 20.14**

Body flattened. Dorsal, tail and anal fins continuous. Sucker on ventral surface. Head flattened, lips thick. Flap of tissue in front of eyes. Body without scales. Pink to red in colour; conspicuous blue spot on dorsal surface just behind each eye. Length up to 65 mm.

Figure 20.14 *Lepadogaster purpurea.*

L. purpurea is found on the south and west coasts of Britain, north to Scotland. It extends south to the Canary islands and Madeira and into the Mediterranean. Although most common on sheltered rocky shores where it is found clinging to the undersurface of stones and among seaweed at low water, it has also been recorded as high as the middle shore. It is carnivorous on the wide range of invertebrates found on sheltered rocky shores. On the west coast of Ireland, longevity of up to four years has been recorded, the fish reaching maturity in their second year at a body length of about 39 mm. Breeding occurs during summer. Yellow-coloured eggs are laid in sheets under stones on the shore, often as many as 250 eggs in each sheet. They are guarded by one of the parents and as development proceeds the colour changes through orange to olive-green with hatching occurring after about two weeks. Pelagic larvae have been recorded in inshore waters during summer.

Prior to 2002, the clingfish commonly recorded from Britain was referred to as *Lepadogaster lepadogaster*. Some authorities assigned this to the subspecies *purpurea* i.e. **Lepadogaster lepadogaster** subspecies **purpurea**. Recent morphological and molecular studies have established *Lepadogaster purpurea*, as described above, as a distinct species. A second species, **Lepadogaster lepadogaster** (Bonnaterre) is distributed southwards from north-west Spain into the Mediterranean and has substantial overlap in distribution with *L. purpurea*. It can be separated from *L. purpurea* by the absence of the large blue spots on the dorsal surface of the head.

The Connemara clingfish, **Lepadogaster candollii** Risso, has a patchy distribution on the south and west coasts of Ireland and south-west Britain, extending south into the Mediterranean. Body length is up to 100 mm and in this species the dorsal and anal fins are not continuous with the tail fin (cf. *L. purpurea* and *L. lepadogaster*).

Family Gasterosteidae Sticklebacks

Conspicuous row of spines anterior to dorsal fin. Dorsal and anal fins almost exactly opposite. Pelvic fins reduced to one long spine and a small ray, behind the base of the pectoral fins. Without scales but may have bony plates along sides of body.

Body somewhat elongate; 3 dorsal spines, the anterior 2 of which are long and the 3rd small, just in front of the dorsal fin .... **Gasterosteus aculeatus** p. 450

Body long and slender; 15 (14–16) short spines in front of dorsal fin ....
.... **Spinachia spinachia** p. 451

Gasterosteus aculeatus Linnaeus Three-spined stickleback **Fig. 20.15**

Body somewhat elongate; 3 dorsal spines, the anterior 2 of which are long and the 3rd small and just in front of dorsal fin. Dorsal and anal fins opposite, dorsal longer than anal. Pelvic fins reduced to long spine and small ray. Body without scales but may have bony plates along sides. Dorsal surface usually bluish-green, ventral surface silver to white. Breeding males develop red coloration on undersurface and the eyes become bright blue. Length usually up to 60 mm, occasionally longer.

G. aculeatus is common in fresh waters but is also widely distributed throughout north-west Europe and Britain in the intertidal region, particularly in estuaries, feeding on a wide range of invertebrates. When not breeding it is often found in

Figure 20.15 *Gasterosteus aculeatus.*

shoals. Breeding occurs in spring and early summer. The male develops a characteristic breeding coloration of red belly and throat and constructs a nest out of bits of plant material held together by sticky secretions produced by the kidneys. The nest is built in a depression in the sediment. More than one female may be attracted to it and as many as 1000 eggs laid. After laying, the females are chased away by the male and play no part in caring for the eggs. The nest is guarded by the male which fans the eggs using the pectoral fins; the male also removes dead eggs. The eggs hatch after one to three weeks, the young often remaining close to the nest for a few days. Maturity is generally reached after one year and longevity is up to three years.

Spinachia spinachia (Linnaeus) Fifteen-spined stickleback, sea stickleback **Fig. 20.16**
Body very slender. Fifteen (14 or 16) short spines in front of dorsal fin. Dorsal and anal fins opposite. Pelvic fins reduced to small spine and small ray. Body without scales but about 41 bony plates along sides. Dorsal and lateral surfaces green-brown with darker blotches, ventral surface paler. Dorsal and anal fins dark anteriorly. Length up to 150 mm, rarely 200 mm.

Figure 20.16 *Spinachia spinachia.*

S. spinachia is widely distributed in north-west Europe. It is found in rock pools and shallow water among seaweeds extending to depths of about 20 m and feeds on a wide range of crustaceans. Breeding occurs in spring and early summer. The male, which unlike that of the three-spined stickleback (above) does not take on a characteristic breeding coloration, constructs a nest among seaweed. The nest is clear of the ground, up to 80 mm in diameter and made of strands of weed held together by sticky secretions produced by the kidneys. The nest may contain up to 200 eggs which are laid by more than one female. After spawning, the females are chased away by the male and take no part in the care of the eggs, which are protected and aerated by the male. The eggs hatch after about two to three weeks. Longevity is usually one year but some fish are believed to survive for a second year.

Family Syngnathidae Pipefishes and seahorses
Body slender, divided into bony rings along length. Head extended into tube-like snout; mouth small. Dorsal fin conspicuous; other fins reduced or absent.

1 Pectoral, anal and tail fins present 2
   Pectoral, anal and tail fins absent 3

2 Snout long, greater than 1/2 total length of head; 17–21 rings between base of pectoral fin and anus; 38–43 rings between anus and tail fin ...... ***Syngnathus acus*** p. 453

Snout about 1/2 total length of head; 13–17 rings between base of pectoral fin and anus; 37–42 rings between anus and tail .... ***Syngnathus rostellatus*** p. 454

3 Snout short, upturned; dorsal fin with 24–28 rays. Dorsal surface almost black in colour .... ***Nerophis lumbriciformis*** p. 452

Snout straight; dorsal fin with 33–34 rays. Dorsal surface green-brown in colour

.... ***Nerophis ophidion*** p. 452

Nerophis lumbriciformis (Jenyns) Worm pipefish **Fig. 20.17**

Body very slender and elongate, round in section; without pectoral, anal and tail fins. Dorsal fin with 24–28 rays. Snout short, upturned. Body divided into bony rings, 17–19 rings between head and anus, 46–54 behind anus. Dorsal surface dark green, black; ventral surface with paler markings. Length up to 170 mm.

Figure 20.17 *Nerophis lumbriciformis.*

N. lumbriciformis is widely distributed in north-west Europe and is most frequently found on south and west coasts of Britain, particularly at low water on sheltered rocky shores where there is a good cover of seaweed. It is found at depths of up to about 30 m feeding mainly on crustaceans and small fishes. In Britain, breeding has been recorded from April to October while in the southern part of its range in Portugal, breeding has been recorded throughout the year but at reduced levels during the summer. The female transfers up to about 150 eggs to a shallow groove on the undersurface of the abdomen of the male where fertilization and development take place. Young hatch at a body length of about 10 mm and spend some weeks in the plankton before becoming benthic at a length of about 30 mm.

Nerophis ophidion (Linnaeus) Straight-nosed pipefish **Fig. 20.18**

Body very slender and elongate, round in section; without pectoral, anal and tail fins. Dorsal fin with 33–34 rays. Snout about 1/2 total length of head, not upturned. Body divided into bony rings, 28–33 between head and anus. 68–82 rings behind anus. Dorsal surface green-brown, ventral surface paler. Females larger than males and with blue coloration. Length up to 300 mm.

N. ophidion has a widespread but local distribution in north-west Europe. It is predominantly sublittoral to depths of about 15 m but is occasionally found on the lower shore, especially among seaweeds and *Zostera* (p. 75), feeding on crustaceans. It extends into estuaries. Breeding has been recorded from May to August. The female develops a skin fold on the undersurface during the breeding season, making her

Figure 20.18 *Nerophis ophidion.*

appear considerably larger than the male. There is a lengthy courtship period during which the female displays her blue colour and the skin fold. Up to 150 eggs are transferred to a groove on the undersurface of the male and it is here that fertilization and development take place, the embryos receiving oxygen and nutrients from the male. Development takes about a month and on hatching the young, about 12 mm in body length, are pelagic for up to four months before taking up a benthic habit. Eggs carried by the male at any one time are from a single female but more than one brood may be carried during the breeding season.

Syngnathus acus Linnaeus Greater pipefish **Fig. 20.19**
Body very slender and elongate; pectoral, dorsal, anal and tail fins present. Dorsal fin with 36–45 rays. Snout long, greater than 1/2 length of head; in profile, head raised over eyes with hump behind eyes. Body divided into bony rings; 17–21 rings between base of pectoral fin and anus, 38–43 rings between anus and tail fin. Dorsal surface brown with darker markings. Ventral surface lighter. Length up to 460 mm.

Figure 20.19 *Syngnathus acus.*

S. acus is widely distributed in north-west Europe and in Britain, where it is found in shallow water among seaweed and occasionally in rock pools on the lower shore and to depths of about 90 m feeding on small crustaceans and larval fishes. It is also found on sand and mud, extending into the mouths of estuaries. Breeding occurs in spring and summer. Males are mature at about 300 mm body length and about 400 eggs are deposited by the female in a brood chamber on the ventral surface of the tail of the male. The brood chamber is formed of lateral folds of skin which hold the eggs in place. Here, fertilization takes place and after about five weeks young emerge, with a body length of about 25 mm. There is no planktonic stage.

Syngnathus rostellatus Nilsson Nilsson's pipefish, lesser pipefish **Fig. 20.20**
Body slender, elongate. Pectoral, dorsal, anal and tail fins present. Dorsal fin with 36–45 rays. Snout about 1/2 length of head. Body divided into bony rings; 13–17 rings between base of pectoral fin and anus, 37–42 rings between anus and tail. Dorsal surface brown with darker markings. Ventral surface lighter. Length up to 170 mm.

Figure 20.20 *Syngnathus rostellatus.*

S. rostellatus is widely distributed in north-west Europe over sandy and muddy deposits from low water to depths of 18 m. It extends into estuaries, where it is sometimes common, feeding on small crustaceans. Breeding occurs in summer; males are mature at about 100 mm body length and eggs are deposited by the female in a brood chamber on the ventral surface of the tail of the male. Young emerge from the brood chamber at a body length of about 14 mm and have a pelagic phase of a few weeks before assuming a benthic habit.

A third species, ***Syngnathus typhle*** Linnaeus, the deep-snouted pipefish, has a wide but patchy distribution in north-west Europe and in some areas, particularly where *Zostera* (p. 75) is present, it may be common. As suggested by the common name, it can be identified by its deep, laterally flattened snout which is more than half the length of the head.

Family Cottidae Bullheads, sea-scorpions
Stout-bodied, head large. Two dorsal fins, the anterior one spiny. Pelvic fins with one spine and two to five rays. Spines just in front of operculum. Without scales, often with spines, prickles or bony plates along lateral line.

Opercular membranes join to form single membrane under body. Largest spine on the operculum is shorter than diameter of eye .... ***Myoxocephalus scorpius*** p. 455

Opercular membranes join sides of body and do not form a single membrane on ventral surface. Largest spine on the operculum is greater than diameter of eye
.... ***Taurulus bubalis*** p. 455

Myoxocephalus scorpius (Linnaeus) (*Cottus scorpius*) Bull-rout, short-spined
sea-scorpion **Fig. 20.21a**

*Stout-bodied, 2 dorsal fins. Pectoral fins large; pelvic fins with long rays. Head
large. Two conspicuous spines in front of operculum, upper one the longer but
shorter than diameter of eye. Opercular membranes extend under body and join
to form a single membrane. Without scales; small spines above and below lateral
line. Colour varied, dorsal surface greenish-brown with darker patches; ventral
surface red to orange; white spots on ventral and lateral surfaces. Length up to
about 300 mm.*

M. scorpius is a northern species extending from Greenland and Iceland to the north-
west coast of France. It is found all around Britain on rocky, sandy and muddy
sediments to depths of about 60 m only occurring intertidally in the north. It feeds
on a variety of invertebrates, predominantly crustaceans, as well as small fishes.
Breeding occurs from December to March. Orange-coloured eggs are laid in clumps,
up to 80 mm across, in crevices and among seaweed on the lower shore or the shallow
sublittoral. These are guarded by the male which guards the eggs laid by more than
one female. Eggs take from 5 to 12 weeks to hatch and the larvae are found in the
plankton in spring, becoming benthic at a length of 15–20 mm. Maturity is reached in
the second year and longevity is believed to be about six years.

(a)

(b)

Figure 20.21 (a) *Myoxocephalus scorpius.* (b) *Taurulus bubalis.*

Taurulus bubalis (Euphrasen) (*Cottus bubalis*) Sea-scorpion, long-spined
sea-scorpion **Fig. 20.21b**

*Stout-bodied, 2 dorsal fins. Pectoral fins large; pelvic fins with 3 long rays. Head broad.
Conspicuous spines in front of operculum, upper one long, longer than diameter of eye.
Opercular membranes join sides of body. Without scales; small spines along lateral line.
Colour varied, greenish-brown to shades of red. Length up to 180 mm.*

T. bubalis is widely distributed in north-west Europe where it is found among seaweed on rocky shores. It occurs all around Britain and may be common on some shores, extending to depths of 30 m. It feeds on a variety of invertebrates and small fishes. Breeding occurs from about January to May. Orange-coloured eggs are laid in clusters on the shore under seaweed and in crevices, and hatch after about six or seven weeks. Larvae have been recorded in the plankton during spring and early summer, becoming benthic at a body length of about 12 mm. Sexual maturity is reached after two years at a body length of 100–180 mm.

Family Cyclopteridae Lump fishes and sea-snails
Body may be rounded and laterally compressed with rows of bony tubercles along sides – the lumpsuckers; or tadpole-like in shape with loose, prickly skin – the sea-snails. One or two dorsal fins. Pelvic fins modified to form a ventral sucker. No scales.

Cyclopterus lumpus Linnaeus Lumpsucker **Fig. 20.22**
Body thick, rounded, very distinctive in appearance. Two dorsal fins, but in adult fish anterior dorsal is reduced to tubercles. Three, sometimes 4 widely spaced rows of bony plates and tubercles along sides of body. Large sucker on ventral surface. Dorsal surface brown to greyish, paler below. Males develop red-orange colours on ventral surface during breeding season. Length up to 600 mm.

Figure 20.22 *Cyclopterus lumpus*. Note that anterior dorsal fin is reduced to tubercles in adult fish.

C. lumpus is widely distributed in north-west Europe where it has been recorded on rocky ground from the lower shore to depths of 200 m, feeding on ctenophores, crustaceans, polychaetes and other fishes. The lumpsucker migrates into shallow water to spawn and in northern parts of Britain it spawns on the lower shore. Clusters of many thousands of bluish-red to yellow-coloured eggs, up to 300 mm across, are attached to rocks and seaweed in spring and early summer and are guarded by the male. The male fans the eggs with the pectoral fins and this presumably aerates the egg mass. Eggs take several weeks to hatch and the larvae have a length of about 7 mm. Soon after hatching they attach to pieces of seaweed and are carried into the plankton before eventually settling on the bottom. It has been suggested that males breed for the first time when they are two to three years of age, females when three to four years.

Liparis montagui (Donovan) Montagu's sea-snail **Fig. 20.23**
*Rather thick and rounded anteriorly. Long dorsal fin with 28–30 rays, anterior rays
shorter than posterior. Anal fin long, 22–26 rays barely extending to the base of the
tail fin. Dorsal and anal fins separate from tail fin. Pelvic fins fused to form ventral
sucker. Without scales. Skin loose and prickly. Colour varied, brown, yellowish,
reddish-brown, green. Length up to about 60 mm, occasionally longer.*

Sucker

Figure 20.23 *Liparis montagui*, lateral and ventral views.

L. montagui is widely distributed on rocky shores in north-west Europe extending into
the sublittoral to depths of about 30 m. In Britain, where it is commonest in the north, it
is found on the middle and lower shore on the undersurface of stones and the holdfasts
of laminarians, attached, often firmly, by the ventral sucker. It feeds on a variety of
crustaceans. Breeding occurs in spring when red-, pink- or orange-coloured eggs are
laid in small clusters on red seaweeds and in crevices on the lower shore. They take
about six weeks to hatch and the pelagic larvae have been recorded in the plankton
during spring and summer. Young are found on the shore when they are about 12 mm in
length. Maturity is reached in the second year and longevity is three years.

The sea-snail *Liparis liparis* (Linnaeus) is found from shallow water to depths
of about 300 m around northern Britain and into Scandinavia, but is rare in the south
and south-west. It is larger than *Liparis montagui* (up to 120 mm in length) and has
more fin rays (27–29) in the anal fin. The anal fin of *L. liparis* overlaps about half of
the tail fin.

Family Moronidae Sea basses
Body elongate, two dorsal fins separate. Anterior with 8–10 spines, posterior with one
spine and 11–14 rays; anal fin with three spines and 10–12 rays. Scales large.

Dicentrarchus labrax (Linnaeus) European sea bass **Fig. 20.24**
*Two dorsal fins, anterior with 8 or 9 spines and separate from posterior dorsal.
Posterior dorsal with 1 spine and 12 or 13 rays. Scales large, squarish, 66–74 along*

Figure 20.24 *Dicentrarchus labrax.*

lateral line. Row of forward pointing spines anterior to operculum. Grey or blue-green in colour with silvery sides. Length up to 1 m.

D. *labrax* is widely distributed in north-west Europe, extending south to the Canaries and the Mediterranean. It is found as far north as Norway and around Britain it is common on south and west coasts, extending along the English Channel into the southern North Sea but is much less common to the north. The sea bass migrates northwards and inshore in spring, moving south and into deeper water with the onset of winter. It feeds on a wide range of crustaceans and young fishes. Around Britain maturity is reached between four and six years of age at a body length of about 350 mm. Spawning takes place offshore in spring and early summer and eggs and larvae can be taken in the plankton. Young fish move into shallow waters and are found in rock pools and the drainage channels of estuaries, often in large numbers, during the summer when they are two to three months old and about 40 to 50 mm in length. They stay in the estuaries through the summer, migrating into deeper waters in winter, returning inshore the following spring. They migrate into deeper waters with age but it is now well established that the estuaries on the west coast of Britain are important nursery areas for young bass. The sea bass is heavily fished and commands a high market price. It is much prized by sea anglers. Length of life is 20 years or more.

Family Mugilidae Grey mullets
Body elongate with two separate dorsal fins, the anterior with four spines, the posterior with one spine and eight rays. Pelvic fins towards anterior, about halfway between base of pectorals and the anterior dorsal fin. Scales large.

Chelon labrosus (Risso) Thick-lipped grey mullet **Fig. 20.25**
Two dorsal fins, clearly separated; anterior dorsal with 4 spines. Thick upper lip, at broadest point about equal to 1/2 diameter of eye; teeth very small. Scales large. Bluish-grey or dark green in colour; silvery sides with several dark grey lines along length. Up to 750 mm body length.

C. *labrosus* is widely distributed in the inshore waters of north-west Europe and common around Britain off harbour mouths and estuaries. It penetrates well into estuaries, often in shoals, with the incoming tide and feeds over mud flats, taking mouthfuls of mud from which it removes diatoms, algae, invertebrates and detritus by filtering through the gills. Spawning occurs during spring and summer. The eggs and

Figure 20.25 *Chelon labrosus.*

larvae are pelagic and the young fish migrate inshore and are found in rock pools and drainage channels and creeks in estuaries along with the young of the sea bass (above). They stay here during the summer months returning to coastal waters with the onset of winter.

Family Labridae Wrasses

Body deep, laterally compressed. Lips thick and pronounced; mouth protrusible. Strong teeth. Single dorsal fin. Dorsal and anal fins with anterior spines. Scales large. The number of rows of scales along the lateral line is important in identification. Brightly coloured fishes.

1 Five spines (occasionally 4 or 6) and 6–8 rays in anal fin ....
.... ***Centrolabrus exoletus*** p. 462
Three spines in anal fin **2**

2 41–47 scales along lateral line; dorsal fin with 19–20 spines anteriorly and 9–11 rays posteriorly .... ***Labrus bergylta*** p. 459
31–39 scales along lateral line **3**

3 Dark spot on centre of base of tail fin; 31–37 scales along lateral line ....
.... ***Crenilabrus melops*** p. 460
Dark spot in dorsal position on base of tail fin; 35–39 scales along lateral line
.... ***Ctenolabrus rupestris*** p. 461

Labrus bergylta Ascanius Ballan wrasse **Fig. 20.26a**
Body deep, laterally compressed. Large dorsal fin with 19–20 spines anteriorly and 9–11 rays posteriorly. Anal fin with 3 spines and 8–12 rays. Lips thick and pronounced. Scales large, usually with pale centre and dark posterior margin; 41–47 scales along lateral line. Region in front of operculum not toothed. Colour varied, green, brown, sometimes reddish brown. Length up to 600 mm, much smaller intertidally.

Figure 20.26 (a) *Labrus bergylta*. (b) *Crenilabrus melops* (after Du Heaume, in Wheeler, 1969).

L. bergylta is widely distributed in north-west Europe and Britain and is often common among seaweed in shallow water off rocky shores, extending to depths of about 20 m. It is also found in rock pools on the lower shore and feeds mainly on crustaceans and mussels. The ballan wrasse undergoes a sex change. The fishes mature first as female then many, but not all, later change sex to male. On Irish Sea coasts, females spawn when they are five or six years old and sex change occurs at any time thereafter. Breeding takes place in spring and summer. The eggs are believed to be laid in depressions excavated by the males. Larvae hatch at a body length of about 4 mm and have been recorded in the plankton off Plymouth from May to July. Growth is slow and specimens more than 20 years old have been recorded.

Crenilabrus melops (Linnaeus) (*Symphodus melops*) Corkwing wrasse **Fig. 20.26b**
Body deep, laterally compressed. Large dorsal fin with 14–17 spines anteriorly and 8–10 rays posteriorly. Anal fin with 3 spines and 8–11 rays. Scales large, 31–37 along lateral line. Lips thick and pronounced. Region in front of operculum finely toothed. Body colour and markings vary, depending on background and age of specimen; juveniles generally olive-green or blue-green; mature males darker, often reddish-brown to deep purple-brown with bluish lines on lower part of head; mature females pale brown with dark brown lines on head. Crescent-shaped dark patch just posterior to eye; dark spot on centre of base of tail fin. Length usually up to 150 mm but can be 250 mm.

C. melops is widely distributed in north-west Europe. Found among seaweed, it may be common in rock pools on the lower shore, particularly on the west coast of Britain, feeding on a wide range of molluscs and crustaceans. It extends into the sublittoral to depths of about 50 m. In spring the male builds a nest of seaweed in a crevice on the lower shore or shallow sublittoral, in preparation for spawning in summer. After a complex courtship display, eggs are laid and fertilized in the nest which is then guarded and fanned by the male. In a single season some males build a second nest and rear another batch of eggs. The eggs hatch after about two weeks and larvae are found in the plankton off Plymouth from June to August. On English Channel coasts both sexes reach maturity in the second or third year, but the nest-building behaviour of the males apparently does not begin until they are in their fourth year. Longevity is believed to be up to nine years.

Ctenolabrus rupestris (Linnaeus) Goldsinny **Fig. 20.27a**

Body deep, laterally compressed. Large dorsal fin with 16–19 spines anteriorly and 7–10 rays posteriorly. Anal fin with 3 (occasionally 4) spines and 6–9 rays. Scales large, 35–39 along lateral line. Colour brown, green, reddish. Dark spot in dorsal position on base of tail fin and black patch on anterior of dorsal fin. Length up to 180 mm, smaller intertidally.

Figure 20.27 (a) *Ctenolabrus rupestris*. (b) *Centrolabrus exoletus*. ((a) and (b) modified after Du Heaume, in Wheeler, 1969.)

C. rupestris is very widely distributed in north-west Europe and locally abundant in some areas around Britain but is not common in the eastern English Channel and southern North Sea. It is the most widespread and abundant wrasse around Ireland. It is found among seaweed off rocky shores to depths of about 50 m. Breeding occurs in spring and summer and fish of up to 16 years of age have been recorded off the west coast of Scotland.

Centrolabrus exoletus (Linnaeus) Rock cook **Fig. 20.27b**
Body deep, laterally compressed. Large dorsal fin with 18–20 spines anteriorly and 5–7 rays posteriorly. Anal fin with 5 (4–6) spines and 6–8 rays. Scales large, 32–35 along lateral line. Mouth small. Colour greenish, brown; ventral surface paler; tail fin with 1 or 2 brownish bands. Length up to 165 mm.

C. exoletus is widely distributed in shallow waters around north-west Europe. It is found over rocks and *Zostera* beds (p. 75) to depths of about 25 m, feeding on small invertebrates. Breeding has been recorded in spring and summer when males are believed to construct nests of seaweed. Longevity of up to nine years has been recorded on the west coast of Scotland.

Family Zoarcidae Eelpouts
Body elongate, with relatively long head. Dorsal fin long and continuous with tail fin and long anal fin. Pelvic fins anterior to pectoral fins. Scales very small.

Zoarces viviparus (Linnaeus) Viviparous blenny, eelpout **Fig. 20.28**
Body elongate. Dorsal fin long, continuous with tail and anal fin; just anterior to tail the fin rays are short. Pectoral fins large; pelvic fins small. Head large, lips thick. Scales small. Brownish colour, paler ventrally, with number of dark patches along body. Usually up to 300 mm but may reach 500 mm in length.

Z. viviparus is a northern species reaching the southern limit of its distribution in the eastern English Channel. Around Britain it is found on North Sea coasts and the west coast of Scotland where it is often common in rock pools and among seaweed on the lower shore, extending to depths of about 40 m. It also occurs on sandy and muddy substrata. During summer it migrates offshore to deeper, cooler waters, and in the southern part of its range it is not common intertidally. It feeds on a wide range of invertebrates, particularly crustaceans and small fishes. As one of the common names implies, *Zoarces* is viviparous. Mating takes place in late summer and autumn and fertilization is internal, following a brief period of copulation when sperm are deposited in the female by means of a tiny papilla on the undersurface of the male.

Figure 20.28 *Zoarces viviparus.*

The eggs develop inside the female; three to four months later, fully developed young are released at about 40 mm body length, generally from December to February. It has been estimated that the larger females carry as many as 300 young. Maturity is reached in the second or third year and longevity is about four years. The eelpout is fished commercially in Scandinavia using baited pots and seine nets.

Family Pholidae Butterfishes
Body long, slender. Dorsal fin long, slender, made up of large number of short spines. Pelvic fins very small, reduced to small spines. Scales tiny.

Pholis gunnellus (Linnaeus) (*Centronotus gunnellus*) Butterfish, gunnel **Fig. 20.29**
Body long, slender and laterally flattened. Dorsal fin long, low, extending to tail fin; anal fin much shorter. Pelvic fins very small, reduced to small spines just below base of pectoral fins. Scales very small. Brownish in colour with darker patches; about 12 black spots, ringed with white, along base of dorsal fin. Length up to 250 mm.

Figure 20.29 *Pholis gunnellus.*

P. gunnellus is a northern species reaching the southern limit of its distribution on the north-west coast of France. It is common around Britain in rock pools from the middle shore and into the sublittoral to depths of about 100 m. There is a downshore migration during winter. It feeds on a wide variety of invertebrates. Breeding occurs during winter. The eggs, which are laid in clusters on stones, in empty shells and in crevices, are usually guarded by the female and hatch after a few weeks; larvae have been recorded in the plankton up to about May. Young are found inshore when they are about 30 mm in length and maturity is reached at a body length of 100 mm when the fish is in its second year. Longevity is up to five years.

Family Trachinidae Weevers
Body stout, mouth sloping sharply downwards. Eyes situated dorsally. Anterior dorsal fin with large spines, black in colour. Second dorsal fin and anal fin long. Pelvic fins anterior to pectoral fins. Large spine on operculum. Scales small, in conspicuous oblique rows.

Echiichthys vipera (Cuvier) (*Trachinus vipera*) Lesser weever **Fig. 20.30**
*Body thick. Anterior dorsal fin with 5–7 spines, black in colour; posterior dorsal fin long, with 21–24 rays; anal fin long, with 24–26 rays. Margin of pectoral fin rounded. Mouth running sharply downwards. Eyes situated dorsally; no spines near eyes. Long, backwardly directed spine on operculum. Scales small. Dorsal surface yellowish-brown with dark patches; paler ventrally. Length up to 140 mm. **Has poisonous sting.***

E. vipera occurs in the southern half of the North Sea and all around Britain, where it is most abundant in the south and west. It lives on sandy sediments and lies buried

Figure 20.30 *Echiichthys vipera.*

with only the head and dorsal surface visible. Occasionally found on the lower shore when the tide is out, it is most common in shallow water to depths of about 50 m. It feeds on a wide range of molluscs, crustaceans and small fishes. Breeding occurs in summer; the eggs are pelagic and larvae have been recorded in the plankton off Plymouth from June to September. The spines on the operculum and the anterior dorsal fin of the weever are equipped with venom cells which can inflict a severe sting causing pronounced swelling of the tissues. The pain is described as 'intense' and 'excruciating' and may last for up to 24 hours. It is said that some relief can be obtained by immediately bathing the affected area in hot water as this apparently breaks down the toxin.

The great weever, ***Trachinus draco*** Linnaeus, up to 300 mm body length, is found in deeper water than *E. vipera* but is occasionally found in the shallows. It is distinguished from the latter species by the presence of two spines close to the eyes and an indented or straight margin to the pectoral fin. **Like** *E. vipera***, it has a poisonous sting.**

Family Blenniidae Blennies
Body elongate, head rather large. Dorsal fin long and, in the species described in this text, divided into two parts by a depression. Pelvic fins with two long rays; anterior to pectoral fins. Tail fin separate from dorsal and anal fins. Without scales.

1   Flap of tissue across head, just above eyes, with mid-dorsal row of small tentacles
    behind .... ***Coryphoblennius galerita*** p. 466
    Head not as above **2**

2   Pair of branched tentacles between eyes .... ***Parablennius gattorugine*** p. 466
    Without tentacles between eyes .... ***Lipophrys pholis*** p. 464

Lipophrys pholis (Linnaeus) (*Blennius pholis*) Shanny **Fig. 20.31a**
Dorsal fin long, divided into 2 by slight depression; anterior part with 11–13 rays, posterior part with 18–20 rays. Anal fin long. Pelvic fins each with 2 long

Figure 20.31 (a) *Lipophrys pholis*. (b) *Parablennius gattorugine*. (c) *Coryphoblennius galerita*.
((b) After Du Heaume, in Wheeler, 1969.)

rays; anterior to the pectoral fins. Body without scales. Without tentacles on head.
Colour varied, green, brown, olive or yellowish with black markings. Length up to
160 mm.

L. pholis is widely distributed on rocky shores, less frequently on sand and mud,
in north-west Europe but has a patchy distribution in the southern North Sea. It
is essentially a rock pool species, occasionally found on the upper shore and is
highly adapted to conditions in rock pools withstanding low salinity and surviving
in moist situations under stones. Small specimens are generally found under
stones and in pools on the upper shore. *Lipophrys* forages over the shore when
the tide is in, often returning to the same rock pool, which it may occupy for
many weeks, when the tide ebbs. This has been described as homing behaviour. It
feeds on a wide range of crustaceans and molluscs including barnacles, littorinids
and the small bivalve *Lasaea adansoni* (p. 266). The young feed extensively on
the thoracic appendages of barnacles, which are nipped off by the shanny as
the barnacles feed. Larger individuals often have considerable amounts of seaweed
in the gut. There is a downshore migration in winter. Around Britain, breeding
occurs in spring and summer but this shifts to winter and spring in more
southerly localities. The eggs are laid in clusters under stones or in narrow crevices
where they are guarded by the male which fans them with its tail. A single
female lays up to eight batches of eggs per season and these hatch after about
six to eight weeks. The larger females have been estimated to produce as many
as 8000 eggs. Larvae are common in the plankton during summer, young
appearing on the shore at a length of about 20 mm. Maturity is reached after

two or three years and although specimens of 14 years of age have been recorded, few live more than five years.

During the winter months, specimens of *L. pholis* on the west coast of Britain are sometimes parasitized by the ectoparasitic marine leech **Oceanobdella blennii** (Knight-Jones). Small leeches up to about 4 mm in length are found under the operculum and larger specimens (up to 14 mm) behind the pectoral fin. The mature leeches leave the host in spring and lay egg cocoons on the shore. The eggs develop slowly during summer and hatch during the following autumn/winter when the young leeches seek out a host.

Parablennius gattorugine (Linnaeus) (*Blennius gattorugine*) Tompot blenny
Fig. 20.31b

Dorsal fin long, divided into 2 parts by slight depression; anterior part with 12–14 rays, posterior part with 17–20 longer rays. Anal fin long. Pelvic fins each with 2 long rays; anterior to the pectoral fins. Body without scales. A pair of branched tentacles between the eyes. Colour yellowish, olive, brown, with number of darker vertical bands. Length usually up to about 150 mm, may reach 300 mm.

P. gattorugine is a southern species extending into the English Channel and along the west coast of Britain as far north as Scotland. It has not been recorded from North Sea coasts. Although found occasionally in rock pools on the lower shore, it is more common in the sublittoral, extending to depths of about 35 m. It feeds on a wide range of cnidarians, molluscs, crustaceans, polychaetes and fragments of seaweed. Breeding occurs in spring and summer when purplish-black eggs are laid in crevices and guarded by the male. Larvae have been recorded in the plankton in summer. On the west coast of Ireland maturity is reached after one year; longevity is believed to be up to about nine years.

Coryphoblennius galerita (Linnaeus) (*Blennius montagui*) Montagu's blenny
Fig. 20.31c

Dorsal fin long, divided into anterior and posterior parts by depression. Flap of tissue across head, just above the eyes; mid-dorsal row of small tentacles behind the flap. Anal fin long. Pelvic fins each with 2 long rays; anterior to pectoral fins. No scales. Green-brown colour with dark patches; light-coloured spots on head and body. Length up to 85 mm.

C. galerita is found on rocky shores on the Atlantic coast of France and south-west Britain, extending to the west coast of Ireland. It is found on the middle shore, particularly in pools with the red seaweed *Corallina* (p. 43). It feeds on barnacles, nipping off the thoracic appendages as the barnacle feeds, and other small crustaceans. Eggs are laid in crevices from May to August and are guarded by the males which are territorial.

Family Ammodytidae Sand eels

Body elongate. Tail fin forked. Lower jaw projecting beyond upper jaw. Dorsal fin long. No pelvic fins. Scales tiny.

Sand eels feature prominently in the diet of many of the most common seabirds and a wide range of commercial fish such as the cod, haddock and mackerel. They are also preyed on heavily by seals and porpoises. They are fished commercially and have been subjected to very heavy fishing pressure.

1 Upper jaw not extensible **2**
Upper jaw extensible (Fig. 20.32d) **3**

2 Black spot in front of eye .... *Hyperoplus lanceolatus* p. 468
Without black spot in front of eye .... *Hyperoplus immaculatus* p. 469

3 Scales in >>>> pattern on undersurface. Scales on base of tail fin ....
.... *Ammodytes tobianus* p. 468
Scales on undersurface not in >>>> pattern. Scales not present on base
of tail fin .... *Ammodytes marinus* p. 467

Ammodytes marinus Raitt Raitt's sand eel **Fig. 20.32c**
Body elongate, eel-like. Dorsal fin long, with 55–67 rays; anal fin much shorter, 29–33 rays. Tail fin forked. Pectoral fins extend beyond anterior margin of dorsal fin. Pelvic fins absent. Lower jaw protruding beyond upper jaw; upper jaw extensible. Scales small, not in >>>> pattern on undersurface. Scales absent from base of tail fin. Dorsal surface blue-green, silvery below. Length up to 240 mm.

Figure 20.32 (a) *Ammodytes tobianus*, with detail of tail fin (b). (c) Detail of tail fin of *Ammodytes marinus*. (d) Extensible jaw of *Ammodytes* sp. ((b), (c), (d) after Stebbing, in Wheeler, 1978.)

A. marinus is essentially a sublittoral species, swimming in shoals in water of about 20–70 m depth. It is often abundant and juveniles are occasionally found in much shallower water where they burrow into sand and fine gravel. It is generally regarded as a zooplankton feeder and there is some evidence to suggest that in the North Sea it is a serious predator of herring larvae and eggs. In the North Sea spawning occurs in December and January and the larval stage lasts for up to 90 days. Longevity is about four years.

Ammodytes tobianus Linnaeus (*Ammodytes lancea*) Lesser sand eel **Fig. 20.32a,b**
Body elongate, eel-like. Dorsal fin long, with 50–56 rays; anal fin much shorter, with 24–32 rays. Tail fin forked. Pectoral fins extend beyond anterior margin of dorsal fin. Pelvic fins absent. Lower jaw protruding beyond upper jaw; upper jaw extensible. Scales small, in >>>> pattern on undersurface. Scales on base of tail fin. Dorsal surface yellow-green or bluish-green; silvery below. Length up to 200 mm.

A. tobianus is widely distributed and common in north-west Europe and Britain on sandy beaches from the middle shore to depths of about 30 m. It swims in shoals just above the bottom, feeding on a wide variety of organisms, including small fishes. It is often found buried in the sand at low water when the tide is out. There are apparently two spawning groups, one spawning in spring and the other in the autumn. Eggs are attached to sand grains and the larvae are pelagic. Longevity is believed to be four years.

Hyperoplus lanceolatus (Le Sauvage) Greater sand eel **Fig. 20.33a**
Body elongate, eel-like. Dorsal fin long, with 52–61 rays; anal fin much shorter (27–33 rays). Tail fin forked. Pectoral fins do not extend beyond anterior margin of dorsal fin. Pelvic fins absent. Lower jaw protruding beyond upper jaw; upper jaw not extensible. Scales small. Dorsal surface greenish, silver below. Conspicuous black spot in front of eye. Body length up to 440 mm.

H. lanceolatus is widely distributed and common in north-west Europe, occurring sublittorally to depths of 150 m. Young specimens are often found buried on the lower shore on sandy beaches. The greater sand eel feeds on crustaceans and other fauna associated with sandy deposits. Breeding occurs in spring and summer when the fish are about two years old. Longevity is up to five years.

Figure 20.33 (a) *Hyperoplus lanceolatus*. (b) *Hyperoplus immaculatus*.

Hyperoplus immaculatus (Corbin) Corbin's sand eel **Fig. 20.33b**
Body elongate, eel-like. Dorsal fin long, with 59–62 rays; anal fin much shorter (31–43 rays). Tail fin forked. Pectoral fins extend beyond anterior margin of dorsal fin. Pelvic fins absent. Lower jaw protruding beyond upper jaw; upper jaw not extensible. Scales small. Dorsal surface blue-green, silver below; without black spot in front of eye (cf. Hyperoplus lanceolatus, *above). Body length up to 300 mm.*

H. immaculatus is generally found sublittorally in waters of 50–300 m depth, but young occur close inshore on sand and gravel. It takes a wide range of food, including zooplankton and small fish. Breeding occurs in winter.

Family Gobiidae Gobies

Eyes large, protruding, dorsolateral in position. Two dorsal fins; anterior dorsal short with spines, pelvic fins joined ventrally to form a crescent-shaped disc enabling the fish to attach loosely to the substratum. Body with scales. A large family of fishes found on the open coast, in estuaries and fresh water. The majority are benthic and generally of a small size, mostly less than 150 mm body length. Although identification can be difficult, the species are often restricted to a particular habitat and this can be helpful in identification.

1 Large black spot at base of tail fin. Anterior dorsal fin with 7 (rarely 6 or 8) spines
.... ***Gobiusculus flavescens*** p. 471
Without black spot at base of tail fin. Anterior dorsal fin with 6 (occasionally 5–7) spines **2**

2 Upper rays of pectoral fins free distally, extending beyond fin membrane (Fig. 20.34a) **3**
Upper rays of pectoral fins not free distally (Fig. 20.36) **4**

3 Upper rays of pectoral fins short; 32–42 scales in a line from base of pectoral fin to tail fin .... ***Gobius niger*** p. 469
Upper rays of pectoral fins long, branched; 50–57 scales in a line from base of pectoral fin to tail fin .... ***Gobius paganellus*** p. 470

4 (a) 55–75 scales in a line from base of pectoral fin to tail fin ....
.... ***Pomatoschistus minutus*** p. 472
or
(b) 39–52 scales in a line from base of pectoral fin to tail fin ....
.... ***Pomatoschistus microps*** p. 472
or
(c) 36–43 scales in a line from base of pectoral fin to tail fin; 1 or 2 rows of black spots at base of dorsal fin .... ***Pomatoschistus pictus*** p. 473

Gobius niger Linnaeus Black goby **Fig. 20.34a**

Anterior dorsal fin with 6 (sometimes 5–7) long spines, the central ones longest; posterior dorsal fin with 1 spine and 11–13 rays. Eyes large. Upper rays of pectoral fins short, free distally. Pelvic fins united to form crescent-shaped disc; 32–42 scales in

Figure 20.34 (a) *Gobius niger*. (b) *Gobius paganellus*, with ventral view to show pelvic fins joined to form crescent-shaped disc (c).

a line from base of pectoral fin to tail fin. Body various shades of brown with darker patches on sides. Black spot on upper, anterior margin of dorsal fins. Length up to 170 mm.

G. *niger* is widely distributed throughout north-west Europe over sand and mud to depths of about 70 m and extends into estuaries. It feeds on a wide range of invertebrates and young fishes. Breeding occurs in spring and summer. Club-shaped eggs are laid on stones, shells and seaweed and are guarded by the male. Larvae hatch after about two weeks and are found in the plankton during summer, becoming benthic at a length of about 12 mm. Maturity is reached after two or three years and longevity is believed to be about four or five years.

Gobius paganellus Linnaeus Rock goby **Fig. 20.34b,c**
Anterior dorsal fin with 6 spines; posterior dorsal with 1 spine and 13–14 rays. Upper rays of pectoral fins long, branched; free distally, almost reaching the base of the

anterior dorsal fin. Small branched tentacle in front of eye. Pelvic fins united to form crescent-shaped disc; 50–57 scales in a line from base of pectoral fin to tail fin. Blackish-brown in colour; pale yellowish horizontal band along upper margin of anterior dorsal fin; band is bright orange in breeding males. Length up to 170 mm.

G. paganellus is found on the west coast of France, in the English Channel and along the west coast of Britain, and has recently been recorded from the southern North Sea. It is found under stones on rocky shores and in pools on the middle shore and below, being most common among seaweed on sheltered shores. In the sublittoral it occurs to depths of about 15 m and undergoes a downshore migration during winter. It feeds on a wide range of small crustaceans and molluscs. Breeding occurs in spring and summer. The eggs are laid in sheets under stones, but have also been found on a variety of substrata such as shells and the tubes of polychaete worms. Development takes about three weeks; the larvae have been recorded in the plankton during summer, becoming benthic at about 10 mm in length. Around Britain maturity is reached in the second or third year. Although individuals of eight and ten years old have been recorded, most live for only two or three years.

Gobiusculus flavescens (Fabricius) (*Gobius ruthensparri*) Two-spot goby **Fig. 20.35**
Anterior dorsal fin with 7 spines (rarely 6 or 8); posterior dorsal with 1 spine and 9–11 rays. Upper margin of pectoral fins not free distally. Pelvic fins united to form crescent-shaped disc; 35–40 scales in a line from base of pectoral fin to tail fin. Reddish-brown colour, with pale blotches along dorsal surface; large black spot, with yellow edge, at base of tail fin; small dark spot just behind pectoral fin in male; dorsal and tail fins banded. Up to 60 mm in length.

Figure 20.35 *Gobiusculus flavescens.*

G. flavescens is widely distributed in north-west Europe and is found all around Britain in rock pools and sublittorally to depths of 15 m among seaweed, often in shoals, but it is rare on the east coast. It is sometimes locally abundant in the south and west and is also found in sheltered bays and estuaries among *Zostera* beds (p. 75). Planktonic crustaceans form the bulk of its food. Breeding occurs in spring and summer when males attract females with a courtship display and the eggs are laid among the holdfasts of *Sacchoriza polyschides* (p. 59) and in empty barnacle shells. They are guarded by the male. A single male attracts a number of females during the breeding season and may be caring for the eggs from as many as nine females at any one time.

Figure 20.36 *Pomatoschistus microps.*

Pomatoschistus microps (Krøyer) Common goby **Fig. 20.36**
Anterior dorsal fin with 6 spines, posterior dorsal with 1 spine and 8–10 rays. Pelvic fins united to form crescent-shaped disc. Eyes large. Upper rays of pectoral fins not free distally; 39–52 scales in a line from base of pectoral fin to tail fin. Greyish in colour, often with darker blotches across back and faint marks along sides of body; dark area at base of pectoral fins and tail fin; dark spot on anterior dorsal fin of males. Length up to about 65 mm.

P. microps is widely distributed in north-west Europe and Britain on sandy and muddy sediments and has been described as 'the most important small fish of British estuaries'. It tolerates salinities from about 8 to 80 and is often abundant in pools and creeks on mud-flats. It feeds predominantly on small crustaceans such as *Corophium* (p. 343), which are found in such habitats. When winter temperatures fall below 5 °C the goby migrates into deeper water, returning in spring. Breeding occurs in spring and summer when a single female can produce as many as 10 batches of eggs. The pear-shaped eggs are laid under stones and shells but apparently there is a preference for laying under empty bivalve shells. The fish removes sediment from under the shell to create a nest and the eggs are attached by the female to the inner surface of the shell which forms the roof of the nest. The shell is then anchored by being partly covered with sediment. Nest construction is carried out solely by males, which have been seen to turn shells so that the concave side faces downwards. The male guards and fans the eggs until they hatch, generally after about 14 days. Both sexes spawn a number of times during the breeding season. Larvae have been recorded in the plankton during summer and young fishes are found on the shore at a body length of 11–12 mm. Sexual maturity is reached in the first year and longevity is believed to be about two years.

Pomatoschistus minutus (Pallas) (*Gobius minutus*) Sand goby **Fig. 20.37**
Anterior dorsal fin with 6–7 spines, posterior dorsal with 1 spine and 10–12 rays. Pelvic fins united to form crescent-shaped disc. Eyes large. Upper rays of pectoral fins not free distally; 55–75 scales in a line from base of pectoral fin to tail fin. Pale brown with darker markings on sides; dark spot on posterior margin of anterior dorsal fin of males. Length up to 110 mm but generally about 60 mm.

P. minutus is widely distributed in north-west Europe and Britain on sand and mud from the middle shore into the sublittoral, sometimes occurring in shoals. It is also found in the mouths of estuaries. Juveniles are apparently more

Figure 20.37 *Pomatoschistus minutus* (after Day, 1884).

tolerant of reduced salinity than the adults and extend well into estuaries. The diet is predominantly small crustaceans such as *Corophium* (p. 343). There is an offshore migration into deeper water during winter. Breeding occurs in spring and summer. Pear-shaped eggs are laid in empty bivalve shells where they are guarded by the male until they hatch, usually after about two weeks. Larvae are found in the plankton throughout the summer; young become benthic at about 12 mm in length. The sand goby reaches maturity in its first year and lives for two years.

Pomatoschistus pictus (Malm) (*Gobius pictus*) Painted goby **Fig. 20.38**
Anterior dorsal fin with 6 spines, posterior dorsal with 1 spine and 8–9 rays. Pelvic fins united to form crescent-shaped disc. Eyes large. Upper rays of pectoral fins not free distally; 36–43 scales in a line from base of pectoral fin to tail fin. Brownish in colour with paler areas on dorsal surface and sides: 4 dark spots, sometimes double, on each side; 1 or 2 rows of black spots at base of dorsal fins. Length generally up to 60 mm.

Figure 20.38 *Pomatoschistus pictus* (after Du Heaume, in Wheeler, 1969).

P. pictus is widely distributed in north-west Europe on gravel and coarse sand from the lower shore to depths of about 50 m. It is occasionally found in rock pools on the middle and lower shore. In Britain it is most common on south and west coasts, feeding mainly on crustaceans. Breeding occurs during spring and summer. As in *P. microps* (p. 472), pear-shaped eggs are laid in empty bivalve shells and guarded by the male, hatching after about 12 days. Larvae have been recorded in the plankton during summer. Sexual maturity is reached in the second year and longevity is believed to be two, occasionally three, years.

Two species of goby occurring in north-west Europe, *Aphia minuta* (Risso) and *Crystallogobius linearis* (von Düben), are pelagic. They can be recognized by their transparent, laterally compressed bodies.

References

Chordata: Ascidiacea

Alder, J. & Hancock, A. (1905–12). *The British Tunicata. An unfinished monograph*. London: Ray Society.

Berrill, N. J. (1950). *The Tunicata with an account of the British species*. London: Ray Society.

Millar, R. H. (1970). *British ascidians. Tunicata: Ascidiacea. Keys and notes for the identification of the species*. Synopses of the British fauna (New Series), no. 1. London: Academic Press.

Picton, B. E. (1985). *Ascidians of the British Isles. A colour guide*. Ross-on-Wye: Marine Conservation Society.

Chordata: Actinopterygii

Day, F. (1880–4). *The fishes of Great Britain and Ireland*, 2 vols. London: Williams & Norgate.

Kay, P. & Dipper, F. (2009). *A field guide to the marine fishes of Wales and adjacent waters*. Llanfairfechan: Marine Wildlife.

Maitland, P. S. & Herdson, D. (undated). *Key to the marine and freshwater fishes of Britain and Ireland*. Almondsbury: Environment Agency.

Wheeler, A. (1969). *The fishes of the British Isles and north-west Europe*. London: Macmillan.

Wheeler, A. (1978). *Key to the fishes of northern Europe*. London: Warne.

Wheeler, A. (1994). Field key to the shore fishes of the British Isles. *Field Studies*, 8, 481–521.

Whitehead, P. J. P., Bauchot, M. L., Hureau, J. C., Nielson, J. & Tortonese, E. eds. (1984–6). *Fishes of the north-eastern Atlantic and the Mediterranean*, Vol. I (1984), Vol. II (1986), Vol. III (1986). Paris: UNESCO.

Bibliography

The following references are a selection of those used in preparing the text. Those listed under 'Introduction' include a range of identification guides and marine ecology texts that are a valuable source of additional/background reading. Those under specific phylum/group headings are a selection of those from which the biological notes have been prepared.

INTRODUCTION

Admiralty Tide Tables. Vol. 1. *European waters*. Published annually by the Hydrographer of the Navy. Available from agents for the sale of British Admiralty Publications.

Baker, J. M. & Wolff, W. J. eds. (1987). *Biological surveys of estuaries and coasts*. Estuarine & Brackish-Water Sciences Association Handbook. Cambridge: Cambridge University Press.

Ballantine, W. J. (1961). A biologically defined exposure scale for the comparative description of rocky shores. *Field Studies*, **1**, 1–19.

Barnes, R. S. K. (1994). *The brackish-water fauna of northwestern Europe*. Cambridge: Cambridge University Press.

Barnes, R. S. K. & Hughes, R. N. (1999). *An introduction to marine ecology*, 3rd edn. Oxford: Blackwell Science.

Barnes, R. S. K., Calow, P., Olive, P. J. W., Golding, D. W. & Spicer, J. I. (2001). *The invertebrates: a synthesis*, 3rd edn. Oxford: Blackwell Science.

Barrett, J. H. & Yonge, C. M. (1958). *Collins pocket guide to the sea shore*. London: Collins.

Brattegard, T. & Holthe, T. eds. (1997). *Distribution of marine, benthic macro-organisms in Norway. A tabulated catalogue*. Trondheim: Directorate for Nature Management.

Brusca, R. C. & Brusca, G. J. (2003). *Invertebrates*, 2nd edn. Sunderland, MA: Sinauer Associates.

Campbell, A. C. (1976). *The* Country Life *guide to the seashore and shallow seas of Britain and Europe*. Feltham: Country Life Books.

Campbell, N. A. & Reeece, J. B. (2008). *Biology*, 8th edn. San Francisco, CA: Pearson Benjamin Cummings.

Costello, M. J., Emblow, C. S. & White, R. eds. (2001). *European register of marine species. A check-list of the marine species in Europe and a bibliography of guides to their identification*, Patrimoines naturels, no. 50. Paris: Muséum National d'Histoire Naturelle. (http://erms.biol.soton.ac.uk)

Gibson, R., Hextall, B. & Rogers, A. (2001). *Photographic guide to the sea and shore life of Britain and north-west Europe*. Oxford: Oxford University Press.

Hayward, P. J. (2004). *A natural history of the seashore*. London: HarperCollins.

Hayward, P. J. & Ryland, J. S. eds. (1990). *The marine fauna of the British Isles and north-west Europe*, Vols. 1 and 2. Oxford: Clarendon Press.

(1995). *Handbook of the marine fauna of north-west Europe*. Oxford: Oxford University Press.

Hayward, P., Nelson-Smith, T. & Shields, C. (1996). *Collins Pocket Guide: Sea shore of Britain & Northern Europe*. London: HarperCollins.

Howson, C. M. & Picton, B. E. eds. (1997). *The species directory of the marine fauna and flora of the British Isles and surrounding seas*. Belfast and Ross-on-Wye: Ulster Museum and The Marine Conservation Society.

International Commission on Zoological Nomenclature (1999). *International code of zoological nomenclature*, 4th edn. London: The International Trust for Zoological Nomenclature.

Jeffrey, C. (1989). *Biological nomenclature*, 3rd edn. London: Edward Arnold.

Lewis, J. R. (1964). *The ecology of rocky shores*. London: English Universities Press.

Lincoln, R. J. & Sheals, J. G. (1979). *Invertebrate animals. Collection and preservation*. London and Cambridge: British Museum (Natural History) and Cambridge University Press.

Little, C., Williams, G. A. & Trowbridge, C. D. (2009). *The biology of rocky shores*, 2nd edn. Oxford: Oxford University Press.

Marine biodiversity and ecosystem functioning EU network of excellence (MarBEF). (http://www.marbef.org/modules.php?name)

McLachlan, A. & Brown, A. C. (2006). *The ecology of sandy shores*, 2nd edn. London: Academic Press.

McLusky, D. S. (1989). *The estuarine ecosystem*, 2nd edn. London: Chapman & Hall.

McNeill, J., Barrie, F. R., Burdet, H. M. *et al*. (2006). *International code of botanical nomenclature (Vienna Code)*. Regnum Vegetabile Vol. 146. Ruggell: A. R. G. Ganter Verlag.

Moore, J. (2006). *An introduction to the invertebrates*, 2nd edn. Cambridge: Cambridge University Press.

Nichols, D. ed. (1999). *Safety in biological fieldwork. Guidance notes for codes of practice*, 4th edn. London: Institute of Biology.

Raffaelli, D. & Hawkins, S. (1999). *Intertidal ecology*, 2nd edn. Dordrecht, the Netherlands: Kluwer Academic.

Sims, R. W., Freedman, P. & Hawksworth, D. L. (1988). *Key works to the fauna and flora of the British Isles and north-western Europe*, 5th edn. Systematics Association Special Volume No. 33. Oxford: Clarendon Press.

Smith, J. E. ed. (1968). *'Torrey Canyon' pollution and marine life*. London: Cambridge University Press.

Yonge, C. M. (1949). *The sea shore*. London: Collins.

SEAWEEDS

Bartsch, I., Wiencke, C., Bischof, K. *et al*. (2008). The genus *Laminaria sensu lato*: recent insights and developments. *European Journal of Phycology*, **43**, 1–86.

Blomster, J., Maggs, C. A. & Stanhope, M. J. (1998). Molecular and morphological analysis of *Enteromorpha intestinalis* and *E. compressa* (Chlorophyta) in the British Isles. *Journal of Phycology*, **34**, 319–40.

Boney, A. D. (1966). *A biology of marine algae*. London: Hutchinson Educational.

Brawley, S. H. (1992). Fertilization in natural populations of the dioecious brown alga *Fucus ceranoides* and the importance of the polyspermy block. *Marine Biology*, **113**, 145–57.

Chapman, A. R. O. (1986). Population and community ecology of seaweeds. *Advances in Marine Biology*, **23**, 1–161.

Critchley, A. T. (1983). *Sargassum muticum*: a morphological description of European material. *Journal of the Marine Biological Association of the United Kingdom*, **63**, 813–24.

Critchley, A. T., Farnham, W. F. & Morrell, S. L. (1983). A chronology of new European sites of attachment for the invasive brown alga, *Sargassum muticum*, 1973–1981. *Journal of the Marine Biological Association of the United Kingdom*, **63**, 799–811.

Farrell, P. & Fletcher, R. L. (2006). An investigation of dispersal of the introduced brown alga *Undaria pinnatifida* (Harvey) Suringar and its competition with some species on the man-made structures of Torquay Marina (Devon, UK). *Journal of Experimental Marine Biology and Ecology*, **334**, 236–43.

Fletcher, R. L. & Farrell, P. (1999). Introduced brown algae in the North East Atlantic, with particular respect to *Undaria pinnatifida* (Harvey) Suringar. *Helgoländer Meeresuntersuchungen*, **52**, 259–75.

Geiselman, J. A. & McConnell, O. J. (1981). Polyphenols in brown algae *Fucus vesiculosus* and *Ascophyllum nodosum*: chemical defenses against the marine herbivorous snail. *Littorina littorea*. *Journal of Chemical Ecology*, **7**, 1115–33.

Gibb, D. C. (1957). The free-living forms of *Ascophyllum nodosum* (L.) Le Jol. *Journal of Ecology*, **45**, 49–83.

Grahame, J. & Hanna, F. S. (1989). Factors affecting the distribution of the epiphytic fauna of *Corallina officinalis* (L.) on an exposed rocky shore. *Ophelia*, **30**, 113–29.

Harries, D. B., Harrow, S., Wilson, J. R., Mair, J. M. & Donnan, D. W. (2007). The establishment of the invasive alga *Sargassum muticum* on the west coast of Scotland: a preliminary assessment of community effects. *Journal of the Marine Biological Association of the United Kingdom*, **87**, 1057–67.

Hayden, H. S., Blomster, J., Maggs, C. A. *et al.* (2003). Linnaeus was right all along: *Ulva* and *Enteromorpha* are not distinct genera. *European Journal of Phycology*, **38**, 277–94.

Kitching, J. A. (1987). The flora and fauna associated with *Himanthalia elongata* (L.) S. F. Gray in relation to water current and wave action in the Lough Hyne marine nature reserve. *Estuarine, Coastal and Shelf Science*, **25**, 663–76.

Lane, C. E., Mayes, C., Druehl, L. D. & Saunders, G. W. (2006). A multi-gene molecular investigation of the kelp (Laminariales, Phaeophyceae) supports substantial taxonomic re-organization. *Journal of Phycology*, **42**, 493–512.

Lobban, C. S., Harrison, P. J. & Duncan, M. J. (1985). *The physiological ecology of seaweeds*. Cambridge: Cambridge University Press.

Monteiro, C. A., Engelen, A. H. & Santos, R. O. P. (2009). Macro- and mesoherbivores prefer native seaweeds over the invasive brown seaweed *Sargassum muticum*: a potential regulating role on invasions. *Marine Biology*, **156**, 2505–15.

Nam, K. W., Maggs, C. A., McIvor, L. & Stanhope, M. J. (2000). Taxonomy and phylogeny of *Osmundea* (Rhodomelaceae, Rhodophyta) in Atlantic Europe. *Journal of Phycology*, **36**, 759–72.

Norton, T. A. & Burrows, E. M. (1969). Studies on marine algae of the British Isles. 7. *Saccorhiza polyschides* (Lightf.) Batt. *British Phycological Journal*, **4**, 19–53.

Oswald, R. C., Telford, N., Seed, R. & Happey-Wood, C. M. (1984). The effect of encrusting bryozoans on the photosynthetic activity of *Fucus serratus* L. *Estuarine, Coastal and Shelf Science*, **19**, 697–702.

Roberts, M. (1970). Studies on marine algae of the British Isles. 8. *Cystoseira tamariscifolia* (Hudson) Papenfuss. *British Phycological Journal*, **5**, 201–10.

Saunders, G. W. & Lehmkuhl, K. V. (2005). Molecular divergence and morphological diversity among four cryptic species of *Plocamium* (Plocamiales, Florideophyceae) in northern Europe. *European Journal of Phycology*, **40**, 293–312.

Scott, G. W., Hull, S. L., Hornby, S. E., Hardy, F. G. & Owens, N. J. P. (2001). Phenotypic variation in *Fucus spiralis* (Phaeophyceae): morphology, chemical phenotype and their relationship to the environment. *European Journal of Phycology*, **36**, 43–50.

South, G. R. & Burrows, E. M. (1967). Studies on marine algae of the British Isles. 5. *Chorda filum* (L.) Stackh. *British Phycological Bulletin*, **3**, 379–402.

Stengel, D. B. & Dring, M. J. (1997). Morphology and *in situ* growth rates of plants of *Ascophyllum nodosum* (Phaeophyta) from different shore levels and responses of plants to vertical transplantation. *European Journal of Phycology*, **32**, 193–202.

Stengel, D. B., Wilkes R. J. & Guiry, M. D. (1999). Seasonal growth and recruitment of *Himanthalia elongata* (Fucales, Phaeophycota) in different habitats on the Irish west coast. *European Journal of Phycology*, **34**, 213–21.

Tan, I. H., Blomster, J., Hansen, G. *et al.* (1999). Molecular phylogenetic evidence for a reversible morphogenetic switch controlling the gross morphology of two common genera of green seaweeds, *Ulva* and *Enteromorpha*. *Molecular Biology and Evolution*, **16**, 1011–18.

Trowbridge, C. D. (2001). Coexistence of introduced and native congeneric algae: *Codium fragile* and *C. tomentosum* on Irish rocky intertidal shores. *Journal of the Marine Biological Association of the United Kingdom*, **81**, 931–7.

Trowbridge, C. D., Farnham, W. F. & White, L. F. (2004). Thriving populations of the native macroalga *Codium tomentosum* on Guernsey rocky shores. *Journal of the Marine Biological Association of the United Kingdom*, **84**, 873–7.

Walker, R. H., Brodie, J., Russell, S. & Irvine, L. M. (2009). Biodiversity of coralline algae in the Northeastern Atlantic including *Corallina caespitosa* sp. nov. (Corallinoideae, Rhodophyta). *Journal of Phycology*, **45**, 287–97.

LICHENS

Boney, A. D. (1961). A note on the intertidal lichen *Lichina pygmaea* Ag. *Journal of the Marine Biological Association of the United Kingdom*, **41**, 123–6.

Fletcher, A. (1973). The ecology of marine (littoral) lichens on some rocky shores of Anglesey. *Lichenologist*, **5**, 368–400.

(1973). The ecology of maritime (supralittoral) lichens on some rocky shores of Anglesey. *Lichenologist*, **5**, 401–22.

ANTHOPHYTA

Cox, P. A., Laushman, R. H. & Ruckelshaus, M. H. (1992). Surface and submarine pollination in the sea grass *Zostera marina*. *Botanical Journal of the Linnean Society*, **109**, 281–91.

Portig, A. A., Mathers, R. G., Montgomery, W. I. & Govier, R. N. (1994). The distribution and utilisation of *Zostera* species in Strangford Lough, Northern Ireland. *Aquatic Botany*, **47**, 317–28.

Ralph, P. J. & Short, F. T. (2002). Impact of the wasting disease pathogen, *Labyrinthula zosterae*, on the photobiology of eelgrass *Zostera marina*. *Marine Ecology Progress Series*, **226**, 265–71.

Thompson, J. D. (1991). The biology of an invasive plant: what makes *Spartina anglica* so successful? *BioScience*, **41**, 393–401.

Turk, S. M. (1986). The three species of eelgrass (*Zostera*) on the Cornish coast. *Cornish Studies*, **14**, 15–22.

Vergeer, L. H. T. & den Hartog, C. (1991). Occurrence of wasting disease in *Zostera noltii*. *Aquatic Botany*, **40**, 155–63.

Wilson, D. P. (1949). The decline of *Zostera marina* L. at Salcombe and its effects on the shore. *Journal of the Marine Biological Association of the United Kingdom*, **28**, 395–412.

PORIFERA

Barucca, M., Azzini, F., Bavestrello, G. *et al.* (2007). The systematic position of some boring sponges (Demospongiae, Hadromerida) studied by molecular analysis. *Marine Biology*, **151**, 529–35.

Hooper, J. N. A. & van Soest, R. W. M. (2002). *Systema Porifera. A guide to the classification of sponges*, Vols. 1 and 2. New York: Kluwer Academic/Plenum Publishers.

Howson, C. M. & Chambers, S. J. (1999). *Ophlitaspongia* and *Ophlitaspongia papilla* reinstated, and a new species of *Ophlitaspongia* described (Porifera: Demospongiae: Microcionidae). *Journal of the Marine Biological Association of the United Kingdom*, **79**, 609–20.

Peattie, M. E. & Hoare, R. (1981). The subittoral ecology of the Menai Strait. II. The sponge *Halichondria panicea* (Pallas) and its associated fauna. *Estuarine, Coastal and Shelf Science*, **13**, 621–35.

Picton, B. E. & Goodwin, C. E. (2007). Sponge biodiversity of Rathlin Island, Northern Ireland. *Journal of the Marine Biological Association of the United Kingdom*, **87**, 1441–58.

Soest, R. W. M. van, Kluijver, M. J. de, Bragt, P. H. van *et al.* (2007). Sponge invaders in Dutch coastal waters. *Journal of the Marine Biological Association of the United Kingdom*, **87**, 1733–48.

Solé-Cava, A. M. & Thorpe, J. P. (1986). Genetic differentiation between morphotypes of the marine sponge *Suberites ficus* (Demospongiae: Hadromerida). *Marine Biology*, **93**, 247–53.

Stefaniak, L. M., McAtee, J. & Shulman, M. J. (2005). The costs of being bored: effects of a clionid sponge on the gastropod *Littorina littorea* (L). *Journal of Experimental Marine Biology and Ecology*, **327**, 103–14.

Stone, A. R. (1970). Growth and reproduction of *Hymeniacidon perleve* (Montagu) (Porifera) in Langstone Harbour, Hampshire. *Journal of Zoology (London)*, **161**, 443–59.

Vacelet, J. & Boury-Esnault, N. (1995). Carnivorous sponges. *Nature*, **373**, 333–5.

CNIDARIA

Arai, M. N. (1997). *A functional biology of Scyphozoa*. London: Chapman & Hall.

Bell, J. J. & Turner, J. R. (2000). Factors influencing the density and morphometrics of the cup coral *Caryophyllia smithii* in Lough Hyne. *Journal of the Marine Biological Association of the United Kingdom*, **80**, 437–41.

Brace, R. C. & Quicke, D. L. J. (1985). Further analysis of individual spacing within aggregations of the anemone, *Actinia equina*. *Journal of the Marine Biological Association of the United Kingdom*, **65**, 35–53.

(1986). Dynamics of colonization by the beadlet anemone, *Actinia equina*. *Journal of the Marine Biological Association of the United Kingdom*, **66**, 21–47.

Brace, R. C. & Sauter, S.-J. (1991). Experimental habituation of aggression in the sea anemone, *Actinia equina*. *Hydrobiologia*, **216/217**, 533–7.

Bucklin, A. (1985). Biochemical genetic variation, growth and regeneration of the sea anemone, *Metridium*, of British shores. *Journal of the Marine Biological Association of the United Kingdom*, **65**, 141–57.

Carter, M. A. & Miles, J. (1989). Gametogenic cycles and reproduction in the beadlet sea anemone *Actinia equina* (Cnidaria: Anthozoa). *Biological Journal of the Linnean Society*, **36**, 129–55.

Dawson, M. N. & Jacobs, D. K. (2001). Molecular evidence for cryptic species of *Aurelia aurita* (Cnidaria, Scyphozoa). *Biological Bulletin*, **200**, 92–6.

Denitto, F., Miglietta, M. P. & Boero, F. (2007). Life cycle of *Bougainvillia nana* (Cnidaria: Hydrozoa: Bougainvilliidae) from Italy, including a discussion of *Bougainvillia muscus* in the Mediterranean Sea. *Journal of the Marine Biological Association of the United Kingdom*, **87**, 853–7.

Donaghue, A. M., Quicke, D. L. J. & Brace, R. C. (1985). Biochemical–genetic and acrorhagial characteristics of pedal disc colour phenotypes of *Actinia equina*. *Journal of the Marine Biological Association of the United Kingdom*, **65**, 21–33.

Dorsett, D. A. & Turner, R. (1986). Colour morph dominance and avoidance responses of *Anemonia sulcata*. *Marine Behaviour and Physiology*, **12**, 115–23.

Gashout, S. E. & Ormond, R. F. G. (1979). Evidence for parthenogenetic reproduction in the sea anemone *Actinia equina* L. *Journal of the Marine Biological Association of the United Kingdom*, **59**, 975–87.

Hartnoll, R. G. (1975). The annual cycle of *Alcyonium digitatum*. *Estuarine and Coastal Marine Science*, **3**, 71–8.

Haylor, G. S., Thorpe, J. P. & Carter, M. A. (1984). Genetic and ecological differentiation between sympatric colour morphs of the common intertidal sea anemone *Actinia equina*. *Marine Ecology Progress Series*, **16**, 281–9.

Heuwinkel, U. B. (1988). Brooding behaviour in the marine hydropolyp *Myriothela cocksi* (Vigors) by means of special tentacles. *Acta Biologica Benrodis*, **1**, 57–67.

Hiscock, K. & Howlett, R. M. (1976). The ecology of *Caryophyllia smithi* Stokes and Broderip on south-western coasts of the British Isles. In *Underwater research*, ed. A. Drew, J. N. Lythgoe & J. D. Woods, pp. 319–34. London: Academic Press.

Holst, S., Sötje, I., Tiemann, H. & Jarms, G. (2007). Life cycle of the rhizostome jellyfish *Rhizostoma octopus* (L.) (Scyphozoa, Rhizostomeae), with studies on cnidocysts and statoliths. *Marine Biology*, **151**, 1695–710.

Hughes, R. G. (1983). The life-history of *Tubularia indivisa* (Hydrozoa: Tubulariidae) with observations on the status of *T. ceratogyne*. *Journal of the Marine Biological Association of the United Kingdom*, **63**, 467–79.

Lilley, M. K. S., Houghton, J. D. R. & Hays, G. C. (2009). Distribution, extent of inter-annual variability and diet of the bloom-forming jellyfish *Rhizostoma* in European waters. *Journal of the Marine Biological Association of the United Kingdom*, **89**, 39–48.

Manton, S. M. (1944). On the hydrorhiza and claspers of the hydroid *Myriothela cocksi* (Vigurs). *Journal of the Marine Biological Association of the United Kingdom*, **25**, 143–50.

Marfenin, N. N. & Belorustseva, S. A. (2008). How the distribution of colonies of the hydroid *Laomedea flexuosa* is limited to a narrow belt along the lower littoral zone. *Journal of the Marine Biological Association of the United Kingdom*, **88**, 1559–66.

Orr, J., Thorpe, J. P. & Carter, M. A. (1982). Biochemical genetic confirmation of the asexual reproduction of brooded offspring in the sea anemone *Actinia equina*. *Marine Ecology Progress Series*, **7**, 227–9.

Perrin, M. C., Thorpe, J. P. & Solé-Cava, A. M. (1999). Population structuring, gene dispersal and reproduction in the *Actinia equina* species group. *Oceanography and Marine Biology. An Annual Review*, **37**, 129–52.

Pyefinch, K. A. & Downing, F. S. (1949). Notes on the general biology of *Tubularia larynx* Ellis & Solander. *Journal of the Marine Biological Association of the United Kingdom*, **28**, 21–43.

Quicke, D. L. J., Donoghue, A. M., Keeling, T. F. & Brace, R. C. (1985). Littoral distributions and evidence for differential post-settlement selection of the morphs of *Actinia equina*. *Journal of the Marine Biological Association of the United Kingdom*, **65**, 1–20.

Solé-Cava, A. M. & Thorpe, J. P. (1992). Genetic divergence between colour morphs in populations of the common intertidal sea anemones *Actinia equina* and *A. prasina* (Anthozoa: Actiniaria) in the Isle of Man. *Marine Biology*, **112**, 243–52.

Solé-Cava, A. M., Thorpe, J. P. & Kaye, J. G. (1985). Reproductive isolation with little genetic divergence between *Urticina* (= *Tealia*) *felina* and *U. eques* (Anthozoa: Actiniaria). *Marine Biology*, **85**, 279–84.

Stepanjants, S. D. (1998). *Obelia* (Cnidaria, Medusozoa, Hydrozoa): phenomenon, aspects of investigations, perspectives for utilization. *Oceanography and Marine Biology. An Annual Review*, **36**, 179–215.

Turk, S. M. (1982). Influx of warm-water oceanic drift animals into the Bristol and English Channels, summer 1981. *Journal of the Marine Biological Association of the United Kingdom*, **62**, 487–9.

Turner, V. L. G., Lynch, S. M., Paterson, L., León-Cortés, J. L. & Thorpe, J. P. (2003). Aggression as a function of genetic relatedness in the sea anemone *Actinia equina* (Anthozoa: Actiniaria). *Marine Ecology Progress Series*, **247**, 85–92.

Watts, P. C. & Thorpe, J. P. (1998). Phenotypic identification of three genetically differentiated morphs of the intertidal beadlet anemone *Actinia equina* (Anthozoa: Cnidaria). *Journal of the Marine Biological Association of the United Kingdom*, **78**, 1365–8.

Williams, G. B. (1965). Observations on the behaviour of the planulae larvae of *Clava squamata*. *Journal of the Marine Biological Association of the United Kingdom*, **45**, 257–73.

CTENOPHORA

Fraser, J. H. (1970). The ecology of the ctenophore *Pleurobrachia pileus* in Scottish waters. *Journal du Conseil International pour l'Exploration de la Mer*, **33**, 149–68.

Haddock, S. H. D. & Case, J. F. (1995). Not all ctenophores are bioluminescent: *Pleurobrachia*. *Biological Bulletin*, **189**, 356–62.

Harbison, G. R. (1985). On the classification and evolution of the Ctenophora. In *The origins and relationships of lower invertebrates*, ed. S. C. Morris, J. D. George, R. Gibson & H. M. Platt, Systematics Association Special Volume no. 28, pp. 78–100. Oxford: Clarendon Press.

Swanberg, N. (1974). The feeding behaviour of *Beroe ovata*. *Marine Biology*, **24**, 69–76.

PLATYHELMINTHES

Baguñà, J. & Riutort, M. (2004). Molecular phylogeny of the Platyhelminthes. *Canadian Journal of Zoology*, **82**, 168–93.

Boaden, P. J. S. (1995). Where Turbellaria? Concerning knowledge and ignorance of marine turbellarian ecology. *Hydrobiologia*, **305**, 91–9.

Faubel, A. & Warwick, R. M. (2005). The marine flora and fauna of the Isles of Scilly: free-living Platyhelminthes ('Turbellaria'). *Journal of Natural History*, **39**, 1–45.

Lang, A. (1884). Die Polycladen (Seeplanarian) des Golfes von Neapel und der angrenzeden Meeresabschnitte. Eine monographie. *Fauna und flora des Golfes von Neapel*, **11**, ix–688.

Sluys, R. (1987). The taxonomy of three species of the genus *Procerodes* (Platyhelminthes: Tricladida: Maricola). *Journal of the Marine Biological Association of the United Kingdom*, **67**, 373–84.

NEMERTEA

Bartsch, I. (1973). Zur Nahrungsaufnahme von *Tetrastemma melanocephalum* (Nemertini). *Helgoländer wissenschaftliche Meeresuntersuchungen*, **25**, 326–31.

Gibson, R. (1995). Nemertean genera and species of the world: an annotated checklist of original names and descriptions, citations, synonyms, current taxonomic status, habitats and recorded zoogeographic distribution. *Journal of Natural History*, **29**, 271–562.

McDermott, J. J. & Roe, P. (1985). Food, feeding behaviour and feeding ecology of nemerteans. *American Zoologist*, **25**, 113–25.

Rogers, A. D., Thorpe, J. P., Gibson, R. & Norenburg, J. L. (1998). Genetic differentiation of populations of the common intertidal nemerteans *Lineus ruber* and *Lineus viridis* (Nemertea, Anopla). *Hydrobiologia*, **365**, 1–11.

Sundberg, P. (1984). Multivariate analysis of polymorphism in the hoplonemertean *Oerstedia dorsalis* (Abildgaard, 1806). *Journal of Experimental Marine Biology and Ecology*, **78**, 1–22.

Sundberg, P. & Andersson, S. (1995). Random amplified polymorphic DNA (RAPD) and intraspecific variation in *Oerstedia dorsalis* (Hoplonemertea, Nemertea). *Journal of the Marine Biological Association of the United Kingdom*, **75**, 483–90.

Sundberg, P. & Janson, K. (1988). Polymorphism in *Oerstedia dorsalis* (Abildgaard, 1806) revisited: electrophoretic evidence for a species complex. *Hydrobiologia*, **156**, 93–8.

Sundberg, P., Gibson, R. & Strand, M. (2007). Swedish nemerteans (phylum Nemertea), with description of a new hoplonemertean genus and species. *Journal of Natural History*, **41**, 2287–99.

Vernet, G. & Bierne, T. (1993). The influence of light and sea water temperature on the reproductive cycle of *Lineus ruber* (Heteronemertea). *Hydrobiologia*, **266**, 267–71.

PRIAPULA

Hammond, R. A. (1970). The burrowing of *Priapulus caudatus*. *Journal of Zoology (London)*, **162**, 469–80.

Lang, K. (1948). Contribution to the ecology of *Priapulus caudatus* Lam. *Arkiv för Zoologi*, **41**, 1–12.

ANNELIDA

Bass, N. R. & Brafield, A. E. (1972). The life-cycle of the polychaete *Nereis virens*. *Journal of the Marine Biological Association of the United Kingdom*, **52**, 701–26.

Bedford, A. P. & Moore, P. G. (1985). Macrofaunal involvement in the sublittoral decay of kelp debris: the polychaete *Platynereis dumerilii* (Audouin and Milne-Edwards) (Annelida: Polychaeta). *Estuarine, Coastal and Shelf Science*, **20**, 117–34.

Bentley, M. G., Olive, P. J. W., Garwood, P. R. & Wright, N. H. (1984). The spawning and spawning mechanism of *Nephtys caeca* (Fabricius, 1780) and *Nephtys hombergi* Savigny, 1818 (Annelida: Polychaeta). *Sarsia*, **69**, 63–8.

Bhaud, M. R. (1991). Larval release from the egg mass and settlement of *Eupolymnia nebulosa* (Polychaeta, Terebellidae). *Bulletin of Marine Science*, **48**, 420–31.

Böggemann, M. & Fiege, D. (2001). Description of seven new species of the genus *Glycera* Savigny, 1818 (Annelida: Polychaeta: Glyceridae). *Ophelia*, **54**, 29–49.

Bonse, S., Schmidt, H., Eibye-Jacobsen, D. & Westheide, W. (1996). *Eulalia viridis* (Polychaeta: Phyllodocidae) is a complex of two species in northern Europe: results from biochemical and morphological analyses. *Cahiers de Biologie Marine*, **37**, 33–48.

Bosence, D. W. J. (1979). The factors leading to aggregation and reef formation in *Serpula vermicularis* L. In *Biology and systematics of colonial organisms*, ed. G. Larwood & B. R. Rosen, Systematics Association Special Volume no. 11, pp. 299–318. London: Academic Press.

Breton, S., Dufresne, F., Desrosiers, G. & Blier, P. U. (2003). Population structure of two northern hemisphere polychaetes, *Neanthes virens* and *Hediste diversicolor* (Nereididae), with different life-history traits. *Marine Biology*, **42**, 707–15.

Cadman, P. S. & Nelson-Smith, A. (1993). A new species of lugworm: *Arenicola defodiens* sp. nov. *Journal of the Marine Biological Association of the United Kingdom*, **73**, 213–23.

Callaway, R. (2003). Juveniles stick to adults: recruitment of the tube-dwelling polychaete *Lanice conchilega* (Pallas, 1766). *Hydrobiologia*, **503**, 121–30.

Chapman, N. D., Moore, C. G., Harries, D. B. & Lyndon, A. R. (2007). Recruitment patterns of *Serpula vermicularis* L. (Polychaeta, Serpulidae) in Loch Creran, Scotland. *Estuarine, Coastal and Shelf Science*, **73**, 598–606.

Cotter, E., O'Riordan, R. M. & Myers, A. A. (2003). A histological study of reproduction in the serpulids *Pomatoceros triqueter* and *Pomatoceros lamarckii* (Annelida: Polychaeta). *Marine Biology*, **142**, 905–14.

Cram, A. & Evans, S. M. (1980). Shell entry behaviour in the commensal ragworm *Nereis fucata*. *Marine Behaviour and Physiology*, **7**, 57–64.

Crisp, D. J. & Ekaratne, K. (1984). Polymorphism in *Pomatoceros*. *Zoological Journal of the Linnean Society*, **80**, 157–75.

Dales, R. P. (1950). The reproduction and larval development of *Nereis diversicolor* O. F. Müller. *Journal of the Marine Biological Association of the United Kingdom*, **29**, 321–60.

(1955). Feeding and digestion in terebellid polychaetes. *Journal of the Marine Biological Association of the United Kingdom*, **34**, 55–79.

Dales, R. P. (1957). The feeding mechanism and structure of the gut of *Owenia fusiformis* Delle Chiaje. *Journal of the Marine Biological Association of the United Kingdom*, **36**, 81–9.

Daly, J. M. (1972). The maturation and breeding biology of *Harmothoë imbricata* (Polychaeta: Polynoidae). *Marine Biology*, **12**, 53–66.

(1973). Behavioural and secretory activity during tube construction by *Platynereis dumerilii* Aud. & M. Edw. (Polychaeta: Nereidae). *Journal of the Marine Biological Association of the United Kingdom*, **53**, 521–9.

(1978). The annual cycle and the short term periodicity of breeding in a Northumberland population of *Spirorbis spirorbis* (Polychaeta: Serpulidae). *Journal of the Marine Biological Association of the United Kingdom*, **58**, 161–76.

Daly, J. M. & Golding, D. W. (1977). A description of the spermatheca of *Spirorbis spirorbis* (L.) (Polychaeta: Serpulidae) and evidence for a novel mode of sperm transmission. *Journal of the Marine Biological Association of the United Kingdom*, **57**, 219–27.

Daro, M. H. & Polk, P. (1973). The autecology of *Polydora ciliata* along the Belgian coast. *Netherlands Journal of Sea Research*, **6**, 130–40.

Dauer, D. M. & Ewing, M. R. (1991). Functional morphology and feeding behaviour of *Malacoceros indicus* (Polychaeta: Spionidae). *Bulletin of Marine Science*, **48**, 395–400.

Dauvin, J.-C., Dewarumez, J.-M. & Gentil, F. (2003). Liste actualisée des espèces d'Annélides Polychètes présentes en Manche. *Cahiers de Biologie Marine*, **44**, 67–95.

Dean, D., Chapman, S. R. & Chapman, C. S. (1987). Reproduction and development of the sabellid polychaete *Myxicola infundibulum*. *Journal of the Marine Biological Association of the United Kingdom*, **67**, 431–9.

de Silva, P. H. D. H. (1967). Studies on the biology of Spirorbinae (Polychaeta). *Journal of Zoology (London)*, **152**, 269–79.

Dorsett, D. A. (1961). The reproduction and maintenance of *Polydora ciliata* (Johnst.) at Whitstable. *Journal of the Marine Biological Association of the United Kingdom*, **41**, 383–96.

(1961). The behaviour of *Polydora ciliata* (Johnst.): tube building and burrowing. *Journal of the Marine Biological Association of the United Kingdom*, **41**, 577–90.

Emson, R. H. (1977). The feeding and consequent role of *Eulalia viridis* (O. F. Müller) (Polychaeta) in intertidal communities. *Journal of the Marine Biological Association of the United Kingdom*, **57**, 93–6.

Enders, H. E. (1909). A study of the life-history and habits of *Chaetopterus variopedatus*, Renier et Claparede. *Journal of Morphology*, **20**, 479–531.

Fauchald, K. & Jumars, P. A. (1979). The diet of worms: a study of polychaete feeding guilds. *Oceanography and Marine Biology. An Annual Review*, **17**, 193–284.

Fauchald, K. & Rouse, G. (1997). Polychaete systematics: past and present. *Zoologica Scripta*, **26**, 71–138.

Fischer, A. & Pfannenstiel, H.-D. eds. (1984). *Polychaete reproduction: progress in comparative reproductive biology*. Stuttgart: Gustav Fischer.

Ford, E. & Hutchings, P. (2005). An analysis of morphological characters of *Owenia* useful to distinguish species: description of three new species of *Owenia* (Oweniidae: Polychaeta) from Australian waters. *Marine Ecology*, **26**, 181–96.

George, J. D. (1964). The life history of the cirratulid worm, *Cirriformia tentaculata*, on an intertidal mud flat. *Journal of the Marine Biological Association of the United Kingdom*, **44**, 47–65.

Gibbs, P. E. (1968). Observations on the population of *Scoloplos armiger* at Whitstable. *Journal of the Marine Biological Association of the United Kingdom*, **48**, 225–54.

(1971). A comparative study of reproductive cycles in four polychaete species belonging to the family Cirratulidae. *Journal of the Marine Biological Association of the United Kingdom*, **51**, 745–69.

Gilpin-Brown, J. B. (1959). The reproduction and larval development of *Nereis fucata* (Savigny). *Journal of the Marine Biological Association of the United Kingdom*, **38**, 65–80.

(1969). Host-adoption in the commensal polychaete *Nereis fucata*. *Journal of the Marine Biological Association of the United Kingdom*, **49**, 121–7.

Grant, A. (1989). The reproductive cycle of *Platynereis dumerilii* (Audouin & Milne-Edwards) (Polychaeta: Nereidae) from the Firth of Clyde. *Sarsia*, **74**, 79–84.

Grassle, J. P. (1984). Speciation in the genus *Capitella* (Polychaeta, Capitellidae). In *Polychaete reproduction: progress in comparative reproductive biology*, ed. A. Fischer & H.-D. Pfannenstiel, pp. 293–8. Stuttgart: Gustav Fischer.

Grémare, A. (1986). A comparative study of reproductive energetics in two populations of the terebellid polychaete *Eupolymnia nebulosa* Montagu with different reproductive modes. *Journal of Experimental Marine Biology and Ecology*, **96**, 287–302.

Gruet, Y. & Lassus, P. (1983). Contribution à l'étude de la biologie reproductive d'une population naturelle de l'annélide polychète *Sabellaria alveolata* (Linné). *Annales de l'Institut Océanographique, Paris*, **59**, 127–40.

Gudmundsson, H. (1985). Life history pattern of polychaete species of the family Spionidae. *Journal of the Marine Biological Association of the United Kingdom*, **65**, 93–111.

Guérin, J.-P. & Kerambrun, P. (1984). Role of reproductive characters in the taxonomy of spionids and elements of speciation in the 'Malacoceros fuliginosus complex'. In *Polychaete reproduction: progress in comparative reproductive biology*, ed. A. Fischer & H.-D. Pfannenstiel, pp. 317–33. Stuttgart: Gustav Fischer.

Hardege, J. D. & Bentley, M. G. (1997). Spawning synchrony in *Arenicola marina*: evidence for sex pheromonal control. *Proceedings of the Royal Society of London Series B–Biological Sciences*, **264**, 1041–7.

Hateley, J. G., Grant, A., Taylor, S. M. & Jones, N. V. (1992). Morphological and other evidence on the degree of genetic differentiation between populations of *Nereis diversicolor*. *Journal of the Marine Biological Association of the United Kingdom*, **72**, 365–81.

Holthe, T. (1992). Identification of Annelida Polychaeta from northern European and adjacent Arctic waters. *Gunnerus*, **66**, 1–30.

Howie, D. I. D. (1984). The reproductive biology of the lugworm, *Arenicola marina* L. In *Polychaete reproduction: progress in comparative reproductive biology*, ed. A. Fischer & H.-D. Pfannenstiel, pp. 247–63. Stuttgart: Gustav Fischer.

Irlinger, J. P., Gentil, F. & Quintino, V. (1991). Reproductive biology of the polychaete *Pectinaria koreni* (Malmgren) in the Bay of Seine (English Channel). *Ophelia* (Supplement), **5**, 343–50.

Joyner, A. (1962). Reproduction and larval life of *Nerine cirratulus* (Delle Chiaje) Family Spionidae. *Proceedings of the Zoological Society of London*, **138**, 655–66.

Knight-Jones, P. & Knight-Jones, E. W. (1977). Taxonomy and ecology of British Spirorbidae (Polychaeta). *Journal of the Marine Biological Association of the United Kingdom*, **57**, 453–99.

Knight-Jones, P. & Perkins, T. H. (1998). A revision of *Sabella*, *Bispira* and *Stylomma* (Polychaeta: Sabellidae). *Zoological Journal of the Linnean Society*, **123**, 385–467.

Knight-Jones, P. & Thorp, C. H. (1984). The opercular brood chambers of Spirorbidae. *Zoological Journal of the Linnean Society*, **80**, 121–33.

Knight-Jones, P., Knight-Jones, E. W. & Kawahara, T. (1975). A review of the genus *Janua*, including *Dexiospira* (Polychaeta: Spirorbinae). *Zoological Journal of the Linnean Society*, **56**, 91–129.

Kristensen, E. (1984). Life cycle, growth and production in estuarine populations of the polychaetes, *Nereis virens* and *N. diversicolor*. *Holarctic Ecology*, **7**, 249–56.

Kruse, I., Strasser, M. & Thiermann, F. (2004). The role of ecological divergence in speciation between intertidal and subtidal *Scoloplos armiger* (Polychaeta, Orbiniidae). *Journal of Sea Research*, **51**, 53–62.

Kupriyanova, E. K., Nishi, E., Ten Hove, H. A. & Rzhavsky, A. V. (2001). Life-history patterns in serpulimorph polychaetes: ecological and evolutionary perspectives. *Oceanography and Marine Biology. An Annual Review*, **39**, 1–101.

Lewis, D. B. (1968). Some aspects of the ecology of *Fabricia sabella* (Ehr.) (Annelida, Polychaeta). *Journal of the Linnean Society of London, Zoology*, **47**, 515–26.

(1968). Feeding and tube-building in the Fabriciinae (Annelida, Polychaeta). *Proceedings of the Linnean Society of London*, **179**, 37–49.

Luttikhuizen, P. C. & Dekker, R. (2010). Pseudo-cryptic species *Arenicola defodiens* and *Arenicola marina* (Polychaeta: Arenicolidae) in Wadden Sea, North Sea and Skagerak: morphological and molecular variation. *Journal of Sea Research*, **63**, 17–23.

Maltagliati, F., Casu, M. & Castelli, A. (2004). Morphological and genetic evidence supports the existence of two species in the genus *Ophelia* (Annelida, Polychaeta) from the western Mediterranean. *Biological Journal of the Linnean Society*, **83**, 101–13.

Mettam, C. (1980). On the feeding habits of *Aphrodita aculeata* and commensal polynoids. *Journal of the Marine Biological Association of the United Kingdom*, **60**, 833–4.

Morgan, T. S., Rogers, A. D., Paterson, G. L. J., Hawkins, L. E. & Sheader, M. (1999). Evidence for poecilogony in *Pygospio elegans* (Polychaeta: Spionidae). *Marine Ecology Progress Series*, **178**, 121–32.

Newell, G. E. (1949). The later larval life of *Arenicola marina* L. *Journal of the Marine Biological Association of the United Kingdom*, **28**, 635–9.

Nicol, E. A. T. (1931). The feeding mechanism, formation of the tube, and physiology of digestion in *Sabella pavonina*. *Transactions of the Royal Society of Edinburgh*, **56**, 537–98.

Nicolaidou, A. (1988). Notes on the behaviour of *Pectinaria koreni*. *Journal of the Marine Biological Association of the United Kingdom*, **68**, 55–9.

Nygren, A. (1999). Phylogeny and reproduction in Syllidae (Polychaeta). *Zoological Journal of the Linnean Society*, **126**, 365–86.

O'Connor, B. D. S. (1987). The Glyceridae (Polychaeta) of the North Atlantic and Mediterranean, with descriptions of two new species. *Journal of Natural History*, **21**, 167–89.

Olive, P. J. W. (1970). Reproduction of a Northumberland population of the polychaete *Cirratulus cirratus*. *Marine Biology*, **5**, 259–73.

(1975). Reproductive biology of *Eulalia viridis* (Müller) (Polychaeta: Phyllodocidae) in the north eastern U.K. *Journal of the Marine Biological Association of the United Kingdom*, **55**, 313–26.

(1978). Reproduction and annual gametogenic cycle in *Nephtys hombergii* and *N. caeca* (Polychaeta: Nephtyidae). *Marine Biology*, **46**, 83–90.

Olive, P. J. W. & Garwood, P. R. (1981). Gametogenic cycle and population structure of *Nereis (Hediste) diversicolor* and *Nereis (Nereis) pelagica* from north-east England. *Journal of the Marine Biological Association of the United Kingdom*, **61**, 193–213.

Olive, P. J. W. & Pillai, G. (1983). Reproductive biology of the polychaete *Kefersteinia cirrata* Keferstein (Hesionidae). II. The gametogenic cycle and evidence for photoperiodic control of oogenesis. *International Journal of Invertebrate Reproduction*, **6**, 307–15.

Olive, P. J. W., Garwood, P. R., Bentley, M. G. & Wright, N. (1981). Reproductive success, relative abundance and population structure of two species of *Nephtys* in an estuarine beach. *Marine Biology*, **63**, 189–96.

Olivier, F., Desroy, N. & Retière, C. (1996). Habitat selection and adult-recruit interactions in *Pectinaria koreni* (Malmgren) (Annelida: Polychaeta) post-larval populations: results of flume experiments. *Journal of Sea Research*, **36**, 217–26.

Osborne, K., Rouse, G. W., Goffed, S. & Robison, B. H. (2007). Description and relationships of *Chaetopterus pugaporcinus* an unusual pelagic polychaete (Annelida, Chaetopteridae). *Biological Bulletin*, **212**, 40–54.

Oyenekan, J. A. (1983). Production and population dynamics of *Capitella capitata*. *Archiv für Hydrobiologie*, **98**, 115–26.

Pearson, M. & Pearson, T. H. (1991). Variation in populations of *Capitella capitata* (Fabricius, 1780) (Polychaeta) from the west coast of Scotland. *Ophelia* (Supplement), **5**, 363–70.

Pearson, T. H. (1969). *Scionella lornensis* sp. nov., a new terebellid (Polychaeta: Annelida) from the west coast of Scotland, with notes on the genus *Scionella* Moore, and a key to the genera of the Terebellidae recorded from European waters. *Journal of Natural History*, **3**, 509–16.

Plate, S. & Husemann, E. (1991). An alternative mode of larval development in *Scoloplos armiger* (O. F. Müller, 1776) (Polychaeta, Orbiniidae). *Helgoländer Meeresuntersuchungen*, **45**, 487–92.

Prevedelli, D., N'Siala, G. M., Ansaloni, I. & Simonini, R. (2007). Life cycle of *Marphysa sanguinea* (Polychaeta: Eunicidae) in the Venice lagoon. (Italy). *Marine Ecology*, **28**, 384–93.

Rainer, S. F. (1991). The genus *Nephtys* (Polychaeta: Phyllodocida) of northern Europe: a review of species, including the description of *N. pulchra* sp. n. and a key to the Nephtyidae. *Helgoländer Meeresuntersuchungen*, **45**, 65–96.

Rouse, G. W. & Fauchald K. (1997). Cladistics and polychaetes. *Zoologica Scripta*, **26**, 139–204.

Scaps, P., Rouabah, A. & Leprêtre, A. (2000). Morphological and biochemical evidence that *Perinereis cultrifera* (Polychaeta: Nereididae) is a complex of species. *Journal of the Marine Biological Association of the United Kingdom*, **80**, 735–6.

Schiedges, K.-L. (1980). Morphological and systematic studies of an *Autolytus* population (Polychaeta, Syllidae, Autolytinae) from the Oosterschelde Estuary. *Netherlands Journal of Sea Research*, **14**, 208–19.

Stephenson, T. A. (1944). *Seashore life and pattern*. London: King Penguin Books.

Thorson, G. (1946). Reproduction and larval development of Danish marine bottom invertebrates, with special reference to the planktonic larvae in the Sound (Øresund). *Meddelelser fra Kommissionen for Danmarks Fiskeri-og Havundersøgelser (Serie: Plankton)* **4**, 1–523.

Tsutsumi, H. & Kikuchi, T. (1984). Study on the life history of *Capitella capitata* (Polychaeta: Capitellidae) in Amakusa, south Japan including a comparison with other geographical regions. *Marine Biology*, **80**, 315–21.

Upton, B. (1953). The escape from the cocoon of *Scoloplos armiger*. *British Journal of Animal Behaviour*, **1**, 87.

Vedel, A. & Riisgård, H. A. (1993). Filter-feeding in the polychaete *Nereis diversicolor*: growth and biogenetics. *Marine Ecology Progress Series*, **100**, 145–52.

Warren, L. M. (1977). The ecology of *Capitella capitata* in British waters. *Journal of the Marine Biological Association of the United Kingdom*, **57**, 151–9.

Watson, G. J., Cadman, P. S., Paterson, L. A., Bentley, M. G. & Auckland, M. F. (1998). Control of oocyte maturation, sperm activation and spawning in two lugworm species: *Arenicola marina* and *A. defodiens*. *Marine Ecology Progress Series*, **175**, 167–76.

Willcox, M. S. & Nickell, T. D. (1998). Field evidence of poecilogony in *Capitella capitata*. *Ophelia*, **49**, 141–5.

Wilson, D. P. (1948). The larval development of *Ophelia bicornis* Savigny. *Journal of the Marine Biological Association of the United Kingdom*, **27**, 540–53.

(1971). *Sabellaria* colonies in Duckpool, North Cornwall, 1961–70. *Journal of the Marine Biological Association of the United Kingdom*, **51**, 509–80.

MOLLUSCA: POLYPLACOPHORA

Matthews, G. (1953). A key for use in the identification of British chitons. *Proceedings of the Malacological Society of London*, **29**, 241–8.

MOLLUSCA: GASTROPODA

Barnes, R. S. K. (1991). On the distribution of the northwest European species of the gastropod *Hydrobia*, with particular reference to *H. neglecta*. *Journal of Conchology*, **34**, 59–62.

Barnes, R. S. K. (2005). Interspecific competition and rarity in mudsnails: feeding interactions between and within *Hydrobia acuta neglecta* and sympatric *Hydrobia* species. *Aquatic Conservation: Marine and Freshwater Ecosystems*, **15**, 485–93.

Barnett, P. R. O., Hardy, B. L. S. & Watson, J. (1980). Substratum selection and egg-capsule deposition in *Nassarius reticulatus* (L.). *Journal of Experimental Marine Biology and Ecology*, **45**, 95–103.

Birchenough, A. C., Evans, S. M., Moss, C. & Welch, R. (2002). Recolonisation and recovery of populations of dogwhelks *Nucella lapillus* (L.) on shores formerly subject to severe TBT contamination. *Marine Pollution Bulletin*, **44**, 652–9.

Boyle, P. R. (1977). The physiology and behaviour of chitons (Mollusca: Polyplacophora). *Oceanography and Marine Biology. An Annual Review*, **15**, 461–509.

Buschbaum, C., Buschbaum, G., Schrey, I. & Thieltges, D. W. (2007). Shell-boring polychaetes affect gastropod shell strength and crab predation. *Marine Ecology Progress Series*, **329**, 123–30.

Byers, B. A. (1990). Shell colour polymorphism associated with substrate colour in the intertidal snail *Littorina saxatilis* Olivi (Prosobranchia: Littorinidae). *Biological Journal of the Linnean Society*, **40**, 3–10.

Camacho-García, Y. E. & Gosliner, T. M. (2008). Systematic revision of *Jorunna* Berg, 1876 (Nudibranchia: Discodorididae) with a morphological phylogenetic analysis. *Journal of Molluscan Studies*, **74**, 143–81.

Carefoot, T. H. (1987). *Aplysia*: its biology and ecology. *Oceanography and Marine Biology. An Annual Review*, **25**, 167–284.

Chia, F. S. (1971). Oviposition, fecundity and larval development of three saccoglossan opistho-branchs from the Northumberland coast, England. *Veliger*, **13**, 319–25.

Chipperfield, P. N. J. (1951). The breeding of *Crepidula fornicata* (L.) in the River Blackwater, Essex. *Journal of the Marine Biological Association of the United Kingdom*, **30**, 49–71.

Claverie, T. & Kamenos, N. A. (2008). Spawning aggregations and mass movements in subtidal *Onchidoris bilamellata* (Mollusca: Opisthobranchia). *Journal of the Marine Biological Association of the United Kingdom*, **88**, 157–9.

Crothers, J. H. (1985). Dog-whelks: an introduction to the biology of *Nucella lapillus* (L.). *Field Studies*, **6**, 291–360.

Dussart, G. B. J. (1977). The ecology of *Potamopyrgus jenkinsi* (Smith) in north-west England with a note on *Marstoniopsis scholtzi* (Schmidt). *Journal of Molluscan Studies*, **43**, 208–16.

Fish, J. D. (1972). The breeding cycle and growth of open coast and estuarine populations of *Littorina littorea*. *Journal of the Marine Biological Association of the United Kingdom*, **52**, 1011–19.

(1979). The rhythmic spawning behaviour of *Littorina littorea* (L.). *Journal of Molluscan Studies*, **45**, 172–7.

Fish, J. D. & Fish, S. (1974). The breeding cycle and growth of *Hydrobia ulvae* in the Dovey Estuary. *Journal of the Marine Biological Association of the United Kingdom*, **54**, 685–97.

(1977). The veliger larva of *Hydrobia ulvae* with observations on the veliger of *Littorina littorea* (Mollusca: Prosobranchia). *Journal of Zoology (London)*, **182**, 495–503.

(1981). The early life-cycle stages of *Hydrobia ventrosa* and *Hydrobia neglecta* with observations on *Potamopyrgus jenkinsi*. *Journal of Molluscan Studies*, **47**, 89–98.

Fish, J. D. & Sharp, L. (1985). The ecology of the periwinkle, *Littorina neglecta* Bean. In *The ecology of rocky coasts*, ed. P. G. Moore & R. Seed, pp. 143–56. London: Hodder & Stoughton.

Fish, J. D., Fish, S. & Foley, H. (2000). The biology of mud snails with particular reference to *Hydrobia ulvae*. In *British salmarshes*, ed. B. R. Sherwood, B. G. Gardiner & T. Harris, pp. 165–79. Cardigan: Forrest Text.

Fretter, V. & Graham, A. (1954). Observations on the opisthobranch mollusc *Acteon tornatilis* (L.). *Journal of the Marine Biological Association of the United Kingdom*, 33, 565–85.

(1994). *British prosobranch molluscs: their functional anatomy and ecology*, revised and updated edition. London: Ray Society.

Fretter, V. & Manly, R. (1977). Algal associations of *Tricolia pullus*, *Lacuna vincta* and *Cerithiopsis tubercularis* (Gastropoda) with special reference to the settlement of their larvae. *Journal of the Marine Biological Association of the United Kingdom*, 57, 999–1017.

Garwood, P. R. & Kendall, M. A. (1985). The reproductive cycles of *Monodonta lineata* and *Gibbula umbilicalis* on the coast of mid-Wales. *Journal of the Marine Biological Association of the United Kingdom*, 65, 993–1008.

Goodwin, B. J. (1978). The growth and breeding cycle of *Littorina obtusata* (Gastropoda: Prosobranchiata) from Cardigan Bay. *Journal of Molluscan Studies*, 44, 231–42.

(1979). The egg mass of *Littorina obtusata* and *Lacuna pallidula* (Gastropoda: Prosobranchia). *Journal of Molluscan Studies*, 45, 1–11.

Goodwin, B. J. & Fish, J. D. (1977). Inter- and intraspecific variation in *Littorina obtusata* and *L. mariae* (Gastropoda: Prosobranchia). *Journal of Molluscan Studies*, 43, 241–54.

Graham, A. & Fretter, V. (1947). The life history of *Patina pellucida* (L.). *Journal of the Marine Biological Association of the United Kingdom*, 26, 590–601.

Grahame, J. (1970). Shedding of the penis in *Littorina littorea*. *Nature*, 221, 976.

Hall, S. J. & Todd, C. D. (1986). Growth and reproduction in the aeolid nudibranch *Aeolidia papillosa* (L.). *Journal of Molluscan Studies*, 52, 193–205.

Hannaford Ellis, C. (1979). Morphology of the oviparous rough winkle, *Littorina arcana* Hannaford Ellis, 1978, with notes on the taxonomy of the *L. saxatilis* species-complex (Prosobranchia: Littorinidae). *Journal of Conchology*, 30, 43–56.

Havenhand, J. N. & Todd, C. D. (1988). Physiological ecology of *Adalaria proxima* (Alder *et* Hancock) and *Onchidoris muricata* (Müller) (Gastropoda: Nudibranchia). 1. Feeding, growth and respiration. *Journal of Experimental Marine Biology and Ecology*, 118, 151–72. 2. Reproduction. *Journal of Experimental Marine Biology and Ecology*, 118, 173–89.

Hayashi, I. (1980). Structure and growth of a shore population of the ormer, *Haliotis tuberculata*. *Journal of the Marine Biological Association of the United Kingdom*, 60, 431–7.

Holmes, S. (1997). Notes on gamete release and fertilization in *Calliostoma zizyphinum* (L.) (Gastropoda; Trochidae). *Journal of Molluscan Studies*, 63, 471–3.

Holmes, S. P., Sturgess, C. J., Cherrill, A. & Davies, M. S. (2001). Shell wiping in *Calliostoma zizyphinum*: the use of pedal mucus as a provendering agent and its contribution to daily energetic requirements. *Marine Ecology Progress Series*, 212, 171–81.

Hughes, R. N. (1980). Population dynamics, growth and reproductive rates of *Littorina nigrolineata* Gray from a moderately sheltered locality in North Wales. *Journal of Experimental Marine Biology and Ecology*, 44, 211–28.

Hughes, R. N. & Roberts, D. J. (1980). Growth and reproductive rates of *Littorina neritoides* (L.) in North Wales. *Journal of the Marine Biological Association of the United Kingdom*, 60, 591–9.

Johannesson, K. (2003). Evolution in *Littorina*: ecology matters. *Journal of Sea Research*, 49, 107–17.

Jørgensen, A. (2002). Morphological variation in *Lacuna parva* (Gastropoda: Littorinidae) from different European populations. *Helgoland Marine Research*, 56, 149–58.

Kennedy, J. J. (1995). The courtship, pseudo-copulation behaviour and spermatophore of *Turritella communis* Risso 1826 (Prosobranchia: Turritellidae). *Journal of Molluscan Studies*, 61, 421–34.

Kingsley-Smith, P. R., Richardson, C. A. & Seed, R. (2003). Size-related and seasonal patterns of egg collar production in *Polinices pulchellus* (Gastropoda: Naticidae) Risso 1826. *Journal of Experimental Marine Biology and Ecology*, 295, 191–206.

Lafuente, I. (1996). Seasonal variation in the offspring development of *Onchidoris bilamellata* (Gastropoda: Nudibranchia). *Journal of the Marine Biological Association of the United Kingdom*, **76**, 243–6.

Lebour, M. V. (1933). The eggs and larvae of *Turritella communis* Lamarck and *Aporrhais pes-pelicani* (L.). *Journal of the Marine Biological Association of the United Kingdom*, **18**, 499–506.

Lindberg, D. R. (1986). Name changes in the "Acmaeidae". *Veliger*, **29**, 142–8.

Macan, T. T. (1977). *A key to the British fresh- and brackish-water gastropods with notes on their ecology*, Scientific Publications no. 13, 4th edn. Ambleside: Freshwater Biological Association.

McGrath, D. (1992). Recruitment and growth of the blue-rayed limpet, *Helcion pellucidum* (L.), in south east Ireland. *Journal of Molluscan Studies*, **58**, 425–31.

(2001). Inter-algal movement of marked blue-rayed limpets, *Patella pellucida* L., between kelps on the lower shore. *Journal of Molluscan Studies*, **67**, 398–400.

Mieszkowska, N., Kendall, M. A., Hawkins, S. J. *et al.* (2006). Changes in the range of some common rocky shore species in Britain: a response to climate change? *Hydrobiologia*, **555**, 241–51.

Mieszkowska, N., Hawkins, S. J., Burrows, M. T. & Kendall, M. A. (2007). Long-term changes in the geographic distribution and population structures of *Osilinus lineatus* (Gastropoda: Trochidae) in Britain and Ireland. *Journal of the Marine Biological Association of the United Kingdom*, **87**, 537–45.

Morton, J. E. (1954). The crevice faunas of the upper intertidal zone at Wembury. *Journal of the Marine Biological Association of the United Kingdom*, **33**, 187–224.

(1955). The functional morphology of *Otina otis*, a primitive marine pulmonate. *Journal of the Marine Biological Association of the United Kingdom*, **34**, 113–50.

Muus, B. J. (1963). Some Danish Hydrobiidae with the description of a new species, *Hydrobia neglecta*. *Proceedings of the Malacological Society of London*, **35**, 131–8.

Nakano, T. & Ozawa, T. (2004). Phylogeny and historical biogeography of limpets of the order Patellogastropoda based on mitochondrial DNA sequences. *Journal of Molluscan Studies*, **70**, 31–41.

(2007). Worldwide phylogeography of limpets of the order Patellogastropoda: molecular, morphological and palaeontological evidence. *Journal of Molluscan Studies*, **73**, 79–99.

Orton, J. H. & Southward, A. J. (1961). Studies on the biology of limpets. IV. The breeding of *Patella depressa* Pennant on the north Cornish coast. *Journal of the Marine Biological Association of the United Kingdom*, **41**, 653–62.

Patil, A. M. (1958). The occurrence of a male of the prosobranch *Potamopyrgus jenkinsi* (Smith) var. *carinata* Marshall in the Thames at Sonning, Berkshire. *Annals and Magazine of Natural History*, **1**, 232–40.

Phorson, J. E. (1990). Some observations on the development of juvenile shells of *Trivia*. *Conchologists' Newsletter*, **113**, 279–83.

Potts, G. W. (1970). The ecology of *Onchidoris fusca* (Nudibranchia). *Journal of the Marine Biological Association of the United Kingdom*, **50**, 269–92.

Power, A. J. & Keegan, B. F. (2001). Seasonal patterns in the reproductive activity of the red whelk, *Neptunea antiqua* (Mollusca: Prosobranchia) in the Irish Sea. *Journal of the Marine Biological Association of the United Kingdom*, **81**, 243–50.

Reimchen, T. E. (1979). Substratum heterogeneity, crypsis, and colour polymorphism in an intertidal snail (*Littorina mariae*). *Canadian Journal of Zoology*, **57**, 1070–85.

Richard, J., Huet, M., Thouzeau, G. & Paulet, Y.-M. (2006). Reproduction of the invasive slipper limpet, *Crepidula fornicata*, in the Bay of Brest, France. *Marine Biology*, **149**, 789–801.

Richardson, C. A., Kingsley-Smith, P. R. & Seed, R. (2005). Age and growth of the naticid gastropod *Polinices pulchellus* (Gastropoda: Naticidae) based on length frequency analysis and statolith growth rings. *Marine Biology*, **148**, 319–26.

Richardson, C. A., Saurel, C., Barroso, C. M. & Thain, J. (2005). Evaluation of the age of the red whelk *Neptunea antiqua* using statoliths, opercula and element ratios in the shell. *Journal of Experimental Marine Biology and Ecology*, **325**, 55–64.

Ridgway, S. A., Reid, D. G., Taylor, J. D., Branch, G. M. & Hodgson, A. N. (1998). A cladistic phylogeny of the family Patellidae (Mollusca: Gastropoda). *Philosophical Transactions of the Royal Society of London Series B*, **353**, 1645–71.

Seed, R. (1979). Distribution and shell characteristics of the painted topshell *Calliostoma zizyphinum* (L.) (Prosobranchia: Trochidae) in County Down, N. Ireland. *Journal of Molluscan Studies*, **45**, 12–18.

Silva, C. M. da, Landau, B. M., Domènech, R. & Martinell, J. (2006). Pliocene Atlanto-Mediterranean biogeography of *Patella pellucida* (Gastropoda, Patellidae): palaeoceanographic implications. *Palaeogeography, Palaeoclimatology, Palaeoecology*, **233**, 225–34.

Small, M. P. & Gosling, E. M. (2001). Population genetics of a snail species complex in the British Isles: *Littorina saxatilis* (Olivi), *L. neglecta* Bean and *L. tenebrosa* (Montagu), using SSCP analysis of cytochrome *b* gene fragments. *Journal of Molluscan Studies*, **67**, 69–80.

Smith, D. A. S. (1973). The population biology of *Lacuna pallidula* (da Costa) and *Lacuna vincta* (Montagu) in north-east England. *Journal of the Marine Biological Association of the United Kingdom*, **53**, 493–520.

Smith, S. T. (1967). The ecology and life-history of *Retusa obtusa* (Montagu) (Gastropoda, Opisthobranchia). *Canadian Journal of Zoology*, **45**, 397–405.

(1967). The development of *Retusa obtusa* (Montagu) (Gastropoda, Opisthobranchia). *Canadian Journal of Zoology*, **45**, 737–64.

Son, M. H. & Hughes, R. N. (2000). Relationship between imposex and morphological variation of the shell in *Nucella lapillus* (Gastropoda: Thaididae). *Estuarine, Coastal and Shelf Science*, **50**, 599–606.

Southgate, T. (1982). Studies on an intertidal population of *Rissoa parva* (Gastropoda: Prosobranchia) in south-west Ireland. *Journal of Natural History*, **16**, 183–94.

Spence, S. K., Bryan, G. W., Gibbs, P. E. *et al.* (1990). Effects of TBT contamination on *Nucella* populations. *Functional Ecology*, **4**, 425–32.

Stefaniak, L. M., McAtee, J. & Shulman, M. J. (2005). The costs of being bored: effects of a clionid sponge on the gastropod *Littorina littorea* (L). *Journal of Experimental Marine Biology and Ecology*, **327**, 103–14.

Steneck, R. S. & Watling, L. (1982). Feeding capabilities and limitation of herbivorous molluscs: a functional group approach. *Marine Biology*, **68**, 299–319.

Tallmark, B. (1980). Population dynamics of *Nassarius reticulatus* (Gastropoda, Prosobranchia) in Gullmar Fjord, Sweden. *Marine Ecology Progress Series*, **3**, 51–62.

Thompson, G. B. (1979). Distribution and population dynamics of the limpet *Patella aspera* (Lamarck) in Bantry Bay. *Journal of Experimental Marine Biology and Ecology*, **40**, 115–35.

Thompson, T. E. & Slinn, D. J. (1959). On the biology of the opisthobranch *Pleurobranchus membranaceus*. *Journal of the Marine Biological Association of the United Kingdom*, **38**, 507–24.

Todd, C. D. (1979). The population ecology of *Onchidoris bilamellata* (L.) (Gastropoda: Nudibranchia). *Journal of Experimental Marine Biology and Ecology*, **41**, 213–55.

(1981). The ecology of nudibranch molluscs. *Oceanography and Marine Biology. An Annual Review*, **19**, 141–234.

Underwood, A. J. (1972). Observations on the reproductive cycles of *Monodonta lineata*, *Gibbula umbilicalis* and *G. cineraria*. *Marine Biology*, **17**, 333–40.

Vahl, O. (1971). Growth and density of *Patina pellucida* (L.) (Gastropoda: Prosobranchia) on *Laminaria hyperborea* (Gunnerus) from western Norway. *Ophelia*, **9**, 31–50.

Valdés, Á. (2002). A phylogenetic analysis and systematic revision of the cryptobranch dorids (Mollusca, Nudibranchia, Anthobranchia). *Zoological Journal of the Linnean Society*, **136**, 535–636.

Waite, M. E., Waldock, M. J., Thain, J. E., Smith, D. J. & Milton, S. M. (1991). Reductions in TBT concentrations in UK estuaries following legislation in 1986 and 1987. *Marine Environmental Research*, **32**, 89–111.

Wallace, C. (1985). On the distribution of the sexes of *Potamopyrgus jenkinsi* (Smith). *Journal of Molluscan Studies*, **51**, 290–6.

Warén, A. (1996). Ecology and systematics of the north European species of *Rissoa* and *Pusillina* (Prosobranchia: Rissoidae). *Journal of the Marine Biological Association of the United Kingdom*, **76**, 1013–59.

Weber, L. I. & Hawkins, S. J. (2005). *Patella aspera* and *P. ulyssiponensis*: genetic evidence of speciation in the North-east Atlantic. *Marine Biology*, **147**, 153–62.

Wigham, G. D. (1975). The biology and ecology of *Rissoa parva* (da Costa) (Gastropoda: Prosobranchia). *Journal of the Marine Biological Association of the United Kingdom*, **55**, 45–67.

(1975). Environmental influences upon the expression of shell form in *Rissoa parva* (da Costa) (Gastropoda: Prosobranchia). *Journal of the Marine Biological Association of the United Kingdom*, **55**, 425–38.

Wilding, C. S., Grahame, J. & Mill, P. J. (2000). Nuclear DNA restriction site polymorphisms and the phylogeny and population structure of an intertidal snail species complex (*Littorina*). *Hereditas*, **133**, 9–18.

Wilke, T. & Pfenninger, M. (2002). Separating historic events from recurrent processes in cryptic species: phylogeography of mud snails (*Hydrobia* spp.). *Molecular Ecology*, **11**, 1439–51.

Wilke, T., Rolán E. & Davis, G. M. (2000). The mudsnail genus *Hydrobia* s.s. in the northern Atlantic and western Mediterranean: a phylogenetic hypothesis. *Marine Biology*, **137**, 827–33.

Williams, G. A., Little, C., Morritt, D. *et al.* (1999). Foraging in the limpet *Patella vulgata*: the influence of rock slope on the timing of activity. *Journal of the Marine Biological Association of the United Kingdom*, **79**, 881–9.

Williams, G. R. (1990). The comparative ecology of the flat periwinkles, *Littorina obtusata* (L.) and *L. mariae* Sacchi et Rastelli. *Field Studies*, **7**, 469–82.

Williams, S. I. & Walker, D. I. (1999). Mesoherbivore–macroalgal interactions: feeding ecology of saccoglossan sea slugs (Mollusca, Opisthobranchia) and their effects on their food algae. *Oceanography and Marine Biology. An Annual Review*, **37**, 87–128.

Williamson, P. & Kendall, M. A. (1981). Population age structure and growth of the trochid *Monodonta lineata* determined from shell rings. *Journal of the Marine Biological Association of the United Kingdom*, **61**, 1011–26.

Yonge, C. M. (1937). The biology of *Aporrhais pes-pelecani* (L.) and *A. serresiana* (Mich). *Journal of the Marine Biological Association of the United Kingdom*, **21**, 687–703.

(1946). On the habits of *Turritella communis* Risso. *Journal of the Marine Biological Association of the United Kingdom*, **26**, 377–80.

Yonge, C. M. & Thompson, T. E. (1976). *Living marine molluscs*. London: Collins.

MOLLUSCA: BIVALVIA

Allen, J. A. (1969). Observations on size composition and breeding of Northumberland populations of *Zirphaea crispata* (Pholadidae: Bivalvia). *Marine Biology*, **3**, 269–75.

Ansell, A. D. (1961). Reproduction, growth and mortality of *Venus striatula* (da Costa) in Kames Bay, Millport. *Journal of the Marine Biological Association of the United Kingdom*, **41**, 191–215.

(1961). The functional morphology of the British species of Veneracea (Eulamellibranchia). *Journal of the Marine Biological Association of the United Kingdom*, **41**, 489–515.

Ansell, A. D., Harvey, R. & Günther, C.-P. (1999). Recovery from siphon damage in *Donax vittatus* (Da Costa) (Bivalvia: Donacidae). *Journal of Molluscan Studies*, **65**, 223–32.

Armonies, W. & Reise, K. (1999). On the population development of the introduced razor clam *Ensis americanus* near the island of Sylt (North Sea). *Helgoländer Meeresunter-suchungen*, **52**, 291–300.

Bayne, B. L. ed. (1976). *Marine mussels: their ecology and physiology.* International Biological Programme Publication no. 10. Cambridge: Cambridge University Press.

Board, P. A. (1970). Some observations on the tunnelling of shipworms. *Journal of Zoology (London)*, **161**, 193–201.

Bodger, P. M. & Allen, J. A. (2008). The ecology and life cycle of a population of *Modiolarca tumida* (Hanley, 1843) (Bivalvia: Mytilidae) of the coast off north-eastern England. *Journal of Molluscan Studies*, **74**, 97–101.

Boyden, C. R. (1971). A comparative study of the reproductive cycles of the cockles *Cerasto-derma edule* and *C. glaucum. Journal of the Marine Biological Association of the United Kingdom*, **51**, 605–22.

Boyden, C. R. & Russell, P. J. C. (1972). The distribution and habitat range of the brackish water cockle (*Cardium (Cerastoderma) glaucum*) in the British Isles. *Journal of Animal Ecology*, **41**, 719–34.

Butler, P. (2007). The tree of the sea: the shell of the bivalve mollusc *Arctica islandica* as a natural archive of the marine environment. *Porcupine Marine Natural History Society Newsletter*, **22**, 39–47.

Cardoso, J. F. M. F., Witte, J. I. J. & Van der Veer, H. W. (2007). Habitat related growth and reproductive investment in estuarine waters, illustrated for the tellinid bivalve *Macoma balthica* (L.) in the western Dutch Wadden Sea. *Marine Biology*, **152**, 1271–82.

Conan, G. & Shafee, M. S. (1978). Growth and biannual recruitment of the Black Scallop *Chlamys varia* (L.) in Lanveoc area, Bay of Brest. *Journal of Experimental Marine Biology and Ecology*, **35**, 59–71.

Cote, I. M. & Jeinikar, E. (1999). Predator-induced clumping behaviour in mussels (*Mytilus edulis* Linnaeus). *Journal of Experimental Marine Biology and Ecology*, **235**, 201–11.

Da Costa, F., Darriba, S. & Martínez-Patiño, D. (2008). Embryonic and larval development of *Ensis arcuatus* (Jeffreys, 1865) (Bivalvia: Pharidae). *Journal of Molluscan Studies*, **74**, 103–9.

Darriba, S., San Juan, F. & Guerra, A. (2004). Reproductive cycle of the razor clam *Ensis arcuatus* (Jeffreys, 1865) in northwest Spain and its relation to environ-mental conditions. *Journal of Experimental Marine Biology and Ecology*, **311**, 101–15.

(2005). Gametogenic cycle of *Ensis siliqua* (Linnaeus, 1758) in the Ría de Corcubión, northwestern Spain. *Journal of Molluscan Studies*, **71**, 47–51.

De Montaudouin, X., Bachelet, G. & Sauriau, P.-G. (2003). Secondary settlement of cockles *Cerastoderma edule* as a function of current velocity and substratum: a flume study with benthic juveniles. *Hydrobiologia*, **503**, 103–16.

Dixon, D. R. & Flavell, N. (1986). A comparative study of the chromosomes of *Mytilus edulis* and *Mytilus galloprovincialis. Journal of the Marine Biological Association of the United Kingdom*, **66**, 219–28.

Duval, D. M. (1962). Observations on the annual cycle of *Barnea candida* (Class Lamellibran-chiata, Family Pholadidae). *Proceedings of the Malacological Society of London*, **35**, 101–2.

Eales, N. B. (1950). *The littoral fauna of Great Britain.* Cambridge: Cambridge University Press.

Fahy, E. & Gaffney, J. (2001). Growth statistics of an exploited razor clam (*Ensis siliqua*) bed at Gormanstown, Co. Meath, Ireland. *Hydrobiologia*, **465**, 139–51.

Farrell, E. D. & Crowe, T. P. (2007). The use of byssus threads by *Mytilus edulis* as an active defence against *Nucella lapillus. Journal of the Marine Biological Association of the United Kingdom*, **87**, 559–64.

Forester, A. J. (1979). The association between the sponge *Halichondria panicea* (Pallas) and scallop *Chlamys varia* (L.): a commensal–protective mutualism. *Journal of Experimental Marine Biology and Ecology*, **36**, 1–10.

Gage, J. (1966). The life-histories of the bivalves *Montacuta substriata* and *M. ferruginosa*, 'commensals' with spatangoids. *Journal of the Marine Biological Association of the United Kingdom*, **46**, 499–511.

Gardner, J. P. A. (1992). *Mytilus galloprovincialis* (Lmk) (Bivalvia, Mollusca): the taxonomic status of the Mediterranean mussel. *Ophelia*, **35**, 219–43.

Green, J. (1957). The growth of *Scrobicularia plana* (da Costa) in the Gwendraeth estuary. *Journal of the Marine Biological Association of the United Kingdom*, **36**, 41–7.

Hiddink, J. G. (2003). Modelling the adaptive value of intertidal migration and nursery use in the bivalve *Macoma balthica*. *Marine Ecology Progress Series*, **252**, 173–85.

Hodgson, A. N. (1982). Studies on wound healing, and an estimation of the rate of regeneration, of the siphon of *Scrobicularia plana* (da Costa). *Journal of Experimental Marine Biology and Ecology*, **62**, 117–28.

Holme, N. A. (1954). The ecology of British species of *Ensis*. *Journal of the Marine Biological Association of the United Kingdom*, **33**, 145–72.

(1959). The British species of *Lutraria* (Lamellibranchia), with a description of *L. angustior* Philippi. *Journal of the Marine Biological Association of the United Kingdom*, **38**, 557–68.

(1961). Shell form in *Venerupis rhomboides*. *Journal of the Marine Biological Association of the United Kingdom*, **41**, 705–22.

Hughes, R. N. (1969). A study of feeding in *Scrobicularia plana*. *Journal of the Marine Biological Association of the United Kingdom*, **49**, 805–23.

(1971). Reproduction of *Scrobicularia plana* da Costa (Pelecypoda: Semelidae) in North Wales. *Veliger*, **14**, 77–81.

Hunter, W. R. (1948). The structure and behaviour of *Hiatella gallicana* (Lamarck) and *H. arctica* (L.), with special reference to the boring habit. *Proceedings of the Royal Society of Edinburgh Series B*, **63**, 271–89.

Ivell, R. (1979). The biology and ecology of a brackish lagoon bivalve, *Cerastoderma glaucum* Bruguière in an English lagoon, the Widewater, Sussex. *Journal of Molluscan Studies*, **45**, 383–400.

Jansen, J. M., Pronker, A. E., Bonga, S. W. & Hummel, H. (2007). *Macoma balthica* in Spain, a few decades back in climate history. *Journal of Experimental Marine Biology and Ecology*, **344**, 161–9.

Jensen, A. C., Humphreys, J., Caldow, R. W. G., Grisley, C. & Dyrynda, P. E. J. (2004). Naturalization of the Manila clam (*Tapes philippinarum*), an alien species, and establishment of a clam fishery within Poole Harbour, Dorset. *Journal of the Marine Biological Association of the United Kingdom*, **84**, 1069–73.

Johannessen, O. H. (1973). Population structure and individual growth of *Venerupis pullastra* (Montagu) (Lamellibranchia). *Sarsia*, **52**, 97–116.

Kingston, P. F. (1974). Studies on the reproductive cycles of *Cardium edule* and *C. glaucum*. *Marine Biology*, **28**, 317–23.

Kristensen, E. S. (1979). Observations on growth and life cycle of the shipworm *Teredo navalis* L. (Bivalvia, Mollusca) in the Isefjord, Denmark. *Ophelia*, **18**, 235–42.

Lane, D. J. W., Beaumont, A. R. & Hunter, J. R. (1985). Byssus drifting and the drifting threads of the young post-larval mussel *Mytilus edulis*. *Marine Biology*, **84**, 301–8.

Langston, W. J., Burt, G. R. & Chesman, B. S. (2007). Feminisation of male clams *Scrobicularia plana* from estuaries in southwest UK and its induction by endocrine-disrupting chemicals. *Marine Ecology Progress Series*, **333**, 173–84.

Macdonald, S. A. & Thomas, M. L. H. (1980). Age determination of the soft-shell clam *Mya arenaria* using shell internal growth lines. *Marine Biology*, **58**, 105–9.

Macleod, J. A. A., Thorpe, J. P. & Duggan, N. A. (1985). A biochemical genetic study of population structure in queen scallop (*Chlamys opercularis*) stocks in the Northern Irish Sea. *Marine Biology*, **87**, 77–82.

Mason, J. (1958). The breeding of the scallop, *Pecten maximus* (L.), in Manx waters. *Journal of the Marine Biological Association of the United Kingdom*, **37**, 653–71.

(1972). The cultivation of the European mussel, *Mytilus edulis* Linnaeus. *Oceanography and Marine Biology. An Annual Review*, **10**, 437–60.

McGrath, D. & O'Foighil, D. (1986). Population dynamics and reproduction of hermaphroditic *Lasaea rubra* (Montagu) (Bivalvia, Galeommatacea). *Ophelia*, **25**, 209–19.

McGrath, D., King, P. A. & Gosling, E. M. (1988). Evidence for the direct settlement of *Mytilus edulis* larvae on adult mussel beds. *Marine Ecology Progress Series*, **47**, 103–6.

Metaxatos, A. (2004). Population dynamics of the venerid bivalve *Callista chione* (L.) in a coastal area of the eastern Mediterranean. *Journal of Sea Research*, **52**, 293–305.

Morton, B. (1978). Feeding and digestion in shipworms. *Oceanography and Marine Biology. An Annual Review*, **16**, 107–44.

Morton, J. E., Boney, A. D. & Corner, E. D. S. (1957). The adaptations of *Lasaea rubra* (Montagu), a small intertidal lamellibranch. *Journal of the Marine Biological Association of the United Kingdom*, **36**, 383–405.

Moura, P., Gasper, M. B. & Monteiro, C. C. (2008). Gametogenic cycle of the smooth clam *Callista chione* on the south-western coast of Portugal. *Journal of the Marine Biological Association of the United Kingdom*, **88**, 161–7.

Palmer, D. W. (2004). Growth of the razor clam *Ensis directus*, an alien species in the Wash on the east coast of England. *Journal of the Marine Biological Association of the United Kingdom*, **84**, 1075–6.

Petraitis, P. S. (1987). Immobilization of the predatory gastropod, *Nucella lapillus*, by its prey, *Mytilus edulis*. *Biological Bulletin*, **172**, 307–14.

Pinn, E. H., Richardson, C. A., Thompson, R. C. & Hawkins, S. J. (2005). Burrow morphology, biometry, age and growth of piddocks (Mollusca: Bivalvia: Pholadidae) on the south coast of England. *Marine Biology*, **147**, 943–53.

Quayle, D. B. (1951). The rate of growth of *Venerupis pullastra* (Montagu) at Millport, Scotland. *Proceedings of the Royal Society of Edinburgh Series B*, **64**, 384–406.

Salas, C., Tirado, C. & Manjón-Cabeza, M. E. (2001). Sublethal foot-predation on Donacidae (Mollusca: Bivalvia). *Journal of Sea Research*, **46**, 43–56.

Scolding, J. W. S., Richardson, C. A. & Luckenbach, M. J. (2007). Predation of cockles (*Cerastoderma edule*) by the whelk (*Buccinum undatum*) under laboratory conditions. *Journal of Molluscan Studies*, **73**, 333–7.

Seed, R. (1980). Reproduction and growth in *Anomia ephippium* (L.) (Bivalvia: Anomiidae) in Strangford Lough, Northern Ireland. *Journal of Conchology*, **30**, 239–45.

Seed, R. & Brown, R. A. (1977). A comparison of the reproductive cycles of *Modiolus modiolus* (L.), *Cerastoderma* (= *Cardium*) *edule* (L.), and *Mytilus edulis* L. in Strangford Lough, Northern Ireland. *Oecologia (Berlin)*, **30**, 173–88.

Seed, R. & O'Connor, R. J. (1980). Shell shape and seasonal changes in population structure in *Lasaea rubra* (Bivalvia: Erycinidae). *Journal of Molluscan Studies*, **46**, 66–73.

Shumway, S. E. & Parsons, G. J. (2006). *Scallops: biology, ecology and aquaculture*, 2nd edn. Developments in aquaculture and fisheries science, no. 35. New York: Elsevier.

Smith, S. M. & Heppell, D. (1991). *Checklist of British marine Mollusca*. National Museums of Scotland Information Series, no. 11. Edinburgh: National Museums of Scotland.

Thompson, I., Jones, D. S. & Dreibelbis, D. (1980). Annual internal growth banding and life history of the ocean quahog *Arctica islandica* (Mollusca: Bivalvia). *Marine Biology*, **57**, 25–34.

Thorarinsdóttir, G. G. (2000). Annual gametogenic cycle in ocean quahog, *Arctica islandica* from north-western Iceland. *Journal of the Marine Biological Association of the United Kingdom*, **80**, 661–6.

Urk, R. M. van (1987). *Ensis americanus* (Binney) (Syn. *E. directus* auct. non Conrad) a recent introduction from Atlantic North America. *Journal of Conchology*, **32**, 329–33.

Witbaard, R., Duineveld, G. C. A. & de Wilde, P. A. W. J. (1999). Geographical differences in growth rates of *Arctica islandica* (Mollusca: Bivalvia) from the North Sea and adjacent waters. *Journal of the Marine Biological Association of the United Kingdom*, **79**, 907–15.

Yonge, C. M. (1959). On the structure, biology and systematic position of *Pharus legumen* (L.). *Journal of the Marine Biological Association of the United Kingdom*, **38**, 277–90.

(1960). *Oysters*. London: Collins.

Zwarts, L. (1986). Burying depth of the benthic bivalve *Scrobicularia plana* (da Costa) in relation to siphon-cropping. *Journal of Experimental Marine Biology and Ecology*, **101**, 25–39.

ARTHROPODA

Allen, J. A. (1967). *The fauna of the Clyde Sea area. Crustacea: Euphausiacea and Decapoda.* Millport: Scottish Marine Biological Association.

Bachelet, G., Dauvin, J.-C. & Sorbe, J.-C. (2003). An updated checklist of marine and brackish water Amphipoda (Crustacea: Peracarida) of the southern Bay of Biscay (N.E. Atlantic). *Cahiers de Biologie Marine*, **44**, 121–51.

Barnes, H. & Stone, R. L. (1973). The general biology of *Verruca stroemia* (O. F. Müller). *Journal of Experimental Marine Biology and Ecology*, **12**, 279–97.

Bellan-Santini, D. (2005). European species of *Haustorius* (Crustacea: Amphipoda: Haustoriidae), with description of a new Mediterranean species. *Journal of Natural History*, **39**, 1101–10.

Brian, J. V., Fernandes, T., Ladle, R. J. & Todd, P. A. (2006). Patterns of morphological and genetic variability in UK populations of the shore crab, *Carcinus maenas* Linnaeus, 1758 (Crustacea: Decapoda: Brachyura). *Journal of Experimental Marine Biology and Ecology*, **329**, 47–54.

Bruce, N. L. & Holdich, D. M. (2002). Revision of the isopod crustacean genus *Campecopea* (Flabellifera: Sphaeromatidae) with discussion of the phylogenetic significance of dorsal processes. *Journal of the Marine Biological Association of the United Kingdom*, **82**, 51–68.

Burrows, M. T., Hawkins, S. J. & Southward, A. J. (1992). A comparison of reproduction in co-occurring chthamalid barnacles, *Chthamalus stellatus* (Poli) and *Chthamalus montagui* Southward. *Journal of Experimental Marine Biology and Ecology*, **160**, 229–49.

Cheng, L. & Frank, J. H. (1993). Marine insects and their reproduction. *Oceanography and Marine Biology. An Annual Review*, **31**, 479–506.

Choy, S. C. (1986). Natural diet and feeding habits of the crabs *Liocarcinus puber* and *L. holsatus* (Decapoda, Brachyura, Portunidae). *Marine Ecology Progress Series*, **31**, 87–99.

Clark, P. F. (2006). *Eriocheir sinensis* H. Milne Edwards: 1853 or 1854 – Grapsidae or Varunidae? *Aquatic Invasions*, **1**, 17–27.

Conlan, K. E. (1989). Delayed reproduction and adult dimorphism in males of the amphipod genus *Jassa* (Corophioidea: Ischyroceridae): an explanation for systematic confusion. *Journal of Crustacean Biology*, **9**, 601–25.

(1990). Revision of the crustacean amphipod genus *Jassa* Leach (Corophioidea: Ischyroceridae). *Canadian Journal of Zoology*, **68**, 2031–75.

Cook, E. J., Jahnke, M., Kerckhof, F. *et al.* (2007). European expansion of the introduced amphipod *Caprella mutica* Schurin 1935. *Aquatic Invasions*, **2**, 411–21.

Costa, F. O., Henzler, C. M., Lunt, D. H., Whiteley, N. M. & Rock, J. (2009). Probing marine *Gammarus* (Amphipoda) taxonomy with DNA barcodes. *Systematics and Biodiversity*, **7**, 365–79.

Costello, M. J., Holmes, J. M. C., McGrath, D. & Myers, A. A. (1989). *A review and catalogue of the Amphipoda (Crustacea) in Ireland*. Irish Fisheries Investigations Series B (Marine), no. 33.

Crisp, D. J., Southward, A. J. & Southward, E. C. (1981). On the distribution of the intertidal barnacles *Chthamalus stellatus*, *Chthamalus montagui* and *Euraphia depressa*. *Journal of the Marine Biological Association of the United Kingdom*, **61**, 359–80.

Crothers, J. H. (1967). The biology of the shore crab, *Carcinus maenas* (L.). 1. The background-anatomy, growth and life-history. *Field Studies*, **2**, 407–34.

  (1968). The biology of the shore crab, *Carcinus maenas* (L.). 2. The life of the adult crab. *Field Studies*, **2**, 579–614.

Davies, L. & Richardson, J. (1970). Distribution in Britain and habitat requirements of *Petrobius maritimus* (Leach) and *P. brevistylis* Carpenter (Thysanura). *Entomologist*, **103**, 97–114.

Dias, N. & Sprung, M. (2004). Population dynamics and production of the amphipod *Orchestia gammarellus* (Talitridae) in a Ria Formosa saltmarsh (southern Portugal). *Crustaceana*, **76**, 1123–41.

D'Udekem d'Acoz, C. & Vader, W. (2005). New records of West and South African *Bathyporeia*, with the description of four new species and a key to all species of the genus (Crustacea, Amphipoda). *Journal of Natural History*, **39**, 2759–94.

Elwood, R. W. & Stewart, A. (1987). Reproduction in the littoral hermit crab *Pagurus bernhardus*. *Irish Naturalists' Journal*, **22**, 252–5.

Fish, J. D. & Fish, S. (1972). The swimming rhythm of *Eurydice pulchra* Leach and a possible explanation of intertidal migration. *Journal of Experimental Marine Biology and Ecology*, **8**, 195–200.

Fish, J. D. & Mills, A. (1979). The reproductive biology of *Corophium volutator* and *C. arenarium* (Crustacea: Amphipoda). *Journal of the Marine Biological Association of the United Kingdom*, **59**, 355–68.

Fish, J. D. & Preece, G. S. (1970). The annual reproductive patterns of *Bathyporeia pilosa* and *Bathyporeia pelagica* (Crustacea: Amphipoda). *Journal of the Marine Biological Association of the United Kingdom*, **50**, 475–88.

Fish, S. (1970). The biology of *Eurydice pulchra* (Crustacea: Isopoda). *Journal of the Marine Biological Association of the United Kingdom*, **50**, 753–68.

Guerra-Garcia, J. M. (2002). Re-descriptions of *Caprella linearis* (Linnaeus, 1767) and *C. septentrionalis* Kröyer, 1838 (Crustacea: Amphipoda: Caprellidea) from Scotland, with an ontogenetic comparison between the species and a study of the clinging behaviour. *Sarsia*, **87**, 216–35.

Haahtela, I. & Naylor, E. (1965). *Jaera hopeana*, an intertidal isopod new to the British fauna. *Journal of the Marine Biological Association of the United Kingdom*, **45**, 367–71.

Hartnoll, R. G. (1963). The biology of Manx spider crabs. *Proceedings of the Zoological Society of London*, **141**, 423–96.

  (1972). The biology of the burrowing crab, *Corystes cassivelaunus*. *Bijdragen tot de Dierkunde*, **42**, 139–55.

Harvey, C. E. (1968). Distribution and seasonal population changes of *Campecopea hirsuta* (Isopoda: Flabellifera). *Journal of the Marine Biological Association of the United Kingdom*, **48**, 761–7.

  (1969). Breeding and distribution of *Sphaeroma* (Crustacea: Isopoda) in Britain. *Journal of Animal Ecology*, **38**, 399–406.

Healey, B. & O'Neill, M. (1984). The life cycle and population dynamics of *Idotea pelagica* and *I. granulosa* (Isopoda: Valvifera) in south-east Ireland. *Journal of the Marine Biological Association of the United Kingdom*, **64**, 21–33.

Heath, D. J. (1975). Geographical variation in populations of the polymorphic isopod, *Sphaeroma rugicauda*. *Heredity*, **35**, 99–107.

Herborg, L.-M., Rushton, S. P., Clare, A. S. & Bentley, M. G. (2003). Spread of the Chinese mitten crab (*Eriocheir sinensis* H. Milne Edwards) in continental Europe: analysis of a historical data set. *Hydrobiologia*, **503**, 21–8.

Hewett, C. J. (1974). Growth and moulting in the common lobster (*Homarus vulgaris* Milne-Edwards). *Journal of the Marine Biological Association of the United Kingdom*, **54**, 379–91.

Høeg, J. T. (1984). Size and settling behaviour in male and female cypris larvae of the parasitic barnacle *Sacculina carcini* Thompson (Crustacea: Cirripedia: Rhizocephala). *Journal of Experimental Marine Biology and Ecology*, **76**, 145–56.

— (1995). The biology and life cycle of the rhizocephala (Cirripedia). *Journal of the Marine Biological Association of the United Kingdom*, **75**, 517–50.

Holdich, D. M. (1968). Reproduction, growth and bionomics of *Dynamene bidentata* (Crustacea: Isopoda). *Journal of Zoology (London)*, **156**, 137–53.

Howard, A. E. & Bennett, D. B. (1979). The substrate preference and burrowing behaviour of juvenile lobsters (*Hommarus gammarus* (L.)). *Journal of Natural History*, **13**, 433–8.

Ingle, R. W. (1986). The Chinese mitten crab *Eriocheir sinensis* H. Milne Edwards: a contentious immigrant. *London Naturalist*, **65**, 101–5.

Jacobs, B. J. M. (1987). A taxonomic revision of the European, Mediterranean and NW African species generally placed in *Sphaeroma* Bosc, 1802 (Isopoda: Flabellifera: Sphaeromatidae). *Zoologische Verhandelingen*, **238**, 3–71.

Jarvis, J. H. & King, P. E. (1972). Reproduction and development in the pycnogonid *Pycnogonum littorale*. *Marine Biology*, **13**, 146–54.

Jones, L. T. (1963). The geographical and vertical distribution of British *Limnoria* (Crustacea: Isopoda). *Journal of the Marine Biological Association of the United Kingdom*, **43**, 589–603.

Jones, M. B. & Naylor, E. (1971). Breeding and bionomics of the British members of the *Jaera albifrons* group of species (Isopoda: Asellota). *Journal of Zoology (London)*, **165**, 183–99.

King, P. E. & Jarvis, J. H. (1970). Egg development in a littoral pycnogonid *Nymphon gracile*. *Marine Biology*, **7**, 294–304.

Krang, A.-S. & Baden, S. P. (2004). The ability of the amphipod *Corophium volutator* (Pallas) to follow chemical signals from con-specifics. *Journal of Experimental Marine Biology and Ecology*, **310**, 195–206.

Lauckner, G. (1983). Diseases of Mollusca: Bivalvia. In *Diseases of marine animals*, Vol. II, ed. O. Kinne, pp. 477–961. Hamburg: Biologische Anstalt Helgoland.

Laval, P. (1980). Hyperiid amphipods as crustacean parasitoids associated with gelatinous zooplankton. *Oceanography and Marine Biology. An Annual Review*, **18**, 11–56.

Linnane, A., Ball, B., Mercer, J. P. *et al.* (2001). Searching for the early benthic phase (EBP) of the European lobster: a trans-European study of cobble fauna. *Hydrobiologia*, **465**, 63–72.

Manjón-Cabeza, M. E. & Garcia Raso, J. E. (2000). Morphological reproductive aspects of males of *Diogenes pugilator* (Roux, 1829) (Crustacea, Decapoda, Anomura) from southern Spain. *Sarsia*, **85**, 195–202.

— (2000). Reproductive aspects of females of the hermit crab *Diogenes pugilator* (Crustacea: Decapoda: Anomura) from southern Spain. *Journal of the Marine Biological Association of the United Kingdom*, **80**, 85–93.

Martin, J. W. & Davis, G. E. (2001). An updated classification of the recent Crustacea. *Natural History Museum of Los Angeles County. Science Series*, no. 39.

Mauchline, J. (1971). The biology of *Neomysis integer* (Crustacea, Mysidacea). *Journal of the Marine Biological Association of the United Kingdom*, **51**, 347–54.

— (1971). The biology of *Praunus flexuosus* and *P. neglectus* (Crustacea, Mysidacea). *Journal of the Marine Biological Association of the United Kingdom*, **51**, 641–52.

— (1980). The biology of mysids. *Advances in Marine Biology*, **18**, 1–369.

McGrath, D., Costello, M. J. & Emblow, C. (2000). The hermit crab *Diogenes pugilator* (Roux, 1829) in Irish waters. *Biology and Environment: Proceedings of the Royal Irish Academy Series B*, **100**, 115–18.

Metz, P. (1967). On the relations between *Hyperia galba* Montagu (Amphipoda, Hyperiidae) and its host *Aurelia aurita* in the Isefjord area (Sjaelland, Denmark). *Videnskabelige Meddelelser fra Dansk naturhistorisk Forening i Kjøbenhavn*, **130**, 85–108.

Moore, P. G. & Beare, D. J. (1993). Taxonomic confusion in the genus *Pontocrates* (Crustacea: Amphipoda) and the presence of *P. arcticus* in Britain. *Journal of the Marine Biological Association of the United Kingdom*, **73**, 609–15.

Moore, P. G. & Francis, C. H. (1986). Notes on breeding periodicity and sex ratio of *Orchestia gammarellus* (Pallas) (Crustacea: Amphipoda) at Millport, Scotland. *Journal of Experimental Marine Biology and Ecology*, **95**, 203–9.

Myers, A. A. & Lowry, J. K. (2003). A phylogeny and a new classification of the Corophiidea Leach, 1814 (Amphipoda). *Journal of Crustacean Biology*, **23**, 443–85.

Neumann, V. (1996). Comparative investigations on the systematics and taxonomy of European *Maja* species (Decapoda, Brachyura, Majidae). *Crustaceana*, **69**, 821–52.

(1998). A review of the *Maja squinado* (Crustacea: Decapoda: Brachyura) species-complex with a key to the eastern Atlantic and Mediterranean species of the genus. *Journal of Natural History*, **32**, 1667–84.

Norman, C. P. & Jones, M. B. (1993). Reproductive ecology of the velvet swimming crab, *Necora puber* (Brachyura: Portunidae) at Plymouth. *Journal of the Marine Biological Association of the United Kingdom*, **73**, 379–89.

Oh, C.-W., Hartnoll, R. G. & Nash, R. D. M. (2001). Feeding ecology of the common shrimp *Crangon crangon* in Port Erin Bay, Isle of Man, Irish Sea. *Marine Ecology Progress Series*, **214**, 211–23.

O'Riordan, R. M., Myers, A. A. & Cross, T. F. (1992). Brooding in the intertidal barnacles *Chthamalus stellatus* (Poli) and *Chthamalus montagui* Southward in south-western Ireland. *Journal of Experimental Marine Biology and Ecology*, **164**, 135–45.

Pike, R. B. & Williamson, D. I. (1959). Observations on the distribution and breeding of British hermit crabs and the stone crab (Crustacea: Diogenidae, Paguridae and Lithodidae). *Proceedings of the Zoological Society of London*, **132**, 551–67.

Rainbow, P. S. (1984). An introduction to the biology of British littoral barnacles. *Field Studies*, **6**, 1–51.

Rangeley, R. W. & Thomas, M. L. H. (1987). Predatory behaviour of juvenile shore crab *Carcinus maenas* (L.). *Journal of Experimental Marine Biology and Ecology*, **108**, 191–7.

Reuschel, S. & Schubart, C. D. (2006). Phylogeny and geographic differentiation of Atlanto-Mediterranean species of the genus *Xantho* (Crustacea: Brachyura: Xanthidae) based on genetic and morphometric analyses. *Marine Biology*, **148**, 853–66.

Salemaa, H. (1986). Breeding biology and microhabitat utilization of the intertidal isopod *Idotea granulosa* Rathke in the Irish Sea. *Estuarine, Coastal and Shelf Science*, **22**, 335–55.

Sheader, M. (1977). The breeding biology of *Idotea pelagica* (Isopoda: Valvifera) with notes on the occurrence and biology of its parasite *Clypeoniscus hanseni* (Isopoda: Epicaridea). *Journal of the Marine Biological Association of the United Kingdom*, **57**, 659–74.

Smaldon, G. (1972). Population structure and breeding biology of *Pisidia longicornis* and *Porcellana platycheles*. *Marine Biology*, **17**, 171–9.

Upton, N. P. D. (1987). Asynchronous male and female life cycles in the sexually dimorphic, harem-forming isopod *Paragnathia formica* (Crustacea: Isopoda). *Journal of Zoology (London)*, **212**, 677–90.

(1987). Gregarious larval settlement within a restricted intertidal zone and sex differences in subsequent mortality in the polygynous saltmarsh isopod *Paragnathia formica* (Crustacea: Isopoda). *Journal of the Marine Biological Association of the United Kingdom*, **67**, 663–78.

Walker, G. (1987). Further studies concerning the sex ratio of the larvae of the parasitic barnacle, *Sacculina carcini* Thompson. *Journal of Experimental Marine Biology and Ecology*, **106**, 151–63.

Weber, L. I., Hartnoll, R. G. & Thorpe, J. P. (2000). Genetic divergence and larval dispersal in two spider crabs (Crustacea: Decapoda). *Hydrobiologia*, **420**, 211–19.

Wildish, D. J. (1987). Estuarine species of *Orchestia* (Crustacea: Amphipoda: Talitroidea) from Britain. *Journal of the Marine Biological Association of the United Kingdom*, **67**, 571–83.

Williams, J. A. (1978). The annual pattern of reproduction of *Talitrus saltator* (Crustacea: Amphipoda: Talitridae). *Journal of Zoology (London)*, **184**, 231–44.

Willis, K. J., Cook, E. J., Lozano-Fernandez, M. & Takeuchi, I. (2004). First record of the alien caprellid amphipod, *Caprella mutica*, for the UK. *Journal of the Marine Biological Association of the United Kingdom*, **84**, 1027–8.

Willows, R. I. (1987). Population dynamics and life history of two contrasting populations of *Ligia oceanica* (Crustacea: Oniscidea) in the rocky supralittoral. *Journal of Animal Ecology*, **56**, 315–30.

Witteveen, J., Verhoef, H. A. & Huipen, T. E. A. M. (1988). Life history strategy and egg diapause in the intertidal collembolan *Anurida maritima*. *Ecological Entomology*, **13**, 443–51.

Wyer, D. & King, P. E. (1974). Feeding in British littoral pycnogonids. *Estuarine and Coastal Marine Science*, **2**, 177–84.

SIPUNCULA AND ECHIURA

Cutler, E. B. (1994). The Sipuncula. *Their systematics, biology, and evolution*. Ithaca, NY: Cornell University Press.

Grassé, P.-P. ed. (1959). *Traité de zoologie. Anatomie, systématique, biologie*, Vol. 5. Paris: Masson.

Greef, R. (1879). Die Echiuren (*Gephyrea armata*). *Nova Acta Academiae Caesareae Leopoldino-Carolinae*, **41**, 1–172.

Kristensen, J. H. (1969). Irrigation in the sipunculid *Phascolion strombi* (Mont.). *Ophelia*, **7**, 101–12.

(1970). Fauna associated with the sipunculid *Phascolion strombi* (Montagu), especially the parasitic gastropod *Menestho diaphana* (Jeffreys). *Ophelia*, **7**, 257–76.

Nickell, L. A. & Atkinson, R. J. A. (1994). Observations on the behaviour of *Thalassema thalassemum* (Echiura: Echiuridae). *Journal of the Marine Biological Association of the United Kingdom*, **74**, 963–6.

Rietsch, M. (1886). Étude sur les Gephyriens armés ou Échiurens. *Recueil Zoologique Suisse*, **3**, 313–515.

Stephen, A. C. & Edmonds, S. J. (1972). *The phyla Sipuncula and Echiura*. London: British Museum (Natural History).

BRYOZOA

Cancino, J. M. & Hughes, R. N. (1987). The effects of water flow on growth and reproduction of *Celleporella hyalina* (L.) (Bryozoa: Cheilostomata). *Journal of Experimental Marine Biology and Ecology*, **112**, 109–30.

Eggleston, D. (1972). Patterns of reproduction in the marine Ectoprocta of the Isle of Man. *Journal of Natural History*, **6**, 31–8.

Hayward, P. J. (1980). Invertebrate epiphytes of coastal marine algae. In *The shore environment*, Vol. 2. *Ecosystems*, ed. J. H. Price, D. E. G. Irvine & W. F. Farnham, Systematics Association Special Volume no. 17b, pp. 761–87. London: Academic Press.

Hughes, D. J. (1987). Gametogenesis and embryonic brooding in the cheilostome bryozoan *Celleporella hyalina*. *Journal of Zoology, London*, **212**, 691–711.

Owrid, G. M. A. & Ryland, J. S. (1991). Sexual reproduction in *Alcyonidium hirsutum* (Bryozoa: Ctenostomata). *Bulletin de la Société des Sciences Naturelles de l'Ouest de la France*, **1**, 317–26.

Porter, J. S., Hayward, P. J. & Spencer Jones, M. E. (2001). The identity of *Alcyonidium diaphanum* (Bryozoa: Ctenostomatida). *Journal of the Marine Biological Association of the United Kingdom*, **81**, 1001–8.

Porter, J.S., Ellis, J.R., Hayward, P.J., Rogers, S.I. & Callaway, R. (2002). Geographic variation in the abundance and morphology of the bryozoan *Alcyonidium diaphanum* (Ctenstomata: Alcyonidiidae) in UK coastal waters. *Journal of the Marine Biological Association of the United Kingdom*, **82**, 529–35.

Ryland, J. S. (1976). Physiology and ecology of marine bryozoans. *Advances in Marine Biology*, **14**, 285–443.

Seed, R. & Wood, V. (1994). Recruitment and mortality of *Alcyonidium hirsutum* (Fleming) and *Flustrellidra hispida* (Fabricius) (Bryozoa, Ctenostomata) within a *Fucus serratus* L. community. *Cahiers de Biologie Marine*, **35**, 305–26.

Stebbing, A. R. D. (1971). The epizoic fauna of *Flustra foliacea* (Bryozoa). *Journal of the Marine Biological Association of the United Kingdom*, **51**, 283–300.

Temkin, M. H. (1994). Gamete spawning and fertilization in the Gymnolaemate bryozoan *Membranipora membranacea*. *Biological Bulletin*, **187**, 143–55.

Thorpe, J. P. & Ryland, J. S. (1979). Cryptic speciation detected by biochemical genetics in three ecologically important intertidal bryozoans. *Estuarine and Coastal Marine Science*, **8**, 395–8.

PHORONIDA

Cohen, B. L. (2000). Monophyly of brachiopods and phoronids: reconciliation of molecular evidence with Linnaean classification (the subphylum Phoroniformea nov.). *Proceedings of the Royal Society of London Series B–Biological Sciences*, **267**, 225–31.

Dyster, F. D. (1859). Notes on *Phoronis hippocrepia*. *Transactions of the Linnean Society of London*, **22**, 251–6.

ECHINODERMATA

Barker, M. F. & Nichols, D. (1983). Reproduction, recruitment and juvenile ecology of the starfish, *Asterias rubens* and *Marthasterias glacialis*. *Journal of the Marine Biological Association of the United Kingdom*, **63**, 745–65.

Bourgoin, A. & Guillou, M. (1990). Variations in the reproductive cycle of *Acrocnida brachiata* (Echinodermata: Ophiuroidea) according to environment in the Bay of Douarnenez (Brittany). *Journal of the Marine Biological Association of the United Kingdom*, **70**, 57–66.

Brun, E. (1976). Ecology and taxonomic position of *Henricia oculata* Pennant. *Thalassia Jugoslavica*, **12**, 51–64.

Byrne, M. (1990). Annual reproductive cycles of the commercial sea urchin *Paracentrotus lividus* from an exposed intertidal and a sheltered subtidal habitat on the west coast of Ireland. *Marine Biology*, **104**, 275–89.

Carlson, H. R. & Pfister, C. A. (1999). A seventeen-year study of the rose star *Crossaster papposus* population in a coastal bay in southeast Alaska. *Marine Biology*, **133**, 223–30.

Chenuil, A. & Féral, J.-P. (2003). Sequences of mitochondrial DNA suggest that *Echinocardium cordatum* is a complex of several sympatric or hybridizing species: a pilot study. In *Proceedings of the sixth European Conference on Echinoderm Research, Banyuls-sur-Mer, France, 3–7 September 2001*, ed. J.-P. Féral & B. David. Lisse: Swets & Zeitlinger.

Christensen, A. M. (1970). Feeding biology of the sea-star *Astropecten irregularis* Pennant. *Ophelia*, **8**, 1–134.

Comely, C. A. & Ansell, A. D. (1989). The reproductive cycle of *Echinus esculentus* L. on the Scottish west coast. *Estuarine, Coastal and Shelf Science*, **29**, 385–407.

Costelloe, J. (1985). The annual reproductive cycle of the holothurian *Aslia lefevrei* (Dendrochirota: Echinodermata). *Marine Biology*, **88**, 155–65.

Costelloe, J. & Keegan, B. F. (1984). Feeding and related morphological structures in the dendrochirote *Aslia lefevrei* (Holothuroidea: Echinodermata). *Marine Biology*, **84**, 135–42.

Crapp, G. B. & Willis, M. E. (1975). Age determination in the sea urchin *Paracentrotus lividus* (Lamarck), with notes on the reproductive cycle. *Journal of Experimental Marine Biology and Ecology*, **20**, 157–78.

Crump, R. G. & Emson, R. H. (1983). The natural history, life history and ecology of the two British species of *Asterina*. *Field Studies*, **5**, 867–82.

Eales, N. B. (1950). *The littoral fauna of Great Britain*. Cambridge: Cambridge University Press.

Emson, R. H. & Crump, R. G. (1979). Description of a new species of *Asterina* (Asteroidea), with an account of its ecology. *Journal of the Marine Biological Association of the United Kingdom*, **59**, 77–94.

(1984). Comparative studies on the ecology of *Asterina gibbosa* and *A. phylactica* at Lough Ine. *Journal of the Marine Biological Association of the United Kingdom*, **64**, 35–53.

Emson, R. H. & Whitfield, P. J. (1989). Aspects of the life history of a tide pool population of *Amphipholis squamata* (Ophiuroidea) from south Devon. *Journal of the Marine Biological Association of the United Kingdom*, **69**, 27–41.

Feder, H. M. (1981). Aspects of the feeding biology of the brittle star *Ophiura texturata*. *Ophelia*, **20**, 215–35.

Fontaine, A. R. (1965). The feeding mechanisms of the ophiuroid *Ophiocomina nigra*. *Journal of the Marine Biological Association of the United Kingdom*, **45**, 373–85.

Freeman, S. M., Richardson, C. A. & Seed, R. (2001). Seasonal abundance, spatial distribution, spawning and growth of *Astropecten irregularis* (Echinodermata: Asteroidea). *Estuarine, Coastal and Shelf Science*, **53**, 39–49.

Frid, C. L. J. (1992). Foraging behaviour of the spiny starfish *Marthasterias glacialis* in Lough Ine, Co. Cork. *Marine Behaviour and Physiology*, **19**, 227–39.

Gage, J. D. (1990). Skeletal growth bands in brittle stars: microstructure and significance as age markers. *Journal of the Marine Biological Association of the United Kingdom*, **70**, 209–24.

(1991). Skeletal growth zones as age-markers in the sea urchin *Psammechinus miliaris*. *Marine Biology*, **110**, 217–28.

(1992). Growth bands in the sea urchin *Echinus esculentus*: results from tetracycline-mark/recapture. *Journal of the Marine Biological Association of the United Kingdom*, **72**, 257–60.

Hancock, D. A. (1974). Some aspects of the biology of the sunstar *Crossaster papposus* (L.). *Ophelia*, **13**, 1–30.

Jones, M. B. & Smaldon, G. (1989). Aspects of the biology of a population of the cosmopolitan brittlestar *Amphipholis squamata* (Echinodermata) from the Firth of Forth, Scotland. *Journal of Natural History*, **23**, 613–25.

Lahaye, M.-C. & Jangoux, M. (1985). Functional morphology of the podia and ambulacral grooves of the comatulid crinoid *Antedon bifida* (Echinodermata). *Marine Biology*, **86**, 307–18.

La Touche, R. W. (1978). The feeding behaviour of the featherstar *Antedon bifida* (Echinodermata: Crinoidea). *Journal of the Marine Biological Association of the United Kingdom*, **58**, 877–90.

Lawrence, J. (1987). *A functional biology of echinoderms*. London: Croom Helm.

Le Gac, M., Féral, J. P., Poulin, E., Veyret, M. & Chenuil, A. (2004). Identification of allopatric clades in the cosmopolitan ophiuroid species complex *Amphipholis squamata* (Echinodermata). The end of a paradox? *Marine Ecology Progress Series*, **278**, 171–8.

Madsen, F. J. (1987). The *Henricia sanguinolenta* complex (Echinodermata, Asteroidea) of the Norwegian Sea and adjacent waters: a re-evaluation, with notes on related species. *Steenstrupia*, **13**, 201–67.

McKenzie, J. D. (1991). The taxonomy and natural history of north European dendrochirote holothurians (Echinodermata). *Journal of Natural History*, **25**, 123–71.

Minchin, D. (1987). Sea-water temperature and spawning behaviour in the seastar *Marthasterias glacialis*. *Marine Biology*, **95**, 139–43.

Nichols, D. (1994). Reproductive seasonality in the comatulid crinoid *Antedon bifida* (Pennant) from the English Channel. *Philosophical Transactions of the Royal Society of London Series B*, **343**, 113–34.

Nichols, D. & Barker, M. F. (1984). Growth of juvenile *Asterias rubens* L. (Echinodermata: Asteroidea) on an intertidal reef in southwestern Britain. *Journal of Experimental Marine Biology and Ecology*, **78**, 157–65.

(1984). Reproductive and nutritional periodicities in the starfish, *Marthasterias glacialis*, from Plymouth Sound. *Journal of the Marine Biological Association of the United Kingdom*, **64**, 461–70.

(1984). A comparative study of reproductive and nutritional periodicities in two populations of *Asterias rubens* (Echinodermata: Asteroidea) from the English Channel. *Journal of the Marine Biological Association of the United Kingdom*, **64**, 471–84.

Nichols, D., Sime, A. A. T. & Bishop, G. M. (1985). Growth in populations of the sea-urchin *Echinus esculentus* L. (Echinodermata: Echinoidea) from the English Channel and Firth of Clyde. *Journal of Experimental Marine Biology and Ecology*, **86**, 219–28.

Richner, H. & Milinski, M. (2000). On the functional significance of masking behaviour in sea urchins: an experiment with *Paracentrotus lividus*. *Marine Ecology Progress Series*, **205**, 307–8.

Rowe, F. W. E. (1970). A note on the British species of cucumarians, involving the erection of two new nominal genera. *Journal of the Marine Biological Association of the United Kingdom*, **50**, 683–7.

(1971). The marine flora and fauna of the Isles of Scilly Echinodermata. *Journal of Natural History*, **5**, 233–8.

Sloan, N. A. (1980). Aspects of the feeding biology of asteroids. *Oceanography and Marine Biology. An Annual Review*, **18**, 57–124.

Sloan, N. A. & Aldridge, T. H. (1981). Observations on an aggregation of the starfish *Asterias rubens* L. in Morecambe Bay, Lancashire, England. *Journal of Natural History*, **15**, 407–18.

Turon, X., Codina, M., Tarjuelo, I., Uriz, M. J. & Becerro, M. A. (2000). Mass recruitment of *Ophiothrix fragilis* (Ophiuroidea) on sponges: settlement patterns and post-settlement dynamics. *Marine Ecology Progress Series*, **200**, 201–12.

Tuwo, A. & Conand, C. (1992). Reproductive biology of the holothurian *Holothuria forskali* (Echinodermata). *Journal of the Marine Biological Association of the United Kingdom*, **72**, 745–58.

Verling, E., Crook, A. C. & Barnes, D. K. A. (2002). Covering behaviour in *Paracentrotus lividus*: is light important? *Marine Biology*, **140**, 391–6.

Warner, G. F. & Woodley, J. D. (1975). Suspension-feeding in the brittle-star *Ophiothrix fragilis*. *Journal of the Marine Biological Association of the United Kingdom*, **55**, 199–210.

Webb, C. M. & Tyler, P. A. (1985). Post larval development of the common north-west European brittle stars *Ophiura ophiura*, *O. albida* and *Acrocnida brachiata* (Echinodermata: Ophiuroidea). *Marine Biology*, **89**, 281–92.

HEMICHORDATA

Brambell, F. W. R. & Cole, H. A. (1939). *Saccoglossus cambrensis*, sp. n., an enteropneust occurring in Wales. *Proceedings of the Zoological Society of London Series B*, **109**, 211–36.

Burdon-Jones, C. & Patil, A. M. (1960). A revision of the genus *Saccoglossus* (Enteropneusta) in British waters. *Proceedings of the Zoological Society of London Series B*, **134**, 635–45.

CHORDATA: ASCIDIACEA

Berrill, N. J. (1975). Chordata: Tunicata. In *Reproduction of marine invertebrates*, Vol. II, ed. A. C. Giese & J. S. Pearse, pp. 241–82. New York: Academic Press.

Carlisle, D. B. (1954). *Styela mammiculata* n.sp., a new species of ascidian from the Plymouth area. *Journal of the Marine Biological Association of the United Kingdom*, **33**, 329–34.

Davis, M. H. & Davis, M. E. (2007). The distribution of *Styela clava* (Tunicata, Ascidiacea) in European waters. *Journal of Experimental Marine Biology and Ecology*, **342**, 182–4.

Dybern, B. I. (1965). The life cycle of *Ciona intestinalis* (L.) f. *typica* in relation to the environmental temperature. *Oikos*, **16**, 109–31.

(1969). Distribution and ecology of the tunicate *Ascidiella scabra* (Müller) in the Skagerak–Kattegat area. *Ophelia*, **6**, 183–201.

(1969). Distribution and ecology of ascidians in Kviturdvikpollen and Våagsböpollen on the west coast of Norway. *Sarsia*, **37**, 21–40.

Kott, P. (1952). Observations on compound ascidians of the Plymouth area, with descriptions of two new species. *Journal of the Marine Biological Association of the United Kingdom*, **31**, 65–83.

Lafargue, F. & Wahl, M. (1987). The didemnid ascidian fauna of France. *Annales de l'Institute Océanographique, Monaco*, **63**, 1–46.

Lambert, G. (2009). Adventures of a sea squirt sleuth: unravelling the identity of *Didemnum vexillum*, a global ascidian invader. *Aquatic Invasions*, **4**, 5–28.

Millar, R. H. (1954). The annual growth and reproductive cycle of the ascidian *Dendrodoa grossularia* (Van Beneden). *Journal of the Marine Biological Association of the United Kingdom*, **33**, 33–48.

(1960). The identity of the ascidians *Styela mammiculata* Carlisle and *S. clava* Herdman. *Journal of the Marine Biological Association of the United Kingdom*, **39**, 509–11.

(1971). The biology of ascidians. *Advances in Marine Biology*, **9**, 1–100.

Rowe, F. W. E. (1972). The marine flora and fauna of the Isles of Scilly Enteropneusta, Ascidiacea, Thaliacea, Larvacea and Cephalochordata. *Journal of Natural History*, **6**, 207–13.

Svane, I. & Lundälv, T. (1981). Reproductive patterns and populations dynamics of *Ascidia mentula* O. F. Müller on the Swedish west coast. *Journal of Experimental Marine Biology and Ecology*, **50**, 163–82.

Vazquez, E. (1994). A new species of the genus *Aplidium* Savigny, 1816 (Ascidiacea, Polyclinidae) from the Spanish Atlantic coast. *Sarsia*, **79**, 53–7.

CHORDATA: ACTINOPTERYGII

Albert, O. T., Torstensen, E., Bertelsen, B. *et al.* (2002). Age-reading of lumpsucker (*Cyclopterus lumpus*) otoliths: dissection, interpretation and comparison with length frequencies. *Fisheries Research*, **55**, 239–52.

Azzarello, M. Y. (1991). Some questions concerning the Syngnathidae brood pouch. *Bulletin of Marine Science*, **49**, 741–7.

Barron, R. J. C. (1976). The occurrence of the rock goby, *Gobius paganellus* L. 1758 and the two-spot goby, *Chaparrudo flavescens* (Fabricius, 1779) in the Blackwater estuary, Essex (S.E. England). *Journal of Fish Biology*, **8**, 93–5.

Deady, S. & Fives, J. M. (1995). The diet of corkwing wrasse, *Crenilabrus melops*, in Galway Bay, Ireland, and in Dinard, France. *Journal of the Marine Biological Association of the United Kingdom*, **75**, 635–49.

(1995). Diet of ballan wrasse, *Labrus bergylta,* and some comparisons with the diet of corkwing wrasse, *Crenilabrus melops. Journal of the Marine Biological Association of the United Kingdom,* **75**, 651–65.

Dipper, F. (1981). The strange sex lives of British wrasse. *New Scientist,* **90**, 444–5.

Dipper, F. A. & Pullin, R. S. V. (1979). Gonochorism and sex-inversion in British Labridae (Pisces). *Journal of Zoology (London),* **187**, 97–112.

Dipper, F. A., Bridges, C. R. & Menz, A. (1977). Age, growth and feeding in the ballan wrasse *Labrus bergylta* Ascanius 1767. *Journal of Fish Biology,* **11**, 105–20.

Dunne, J. (1977). Littoral and benthic investigations on the west coast of Ireland. VII. (Section A: Faunistic and ecological studies.) The biology of the shanny, *Blennius pholis* L. (Pisces) at Carna, Connemara. *Proceedings of the Royal Irish Academy Section B,* **77**, 207–26.

(1978). Littoral and benthic investigations on the west coast of Ireland. IX. (Section A: Faunistic and ecological studies.) The biology of the rock-goby, *Gobius paganellus* L., at Carna. *Proceedings of the Royal Irish Academy Section B,* **78**, 179–91.

(1981). A contribution to the biology of Montagu's sea snail, *Liparis montagui* Donovan (Pisces). *Irish Naturalists' Journal,* **20**, 217–22.

Dunne, J. & Byrne, P. (1979). Notes on the biology of the tompot blenny, *Blennius gattorugine* Brunnich. *Irish Naturalists' Journal,* **19**, 414–18.

Eschmeyer, W. N. ed. (1998). *Catalog of fishes,* 3 vols. San Francisco, CA: California Academy of Sciences.

Gauld, J. A. & Hutcheon, J. R. (1990). Spawning and fecundity in the lesser sandeel, *Ammodytes marinus* Raitt, in the north-western North Sea. *Journal of Fish Biology,* **36**, 611–13.

Gibson, R. N. (1986). Intertidal teleosts: life in a fluctuating environment. In *The behaviour of teleost fishes,* ed. T. J. Pitcher, pp. 388–408. London: Croom Helm.

Gibson, R. N. & Tong, L. J. (1969). Observations on the biology of the marine leech *Oceanobdella blennii. Journal of the Marine Biological Association of the United Kingdom,* **49**, 433–8.

Healey, M. C. (1972). On the population ecology of the common goby in the Ythan estuary. *Journal of Natural History,* **6**, 133–45.

Henderson, P. A. & Holmes, R. H. A. (1990). Population stability over a ten-year period in the short-lived fish *Liparis liparis* (L.). *Journal of Fish Biology,* **37**, 605–15.

Henriques, M., Lourenço, R., Almada, F. *et al.* (2002). A revision of the status of *Lepadogaster lepadogaster* (Teleostei: Gobiesocidae): sympatric subspecies or a long misunderstood blend of species? *Biological Journal of the Linnean Society,* **76**, 327–38.

Jacquet, T. N. & Raffaelli, D. (1989). The ecological importance of the sand goby *Pomatoschistus minutus* (Pallas). *Journal of Experimental Marine Biology and Ecology,* **128**, 147–56.

Kaiser, M. J. & Croy, M. I. (1991). Population structure of the fifteen-spined stickleback, *Spinachia spinachia* (L.). *Journal of Fish Biology,* **39**, 129–31.

King, P. A. (1989). Littoral and benthic investigations on the west coast of Ireland. XXII. The biology of a population of shore clingfish *Lepadogaster lepadogaster* (Bonaterre, 1788) at Inishbofin, Co. Galway. *Proceedings of the Royal Irish Academy Section B,* **89**, 47–58.

Miller, P. J. (1961). Age, growth and reproduction of the rock goby, *Gobius paganellus* L., in the Isle of Man. *Journal of the Marine Biological Association of the United Kingdom,* **41**, 737–69.

(1975). Age-structure and life-span in the common goby. *Pomatoschistus microps. Journal of Zoology (London),* **177**, 425–48.

Milton, P. (1983). Biology of littoral blenniid fishes on the coast of south-west England. *Journal of the Marine Biological Association of the United Kingdom,* **63**, 223–37.

Monteiro, N., Almada, V. C., Santos, A. M. & Vieira, M. N. (2001). The breeding ecology of the pipefish *Nerophis lumbriciformis* and its relation to latitude and water temperature. *Journal of the Marine Biological Association of the United Kingdom*, **81**, 1031–3.

Nelson, J. S. (2006). *Fishes of the world*, 4th edn. New York: Wiley.

Nolan, C. P. (1985). Observations on the breeding season of the worm pipefish, *Nerophis lumbriciformis* (Jenyns). *Irish Naturalists' Journal*, **21**, 422.

Pawson, M. G. & Pickett, G. D. (1987). The bass (*Dicentrarchus labrax*) and management of its fishery in England and Wales. *Laboratory Leaflet, Ministry of Agriculture, Fisheries and Food Directorate of Fisheries Research, Lowestoft*, No. **59**, 37pp.

Potts, G. W. & Wootton, R. J. eds. (1984). *Fish reproduction: strategies and tactics*. London: Academic Press.

Rosenqvist, G. (1993). Sex role reversal in a pipefish. *Marine Behaviour and Physiology*, **23**, 219–30.

Russell, F. S. (1976). *The eggs and planktonic stages of British marine fishes*. London: Academic Press.

Vestergaard, K. (1976). Nest building behaviour in the common goby *Pomatoschistus microps* (Krøyer) (Pisces, Gobiidae). *Videnskabelige Meddelelser fra Dansk naturhistorisk Forening i Kjøbenhavn*, **139**, 91–108.

Wheeler, A. (1992). A list of the common and scientific names of fishes of the British Isles. *Journal of Fish Biology*, **41**, Supplement A, 1–37.

Wheeler, A. C., Merrett, N. R. & Quigley, D. T. G. (2004). Additional records and notes for Wheeler's (1992) list of the common and scientific names of fishes of the British Isles. *Journal of Fish Biology*, **65**, Supplement B, 1–40.

Glossary

Words in these definitions that are italicized are listed in the glossary.

ACONTIA thread-like structures discharged through mouth and pores in the column of sea-anemones. Armed with nematocysts.

ACRORHAGUS wart-like structure in sea-anemones, often conspicuously coloured, bearing batteries of stinging cells.

ADDUCTOR MUSCLE muscle that closes shell of bivalve mollusc. Adductor muscle scars are seen on inside of shell valves.

AMBULACRUM part of surface of echinoderm with tube-feet.

ANAEROBIC without oxygen.

ANCESTRULA the first-formed *zooid* in a bryozoan colony.

ANNUAL organism living for one year.

ANNULATION ring or ring-like division.

ANTENNA sensory appendage found on the head of some annelids and arthropods.

APOTHECIA saucer-shaped reproductive bodies seen on the surface of lichen *thallus*.

ARISTOTLE'S LANTERN complex feeding structure of sea-urchins; equipped with calcified teeth.

ASEXUAL REPRODUCTION reproduction not involving a sexual process.

ATRIUM space surrounding *pharynx* in the sea-squirts. Communicates with exterior via exhalant *siphon*.

AVICULARIUM specialized bryozoan *zooid*, equipped with movable jaws.

BASAL LACERATION asexual reproduction of sea-anemones by fragmentation of basal disc.

BEAK first-formed part of bivalve shell. Usually pointed.

BENTHIC living in or on the sea bed.

BERRIED bearing eggs, with reference to crustaceans carrying eggs externally, e.g. crabs.

BILATERAL SYMMETRY symmetry of organism that can be cut into two halves, each a mirror image of the other.

BIRAMOUS having two branches or *rami*, e.g. some arthropod appendages.

BLASTOSTYLE hydroid *polyp* specialized for reproduction.

BRACKISH water containing less salt than normal sea water.

BROOD PLATE see *oostegite*.

BROODING the care of eggs, at least during the early stages of development, in a brood chamber which may be internal or external to the body.

BYSSUS fibres produced by the foot of some bivalve molluscs, e.g *Mytilus*. Used for attachment.

CALCAREOUS made of calcium carbonate or chalk.

CARAPACE *exoskeleton* covering the anterior *dorsal* surface of many crustaceans.

CARDINAL TEETH teeth on the hinge line of bivalve molluscs.

CARPOSPOROPHYTE a stage in the life-cycle of red seaweeds. Produces carpospores which grow into the *sporophyte* generation.

CARTILAGE firm, elastic tissue. Not calcified. Found in the skeleton of skates, rays, etc.

CEPHALOTHORAX fusion of head and one or more thoracic segments in crustaceans.

CERATA finger-like outgrowths on the *dorsal* surface of sea-slugs.

CHAETA (SETA) chitinous bristle of annelids and arthropods. Usually referred to as chaetae in annelids and setae in arthropods.

CHELAE pincers on the *pereopods* of decapod crustaceans.

CHELIFORES first pair of appendages on the head of sea-spiders.

CHELIPEDS first pair of *pereopods* of decapod crustaceans.

CHLOROPLAST body in plant cell containing chlorophyll. The site of *photosynthesis*.

CHONDROPHORE depression in which the internal ligament of bivalve shell is attached.

CHROMATOPHORE cell containing pigment granules.

CILIUM very small hair-like structure. Motile, creating water currents which may be used for feeding, locomotion, etc.

CIRRUS tentacle-like appendage.

CLASS unit of classification. A subdivision of a *phylum*.

CNIDOCYTE cell containing capsule-like structure (cnida) with long thread which is discharged on stimulation. Characteristic of *phylum* Cnidaria.

COLUMELLA central pillar of a gastropod shell.

COMMENSAL organism living in close association with another of a different *species*.

CONCEPTACLE flask-shaped invagination opening by tiny pore onto the surface of seaweed *frond*. Contains *gametes*.

CORBULA basket-like structure protecting *gonothecae* in hydroids belonging to the family Aglaeopheniidae.

CRENULATE scalloped or indented margin.

CRUSTOSE crust-like *thallus* of lichens; growth form of red seaweeds.

CUTICLE external, non-cellular layer on the outside of plants and animals.

DACTYLOZOOID *polyp* specialized for the defence of hydrozoan colony.

DEPOSIT FEEDING feeding on bottom deposits and *detritus*.

DETRITUS decomposing plant and animal material.

DIATOMS microscopic, unicellular algae with cell walls of silica. Main constituent of *phytoplankton*.

DICHOTOMOUS BRANCHING repeatedly dividing into two branches.

DIGITATE divided into finger-like segments.

DIOECIOUS having separate sexes.

DIPLOID chromosomes in pairs in nucleus.

DIRECT DEVELOPMENT without a free-living *larva* in the life-cycle.

DISTAL away from point of attachment, e.g. the distal part of a limb.

DORSAL the upper side of a plant or animal.

ECDYSIS the periodic shedding (moulting) of the *cuticle* by arthropods.

ECTOPARASITE organism living as a *parasite* on the outer surface of another organism.

EMERSION the period when an intertidal organism is out of water.

ENDEMIC native to a country.

ENDOPARASITE organism living as a *parasite* within another organism.

ENDOSKELETON an internal skeleton, e.g. the skeleton of fishes.

ENZYME a substance that promotes a chemical reaction without itself being used up in the reaction.

EPIDERMIS outer layer of cells of plants and animals.

EPILITHIC living on the surface of rocks and stones.

EPIPHYTIC living on the surface of a plant, attached but not living as a *parasite*.

EPIPODIAL TENTACLE tentacle on the lateral lobes of the foot of prosobranch molluscs.

EPITOKE specialized polychaete adapted for reproduction.

EPIZOIC living on the surface of an animal, but not living as a *parasite*.

EQUINOCTIAL TIDE tide of large amplitude occurring at time of spring and autumn equinoxes.

EQUIVALVE valves of shell of bivalve mollusc similar in size and shape.

ERRANT animal moving freely on or in substratum, with particular reference to polychaetes.

EULITTORAL ZONE middle region of shore, lying between the upper limit of the laminarian seaweeds and the upper limit of the acorn barnacles.

EXOSKELETON skeleton covering the outside of an animal, e.g. the shell of a crab.

FAMILY unit of classification. A subdivision of an *order*.

FILAMENTOUS thread-like chain of cells with reference to seaweed *thallus*.

FOLIOSE leaf-like *thallus* of lichens.

FOOT muscular organ of molluscs.

FROND that part of a seaweed excluding the *holdfast*.

FRUTICOSE upright or pendant lichen *thallus*, attached at base.

GAMETE reproductive cell, male or female.

GAMETOPHYTE stage in the life-cycle of seaweeds: produces *gametes*.

GASTROZOOID (HYDRANTH) hydroid *polyp* specialized for feeding.

GENUS unit of classification. A subdivision of a *family*.

GIRDLE margin of muscular tissue surrounding valves of chitons.

GNATHOPOD anterior thoracic appendage of amphipods, usually used for feeding and grasping.

GONOTHECA protective cup surrounding the *gonozooid* of a hydroid.

GONOZOOID hydroid *polyp* specialized for reproduction.

HAPLOID single set of unpaired chromosomes in nucleus.

HERBIVORE animal eating plant material.

HERMAPHRODITE organism producing both male and female *gametes*.

HETEROMORPHIC morphologically dissimilar *gametophyte* and *sporophyte* generations of seaweeds.

HOLDFAST attachment organ of seaweeds.

HYBRID offspring resulting from cross between two different *species*.

HYDRANTH see *gastrozooid*.

HYDROTHECA protective cup around the *hydranth* of a hydroid.

INDIRECT DEVELOPMENT process with a free-living larva in the life-cycle.

INEQUIVALVE valves of bivalve mollusc dissimilar in size and/or shape.

INTERSTITIAL FAUNA fauna living in space between sand grains.

INTROVERT narrow, anterior region of sipunculans.

ISOMORPHIC morphologically similar *gametophyte* and *sporophyte* generations of seaweeds.

LANCEOLATE spear-like in shape.

LARVA immature, pre-adult form in which some animals hatch from the egg. Usually different in form and way of life from adult. Often an important dispersal phase for animals.

LATERAL LINE system of sense organs arranged along sides of fish. Sensitive to pressure changes in water.

LATERAL TEETH teeth on hinge line of shell of bivalve mollusc.

LECITHOTROPHIC LARVA *larva* that survives on yolk deposited in the egg.

LITTORAL FRINGE region of shore lying between upper limit of barnacles and upper limit of periwinkles and black lichens, i.e. between *eulittoral zone* and *terrestrial fringe*.

LOPHOPHORE group of ciliated tentacles encircling the mouth in phoronids and bryozoans.

LUNULE depressed area, often heart-shaped, lying in front of the *beaks* in some bivalve molluscs. Usually of different sculpture and/or colour from rest of shell.

MANDIBLE anterior mouthpart of arthropods. The lower jaw of craniates.

MANTLE (PALLIUM) fold of tissue covering whole or part of the body of a mollusc. Secretes the shell. Also used to describe the flaps of tissue inside the *opercular aperture* of barnacles.

MANTLE CAVITY cavity enclosed by *mantle* in which gills and other structures are housed.

MAXILLA anterior mouthpart of arthropods, posterior to the *mandible*. Bone in upper jaw of craniates.

MAXILLIPED posterior crustacean mouthpart.

MEDUSA free-swimming (jellyfish) stage in the life-cycle of hydrozoans and scyphozoans.

MEIOSIS cell division in which the number of chromosomes is halved.

MEMBRANOUS thin sheet of cells, with reference to seaweed *thallus*.

METAMERISM repetition of *segments* of similar structure.

METAMORPHOSIS the process by which a *larva* changes into the adult form.

METAPODIAL TENTACLE tentacle on the posterior lobes of the foot of prosobranch molluscs.

MIDRIB thickened central axis of a seaweed *frond*.

MITOSIS cell division in which the number of chromosomes is not reduced.

MONOSIPHONOUS single row of siphon-like cells, with reference to seaweed *frond*.

NACRE the inner layer of the shell of molluscs. The nacreous layer; mother of pearl layer.

NATANT swimming, with reference to decapods.

NEAP TIDE tide with the smallest difference between the heights of high and low water.

NEMATOPHORE hydroid *polyp* armed with nematocysts, specialized for the defence of a hydroid colony, sometimes protected by a *nematotheca*.

NEMATOTHECA protective cup surrounding *nematophore* of a hydroid.

NEUROPODIUM ventral part of *parapodium* of polychaetes.

NOTOCHORD dorsal, flexible rod between nerve cord and gut; found at some stage in the life-cycle of chordates.

NOTOPODIUM dorsal part of *parapodium* of polychaetes.

OMNIVORE animal eating both plant and animal material.

OOSTEGITE (BROOD PLATE) plate arising from the base of thoracic appendages of females of some crustacean groups. Oostegites form brood chamber in which eggs are held.

OPERCULAR APERTURE opening at the apex of acorn barnacle through which feeding appendages emerge. The aperture is closed by paired plates, the terga and scuta.

OPERCULUM structure which closes the opening of, for example, polychaete tube, molluscan shell, bryozoan *zooid*, etc. Also the gill cover of a fish.

ORDER unit of classification. A subdivision of a *class*.

OSCULUM opening of sponge through which water leaves the body.

OVIGEROUS LEG appendage of sea-spiders, used to carry eggs.

OVIPARITY reproductive process in which the female lays embryos on, for example, seaweeds, rocks and stones.

OVOVIVIPARITY reproductive process in which embryos develop within female but are nourished from yolk of egg. Young emerge from parent.

PALEA broad, flattened *chaeta* of polychaetes.

PALLIAL LINE line on inside of shell of bivalve mollusc, marking position of attachment of *mantle* to shell.

PALLIAL SINUS indentation of *pallial line* marking area occupied by siphons of bivalve mollusc.

PALLIUM see *mantle*.

PALMATE divided like a hand into finger-like lobes.

PALPS paired projections arising from the head of polychaetes and arthropods.

PARAGNATHS chitinous teeth on the *proboscis* of polychaetes.

PARAPODIA paired lateral outgrowths from body *segments* of polychaetes.

PARAPODIAL LOBES lateral outgrowths of the *foot* of some opisthobranch molluscs. May be used in swimming.

PARASITE organism living in or on another organism (the host) from which it obtains nourishment at the expense of the host.

PARTHENOGENESIS development of an egg without fertilization.

PAXILLAE modified spines of asteroid. Have brush-like margin.

PEDICELLARIA pincer-like structure of echinoderms.

PELAGIC swimming or floating in the sea.

PELAGIC DEVELOPMENT development in which there is a *pelagic* stage.

PENTAMEROUS symmetry in which the body can be divided into five parts arranged round centre, e.g. in a starfish.

PERENNIAL plant that continues to grow year after year.

PEREOPOD appendage of thoracic *segment* of crustaceans.

PERIOSTRACUM outer layer of molluscan shell.

PERISARC thin, skeletal structure surrounding hydroid.

PERITHECIUM flask-shaped reproductive body embedded in *thallus* of lichen.

PHARYNGEAL CLEFTS openings leading from the cavity of the *pharynx* to the exterior. Sometimes known as gill slits.

PHARYNX anterior region of alimentary canal.

PHOTOSYNTHESIS synthesis of organic compounds by green plants from water and carbon dioxide, using energy from sunlight.

PHYLUM major division of animal kingdom.

PHYTOPLANKTON plants (usually microscopic) found in *plankton*.

PINNATE branched on either side of central axis, feather-like.

PLANKTON plants and animals which drift in the sea. Most are microscopic.

PLANKTOTROPHIC LARVA larva that feeds and develops in the *plankton*.

PLEOPOD appendage of abdominal *segment* of crustaceans.

PLUMOSE feather-like.

PODIUM see *tube-foot*.

POECILOGONY more than one mode of development in a single *species*.

POLYMORPHISM occurrence in the same population of two or more different forms of the same *species*, e.g. in cnidarians and bryozoans.

POLYP *sedentary* stage in the life-cycle of cnidarians. Usually has single opening, the mouth, surrounded by tentacles.

POLYSIPHONOUS with several rows of siphon-like cells, with reference to seaweed *frond*.

PROBOSCIS specialized, tubular structure, usually associated with the anterior part of the gut of some animals, e.g. the polychaetes.

PROTANDROUS HERMAPHRODITE *hermaphrodite* that matures first as a male.

PROXIMAL towards point of attachment, e.g. the proximal part of limb.

RADIAL SYMMETRY symmetry of an organism in which the parts are symmetrically arranged around a central axis, e.g. a jellyfish.

RADIOLES tentacle-like projections borne on head of some polychaetes and forming crown or fan.

RADULA narrow, ribbon-like structure bearing teeth and used for rasping food. Occurs in some groups of molluscs.

RAMUS branch of appendage of arthropod.

RECEPTACLE specialized reproductive structure on seaweed *frond* bearing *conceptacles*.

REPTANT crawling, with reference to decapods.

RHIZOME underground stem of angiosperms.

ROSTRUM forward projection of the *carapace* or head between the eyes in crustaceans, e.g. prawns.

SALINITY measure of the salt content of water, traditionally expressed as parts per thousand (‰) but now expressed as a ratio against a conductivity standard. The value giving the salinity of a sample does not require a unit and is expressed simply as a number; for example, the salinity of 'normal' sea water is 35.

SCYPHISTOMA *polyp* stage in the life-cycle of a jellyfish.

SEDENTARY not moving freely on or in substratum. Often living in burrow or tube.

SEGMENT unit of body, usually bearing appendages; clearly seen in annelids and arthropods. Term used to describe divisions of arthropod limb.

SEMIDIURNAL TIDE tidal system giving two high waters and two low waters in approximately 24 hours.

SEPTUM partition.

SESSILE living attached to the substratum.

SETA see *chaeta*.

SEXUAL REPRODUCTION reproduction involving fusion of male and female *gametes*.

SIPHON tube carrying water current into or out of body, e.g. in bivalve molluscs, sea-squirts.

SPATULATE having broad end.

SPECIES smallest unit of classification normally used. Organisms in a species are able to breed among themselves and produce fertile offspring.

SPERMATHECA organ in female (or *hermaphroditic* animal) in which sperm is stored.

SPERMATOPHORE packet of sperm.

SPINE stout *seta*.

SPOROPHYTE stage in the life-cycle of seaweeds, producing spores.

SPRING TIDE tide with the greatest difference between heights of high and low water.

STATOCYST organ of balance.

STIPE stalk-like part of seaweed.

STOLON root-like structure which anchors colony to substratum, e.g. in hydroids.

SUBCHELATE claw of *pereopod* closes against part of preceding segment.

SUBLITTORAL lowest zone on the shore, extending below the level of the lowest *spring tides*. Its upper limit is the upper limit of laminarian seaweeds.

SUBMERSION the period when an intertidal organism is covered by water.

SUSPENSION FEEDING feeding on organic particles, e.g. *plankton* or *detritus* suspended in water.

SYMBIOSIS close association between two individuals of different *species*, to the benefit of both.

TAXODONT arrangement of teeth on hinge line of bivalve mollusc in which there are large numbers of small teeth and alternating sockets.

TELSON terminal, usually flattened, region of abdomen, seen in some crustaceans. Telson and *uropods* often form a tail fan.

TENTACULAR CIRRI slender appendages from the anterior segments of some polychaetes.

TERRESTRIAL FRINGE region of shore above the *littoral fringe*.

TEST hard, *calcareous* covering of a sea-urchin. The jelly-like coating of a sea-squirt.

THALLUS the entire body of a seaweed or lichen.

TORSION twisting of the visceral mass through 180° during development of gastropods.

TRIRAMOUS having three branches or *rami*, e.g. some arthropod appendages.

TUBE-FOOT (PODIUM) blind ending projection from the *water vascular system* of echinoderms.

UMBILICUS cavity in the *columella* of a gastropod shell.

UMBO raised area of bivalve shell, just behind the *beak*.

UNIRAMOUS having one branch or *ramus*, e.g. some arthropod appendages.

UROPOD appendage of posterior abdominal segment(s) of crustaceans.

VEGETATIVE REPRODUCTION asexual reproduction in both plants and animals, by the separation of part of the body which develops into a complete organism.

VENTRAL the underside of a plant or animal.

VERTEBRAE small bones or cartilages in chain-like arrangement to form a backbone. Characteristic of vertebrates.

VIBRACULUM specialized *zooid* of bryozoans, equipped with a *seta* which sweeps over surface and keeps it clear of *detritus*.

VISCERAL MASS region of gastropod mollusc in which most internal organs are found.

VIVIPARITY embryos develop within the body of the female. Nutrition provided directly by mother. Young emerge.

WATER VASCULAR SYSTEM fluid-filled canals functioning as a hydraulic system Unique to echinoderms.

ZONATION distribution of shore organisms in zones or bands on the shore.

ZOOID individual member of a colony of animals, e.g. ascidians and bryozoans.

ZOOPLANKTON animals of the *plankton*.

ZYGOTE fertilized egg.

Index